Epilepsy

David W. McCandless

Epilepsy

Animal and Human Correlations

Springer

David W. McCandless
The Chicago Medical School
Department of Cell Biology and Anatomy
Rosalind Franklin University
Chicago, IL, USA

ISBN 978-1-4614-0108-7 e-ISBN 978-1-4614-0361-6
DOI 10.1007/978-1-4614-0361-6
Springer New York Dordrecht Heidelberg London

Library of Congress Control Number: 2011934492

Printed on acid-free paper

Springer is part of Springer Science+Business Media (www.springer.com)

This volume is dedicated to
Jason David McCandless.

Foreword

In April 2011, just a few weeks after our dad submitted the manuscript for this book, he unexpectedly passed away. Dad devoted his career to science and teaching. Over the final few years of his life, he became especially focused on writing medical books. His strong commitment and devotion to sharing his knowledge and contributing to scientific discourse were evident to everyone around him. These books are Dad's legacy and symbolize his dedication to education and the advancement of science. This book on epilepsy was especially meaningful to him. We are particularly grateful to Dr. Richard Wiggins for editing the final proofs of the manuscript. Not only was Dr. Wiggins a colleague and collaborator with our dad, he was also a close friend.

We feel honored that Dad was such an integral part of our lives. He was an outstanding role model and a strong influence on our own educations. We will miss him dearly.

Jeffrey and Steven McCandless

Preface

This volume is intended to be a synopsis of seizure disorders with a goal of describing key studies in animals and humans. The translation of pertinent findings from animal to human studies, and to potential human studies, is emphasized where possible. Specific cogent animal studies/results that deserve exploration in human seizure disorders are identified. The current rate of translation is estimated to be from 7 to 9 years, and the success rate of translation was very recently listed as less than one half. The success rate is defined as results in human studies which were predicted in advance by animal studies. Both the time between animal and human attempts plus the success rate clearly need improvement. A clear cause of delay is a lack of controlled randomized placebo studies in humans once suggestive data are identified in animals.

This epilepsy volume is not intended to be a 3-volume, 3,000-page encyclopedic description of epilepsy. It does not cover every published study relating to epilepsy in animals and humans. Several excellent recent publications filling the need for inclusiveness have been published, and the reader is referred to them.

This volume is designed to facilitate translation and be synoptic in nature. The arrangement of epilepsy chapters is in two parts whenever possible: animal studies, and related human studies. Each chapter has a similar organizational format. The chapters have a brief introductory section, followed by clinical descriptions. Animal studies with a bearing on human clinical issues are presented. The rationale for future human clinical investigations based on animal results is clearly presented when available. Translation is an important consideration in such a complicated field as epilepsy, and its facilitation is critical.

The complicated nature of the study of epilepsy is obvious even at the start by the confusion and controversy regarding even the classification of seizures. Several systems are in use, ranging from those based on genetics, age, metabolism, etc. In this volume, the 1981 classification is used, supplemented by some expansion supported by an updated version from 2006. Classification of seizures is certainly subject to change based on new knowledge gleaned from new technological advances in imaging, diagnostics, new drug development, and new molecular biology concepts.

This volume examines features of animal and human studies related to both simple and complex partial seizures. Partial (focal) seizures that spread and evolve into secondary generalized seizures are described. Attention is paid to those seizure types which produce the most numbers of human epilepsies. Generalized onset seizures will be examined, including absence seizures, as well as clonic, tonic, tonic clonic, etc. Typical as well as atypical seizures are discussed. Some seizure disorders that do not clearly fit into a classification are described, as well as nocturnal epilepsy. Whenever possible, studies in primates receive careful attention. These descriptions are always mindful of translation.

Other areas such as pediatric considerations, status epilepticus, and surgical approaches are included. As stated above, the focus is on the presentation of data from animal studies, which is timely and appropriate for translation to human disorders. The issue of world health concerns is neglected. Epilepsy in third world countries is largely untreated, and so studies applicable to patients in these settings are described. Possible treatments that might be applied or investigated with this in mind receive special attention in this volume.

Therefore, this volume is not intended to be all inclusive, but rather a synopsis of sorts, with a clear focus. If certain studies have been replicated 25 times over the last 10 years, not all will be cited. Studies with no bearing on the stated purpose of this volume are likely omitted. There are other very good sources of this material.

An important feature of this synoptic epilepsy book is to try to provide credence for each quoted study by including salient characteristics of each study. One often sees quotes such as "epilepsy was associated with dementia (followed by references) 32, 34, 36–40, 42, 43, and 48–51". The reader cannot tell if these references are significant or not. In this volume, we have tried to include enough actual information (ages, numbers of patients, epilepsy features, materials and methods, statistics used, etc.) in order for the reader to at least partially evaluate the validity/reliability of the results. We even occasionally state "this is an outstanding study".

In the preparation of this synopsis, we felt it was better to present less papers, but in a way for the reader to judge their significance, than to present 5,000 references which cannot be evaluated. As an aside, many journals now have the Materials and Methods sections at the end of a manuscript, when actually they should be first!

This synopsis of epilepsy – both experimental and clinical – will have appeal to physician assistants, primary care physicians, nurse anesthetists, osteopathic physicians, psychologists, radiologists, psychiatrists, dieticians, neurosurgeons, as well as neurologists will all find value in this book. More and more a team approach is most effective in epilepsy study and treatment. Increasing numbers of patients in the US accessing health care will place additional burdens on already overworked health care providers.

This book is as timely as is possible. As stated above, new technology and drug development are progressing. Thousands of epilepsy-related papers are published annually. It is important to try to identify those results which are of significance. Sources known for publishing high-quality investigations are identified in order to select the best papers.

It is our goal to produce a volume on epilepsy which will provide results of excellent critical investigations which will stimulate further thinking about aspects of this highly interesting, yet complicated subject area. It is our hope that this information, concisely presented, will motivate others to attempt to translate animal results into productive human studies. In this way, it is expected that human epilepsy will be modulated in ways which will reduce the severity and frequency of seizures.

Acknowledgments

The author wishes to thank Jeremy Carter, Candace Larson, and Laura Muncie for important and significant chapter contributions. Jeremy S. Carter and Candace L. Larson contributed a chapter on anesthesia in epilepsy, detailing procedures and methodology surrounding anesthetizing patients undergoing surgery for intractable seizures. Laura Muncie contributed an excellent chapter on diagnostic methodology currently used in evaluating patients prior to surgery to correct epileptic seizures. Jeffrey McCandless provided much valuable input into the Gene Therapy chapter, both in terms of content and organization.

Mrs Cristina Gonzalez spent many hours seeking both references and permissions to use various figures and tables in our book. Mrs Vilmary Friederichs aided in facilitating the completion of this volume and providing encouragement and support. My son, Jeffrey McCandless, NASA Ames Research Center, provided valuable input into the preparation and production of over 30 figures and tables. This consumed many hours.

Ms. Ann Avouris, Senior Editor at Springer Science and Business Media, helped us in many details of producing this book. As I have stated before, she is the most outstanding editor I have ever known. She is always responsive and ready to solve problems. She told me once, that she "loves solving problems", and she is outstanding in that regard.

My wife Sue has spent many years in support of these endeavors, and without her support and patience, this volume, and others, would not have been possible. I owe her a huge debt for her encouragement.

Contents

Part I
Introduction

Chapter 1
World Health Concerns

The World Health Organization (WHO) estimates that there are 50 million people suffering from epilepsy worldwide. In addition, about 75% or more are living in developing countries, sometimes with suboptimal medical care. It is estimated that 2.4 million new cases occur globally each year, and over half have their first seizure in childhood. Probably 70% of epilepsy patients have a measure of treatment success, allowing many to have essentially normal lives. In third world countries, most epileptics receive no treatment for various reasons, including lack of health care, lack of money, and/or lack of knowledge.

Epilepsy has a profound impact on childhood education. Many times, children do not attend school due to parental fears and teacher disdain. They may view epileptic children as disruptive, etc. In schools, there is usually no provision for first aid for a seizure patient. Side effects of anticonvulsant drugs (AEDs) and social stigma frequently lead to school dropouts. This all serves to create an inferiority complex for the epileptic child which lasts forever.

These effects in early schooling carry over into adulthood, where poor schooling may lead to an inability to gain employment. This may lead to a feeling of hopelessness, and worthlessness, leading to a suicide rate many times that of the general population.

In many cases in developing countries, the causes of epilepsy could be prevented. For example, in Ecuador, over 50% of epilepsy cases are caused by neurocysticercosis, which is an infection of the CNS caused by cysticerci of the pork tapeworm. Preventative treatment for those infected with tapeworm is about $0.10 per day. Conversely, the costs of first-line AEDs in some countries can exceed the total wages earned by a family annually. There are variations in costs of AEDs in different countries, and the reasons are not clear. It is obvious that there is a large gap in epilepsy treatment across the globe. A prime example is Zimbabwe where until a few years ago, there was only one fully trained neurologist in the country. He has since moved to the USA. Phenobarbital is often used in developing countries as an AED due to its low cost, whereas it is not frequently used in the USA because of adverse side effects.

D.W. McCandless, *Epilepsy: Animal and Human Correlations*,
DOI 10.1007/978-1-4614-0361-6_1, © Springer Science+Business Media, LLC 2012

The International League Against Epilepsy (ILAE). The International Bureau for Epilepsy (IBE) and the WHO joined together to form a global campaign against epilepsy. This effort is based in part on the startling statistics concerning epilepsy, suggesting it is most likely one of the most universal of all medical/neurological disorders. There is no age, race, social classes, or geographical borders to becoming an epileptic. The estimate of 50 million epileptics is surely an underestimate due to limited diagnostic capabilities in developing countries. In addition, political considerations limit incidence reporting in some countries.

The goal of the combined campaigns is to "bring epilepsy out of the shadows." This would be achieved by increasing awareness that epilepsy is a treatable brain disorder. The effort should improve education about epilepsy and elevate acceptability of epileptics. The needs of patients with epilepsy should be identified and met, and governments should address needs of epileptic patients. A phase 3 objective for IBE and ILAE is to develop greater regional involvement in the campaign.

In terms of underdeveloped countries, the estimate is of the worldwide population of epileptics, about 40 million live in developing countries. The incidence of epilepsy in low income countries may be 190/100,000 people (Placencia et al. 1994). The high incidence rates are likely due to parasitic and infectious diseases not present in developed countries. As many as half of epilepsy cases are untreated worldwide.

Psychosocial issues are also significant. One study shows that those with epilepsy had a lower income, more sick days, and a lower life quality than those without epilepsy (Wiebe et al. 1999). Children with epilepsy performed lower than nonaffected children and even lower than other chronically ill children.

Various factors place restrictions on epilepsy treatment, including economic, political, as well as culture of third world societies. Lack of knowledge is a key element in problems of obtaining treatment, and in compliance. As described earlier, treatment costs may exceed a family's annual income. And there may be a problem with drug supply. Phenobarbital is the favorite first-line AED proposed by WHO. This decision is no doubt based on cost.

Successful approaches in Africa involve a combination of availability of health care workers, availability of free AEDs, and availability of education for the local population regarding epilepsy and its treatment. Follow-ups were initiated, and mobile clinics were instituted in order to make medical help more accessible. The success was measured by the fact that after six months, 56% of patients were seizure free (Watts 1989).

As a part of the global campaign against epilepsy, surveys of epilepsy prevention were conducted regionally, including the Western Pacific region (Li et al. 2005). This region has 37 countries, and about 10 million epileptics reside there, although more than 60% do not have any epileptic care. Even in this area, developed countries have problems of stigma and lack of access to good treatment. The questionnaire sent to governments regarding epilepsy care was only returned from 20 of the 37 countries in the Western Pacific region.

There was a great degree of diversity in the results gained from the Western Pacific region survey. In some ways, for example, prevalence rates (3.8–4.6 per 1,000) were similar to others, but there was a large range in the results. There is a clear need for improvement in delivery of health care to those with none. Large gains can be achieved with just a modest investment.

The importance and need for employment in epileptic patients has been examined (de Boer 2005). People whose epilepsy starts during employment years, especially early, are at a disadvantage as regards employment. The onset of seizures may lead to short term, and perhaps permanent unemployment. The medical opinion is that most epileptic patients should be able to compete with anyone as regards the work place, but epileptics do not perform as well as others.

A project in the Netherlands was undertaken to determine characteristics of education and employment in a group of 1,000 epilepsy patients. Results showed a mean age of 38 years, and an equal distribution of men and women. Complex partial seizures were the most frequent type of seizure, followed by tonic-clonic seizures. Ninety-six percent were on AED treatment, with 32% taking one drug, and 34% taking two different AEDs. Twenty percent were seizure free. Compared to the general Dutch population, the epileptic group had a lower level of education. Forty-four percent were employed, 49% were unemployed or disabled. There was a correlation between the number of AEDs taken and employment: the lower the number of drugs, the higher the number of employed people. There was also an inverse correlation between the seizure severity and the employment rate.

The author notes that two areas need attention. The first is education of the public, especially employers, and second, employment training programs for people with epilepsy. To that end, a work integration program was developed (de Boer 1997) with an objective to facilitate the introduction into the workplace of epileptic patients. This was achieved by bringing together those in government, etc., and sharing each others knowledge. Job placement aims to place the epileptic patient to the proper job.

A brief report from India (Ramaratnam and Narasimha 2005) examined both prevalence and patterns of epilepsy in India. The authors reviewed 21 previous studies which covered 837,000 people, 4,220 of whom were epileptic. This gave a crude prevalence rate of 5.04/1,000. Age standardization revealed a prevalence rate of 5.39/1,000. Onset of seizures was in the first 30 years of life in all cases.

There were significant heterogeneities in results due to criteria such as malaria incidence, denial of epilepsy, prevalence of cysticercosis, etc. The overall number of epileptics in India is about 5.5 million and three-fourths are estimated to live in rural areas. Of these, 75% will probably not receive any treatment. This presents a significant potential burden on the economy and medical resources of the country.

Epilepsy results in a significant burden worldwide. Estimates are based on calculating disability adjusted life years (DALY). One DALY is one lost year of healthy life, and epilepsy contributes about 7 million DALYs (Chisholm 2005). The burden is magnified in places where there is a treatment gap, or lower treatment rate, than seen in developed countries (see Table 1.1).

One concern of WHO is in choosing interventions in treatment which are cost effective. A standardized approach has been developed comparing costs and effects of both current and proposed plans. Diseases are "modeled," and epilepsy was modeled as a disabling chronic condition with an increased risk of premature death.

Table 1.1 Percentage of DALYS lost worldwide

Disorder	Percentage of total DALYs lost worldwide in 1 year
Infectious	23.4
Neuropsychiatric (including epilepsy)	11.5
Trauma	11.3
Cardiovascular	10.3
Respiratory problems	6.2

Adapted from Scott et al. Bull. W. H. O. 79: p. 346, 2001

The classification of epilepsy was based on idiopathic epilepsy, not symptomatic epilepsy, and also based on two or more seizures in the past 5 years. Seventy percent of epilepsy cases are unprovoked, while 30% have a clear cause.

In developing countries, emphasis is on available low cost AEDs for first-line intervention. Four AEDs were compared: phenobarbital, phenytoin, carbamazepine, and valproate. In addition, analysis of costs and efficacy of treatments were performed at a primary health center.

Results showed that there were no measurable differences between the AEDs tested as regards efficacy or effectiveness. DALYs averted per million population ranged from 90 to 350 depending on geographic region. Treatment costs were as little as 70 international dollars (one international dollar equals one US dollar in purchasing power) per year up to 210 I.D. per year for valproate. The most efficient choice was using phenytoin or phenobarbital.

The author comments that there is a low priority and high treatment gap in some world geographic areas regarding epilepsy treatment. This could be averted in some measure by increasing the availability of affordable AEDs. Defects in this overall analysis include not always using the drug of choice for the seizure type. Comorbidity is an unmeasured variable which can increase costs appreciably. Disability weights may have been too low in the estimations of health benefits and averted burdens. The author also states that cost-effectiveness analysis is only one factor in the allocation and priority of health resources for epilepsy. There also needs to be attention paid to psychosocial issues.

The stigma mentioned earlier towards epileptics is a pervasive influence on responses of others at all levels from friends and neighbors to governments. The cost of stigma in terms of reluctance to work, or seek medical/psychological help, is not well studied. There is a growing awareness of this problem (which dates to antiquity), and a recent brief paper addresses some of the problems (Keusch et al. 2006).

There is a strong stigmatization of people with certain disorders. This could be a fear of physical contact, strong feelings about the manner in which some diseases are transmitted, etc. These feelings against certain groups with certain diseases (leprosy) go back thousands of years. The stigma of epilepsy similarly occurred in ancient times.

The recognition of the need to try to deal with, and educate against stigma, motivates health care workers to try to define the cultural disease of stigma. In 2001, an NIH sponsored international conference to examine aspects of stigma. Diseases covered were AIDS, epilepsy, mental illness, addiction, and physical deformities. The major goal was to examine ways stigma could be reduced given its almost endless history.

The view emerged that stigma reduction will occur through action taken according to the condition, country, and culture. A start should be made to define circumstances that amplify fear and discrimination. Various people such as psychologists, behavioral scientists, physicians, health care workers, etc. should work together to implement educational directional concepts in order to develop a better understanding of these afflictions by the general public.

Malaria is a disease in which hundreds of millions of people worldwide suffer. Cerebral malaria is the most severe expression of malaria and carries a morality rate of 10–40% (Molneux 2000). Cerebral malaria results from *Plasmodium falciparum*. The country Mali is endemic for malaria and has a 1.3% prevalence for epilepsy. The present study was undertaken in order to evaluate the role of cerebral malaria in epilepsy in Mali children.

This study is a result of a several year-long examination of risk factors for complicated and severe malaria in children 6 months to 15 years of age. Patients with severe malaria were diagnosed using WHO clinical criteria (Warrell et al. 1990). Malaria was carefully diagnosed, including *P. falciparum* in blood, fever, and coma. Patients were considered as epileptic when they had at least two unprovoked seizures within a time frame greater than 24 h.

Overall results showed 101 children in the cerebral malaria group and 222 patients in the noncerebral malaria group. In the cerebral malaria group, 77% had exhibited convulsions before the onset of coma, and 15% developed status epilepticus. Fifteen percent of the noncerebral malaria patients also developed seizures. Other sequelae included mental retardation, speech delay, oral dyspraxia, and behavior problems.

EEGs were recorded in only two patients who were subsequently confirmed as epileptic. These two patients from the cerebral malaria group showed bilateral centro occipital spikes and age-dependent occipital spikes. CT scans were performed in eight patients, four with epilepsy, and four with other neurological sequelae. Five of the eight showed abnormalities, including atrophy of the affected hemisphere, and diffuse atrophy. In the other neurologic sequelae group, CT results showed one case with dilated ventricles, and two had necrosis of both pallidi and bilateral peri sylvian atrophy in one patient, and another case of a cortico subcortical large zone of atrophy.

The authors note that epilepsy was significantly more frequent in children with cerebral malaria than in noncerebral malaria patients. A nonmalarial group was not a part of the protocol. In a rural tropical context, it is difficult to determine the date of onset of seizures or much else regarding initial seizures such as body temperature. The finding that cerebral malaria is a risk factor for epilepsy has been previously noted (Schmutzhard and Gerstenbrand 1984). Another study (Carter et al. 2004) showed a higher prevalence of epilepsy after the cerebral malaria.

In acute malaria, seizures occur in over half of patients before encephalopathy. Status epilepticus may occur, worsening the prognosis. The authors conclude saying that cerebral malaria dramatically damages the brain, and produces severe cognitive and motor sequelae. Phenobarbital, however, served to exert an easy control on seizures. Phenobarbital is relatively inexpensive, therefore representing an excellent choice in an AED.

Results of the ILAE, IBE, and WHO global campaign against epilepsy survey were published (Dua et al. 2006). Epilepsy care worldwide was evaluated from questionnaires asking for information for a multitude of questions. Data was collected from 160 countries. The questionnaire was translated into multiple languages before distribution. An atlas of epilepsy care in the world was developed in order to present all data in an easy to read form.

Results showed that professional organizations for epilepsy specialists exist in 61% of responding countries; 66% of low income countries have no such organizations. First-line AEDs were listed as phenobarbital (95% of responders), carbamazepine (93%), phenytoin (86%), and valproate (87%). The least expensive by far is phenobarbital. Epilepsy specialists were available in 70% of responding countries. By contrast, 30% did not have an epilepsy specialist for their country. Disability benefits were available in 47% of countries. The high was in high income countries (86%), and only 15% received benefits in low income countries.

Major problems in health care for epileptics included lack of drug supply, low level of community knowledge/awareness of epilepsy, stigma, lack of governmental concern, lack of infrastructure, etc. Social concerns that included help on employment, driving, education opportunities, etc., were identified as problems.

The authors comment that the atlas is a key activity coming from the campaign. It provides a visual way to assess the data. The only possible drawback to the atlas is that only one person in each country provided all the data for that particular country. Responses were mostly "yes" or "no," with limited information which was variable.

The treatment gap should be viewed, state the authors, in the context in which it comes. Areas such as economics, social, political, and cultural all need to be considered. This consideration modifies one's view of limitations. For example, the authors note that epilepsy surgery might be very desirable, but in some countries it is not obtainable at this time. There is also a difference in availability and how is a certain service actually used. In conclusion, these data and resultant atlas are very important, but must be considered carefully and in light of additional factors.

An ILAE report on epilepsy in North America which resulted from the campaign against epilepsy was published (Theodore et al. 2006). In this report, the estimate was that there were over 3 million people in North America with epilepsy. The comment is made that while epilepsy is more common than multiple sclerosis, Parkinson's disease, or autism, research funding is higher in the latter three, and epilepsy stigma extends to epilepsy patients and their family members.

The incidence of epilepsy in the USA, based on the Rochester study (op cit), is 44 per 100,000 people per year (1935–1984) and was 48 per 100,000 in 1975–1984. The age results showed an incidence of 60–70/100,000 under 5 years of age to less than 30/100,000 in older patients (65–74 years old). Since prevalence data include both existing cases plus new cases, the numbers tend to rise. In the Rochester study, prevalence rose from 2.73/1,000 to 6.79/1,000 people 40 years later.

In Canada, the incidence of epilepsy is estimated to be from 40 to 70 per 100,000 people, which would represent about 15,000 new cases per year (Kotsopoulos et al. 2002). In English-speaking Caribbean countries, Jamaica had over 8,000 hospital admissions for epilepsy over a 7-year period (population over 2.6 million).

In terms of etiology and classification in the U.S. Rochester study, about two-thirds of cases were cryptogenic or idiopathic. The remainder had insults such as trauma, stroke, cerebral infections, etc. Similar results were seen in the Texas study (Annegers et al. 1999). Tonic-clonic and complex partial seizures were most common. Mortality rate of epileptics in the USA was 2.3 times that of the general population.

Quality of life was assessed, and showed that there was a significant increase in the perception by the patient, that they were disabled as compared to the general population. Employment and financial differences were significantly different, with epilepsy patients having a mean annual income of $19,000 vs. $32,000 for the general population. In one survey, epileptics reported the fear of having a seizure was the worst aspect of epilepsy. Data in this study put the risk for suicide at about ten times that of nonepilepsy patients.

US public policies have not been successful in dispelling the perception that epilepsy is a stigmatizing condition, and so this attitude prevails in epilepsy patients as well. Epilepsy is a social label as well as a clinical disorder. The US public tends to overestimate the severity of the disorder in terms of both medical problems and social interactions.

In terms of care for epileptics, 41% of members of the American Academy of Neurology state that epilepsy is a focus of their practice. This translates to about 4,000 neurologists for 3 million epileptic patients. Resources reveal about 100 epilepsy centers have procedures such as video EEG, CT, MRI, SPECT, etc. In Canada, there were 45 video EEG beds in 2002.

Health care for epilepsy is generally and widely available in the USA. In one survey, 90% of epilepsy patients were taking AEDs, 56% were receiving monotherapy, 26% were taking two AEDs, 6% were taking three, and 2% were taking four AEDs. About 70% were very satisfied with their medications. The increased use of generic drugs can affect epilepsy care in a significant negative way including increased hospital admissions, and a negative impact on quality of care (Christian-Herman et al. 2004).

The annual cost of epilepsy in the USA is about 12.5 billion dollars. In terms of NIH dollars for research, epilepsy seems to be underfunded as compared to other chronic neurological disorders such as multiple sclerosis and Parkinson's disease. The USA is behind other countries in terms of life expectancy and infant mortality.

The report states that advances in neuroscience research show great promise, but weak links (translation) between advances (genetics and gene therapy) must be strengthened. Results of animal studies of new AEDs should be tried in patients at an accelerated pace whenever possible. Improved access to care and new advances should reduce morbidity and mortality. Translation of clinical trial evidence to community settings is needed.

Finally, there should be cooperation with the psychiatric community since many epilepsy patients have special needs, which psychiatrists can help. In many developing countries, it is psychiatrists who first diagnosis epilepsy due to behavioral problems. There should be interaction between developing countries and developed countries as regards epilepsy. Improved seizure control should be the first goal of all efforts. Secondary goals include mental health and education.

The health care and outcomes to assess epilepsy burden in China has been studied (Ding et al. 2006). In this study, prevalence/incidence have been used in order to assess disease burden in China. The DALY (see Chap. 3) was utilized. Characteristics such as age groups, prevalence, mortality rates, disability, and DALYs were all estimated.

Results showed that epilepsy caused 1.31–1.52 years of life lost per 1,000 general population. Years lived with disability due to epilepsy were 0.46–1.01 per 1,000 population. The greatest number of years of life lost was in the 15- to 29-year-old age group. Over 3.2 DALYs were lost per 1,000 in the 15 to 44-year-old age group. The highest prevalence rate was 12.55 in males in the 60–69 age group. Epilepsy mortality rates in China were between 3 and 7.9 per 100,000 people.

The authors note the overall DALY lost rate was 2.08 per 100,000 population. The authors state that for a complete assessment of the epilepsy disease burden, all DALYs for all diseases should be acquired, and this has not been done. The rate for cerebrovascular disease is 9.86 per 1,000 population, so the epilepsy rate seems reasonable. The DALY rate can be significantly reduced by a cost-effective treatment such as phenobarbital.

Another paper was published using a matched population-based cross-sectional case-controlled survey to examine the relationship between epilepsy and malnutrition in Benin. In sub-Saharan Africa, both epilepsy and malnutrition are health problems which have significant impact on medical and economic problems in Benin.

The prevalence of epilepsy in sub-Saharan Africa is from 10 to 55 per 1,000, whereas it is much lower in industrialized countries. Estimates on malnutrition are that at least one-thirds of the African population is undernourished (Food and Agriculture Organization 2006). A relation between epilepsy and malnutrition was suspected for many years (Hackett et al. 1997).

In this study, patients with epilepsy were matched with control subjects (1–2 ratio) based on age and area of residence. Anthropometric measurements were taken, and clinical examinations were done. Questionnaires were administered.

Results showed the prevalence for epilepsy in the studied region of Benin was 12.7%; 95% CI: 10.8–14.8. The study utilized 131 epileptic patients and 262 controls. The mean age was 25.4. The ages of epileptic cases were 17% children, 18% adolescents, and 65% were adults. Controls were almost identical. Epileptics and controls took a questionnaire asking about occupations, etc., plus questions regarding food. Twenty-two percent of the cases were malnourished versus 9.2% for controls ($p = 0.0006$). Height and weight for height were not significantly different. Four clinical signs of malnutrition (hair depigmentation, curly hair, dry skin, and tooth decay) were all highly significantly evident in epileptic patients as compared to controls.

The authors comment that both epilepsy and malnutrition are major health problems in Benin. Malnutrition was found to be more frequently associated with epileptic patients than with controls. Since epilepsy is stigmatized, often actual patients were reluctant to answer questions and a surrogate respondent was used. There was an association between epilepsy and nutritional status, but no inference about the direction can be made.

If epilepsy begets malnutrition, it could be because of the stigma in which epileptic patients (children) might not be fed as well or as often (benign neglect). Food taboos could also have contributed to malnutrition in Benin. If malnutrition contributed to epilepsy, several mechanisms might be active, such as a reduction in seizure threshold. Biochemical variations such as electrolyte abnormalities, hypoglycemia (and altered energy metabolism), or a decrease in inhibitory amino acid neurotransmitters might contribute. Hippocampal damage from a low protein diet (high frequency in Africa) might play a role in increasing susceptibility to epilepsy.

The authors conclude saying it is difficult to study these phenomena due to the almost complete lack of record keeping. Differently designed studies such as a cohort study are essential to clarify the direction of the association between epilepsy and malnutrition. Animal studies of the neurochemical results associated with malnutrition (hypoglycemia, neurotransmitters) certainly support via potential translation this explanation for the results. Future studies should resolve this issue.

A study of epilepsy, its stigma, and treatment in the African country of Togo have been performed (Balogou et al. 2007). The background is that in sub-Saharan Africa, 80% of epilepsy patients receive no treatment – the treatment gap is 60–98%, and about 70% of epilepsy cases could be controlled. An interesting statistic was that in a northern district of Togo (Kara district) where cysticercosis prevalence was 23%, the epilepsy rate was 16%. In southern Togo (Kloto), cysticercosis is rare, and the epilepsy rate was 12.3%.

The stated purpose of this study was to describe methods aimed at managing epilepsy in an underdeveloped country. The study implemented members of an indigenous tribe called the Batamariba tribe, with a population of 9,750 members. The tribe was primitive in that they relied on traditional healers. Attendance at a health clinic was 1.1% vs. 23% for the rest of Togo.

In the study, questionnaires were completed by over 600 residents to evaluate attitudes, etc. as regards epilepsy. A national campaign was conducted, followed by a detection phase. This consisted of a door-to-door survey looking for epileptic patients. This effort involved multiple neurology interns, neurologists, and a pediatrician, among others. Home follow-up and free drugs were distributed by a health care worker. Phenobarbital was the basic AED.

Results showed that the management of epilepsy was integrated into the daily routine of the identified epileptics. Of the initial 6,250 people participating in the study, 98 were considered to have epilepsy. The prevalence of epilepsy was determined to be 15.7% (95% confidence interval, CI 12.7–19.2). The seizure types were tonic-clonic – 36% and complex partial seizures equaled 33%. Suspected or confirmed causes were found in 76 patients, including 10 with head trauma. In follow-up, 93% of patients treated with phenobarbital had no more seizures, and most of the rest had a reduction of seizures.

The authors comment that their study conclusively shows that epilepsy can effectively be managed in a rural setting in a developing country. The success was based on determining beliefs which originally prevented epileptic residents from seeking help, and a sensitization program resulted in their treatment. The compliance rate was 98% compared to a compliance rate in a study in Tanzania of 90% (Jilek and Jilek-Aall 1970). The study was reproducible since the team of specialists used the same diagnostic criteria throughout the studies.

The present study used EEG methods to better clarify seizures and to minimize false positives. The authors conclude saying the use of the cheapest AED, phenobarbital, was highly successful. The treatment gap decreased from 100% before the study to only 4.1% after 2 years, and to 0% 4 years after program onset.

Another study has examined the prevalence and patterns of epilepsy in Brazil (Noronha et al. 2007). Previous studies in Brazil have shown a prevalence range of from 11.9/1,000 to 21/1,000 people. Similar results are derived from other South American countries. The treatment gap is high in low income countries. The objective of the present study was to assess prevalence and treatment gap in epileptic patients in Brazil, and determine if socioeconomic class played a significant role in these data.

Door-to-door surveys were conducted in three separate Brazilian areas representing three different socioeconomic geographic areas. Clinical epilepsy classification included active epilepsy, adequate AED treatment epilepsy, treatment with monotherapy, treatment with polytherapy, and nontreated epileptic patients. Socioeconomic class was also determined, ranging from 1 to 7, then combined into four classes – A, B, C, and D in decreasing order.

Results showed that of 54,000 people surveyed, 496 were positive for epilepsy. This represented 0.9% of the people surveyed. The minimum lifetime prevalence was 9.2/1,000 people. No differences were found between the social classes in terms of prevalence. Thirty-eight percent of those with active epilepsy had inadequate treatment. Of these, 12% were taking inappropriate medications, 19% were taking no drugs, and in 7%, the treatment was unknown.

The authors note that their study is the first door-to-door epidemiological survey in Brazil. The prevalence of epilepsy in Brazil is similar to other poor countries (Nicoletti et al. 2005). The patient lack of appropriate treatment even extended to more well-to-do patients. The treatment gap was rather equal across socioeconomic groups. The authors further note and conclude that these results indicate a commitment from the government may be inadequate. There should be overall health care management and a decent availability of AEDs. There should also be a campaign to educate the population regarding epilepsy and its treatment.

A recent paper (Co et al. 2007) has examined diagnosis and treatment of neonatal seizures in developing countries. Neonatal seizures can be relatively benign or can be life threatening. One goal of the ILAE/WHO campaign against epilepsy is to investigate the possibility of clinical guidelines and diagnostic/treatment algorithms for seizure patients that could be widely applied. This process has been done, and the present paper examines results. The authors state the guidelines are not a sole source of information for the evaluation of neonatal seizures. They are rather a framework for clinical workup.

The paper suggests several important features necessary for the proposal. First is a complete history. Family history of seizures is important, as are details of pregnancy and delivery. Head trauma, hypoxia, etc., all may be critical. The physical exam of the patient, with emphasis on evidence of trauma, level of consciousness, muscle tone and movement, tendon reflexes, etc. Attention should be paid to possible causes of seizures such as metabolic abnormalities (inborn errors, stroke, etc.), infections, intracranial hemorrhage, etc.

Neonatal infants with continuing seizures require immediate treatment and attention to breathing, and circulation. Abnormal movements have a correlation with electrographic seizures as shown on EEG. AED treatment should begin with a first-line drug such as phenobarbital, phenytoin, diazepam, or lorazepam.

Prognosis is generally tied to the underlying etiology of the seizures. Diffuse brain injury is associated with a poor prognosis. Diffuse brain damage is a feature of generalized myoclonic seizures, tonic seizures, and motor automatisms. Early onset of seizures (under 48 h), status epilepticus, comorbidities, etc., are also predictors of poor prognoses.

The overall view of this report is that there should be consistencies in all phases of epilepsy examination, diagnosis, and treatment. This is essential for consistency in comparing results from different geographic regional studies.

A brief paper (Tran et al. 2008) examines epilepsy in Laos, and represents one of the worst examples of lack of epilepsy care, emphasizing the difficult task of improving diagnosis and treatment. The WHO indicates that at least 80% of epilepsy patients live in areas with poor resources, and never receive treatment. Laos is a small country with a population of about 5.6 million people, mostly living in rural areas. Laos ranks 133rd in the Human Development Index, and has an annual per capita income of US $491. The epilepsy prevalence is 7.7% (Tran et al. 2006). In Laos, there are no guidelines, and many misconceptions and stigma towards epilepsy. Only one half of pharmacies can supply phenobarbital.

A group of medical health workers developed an epilepsy program to provide phenobarbital to epileptic patients. In this program, information was provided to epilepsy patients who were diagnosed on clinical criteria. Ultimately, 46 active cases were identified. Of these, only four had ever received AEDs previously. The daily income of the group was less than US $1.00. Twenty percent were mentally retarded. Seizure frequency ranged from 2 per year to three per week.

Eleven patients did not attend the program, and 16 dropped out. Only ten showed full compliance. Premature death occurred in 11%, from drowning, falls, burns, and sudden deaths. In the group who completed the program, seizure frequency dropped from 3.5 to 0.3 per month, and 70% were seizure free. The major problem was the distance needed to travel for medical accessibility. Phenobarbital was efficient, economical, and well tolerated at the community level.

Low education, low income, and lack of access all may have influenced poor compliance. The mortality rate was strikingly high. Stigma and mental retardation also no doubt contribute to the low success. This study illustrates the highly complicated and difficult problems which can be encountered in trying to provide epilepsy care to developing areas. Nevertheless, there were positive achievements, and not trying is no option.

Chapter 2
Metabolism and Epilepsy

This chapter will examine some aspects of the metabolic (especially energy metabolism) and pharmacologic responses to epileptic seizures. This involves animal models as well as imaging studies in patients. Some results are in conflict, and this is probably linked to the usual differences in methodology, and in definitions and data interpretation.

What's certain is that in seizure activity, there is significant electrical discharge of groups of neurons sufficient to usually result in an uncontrolled motor response. This in turn is associated with a change in cerebral energy metabolism and changes in synaptic function.

Methodological techniques include two basic mechanisms. The first includes directly measuring metabolites and neurotransmitters, mostly in experimental animals. The second method is to estimate energy metabolism using 2-deoxy-glucose PET, and magnetic resonance spectroscopy, both suitable for measuring glutamate turnover. Imaging methods have the advantage of being able to evaluate many cerebral areas simultaneously.

Dozens of studies have been performed looking at the direct measurement of energy metabolites in brains of seizing laboratory animals. There have been studies showing little change in adenosine triphosphate (ATP) or phosphocreatine (PCr), however many studies have shown dramatic alteration in energy metabolites.

One key is to be mindful of the concept that many metabolic encephalopathies have highly regional cerebral effects, and that includes seizures. Studies in which effects on energy metabolism of seizures gained by analyzing whole brain are much less meaningful when compared to regional studies.

Glutamate, the most prominent excitatory neurotransmitter in brain, has turnover which is tightly coupled to glucose metabolism (Magistretti and Pellerin 1996).

The rapid removal of both glutamate and GABA from the synapse is an energy (ATP) requiring process, explaining the coupling between neurotransmitter and glucose metabolism. Compensatory mechanisms such as increased cerebral blood flow during seizures provide increased oxygen and glucose to the stressed brain.

Transporters which serve to remove glutamate and GABA from the synapse are an astrocyte product, and the astrocytes are in close proximity to the synapse. In the

D.W. McCandless, *Epilepsy: Animal and Human Correlations*,
DOI 10.1007/978-1-4614-0361-6_2, © Springer Science+Business Media, LLC 2012

Relationship between ATP and PCr

$$\mathrm{PCr + ADP + H^+ \xrightleftharpoons[\text{Kinase}]{\text{Creatine}} ATP + Cr}$$

Fig. 2.1 Relationship between ATP-PCr. Adapted from Collins, R., Posner, J., and Plum, F. Am. J. Physiol. 218: p. 944, 1970

first few days after birth, the brain is more resistant to metabolic stress because overall utilization of high energy phosphates is lower due to lower relative activity.

As mentioned earlier, glutamate is the primary excitatory neurotransmitter, and GABA is the key inhibitory neurotransmitter. The balance between these two neurotransmitters is thought to be important in maintaining ion channel homeostasis. An increase in synaptic glutamate or a decrease in inhibitory GABA is seen as conducive to seizures. This simple concept underlies the mechanism of action of the majority of all antiepileptic drugs (AEDs).

Early studies of energy metabolism were undertaken to define characteristics of the seizures in relation to changes in high energy phosphates before, during, and after the seizure. Maximal electroshock was utilized as a stimulus for seizure production because of the consistency of the response and ease of producing it. One early study (Collins et al. 1970) looked at mice with electroshock-induced tonic-clonic seizures as regards cerebral energy metabolism.

The question asked was how do mice in tonic-clonic seizures compare when paralyzed and unparalyzed. The unparalyzed mice were not only seizing, but were hypoxic due to inadequate pulmonary ventilation. Mice were shocked through ear electrodes at a level which produced tonic-clonic seizures. Two groups received ventilation of either room air or 100% oxygen. Unparalyzed mice received nothing except the electroshock. Mice were sacrificed in liquid Freon, then the whole brains were dissected and saved frozen for analysis.

Results showed the mouse response to the stimulus consisted of a 15–20 s period of tonic-clonic extension followed by a 30–45 s period of clonic jerks. Postictal depression followed the ictal phase. Respiration ceased for the initial 20 s in unventilated paralyzed mice. The only seizure sign was a body twitch during delivery of the stimulus.

High energy phosphates ATP and PCr dropped significantly from control values in the unparalyzed group of mice. The decrease was greater than 50% in unparalyzed mice, and glucose was similarly decreased. After 25 s, values began to return to normal. By contrast, the paralyzed ventilated room air mice had depletions of only 37% for PCr, and 25% for ATP. Those mice ventilated with 100% oxygen had unchanged cerebral metabolites (see Fig. 2.1).

The authors comment that their results show that if the electroshock mouse model has an adequate oxygen supply, there is a capacity to balance a three- to fourfold increase in electroshock-induced energy demand. The ability of the brain to maintain energy reserves is dependent on a supply of 100% oxygen.

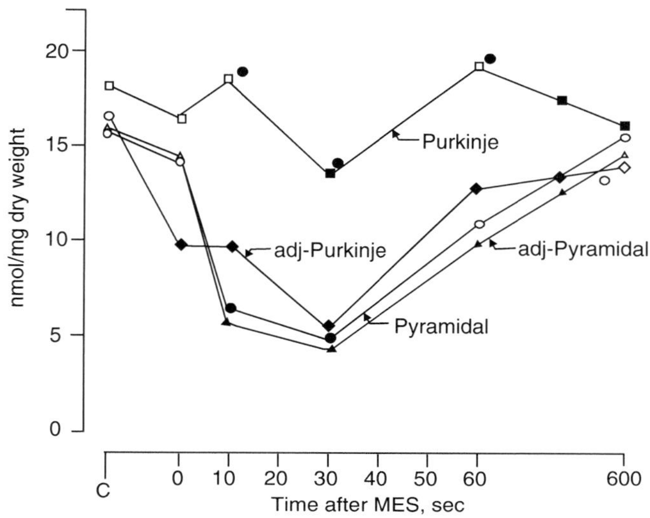

Fig. 2.2 Changes in ATP concentrations in cell bodies of cortex and cerebellum in mice after maximal electroshock. Redrawn from McCandless, D., et al. P.N.A.S. 76: p. 1484

In another report, the effects of maximal electroshock on energy metabolism were examined in single neural cells from the cortex and cerebellum of mice (McCandless et al. 1979). In this study, maximal electroshock was administered to mice via corneal electrodes, frozen at various times after electroshock, and the cortical/cerebellar samples freeze dried and cellular samples prepared and analyzed by elegant methods previously described (Lowry and Passonneau 1972).

Results showed that glucose, PCr, and ATP were decreased in cortical pyramidal neurons, Purkinje cells in the cerebellum, and adjacent neuropil in cortex and cerebellum following electric shock. The largest decrease in metabolism was seen in mice 30 s after maximal electroshock, and at 10 min later, there was an increase in metabolites.

A differential effect was seen in ATP values in Pukinje cells of the cerebellum. In this case, the ATP values were less affected in the cerebellar Purkinje cells than other metabolite levels. It seems that the electroshock is attenuated before it reaches the Purkinje cells. This is in keeping with the concept that the excitable state cannot be sustained in the cerebellum. This difference has not been described before and is only noted by examining single cells (see Fig. 2.2).

Purkinje cells represent the output of the cerebellum, and are inhibitory. The delayed response from the Purkinje cells suggests that when it does become obvious, the inhibitory output modulates the seizure activity coming from pyramidal cells. Thus, the output of the cerebellum is initially low. If the cerebellar output was increased immediately following maximal electroshock, the duration/severity of the seizure response might be lessened. The sparing effect is therefore deleterious to extracerebellar paroxysmal activity.

The 2-deoxy-glucose method (Sokoloff et al. 1977), used for neuroimaging, has been extensively employed to evaluate cerebral activity in an epilepsy mouse model, the E1 mouse. Following seizures in the E1 mouse induced by positional change, 2-deoxy-glucose was highest in brain 5 min after administration. The entire cortex in seizure mice showed a dense radiolabel. The entire hippocampus was similarly darkened by the presence of the isotope.

The results of the 2-deoxy-glucose study represent the total serum level of isotope deposition over the 40 min of the exposure. This includes both the seizure time period and partial recovery. Lightly labeled areas such as thalamic nuclei may represent hypometabolism or decreased circulatory activity. These results are somewhat dissimilar when compared to electroshock and this is due to different mechanisms.

The effects of status epilepticus on cerebral cortical energy metabolism have been described in primates (McCandless et al. 1986). Many studies have focused on single seizures and energy metabolism, while only a few have looked at status epilepticus (Meldrum 1983). Status is of special interest due to the supposed severe cellular damage which can result (Meldrum et al. 1973). The use of primates facilitates translation.

In this study, *Cynamologus fascicolaris* monkeys were immobilized with ketamine and after ether anesthesia, a craniotomy was performed, exposing the cerebral cortex. EEG electrodes were placed on the cortex. A sharpened spoon was used to remove small motor cortex samples prior to bicuculline administration, 20 min after onset of seizures, and 2 h after the previous seizure (2 h 20 min).

Tissue samples were prepared for metabolite analysis as previously described (McCandless et al. 1979). Samples were prepared for electron microscopy by fixation in glutaraldehyde and postfixation in osmium tetroxide. Blood samples were removed for measuring glucose.

Results showed that bicuculline produced sustained epileptiform activity on EEG recordings. At 2 h 20 min, the EEG was characterized by slow, low voltage activity, and occasional spikes. EEG results were normal in appearance until 2 h 20 min, at which time endoplasmic reticulum was dilated, mitochondria were swollen, vesicles were seen, and the cytoplasm was watery and dispersed.

The effects of bicuculline-induced seizures on energy metabolites were pronounced. Glucose samples from the motor cortex were essentially not changed, but PCr was statistically significantly decreased in the two pyramidal cell layers (outer small pyramidal cells and inner large pyramidal cells), and in adjacent white matter 20 min after bicuculline. In contrast, ATP was maintained at normal levels at 20 min, whereas 2 h later, ATP was decreased in all layers of the motor cortex except white matter. Phosphocreatine had returned to normal at 2 h 20 min (see Figs. 2.3 and 2.4).

Many previous studies have examined energy metabolism in experimental models of seizures. Most studies that looked show regional changes in high energy phosphates. But, just like regional changes varying from site to site, so are differences seen between cortical layers from the same site. This emphasizes the importance of measuring labile metabolites in samples as small as possible. The present study shows this still holds true in prolonged status epilepticus.

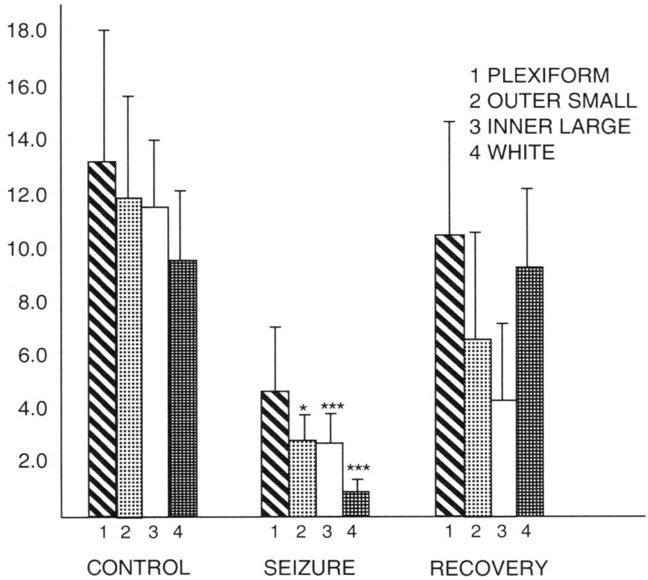

Fig. 2.3 Effect of bicuculine-induced seizures and recovery on regional motor cortex phosphocreatine. Data expressed as nmoles/mg dry weight. * = $P<0.05$; ** = $P<0.01$; *** = $P<0.005$. McCandless, D., et al. Am J. Physiol. 251: p. 778, 1986

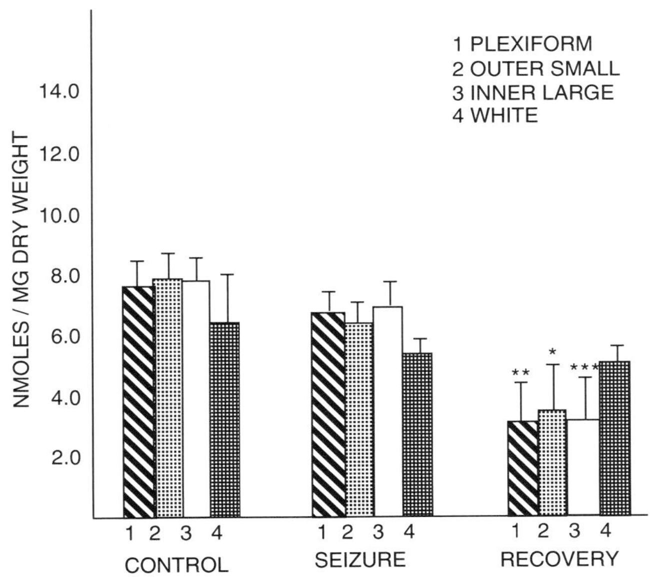

Fig. 2.4 Effect of bicuculline-induced seizures and recovery on regional motor cortex ATP. Layers and significance as in Fig. 2.3. Redrawn from McCandless, D., et al. Am J. Physiol. 251: p. 778, 1986

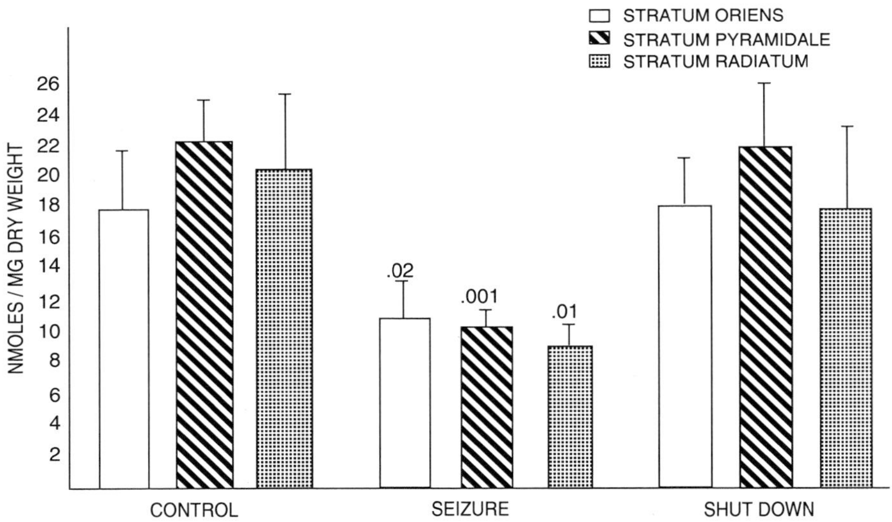

Fig. 2.5 Hippocampal CA1 phosphocreatine during and after seizure activity. Data (mean plus/minus SEM) expressed as nmoles/mg dry weight. Redrawn from DeFrance, J. and McCandless, D. M.B.D. 6: p. 89, 1991

The E.M. changes seen in the present study in mitochondria are in keeping with changes in energy metabolism. These changes are similar to that seen in ischemic cell damage. The possibility that repeated motor cortex sampling might have contributed was examined in a series of rats in which cortical samples were removed over a 2 h 20 min period and metabolism analyzed. There were no significant changes in energy metabolites in this series of rats treated exactly like the primates, except for bicuculline. It is likely single cell analysis might be more revealing than the layer results. Data on metabolite turnover would also be of interest.

The hippocampus displays a postictal depression period in which further seizures cannot be elicited. The hippocampus plays an important role in memory, and so hippocampal postictal depression may underlie postictal amnesia (Penfield 1958). The present study (DeFrance and McCandless 1991) was undertaken to look at energy metabolism in rat brain hippocampus following electrical stimulation.

Results showed that during the electrically induced seizure activity, glucose, ATP, and phosphocreatine were all decreased by about 40% in all three hippocampal CA1 layers. Within 60 s after the seizure, metabolite levels had returned toward normal. At a time when the hippocampus (CA1) was unresponsive, metabolites were at control levels. This suggests the physiological shutdown of the hippocampus was not associated with perturbed energy metabolism. Whatever is the basis of the postictal refractory state, depleted energy metabolites are not the mechanism. These results do not rule out the possibility that some group of single cells is not involved. Testing this would require single cell analysis (see Figs. 2.5 and 2.6).

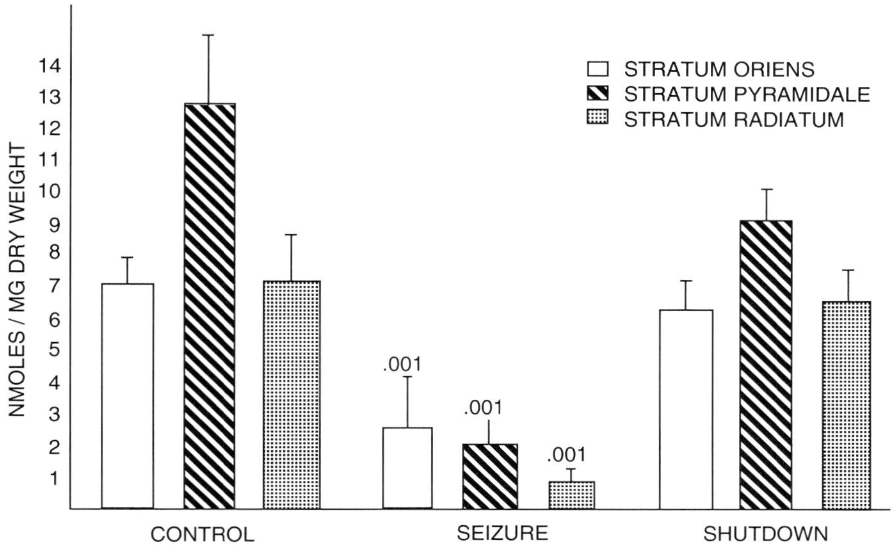

Fig. 2.6 Hippocampal ATP CA1 during and after seizure activity. Data (mean plus/minus SEM) expressed as nmoles/mg dry weight. Redrawn from DeFrance, J. and McCandless, D. M.B.D. 6: p. 89, 1991

Further evidence as regards energy metabolism and seizures can be derived from results of treatment with the ketagenic diet (Bough 2008). The ketogenic diet (and the modified Atkins diet) are very high in fat, low in carbohydrates, and adequate in all other requirements (see Chap. 30). The diet results in increased levels of ketone bodies, which are utilized to affect the depletion of glucose as a result of low carbohydrate intake.

Microarrays are used to identify patterns of gene expression in rats on a ketogenic diet for 3 weeks. Results show 384 transcripts were upregulated. The most frequent group of differentially expressed genes was those associated with energy metabolism, and 21 encoded genes were involved in oxidative phosphorylation (Bough et al. 2006).

Examination of mitochondrial by electron microscopy showed a 46% increase in mitochondrial density in the ketogenic diet rats as compared to control animals. When actual energy metabolites were measured, ketogenic diet rat hippocampus showed little change in ATP, ADP, or AMP, however PCr was elevated. This is in keeping with the concept that the ketone bodies act to "supercharge" the energy capabilities of cerebral tissue. Glutamate was also elevated, but most brain glutamate is used as an energy substrate.

The mechanism of action of the ketogenic diet on seizure control is not clear, but metabolic perturbations may contribute significantly to synaptic instability. The author speculates that the E.M. result showing a 50% increase in mitochondrial density suggests an increase in ATP production. This excess energy is stored in the

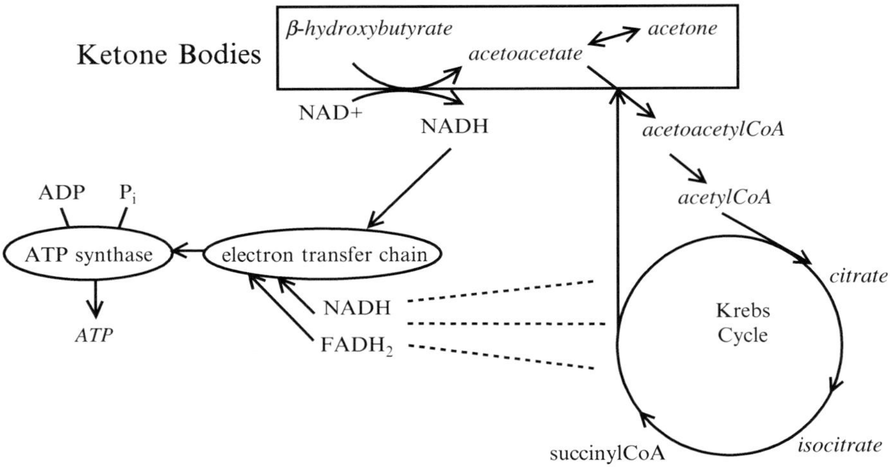

Fig. 2.7 Ketone body metabolism. Adapted from Massino, S., et. al. Curr. Neuropharmacol. 7: p. 260, 2009

form of PCr until needed as ATP. This acts to stabilize the brain electrically leading to a lower sensitivity to seizures. This has also been seen in humans (Pan et al. 1999), (see Fig. 2.7).

Further data supports the concept that positive energy balance, produced by the ketogenic diet can benefit GABA. The creatine kinase enzyme is found mostly in GABAergic neurons (Boero et al. 2003). In this case, the ketogenic diet enhancement of energy metabolism could in turn act to stabilize postsynaptic GABAa receptors. There may also been an effect on glutamate such that its levels are lowered.

The author concludes saying much data exist supporting the idea that the ketogenic diet "dramatically" enhances cerebral energy metabolism leading to increased energy stores, and a stabilization of brain in terms of potential seizure activity. The author states the ketogenic diet is efficacious due to its ability to compensate for other deficits in epileptic foci.

Glycogen may also play a role in seizures in both animal models and in human epilepsies (Cloix et al. 2008). During periods of hypometabolism in epileptic foci, glycogen from astrocytes, or glycogenolysis can be a source of potential energy. Brain biopsies of hippocampus from human epileptic patients have been noted to be high relative to other brain areas (Dalsgaard et al. 2007).

Methionine sulfoximine (MSO) is a compound which resembles glutamate, and may actually cause seizures by acting at the synapse in a way similar to that of glutamate. In this way MSO could induce seizures. MSO has been administered IP as an inducer of seizures in experimental models of epilepsy for over 50 years. Seizures in mice and rats are generalized tonic-clonic seizures. The time frame is a preictal period of several hours, followed by continuous seizures from 24 to 48 h. This is followed by a postictal period.

The preseizure period is characterized by an accumulation of intracerebral glycogen, which in turn signals no correlation with the convulsive period, or a consequence of seizure activity. Accumulation of glycogen is usually localized to astrocytes in the cerebral cortex and cerebellum. With MSO administration, there is a significant accumulation in the cortex and cerebellum of mice and rats. Seizure latency and glycogen accumulation are attributable to genetic factors.

Another paper from the same group (Cloix and Hvéor 2009) examines various aspects of neurotransmission and epilepsy/energy metabolism. Glutamate and GABA are the most common neurotransmitters in the CNS, with glutamate being an excitatory neurotransmitter and GABA being inhibitory. The ionotropic receptors associated with glutamate are AMPA, NMDA, and kainate receptors. As regards GABA receptors, the two are GABAa and GABAb. The former facilitates inward movement of chloride ions, the latter the exit of K ions; both yield an inhibitory action. It is a disequilibrium of these two neurotransmitters which can produce seizures.

Cortical pyramidal cells (large and small) project axons into other brain sites and the spinal cord. These cells are excitatory and neurotransmission is via glutamate. Axons of cortical basket cells contact other cortical neurons and are inhibitory using GABA as a neurotransmitter. These neurons represent the main contributions to EEGs. The EEGs are contributed to by millions of neurons, mostly cortical pyramidal output.

The EEG is composed of by alpha, theta, and delta waves, and is a result of synchronization of groups of neurons (Liberson 1989). Synchronization occurs in seizures and large spikes and waves occur. Synchronization etiology is unclear, as in the "all or none" timing. How thousands of neurons accomplish this feat at the same time requires more of an explanation than decreased inhibition. This area needs more study.

The relationship between astrocytes and neurons no doubt involves a role for astrocytes in the supply of energy to neurons. This occurs especially when the energy potential is not met by blood supply. Two explanations may occur: one is that the astrocyte supplies the neurons with lactate to enter the TCA cycle and the other explanation suggests the astrocytes supply glucose to the neurons.

Glucose enters brain from blood, crossing the BBB in so doing. Three main GLUT proteins facilitate this transport, GLUT1, GLUT2, and GLUT3. GLUT3 is thought to have a higher affinity for glucose as well as a higher capacity for glucose transport. In addition to transport into neurons, it may be that some glucose is transported to astrocytes, then into neurons. Both processes may be occurring simultaneously. Lactate is transported out of cells by MCT transporters.

Adenosine is a neuromodulator with a wide range of actions in the CNS. It is involved in neurochemical responses and morbidity in traumatic brain injury (TBI) (Lusardi 2009), as well as many other neurological perturbations (Boison 2008a). In terms of TBI, epilepsy is common sequelae which has a significant latency. Seizures related to TBI can materialize many years later, even decades, after a TBI.

Recovery from TBI is frequent, especially in terms of gross motor function and intellectual domain, but various sequelae/comorbidities may remain. These include migraine, anxiety, and chronic pain, as well as epilepsy. Adenosine modulation is a key feature of these TBI-related comorbidities.

Posttraumatic epilepsy is a common long latency after effect of TBI. This can occur in both pediatric and adult TBI patients. That the A1 receptor is involved in seizure suppression is shown by results in which an A1 receptor knockout mouse is highly susceptible to status epilepticus. Even in cases in which there are no overt seizures, epileptiform activity is demonstrable. Studies implicate adenosine dysregulation in epileptogenesis (Boison 2008b).

A group of inborn errors of creatine metabolism have been recently described (Schulze 2003). Creatine transport deficiency is one of several of these inborn errors. The diagnosis of this disorder is made by incubating fibroblasts, measuring creatine uptake, and analysis of the SLC6A8 gene in vitro. The phenotype in males includes mental retardation, language delay, autism, and epilepsy.

In a recent retrospective study (Fons et al. 2009), clinical data, seizure types, treatment, etc., were reviewed in seven patients. Results showed an age range from 11 years to 40 years old. All had moderate to severe mental retardation, and six of seven had epilepsy. No genotype–phenotype correlations were obvious.

All patients showed slow EEG background activity, without specific paroxysmal activity. Brain levels of creatine were not associated with epilepsy severity. Imaging studies failed to show any association with brain structural lesions. The authors state that any patient with mental retardation, language disorder, autism, and seizures should be tested for creatine deficiency.

Creatine plays a significant role in forming phosphocreatine, an energy yielding compound. This is a critical role in cerebral energy metabolism. This is another example of an energy metabolism alteration which would appear to directly produce seizures and epileptiform activity.

Another recent paper related to energy metabolism and adenosine has appeared (Masino et al. 2009). Adenosine serves at least two key functions in brain: first as a neuromodulator, and second as a vital component of ATP. In terms of epilepsy and ATP, the efficacious ketogenic diet used to treat refractory epilepsies, has as a mechanism, the providing of ketone bodies which feed the TCA cycle, thereby increasing ATP and stabilizing neurons. This in turn decreases seizures and acts to protect neurons (Masino and Geiger 2008).

Adenosine is a neuromodulator, and the receptor A1 is inhibitory. It is located at least in the hippocampi and cerebral cortices, and when increased, can serve to prevent seizures. In a sense, adenosine as a modulator of brain activity, acts at a "higher" level much as does neuropeptide Y, or ketogenic mechanisms.

The effects of both adenosine and the ketogenic diet on brain metabolism are rather different than that of classical AEDs. Thus, their use in AED refractory epilepsy cases has shown significant beneficial effects. In vitro, ketone bodies increase ATP/ADP ratios, and the ketogenic diet has a similar effect in vivo on energy metabolite ratios in specific brain regions. Hippocampal CA3 neurons can in fact actually regulate their own excitability when glucose is low, and ATP is in sufficient concentration, or higher.

The clinical significance of this can hardly be understated. Evidence is well established that ketone bodies, adenosine, creatine, etc., can decrease seizures and offer cellular protection, and stabilize neurons such that epileptiform activity stops.

Many pediatric patients have a highly significant decrease in seizure frequency or outright cessation of seizures which can be long lasting. This effect is not limited to pediatric patients (Bodenant et al. 2008).

The authors of this paper conclude saying that "……the metabolic relationship among adenosine, a ketogenic diet, and epilepsy could open major new therapeutic applications and avoid peripheral side effects in a way that has eluded receptor-based strategies." This entire area warrants much additional study and translational trials.

A clinical technical problem of significance is the physical resolution of MRI. This problem has been addressed using 18F-FCWAY (18F-trans-4-fluoro-N-2-(4-(2-methoxyphenyl) piperazin-1-Y)ethyl-N-(2 pyridyl) cyclohexane carboxamide PET. This acts to improve resolution such that more specific 5-HT receptor binding is seen.

In this study, the authors (Liew et al. 2009) noted that 30% or more patients with refractory to AEDs temporal lobe epilepsy show no lesions on MRI examination. Foci localization is difficult in these patients, and 18F-FDG PET may not be a viable option. The present study was performed in order to evaluate 18F-FCWAY in refractory epilepsy cases. The goal was localization of otherwise obscure foci in refractory temporal lobe epilepsy cases.

Twelve patients were studied. All had medically refractive temporal lobe epilepsy, and all had MRIs which had resolution insufficient to permit identification of structural abnormalities. Results using 18F-FCWAY showed in 11 of 12 patients an identifiable lateralized epileptogenic area of focus. Nine patients were resected, and five of these were seizure free at one year postoperatively.

The authors comment that both 18F-FCWAY-PET and 18F-FDG-PET were helpful in seeking epileptogenic zones in nonlesioned (by MRI) temporal lobe epilepsy. 18F-FDG-PET could have shown potentially misleading images in two cases. 18F-FCWAY-PET may be a more accurate method than 18F-Fdg-PET.

About one third of all epilepsy patients are refractory to AEDs, and become theoretical candidates for surgery. Many of these are not actually good surgical candidates for a variety of reasons such as multiple foci, and thus better nonsurgical therapies are essential. Recent data has suggested an amplification of glutamate activity may be conducive to the seizure state. The present paper examines this hypothesis (Eid et al. 2008).

It is suspected that excessive extracellular glutamate may contribute to seizure activity and an imbalance in levels of GABA/glutamate. In humans, extracellular hippocampal glutamate increases sixfold during seizures, and this lasts several minutes after the seizure ends. A slowing of the glutamate/glutamine cycle is thought to be the cause of the increased extracellular glutamate.

Glutamate does not cross the BBB readily, so most is synthesized from glucose in astrocytes. The first cell type in this scheme is astrocytes, therefore there are several routes possible, including the TCA cycle. Alpha ketogluterate is one metabolite associated with the TCA cycle, and from which glutamate is formed.

Glutamate in the astrocyte is converted to glutamine (enzyme: glutamine synthetase) then transported to neurons where it again becomes glutamate which can enter extracellular space. Glutamine synthetase is a key enzyme in the glutamate/glutamine cycle. The authors of this paper speculate that an alteration of glutamine synthetase can slow the formation of glutamine, thereby causing excess glutamate, conducive to seizure activity. A loss of astrocytic glutamate transporters could have a similar effect, however the loss of glutamine synthetase is a more likely explanation (Petroff et al. 2002).

A chronic animal model of hippocampal glutamine synthetase deficiency has been developed by daily injections of MSO into the hippocampus. This model is associated with spontaneous recurrent seizures and lowered hippocampal glutamine synthetase activity. The hypothesis is that the loss of hippocampal glutamine synthetase is intimately involved in the pathology of mesial temporal lobe epilepsy.

The mechanism for the glutamine synthetase deficiency may be related to a downregulation due to neuronyl loss in the hippocampus. This has been shown previously (Derouiche et al. 1993). The glutamate/glutamine cycle can be studied using human tissue from patients undergoing temporal lobe resection for epilepsy (Eid et al. 2004).

Results from the above-mentioned studies of four mesial temporal lobe epilepsy patients showed normal glutamate/glutamine cycling in the neocortex and hippocampus. However, when challenged with ammonia, an increased cycling occurred in the cortex, but not in the hippocampus. This finding could be significant even though there are decreased numbers of neurons in the hippocampi of mesial temporal lobe epileptic patients. Studies have shown that astrocytes can be a source of extracellular glutamate (Volterra and Meldolesi 2005). Astrocytes do not have glutamine synthetase in the hippocampi of temporal lobe epileptic patients, therefore accumulation of astrocytic glutamate may impair clearance of the compound from extracellular space. These data, although suggestive, require further investigation.

Another recent paper examines aspects of glutamine synthetase activity (Bidmon et al. 2008). The study authors note that astrocyte specific glutamine synthetase is intimately involved in glutamate recycling as well as GABA metabolism. It is proposed that glutamine synthetase activity alteration is associated with epilepsy, but the exact mechanisms are not clear.

In this study, Wistar rats were maintained in a cage in which seizure activity was measured. Three groups of rats were defined: one group received saline, another group received MSO, and a third received the convulsant pentylenetetrazole. Groups 1 and 3 were sacrificed at 14 days, and group 2 was sacrificed 24 h after treatment.

Results showed rats receiving 2 weeks of pentylenetetrazole injections had an upregulation of heat shock protein-27 (HSP-27) in cortical and hippocampal astrocytes. MSO-treated rats showed significant glutamine synthetase inhibition and HSP-27 induction. Repeated pentylenetetrazole injection induced seizures in all rats, but the phenotype was not identical. There was a small decrease in glutamine synthetase activity after 2 weeks, but a more significant decrease after MSO treatment.

MSO treatment resulted in a more uniform and widespread HSP-27 induction compared to that induced by pentylenetetrazole injections for 14 days. The MSO group showed a significant decrease in glutamate synthetase immunoreactivity in the cortex and hippocampus. There were subregion differences in hippocampal and cortical areas.

Results further showed repeated seizures elicited a stress response in astrocytes, but adversely affected glutamine synthetase by tyrosine nitration and inhibition. The effects were localized to epileptic circuitry and amygdala related structures. The changes were most significant in the entorhinal cortex and dentate gyrus.

Differences between the pentylenetetrazole model and the MSO model include the concept that the pentylenetetrazole model induces seizures in seconds by affecting GABA neurotransmission. MSO, however, is a partial inhibitor of a precursor of GABA and glutamate. Concomitant ammonia increases are not associated with seizures. MSO-induced seizures may be caused by other features such as cerebral energy metabolism and neurotransmitter changes (Cloix and Hevor 1998).

The authors state their findings suggest binding of MSO to the active site of glutamine synthetase and its phosphorylation blocks the enzyme conformation, decreasing activity. Nitration probably has no adverse effect.

The data do support findings of Steffens (Steffens et al. 2005) showing no activity changes in glutamine synthetase in the temporal cortex of human resected tissue. Thus, in the pentylenetetrazole model the authors of the present paper were unable to find any reduction in glutamine synthetase activity. The implication is that the pentylenetetrazole model of tonic-clonic seizures may be excellent for the study of early processes in borderline nonsclerotic tissue. Many questions remain to be resolved in this interesting area.

Chapter 3
Epidemiology of Epilepsy

The epidemiology of epilepsy varies worldwide, with age, sex, comorbidities, overall health, reporting schemes, definitions, etc. It was once stated that "the object of any science is the accumulation of systematized verifiable knowledge, achieved through observation, experimentation, and thought" (Hill 1953).

Epidemiology focuses on the distribution and determinants of disease in groups of individuals who have some characteristics, exposures, or diseases in common. The prevalence of a disease is the total number of subjects who have a certain disease at a particular point in time. This number is factored over the number of subjects at risk. The evaluation of the number of subjects per se over a period of time is the incidence.

Epilepsy incidence ratios have been estimated for several populations, and found to range from a low of 35/100,000 to 124/100,000 in Chile. An estimate for the USA is about 60/100,000 (Shamansky and Glaser 1979; Annegers et al. 1999; Zarrelli et al. 1999). Differences in rates may have to do with definition of epilepsy, methods of data collecting, etc. There are some unexplained variations in incidence of epilepsy over many years, such as decrease in incidence of about 40% from 1935 to 1975 (Hauser et al. 1993) in Rochester Minn. The incidence has started to rise again, and this could be attributable to an increase in survival of low birth weight infants. The reason for the earlier decline is unclear.

Incidence of epilepsy based on age shows the highest rate during the first year of life. This rate may then drop to around 40/100,000 in a Canadian study, then to 20/100,000 in adult years (Camfield et al. 1996). The effect of gender on incidence of epilepsy is another factor often measured. It is frequently stated that this incidence is higher in males, but not statistically significantly higher. This is another epidemiologic/statistical fact: if it is not statistically higher, then it is not higher within some error range (0.05). Comparisons based on race lack reliability. This refers to consistency – will the same data be obtained if the study is repeated.

The etiology of seizures is somewhat more difficult to ascertain. From 60 to 80% of seizures are idiopathic. Those which have a proven etiology include trauma, infection, HUS, etc. Genetic factors play a role in the incidence, as do structural CNS developmental defects and tuberous sclerosis. Equally important is to identify

D.W. McCandless, *Epilepsy: Animal and Human Correlations*,
DOI 10.1007/978-1-4614-0361-6_3, © Springer Science+Business Media, LLC 2012

possible factors which are in fact not causal for epilepsy. For example, "birth trauma" and "complications of pregnancy" have been shown not to be related to epilepsy. Also, febrile seizures per se are not causal, but may reflect preexisting susceptibility.

In terms of incidence, studies of all ages and all seizures, only a handful of population-based studies exist, the Rochester Minn study being one (Zarrelli et al. 1999). In this study, 20% of childhood cases were in nonspecific categories, and another 1/3 were judged to be location-related cryptogenic epilepsies with no further localization.

Overall results show approximate incidences of 2–8/10,000 of West syndrome, 0.6 cases/10,000 of juvenile myoclonic epilepsy, 2–5/1,000 cases of Lennox–Gastaut syndrome, and Dravet syndrome is about 1/40,000.

The overall incidence rates of epilepsy range from about 35 to 95 cases per 100,000, and prevalence rates range from about 3 to 11 cases per 1,000. Many studies show the incidence of epilepsy is decreasing. Overall, it is difficult to compare different studies due to the differences in definitions and in data gathering methods. No doubt political considerations in some countries play a role in reporting disease statistics. Particularly, rare forms and types of epilepsies contribute to the overall picture, with limited reliability.

This paper examines features which have been determined regarding the epidemiology of epilepsy in India (Bharucha 2003). The author notes that while epilepsy is clinically similar in developing countries and in developed countries, the recognition of epilepsy and investigation is different between the two groups. In India, rich areas have medical services equal to any worldwide, whereas in poor rural areas, epilepsy largely goes unrecognized and untreated.

In India, a recent meta-analysis has shown that the overall prevalence rate of epilepsy is 5.6 per 1,000. There is no statistically significant difference between city and rural rates (Sridharan and Murthy 1999). There are no published studies from India on either incidence or prognosis of epilepsy. Unpublished data suggest a mortality rate higher than the usual published rate worldwide of two to three times that of nonepileptic people. This is due to burning, drowning, or accidental death.

Various risk factors may be involved as etiologic predeterminants for epilepsy in developing countries. These may include febrile seizures, head trauma, neurocysticercosis, genetics, etc. A possibly unique epilepsy in South India is termed hot water epilepsy, induced by pouring hot water over the head. The prevalence of hot water epilepsy may range from 1.14 to 2.99 cases per 1,000.

A treatment gap exists in developing countries in which 80% or more epilepsy patients do not receive treatment. This information has stimulated the WHO to initiate the Global Campaign against Epilepsy. The reasons are many for this gap. Stigma is an important negative factor in patients seeking treatment. Costs and lack of knowledge are additional significant factors (Gambhir et al. 1995).

The epidemiology of epilepsy in Europe largely originates from the UK, Nordic, Baltic, and Western Mediterranean countries. Large areas of Europe have no epidemiological data on epilepsy. The studies from Europe are all country based; there is no larger study or the epidemiology of epilepsy in Europe. In many studies,

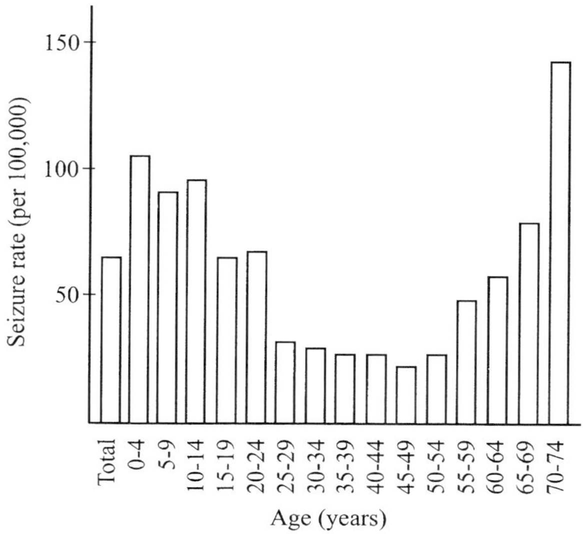

Fig. 3.1 Incidence of epilepsy in relation to age. Adapted from Tallis, R., et al. Age and Aging 20: p. 446, 1991

the prevalence of epilepsy rage from 3.3 to 7.8 per 1,000. In children, the rates were 3.2 to 5.1 per 1,000 inhabitants. Prevalence by gender showed higher numbers of cases in males, but the increase was not statistically significant. The prevalence by age shows a similar incidence for all ages.

In terms of seizure types in Europe, it was found that from 1/3 to 2/3 of patients had partial seizures, 17–60% had generalized seizures, and 2.8% had unclassified seizures (Maremmani et al. 1991).

Incidence studies show that the highest group is children 0–1 years of age. Incidence drops until age 40 at which time it starts to increase again. Incidence numbers for seizure types are similar to prevalence studies, with the most frequent incidence of seizures being partial seizures (see Fig. 3.1).

A population-based study looking to estimate the incidence of first unprovoked seizures in low income people of Northern Manhattan, New York was undertaken (Benn et al. 2008).

Results showed that unprovoked seizures and newly diagnosed epilepsy were identified in 209 people. Fifty-nine percent of cases presented with a single seizure, and 39% presented with newly diagnosed epilepsy. The median age was 24.5 years old in Hispanic patients and 54.6 in non-Hispanics. Hispanics constituted 68% at a public health center. The annual income was below $15,000 in 53%.

The incidence of first unprovoked seizure and newly diagnosed epilepsy in this area of New York was 38.6 cases per 100,000. Age- and sex-adjusted incidence was 41.1 and there were 32 deaths over 4 years, with over half occurring during the first year.

The rates of incidence for this group in terms of first unproved seizures and newly diagnosed epilepsy are somewhat lower than those reported in similar studies. No difference was based on this study as regards incidence by ethnicity.

There was an association between socioeconomic status and increased risk for epilepsy. The mortality rate was greater in this New York area than in the US general population.

A similar study of newly diagnosed single unprovoked seizures was reported from Stockholm (Adelow et al. 2009). Using population-based study of incident cases provides an unbiased assessment of unprovoked seizure and epilepsy incidence. Previous studies have shown incidence rates of 20–80 per 100,000. Previous studies have had relatively small numbers, while in the present study over 1,000 patients met the criteria for inclusion.

Results from this study showed 430 had a single unprovoked seizure, and 585 had recurrent seizures, and were newly diagnosed with epilepsy. Seizures types, using all available information showed 99 patients with generalized seizures and 629 patients with partial seizures. Another 282 had unclassified seizures. The cause of the unprovoked seizures was unknown in 62.4% of seizures. The most commonly identified etiologies were stroke and brain tumors. Genetic factors, hypoxia, and brain injury were other common factors.

The authors comment that the highest incidence was in the first year of life. The overall identification of etiology was 37.6%. The authors comment that in their study the incidence rate was lower than previous studies (MacDonald et. al. 2000). This may be due to an underestimation of incidence in the elderly. This cohort would be useful for long-term follow-up.

Epilepsy is often thought of as a disease of children only, however the actual incidence of epilepsy increase in the elderly, and is a serious health problem (Faught 1999). In the Rochester Minn study (Hauser et al. 1993), data show that the incidence of epilepsy in over 60-year-old patients was 95 per 100,000 in 1935, and rose to 140 per 100,000 by 1944. The incidence of patients surviving stroke has also risen.

The prevalence of people taking anticonvulsant medication aged 20–50 was 5 per 1,000 in the UK, and the rate rose to 7 per 1,000 by age 85 per 1,000. The reasons for the rise could be increased awareness, and an increase in survival from disorders which may precede seizures (stroke).

In the elderly, cerebrovascular disease is the most commonly implicated cause of seizures. Another symptom in the elderly is dementia, and it is associated with epilepsy (Hesdorffer and Hauser 1996). Infections and trauma from falls are associated with epilepsy and are common in elderly patients.

In terms of seizure types, partial onset seizure, predominate in the elderly, contributes 70% of cases. Complex partial seizures can be difficult to diagnose in elderly patients. The seizures may become "briefer and less elaborate" in old age (Tinuper et al. 1996). The effects of seizures in terms of quality of life are likely to be similar in elderly and children, which means try to achieve complete seizure control.

Regarding treatment, reactive epilepsies may not need treatment, for example if a patient has a correctible metabolic encephalopathy, the seizures will stop when the metabolic disturbance is corrected. First seizures not usually treated in children, probably should be treated in the elderly, since 3/4 go on to develop more seizures (Luhdorf et al. 1986). In drug treatment, one must always be aware that pharmacokinetic parameters are likely altered in aged patients. Toxicities such as ataxia and/or cognitive effects are often difficult to distinguish in the elderly.

The author concludes this thoughtful paper saying that physician treating potentially epileptic elderly patients should always be aware of the rising incidence of epilepsy in these patients. Staring spells, confusion, ataxia, etc., all could be semiology of epilepsy. Potential etiology should be looked for, although not determinable in ½ of the cases. Drug treatment should start at low dozes, and slow upward titration. Successful treatment of epilepsy in elderly patients is an intellectual challenge, and when successful, rewarding for both patient and physician.

An interesting study was performed looking at the genetic epidemiology of seizures (Sharma 2005). This study included 199 pairs of twins who were originally recruited for a growth and developmental study.

Results show 66 monozygotic and 133 dizygotic twins. Epilepsy was defined as two or more idiopathic seizures. The concordance rates were higher in monozygotic twins than in dizygotic twins. The case-wise prevalence of epilepsy was similar between the two groups. Twenty percent of affected twin kinships had epileptic first-degree relations. Proband concordance rate in monozygotic twins was four times more than seen in dizygotic twins (0.67 vs. 0.17). These results are consistent with significant genetic susceptibility to epilepsy.

The author notes that prevalence rate between monozygotic twins and dizygotic twins was similar. This is an agreement with a previous study also showing no difference in prevalence between the two groups (Berkovic et al. 1993). Additionally, the author notes the prevalence rate of epilepsy in twins is higher than in the general population. Further, the prevalence rate is higher in underdeveloped countries (WHO 1978). The author observes that after WWII, the prevalence of epilepsy would have dropped since the use of antibiotics and improved healthcare greatly increased. The opposite has occurred, with prevalence rates increased. The author states this must be due to an increase in environmental factors, which influence epilepsy.

Malaria is a common parasitic disease worldwide. This disorder accounts for over 1 million deaths per year (WHO 2004). The parasite *Plasmodium falciparum* is common in sub-Saharan Africa and in south East Asia, and represents the most severe form of malaria parasite. This parasite accounts for at least 100,000 childhood death per annum (mortality rate 20%) and 20,000 more neurological sequelae. Kernicterus actually accounts for a significant number of these morbidities and mortalities (McCandless 2011).

Cerebral malaria, a severe neurologic form of malaria has long been thought of as causing epilepsy. Recent epidemiological studies have examined the link between cerebral malaria and epilepsy (Carter et al. 2004; Ngoungou et al. 2006). In the first study (Carter), 152 children with cerebral malaria were compared to similar numbers of cases of malaria without cerebral malaria, and a group not exposed to malaria.

The prevalence of epilepsy statistically was significantly increased in cerebral malaria, children odds ratio = 4.4; and 95% confidence internal was 1.4–13.7.

The other study (Ngoungou) was an incidence based obtained from a database. Cerebral malaria children (101 patients) were compared to nonexposed children (222 patients). The incidence rate of epilepsy in the cerebral malaria group was 17.0/1,000 person years, 1.8/1,000 person years in the nonexposed group. The age-adjusted risk for developing epilepsy was 14.3 (95% confidence interval = 1.6–132, $p = 0.01$).

This obviously suggests a strong association between cerebral epilepsy and cerebral malaria.

A paper reviewing these studies (Ngoungou and Preux 2008) notes that the methods vary somewhat in design, and may have a bias due to the fact that the source populations were hospital based, and not all subjects in a group could be enrolled. There was also a wide variation in some data. The data are, however, consistent. The association between cerebral malaria and epilepsy was temporally related in that the epilepsy followed onset of cerebral malaria.

The issue of posttraumatic seizures is important due to frequency and severity. The epidemiology of posttraumatic epilepsy has been examined to try to clarify the wide range of reported incidence (Vespa and Nuwer 2000). The significance of posttraumatic head injuries cannot be underestimated. The occurrence of head trauma and resultant seizures has a major effect on cerebral metabolism.

Previous studies over 30 years show an incidence rate of up to 15% in head trauma cases, and that posttraumatic clinical seizure activity occurs within the first 40 h posttrauma (Kollevold 1976). Focal and generalized epilepsy occur with an increased frequency in patients more severely injured (Annegers et al. 1980). Overall, severity of injury and an early age increase the incidence of posttrauma epilepsy.

The incidence of EEG posttraumatic seizures is greater than that of clinical seizures. In one study (Dawson et al. 1951) of 45 patients with severe brain trauma, 24% had EEG evidence of seizures after 1 week following the trauma. The authors of this current paper (Vespa et al. 1997) reported a very similar result (21%) as regards posttrauma epilepsy.

Risk factors include young age (under 5 years of age), diffuse edema, the presence of subdural hematoma, and coma. Penetrating brain injuries such as gunshot victims may have an epilepsy incidence rate of over 40%. The authors (Vespa and Nuwer) note that the identification of posttraumatic epilepsy may be difficult in part because of the low age of many patients. Semiology includes focal clonic activity, eyelid fluttering, lip twitching/smacking, and episodic staring and motor automatisms. Identification is not always easy in a newborn/early infancy setting, or in a stuporous or comatose patient. There have been cases reported of childhood nonconvulsive status epilepticus.

The authors conclude stating that the actual incidence rate of posttraumatic epilepsy may be as high as 20%, and at least one half may only be discerned using EEG data. If seizures occur, as are suspected, then AEDs should be started immediately, and at doses near the high end of the therapeutic range. The prompt and aggressive treatment of status epileptics with benzodiazepines, followed by phenytoin is critical in a good outcome.

The epidemiology of epilepsy and learning disorders has been examined (Lhatoo and Sander 2001). Epilepsy is present in around 0.5–1.0% of the general population in the USA, and learning disabilities are present in about the same numbers (Alberman 1984). As many as ¼ epilepsy patients have learning disorders, and ½ of learning impaired patients have seizures. The authors state the usual reservations, namely that in epidemiology, prevalence in epilepsy is quite difficult to ascertain due to the high variability in screening tests, standardized definitions, and classifications of epilepsy. In addition, adult and elderly patients are underrepresented; the majority of studies are on children.

The prevalence of epilepsy varies in proportion to the prevalence of learning disability. Community studies show a prevalence rate of 6% in children to as high as 24% epilepsy in severe cases of learning disabilities (I.Q. below 50). Epilepsy and learning disabilities are common comorbidities in disorder such as cerebral palsy and cortical dysplasias.

The prognosis for seizure control in patients with learning disorders is poorer than in epilepsy only patients (Rodin 1968). The prognosis of epilepsy depends on the seizure type, for example, in the Lennox–Gastaut syndrome in which seizure control rarely occurs. In one study (Forsgren et al. 1996), language deficit patients with epilepsy were followed. The standardized mortality ratios (SMRs) in patients with both language deficits and epilepsy were higher [SMR, 5.0 (95% CI 3.3–7.5)] than those in children with language deficits alone [SMR, 1.6 (95% CI 1.3–2.0)]. In language deficit patients who have severe epilepsy, seizure-related deaths are most frequent, with sudden unexpected death in epilepsy (SUDEP) the most common category.

The authors state that the association between epilepsy and language deficits is common. In spite of the prevalence of the two conditions (morbidity and mortality), relatively few studies have examined neurological sequelae of language deficits in children. Most treatments for seizures in patients with language deficits are derived from studies on seizures alone, not in studies of patients with seizures plus language deficits.

The authors state that there is a need for well-designed prospective incidence studies of patients with language deficits and epilepsy. Case-controlled studies of etiology and risk factors are needed. And, the authors state that long-term studies using epidemiologic techniques are required. There also needs to be some standardization efforts.

In a paper on epilepsy following stroke (Ferro and Pinto 2004), clinical epidemiology risk factors, etc., are examined. Seizures can be an important initial feature of stroke. Stroke may be the main cause of epilepsy in the elderly (Schreiner et al. 1995), who emphasize the importance of vigorous definitions is critical.

In clinical settings, there is a wide variation in reports of incidence, varying between 1 and 25% of stroke patients. Reasons for the variation include use of nonstandard definitions, case ascertainment, a wide variation in the reports of the seizure witnesses, inclusion in studies of patients with TIAs, who almost never have seizures, etc.

In a large multicenter study of over 2,000 stroke patients, 8.9% of patients had seizures at a 9-month follow-up (Bladin et al. 2000), and the incidence rate was 10.6% in hemorrhagic stroke. Forty percent and fifty-seven percent of seizures occurred in less than 24 h in ischemic hemorrhagic, respectively. Remote seizures (greater than 2 weeks after stroke) occurred at a rate of 3.8% in ischemic and 2.6% in hemorrhagic strokes. In a French study (Giroud et al. 1994), seizures occurred in 5.4% of stroke/TIAA patients with 15 days. In the seizure groups, seizures were the initial symptom in 89% of cases. These authors also reported seizures in 15% of subcortical strokes.

In terms of risk factors, studies have indicated that young patients are at an increased risk of poststroke seizures (Arboix et al. 1997). The potential role of risk factors, such as febrile seizures, previous head trauma, etc., has not been studied in stroke/epilepsy patients. Preexisting medical and/or neurological conditions are, of course, associated with increased incidence of epilepsy in stroke patients. There is a lack of follow-up studies looking for epilepsy incidence after stroke.

Status epilepticus may range from 19 to 27% of stroke patients (Labovitz et al. 2001). Status epilepticus can be the first epileptic symptom of a stroke. Status epilepticus may be nonconvulsive, and there must be a high degree of suspicion and EEG monitoring. Stroke is the most common etiology in the elderly patient for epilepsy.

In terms of pathogenesis, stroke-induced seizures most likely would involve ischemia. One animal study (Marciani et al. 1991) shows that there is spiking and epileptic form discharges in the hippocampus and frontal cortex. The postulate was that in time there was a release of glutamate. Other studies show that there are alterations to the structure and function of GABAergic interneurons (Kessler et al. 2002).

There is scant information on treatment for epilepsy in stroke patients. It is reasonable to start AEDs with one of the three classic choices, phenytoin, carbamazepine, or valproate. The authors state that there is a need for controlled studies on this aspect of epilepsy and stroke. The authors note that studies of epilepsy in stroke are needed to elucidate mechanisms and to define prognoses are critical.

The last several years have had many new developments in the area of genetics in epilepsy. Well over 12 genes potentially involved in seizures have been identified. The present paper (Winawer and Shinnar 2005) examines various concepts and epidemiology of epilepsy.

Epilepsy is a highly heterogeneous disorder which results from many causes, including genetic. Most genetic epilepsy disorders do not follow classical genetic modes of inheritance. Therefore, risk of affliction lies in epidemiology studies against a background of a 1% incidence in the general population. The risk of a child becoming epileptic when born to epileptic parents is 2.4–4.6%. The rate for a monther is about double the rate for then having epileptic offspring (Anderson and Hauser 1997). Not enough date exists to calculate risk in cases of multiple family members with epilepsy. Acquired epilepsies, from head trauma, for example, have no impact on relative's epilepsies.

Some seizure types predict an increased risk for epilepsy in offspring. There is a 4.8% risk of epilepsy in offspring of parents with myoclonic seizures, and about the same increase for offspring of absence seizure parents (Annegers et al. 1982).

Lennox (1951) showed 60 years ago that there was a greater concordance in monozygotic twins than in dizygotic twins. This provided evidence for a genetic component in some epilepsies. This has been confirmed in later studies. Family studies also provide some evidence for genetic factors in epilepsy. Data support evidence for genetic effects on generalized vs. localization-related epilepsy (Winawer et al. 2003).

The authors comment that a thorough history is critical in ascertaining risk for offspring of parents who both have seizures. The etiology of parental epilepsy is important to determine if possible. The overall risk is relatively small except in monozygotic twins.

Associated above, juvenile myoclonic epilepsy has an etiologic genetic factor, and the epidemiology has been examined (Welty 2006). Juvenile myoclonic epilepsy has been first described over 150 years ago, and is a primary generalized, and is idiopathic, often first presenting in the teens. Multiple gene mutations, seizure types are associated with juvenile myoclonic epilepsy (see Chaps. 11 and 12).

From an epidemiological standpoint, juvenile myoclonic epilepsy is one of the more common seizure types with a prevalence rate of from 4.10% of all epileptic patients. The incidence of all epilepsies is from 35 to 124/100,000 persons per year in pediatric ages, and the incidence of juvenile myoclonic epilepsy is 0.5–6.3/100,000 (Hauser 2001).

Forty percent or more of those diagnosed with juvenile myoclonic epilepsy have a family history of seizures. This incidence is higher than other seizure types. Many genetic alterations have been described with at least half on chromosome 6, and related to neuro GABA on glutamate receptor. Otherwise neuropathologic mechanisms are not clear, in part because of the genetic diversity.

Clinical features are described in the juvenile myoclonic epilepsy chapter in this volume. They consist of generalized tonic-clonic seizures in otherwise healthy patient. Seizures frequently occur early in the A.M. EEG recordings are essential in making an accurate diagnosis.

A recent paper (Neville et al. 2007) has examined epidemiology in cases of convulsive status epilepticus in children. Convulsive status epilepticus is a true medical emergency, which requires immediate aggressive treatment in order to avert disaster. The definition used in this study (and in most studies) is continuing seizures or a series of seizures lasting over 30 min without regaining consciousness (see Chap. 23).

Most previous work on epidemiology of status epilepticus has been based on adult status. There were widely based variations reported on incidence, ranging from 4.27/100,000 per year (Chin et al. 2004). In another study (Chin et al. 2006), incidence, occurrence, etiology, and seizure features were described in a cohort of children with status epilepticus.

Results from this study showed that the ascertainment corrected incidence of childhood status epilepticus was 17–23 per 100,000 per year. One third were caused by prolonged febrile seizures. The age-specific incidence reflected a predominance of 0.4 years of age. There was a significant increase in incidence in children of a low socioeconomic background. The mortality rate in other studies ranges from 3.6% to 11%; in the present study the mortality rate was 3.4%.

The authors conclude stating that childhood status epilepticus more common in younger children, and febrile seizures were the most common cause. They advocate IV lorazepam as a first treatment rather than rectal diazepam. The 1-year recurrence rate was 17%.

Another study by the same group (Raspall-Chaure et al. 2007) adds to the previous study on convulsive status epilepticus in children. The authors correctly state in the introduction that one significant shortcoming in understanding the epidemiology of status epilepticus is the lack of consistent definitions of etiologies, seizure types, and overall inappropriate study design.

An example is in actually defining status epilepsy. The ongoing debate, reflected in varying definitions in the literature centers and the length of time of seizures before status epilepticus is declared. In the 1970s, the time was 1 h uninterrupted seizures/lack of consciousness. Later, the time was dropped to 30 min. On the last terminology definition, the ILAE used a definition without any time criterion stated.

It has been shown that there is an association between fever and childhood status epilepticus. Recent ILAE guidelines include febrile childhood status epilepticus in the same category as childhood status epilepticus due to unidentified acute neurological insult. This combination may not be proper in terms of outcome studies. A revised classification should have febrile childhood status epilepticus as a distinct category.

Further incidence data show that problems with estimating length of time of seizure could act to lower actual incidence rates. The crude incidence rate may need adjustment for ascertainment rates. The estimated incidence rate in Europe is 10–38/100,000 per year. The incidence rate is higher in children less than 1 year of age. This may in part be due to the higher propensity for young brains to seize, regardless of cause.

The incidence of childhood status epilepticus is significantly higher in males which could be due to the higher incidence of head injuries in male children. There may also be an association between socioeconomic states and childhood status epilepticus, although children of lower socioeconomic backgrounds may engage in more dangerous play (with resultant head injuries) than other groups. In addition to the mortality rate associated with childhood status epilepticus, morbidity is an important outcome. Studies have shown that etiology is an important predictor of morbidity. Overall morbidity is relatively low (15%) in those who survive childhood status epilepticus. Minor small sequelae may easily be missed, however. One study has shown no cognitive determination as confirmed by I.Q. tests administered before and after the seizure episode (Adachi et al. 2005).

A significant number (16%) of children will have another occurrence of status epilepticus in less than 1 year. The risk of subsequent epilepsy over a 2-year period ranges from 25 to 40% (Maytal et al. 1989). Animal studies support the concept that prolonged febrile seizures may decrease resistance to seizure activity (Dube et al. 2000).

In summary, the authors note that the incidence in childhood status epilepticus has been clarified, and risk factors identified. Further epidemiologic studies are needed in examining the influence of genetic factors, geographic variations, and prehospital pretreatment. These studies should be population based and prospective. Many new well-done studies are needed.

Another paper examines the epidemiology of convulsive and nonconvulsive status epilepticus (Rosenow et al. 2007). The paper's authors state in the introduction that the ILAE defines status in terms of the 30-min course described earlier, but they add that in clinical practices, a tonic seizure lasting more than 5–10 min would be considered as status epilepticus and treated accordingly. The authors continue saying that there are as many types of status epilepticus as there are seizure types. There is also a classification based on the presence or absence of convulsions. The above exemplifies the problems relating to definitions.

This paper reviews recent papers showing incidence rates of status epilepticus of about 10–20/100,000. These data are from recent population-based studies in Europe. Other data from this report were that the incidence of status epilepticus was three times higher in African Americans than in white people (DeLorenzo 2006). As stated earlier, the incidence of status epilepticus is higher in males than females. Noncompulsive status epilepticus constitutes about 25–50% of all status cases.

The epidemiological aspects of Panayiotopoulos syndrome have been published (Durá-Travé et al. 2008). This syndrome falls under a classification to benign childhood occipital epilepsy. There are two types, one with early onset (Panayiotopoulos syndrome) and the other late onset epilepsy (Gastaut type).

Clinical manifestation relates somewhat to the occipital lobe and includes alterations in consciousness, behavioral abnormalities, autonomic changes, eye deviations, blindness, visual hallucinations, etc. The mean duration of seizures was 9.3 min. Twenty-five percent of the patients in this study had at least one seizure once every 30 min.

The study consisted of 37 patients diagnosed with Panayiotopoulos syndrome. There was a female/male ratio of 2:1. Ten patients of the 37 had febrile seizures, or there was a family history of febrile seizures. Mean age of the first seizure was 5.4 years of age; follow-up lasted a mean of 6.0 years. Psychomotor development was normal in all; three had learning difficulties.

EEG findings showed occipital spikes in 28 of the 37 patients. In 50% the spikes were bilateral and synchronous. The remainder had unilateral right-sided spikes. MRI and PET showed no abnormalities associated with epilepsy. AEDs were prescribed in 23 of the patients. After a mean of 3.2 years, treatment was suspended, and after the suspension, and a 3.5-year follow-up (95% CI: 2.0–4.0) there was no relapse.

The authors comment that the age of initial diagnosis matched that of other studies (Ferrie et al. 1997). Eighty-nine percent of children were diagnosed below the age of 9, the rest at an older age, but symptoms and prognosis were similar. The combination of autonomic and behavioral symptoms in a very high percentage of patients could indicate that many cases may go unnoticed.

The authors conclude saying that the prognosis of Panayiotopoulos syndrome is excellent. The syndrome shows a tendency to spontaneous remission after 1–2 years, and psychosocial development continues unimpeded. There is an excellent response to AEDs. This seizure disorder is relatively frequent and should be watched for, even though the semiology is almost unique in epilepsies.

Another paper (Singh and Prabhakar 2008) examines various aspects of the association of CNS infections and epilepsy, including epidemiological approaches. In the USA, at least 25,000 patients are hospitalized with bacterial meningitis or viral encephalitis. The upper incidence is 14/100,000 for the two disorders combined

(Schuchat et al. 1997). A safe figure for the etiology of epilepsy due to CNS infections is probably 3–5% (Annegers et al. 1996).

The occurrence of acute symptomatic seizures resulting from CNS infections increases the risk for late unprovoked seizures. The risk depends no doubt, say the authors, on the underlying mechanism responsible for the seizures associated with the infection. Structural residue in brain parenchyma following the acute phase may increase this risk to 25% (Hesdorffer et al. 1998).

The authors note that certain problems complicate the evaluation of the association of CNS infections and epilepsy. There are, for example, no validated scales for assessing the severity of brain damage in CNS infections. This renders impractical the correlation of infection injury to the incidence of late seizures. The differentiation of provoked and unprovoked seizures may be difficult.

The authors continue saying further studies looking at the association between CNS infections and epilepsy are much needed. Studies should focus on late unprovoked seizures and the major CNS infections, and the remission and intractability rates of the associated epilepsy. Contemporary imaging methods should define anatomical substrates. The authors note that the aim is to estimate the burden of epilepsy due to CNS infections, then reduce the burden of this epilepsy by attempting to prevent infectious disorders. This is accomplished by improving preventative health care.

Part II
Generalized Seizures

Chapter 4
Animal Models of Absence Seizures

Absence seizures are characterized by many brief (up to 30 s) interruptions in consciousness, with characteristic bilaterally synchronous spike wave discharges of about 3 Hz. The phenotype consists of staring spells, and may be accompanied by atonia and automatisms such as lip or mouth movements. The incidence is somewhere between two and eight cases per 100,000, with girls affected more than boys (Panayiotopoulos 1997). Early studies in animals demonstrated that electrical stimulation to the external midline of the brain at a frequency of 3 Hz yielded spike wave discharges (Jasper and Droogleever-Fortuyn 1947). A bit later, it was shown that identical 3-Hz spike wave discharges could be recorded in children with deep recording electrodes in the thalamus (Williams 1953).

Absence seizures in humans are relatively common, although less frequent than complex partial seizures, which may be difficult to differentiate from absence seizures. The typical absence seizure is a rapid interruption in activity. This is followed by several seconds of loss of awareness (consciousness) in which the subject appears to be starring. Automatisms may be a feature, usually in the oral region. As quickly as it starts, the seizure stops, and the patients returns to the activity in which they were previously engaged. It is classified as a generalized seizure because of the loss of consciousness. This seizure type is often called pyknolepsy because of the tendency for the episodes to occur in clusters. The disorder occurs primarily in children, with a usual onset before 10 years of age. Absence seizures usually respond well to AEDs, and frequently resolve spontaneously by mid- to late teens. The EEG is characterized by spike and slow wave paroxysms of 3 Hz. Absence seizures are often confused with complex partial seizures, which share many of the same features. Ultimately, the EEG will serve to differentiate the two in cases which are difficult to distinguish. Many children are accused of "day dreaming" in school when in fact they are experiencing absence seizures.

As regards animal models of seizures, there are some excellent models which have been important in providing data regarding all aspects of absence seizures. Generally, excellent animal models of absence seizures should have overt features consistent with the EEG. A good absence seizure model will display clinical

D.W. McCandless, *Epilepsy: Animal and Human Correlations*,
DOI 10.1007/978-1-4614-0361-6_4, © Springer Science+Business Media, LLC 2012

features in accordance with the approximately 2-year life span of mice and rats; the age of onset of animal models should parallel that of the human disorder. The treatment of animal models with AEDs should have the same response or lack of response as humans. Ideally these criteria for animal models of absence seizures will be met (Loscher and Schmidt 1988).

Genetic models of animal seizures are favored by some since human absence seizures may have a genetic aspect. Results from twin studies suggest that when one twin has seizures, so will the other twin. On the other hand, some seizure states are acquired, so other nongenetic models are also useful.

There are many genetic models of seizures in both mice and rats. Several of these have first been described and/or have been maintained by the Jackson Laboratories in Maine. One such mouse is called the "totterer" mouse, which is ataxic and has motor seizures, and cortical spike wave discharges. This represents two seizure types in one model, each of which responds to different AEDs.

In terms of absence seizures, there is a rat model created by breeding two different single cause epileptic rats, and the progeny have a combination of convulsive and nonconvulsive (absence) seizures. In these rats, the convulsive symptoms responded to phenytoin, while the nonconvulsive aspect responded to ethosuximide (Sasa et al. 1988). This model also has significant EEG features from the hippocampus.

Another rat model of absence seizures, the WAG/Rij model, has also been described (Van Luijtelaar and Coenen 1986). These animals display a phenotype characterized by immobilized behavior, vibrissal twitching, increased breathing, and head tilting. The behavior was recorded by video EEG which showed the immobilization immediately preceded the spike wave discharges. These lasted about 5 s, and occurred up to 18 times per hour. These discharges also occurred during sleep. It was noted that these absence nocturnal discharges also are seen in human patients with absence seizures. Finally the WAG/Rij rat shows circadian rhythms similar to those seen in human patients with absence seizures.

In terms of the pharmacological responses of the WAG/Rij rat, both ethosuximide and trimethadione produced a reduction in the number of spike and wave EEG discharges, whereas the anticonvulsant drugs diphenylhydantoin and carbamazepine actually caused an increase in spike wave discharges, These pharmacological results, similar to those in the WAG/Rij rat and in human absence epilepsy, favor the rat WAG/Rij model as an excellent one (Peeters et al. 1988). Further studies (Coenen et al. 1992) examined the effect on convulsive and nonconvulsive rats of GABA agonists and antagonists. It was found that muscimal, a GABA agonist decreased convulsive and increased nonconvulsive seizures, whereas NMDA, a glutamate agonist increased convulsions, and also increased nonconvulsive seizures. These results support the concept of a differential pharmacological profile of convulsive and nonconvulsive epilepsies.

In terms of genetic rat absence models, other strains such as BN/BiRij, G/Cpb, and B/Cpb all show spike wave discharges on EEG examination to a lesser extent. This suggests that absence epilepsy is more frequent in inbred rodents than was previously believed. As in the rat models of absence epilepsy, there is a brief interruption in the epilepsy patient's ability to maintain contact with the surroundings.

An interesting paper (Hosford and Wang 1997) examined the utility of the ih/ih mouse model to predict the effects of putative AEDs. The ih/ih mouse model was used due to its ability to predict effects of ethosuximide, valproate, phenytoin, etc. In this study, the model was used to estimate the effect of lamotrigene, vigabatrin, tiagabine, gabcepentin, and topiramate. Bilateral electrodes were implanted in the frontal neonatal cortex of 8-week-old mice. All AEDs were injected I.P. on every other day except vigabatrine, which was administered daily.

Results demonstrated that of the AEDs tried, only lamotrigine significantly reduced (65%) seizure frequency. Vigrabatrin and tiagabine significantly increased seizure activity, while the other anticonvulsants had no effect.

The authors conclude that the ih/ih mouse model of human absence epilepsy was a good model for predicting efficacy of AEDs. Previous use by others of a high-dose pentylenetetrazole (a generalized clonic seizure producer) has not always been as efficient as should be in testing potential anti absence epilepsy drugs. The ih/ih model correctly predicted the positive effect of lamotrigine, and the negative (pro absence) effects of both vigabatrin and tiagubine. Thus, the authors state the ih/ih absence model seems much better as compared with the pentyltetrazole mouse model in its ability to assess the putative efficacy or lack thereof of new AEDs designed to help patients with absence seizures.

An interesting paper looking developmental processes and absence seizure activity in the stargazer mouse model yielded interesting results (Qiao and Noebels 1993). In this study, 2-week-old mice (control 57BL/6J and the mutant stargazer stg/stg C3Ble Fe) were anesthetized and electrodes attached to a microminiature connector were implanted over the frontal and parietal cortices. Deep electrodes were implanted in the hippocampus in some animals. Recordings were obtained after surgery from mice while they were freely moving in their cages. Histological sections were examined for mossy fibers. Resultant brain sections were stained for vesicular zinc to determine the extent of recurrent mossy fiber outgrowth. Regional staining was assured using image analysis of neuronyl hippocampal populations. Electron microscopy was also used for localization of zinc-rich terminals.

Results showed that stargazer EEGs spontaneously displayed bilateral symmetrical 6–7 s spike wave discharges in both cortex and hippocampus. The discharges from the two areas were similar. No other abnormal EEG recordings were noted. Spontaneous seizure rate of adult stargazer mice ranges from about 43 to 200 discharges per hour.

Histochemical results showed Timm's silver precipitate present in areas representing the hippocampal mossy fibers. No staining was present in the inner molecular layer of the dentate gyrus. Similar staining results were obtained using selenium. Electron microscopy of mossy fiber sprouting demonstrated Timm's positive particles in mossy fiber terminal elements. They contained mitochondria and neurotransmitter vesicles.

Stargazer mice from postnatal day 14 to 6 months of age were examined to study mossy fiber sprouting and seizure activity. Spike wave discharges in stargazer mice were not recorded at day 15, but were present by day 18. The discharges in young mice were essentially the same as those seen in adult stargazer mice. Within 1 week,

Fig. 4.1 Representative EEG traces in stargazer mice (postnatal day 18). Adapted from Qiao, X., and Noebels, J. J. Neurosci. 13:4622–4635, 2001

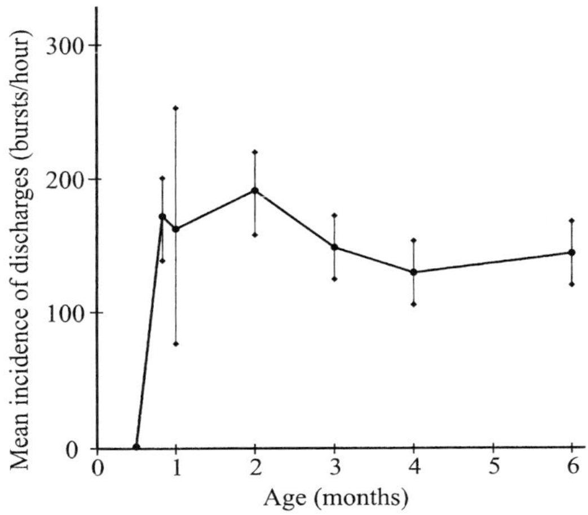

Fig. 4.2 Onset and mean incidence of discharge activity in stargazer mice. Data and source as in Fig. 4.1

the frequency approached 150 spike wave discharges per hour, which stays consistent during further development. The authors state a positive correlation between continuous seizure activity and increased mossy fiber sprouting. In terms of neuronal density (as an indication of cell death), results showed that hilar cells were significantly reduced (16%), whereas cells of CA3 granular and pyramidal cells were not decreased (Figs. 4.1 and 4.2).

The authors conclude saying that, just like other more severe seizure states, this study suggests that spike wave seizure discharges increase the possibility for structural brain damage. Further, continued activity over time, may induce cell death. Indeed, bursts of spike wave discharges may occur many hundreds of times per day (Browne et al. 1983). The structural reorganization of mossy fiber terminals may represent a secondary mechanism for additional epileptogenesis.

Another study (Letts et al. 1998) describes a gene encoding a 36-kD transmembrane protein whose expression was disrupted by two alleles in stargazer mice. The compound stargazin is encoded by four exons and one intron. The spg allele has an insertion of an ETn retrotransposon into intron 2. This is not the only mouse mutant which has an ETn insertion (Steinmeyer et al. 1991). The stargazin mRNA is expressed in all brain areas, but the highest levels are in hippocampus and cerebellum, among others. The stargazer's phenotype, similar to that of lethargic and tottering mice, suggests that stargazin may represent a new gamma subunit for neuronyl Ca2 channels. The discovery of stargazin completes the description of the major subunits for neuronyl Ca^{2+} channels. These include alpha 1, alpha 2 theta, and gamma.

A proper examination of mechanisms of seizures in the lethargic mouse model of absence seizures may shed light on regulatory mechanisms (Hosford et al. 1999). In this paper, the authors note that the lethargic (lh/lh) mouse mutant has spontaneous absence seizures that are similar to those in human absence seizures in that these mice share behavioral, EEG, and anticonvulsant features with afflicted humans. This represents at least partial justification for the lh/lh mouse serving as an absence seizure model.

A couple of mechanisms regulating the synchronized discharges resulting in absence seizures are described. One of these is the activation of GABA B receptors in thalamocortical cells. These receptors are increased in lh/lh mice as compared to littermate controls. Compounds, which activate GABA B receptors in humans, generate absence seizures. This suggests antagonists to GABA B may serve as effective AEDS. In addition, activation of GABA A in the nucleus reticularis thalami seems to act to suppress generation of absence seizures in the ih/ih mouse model.

These data seem to indicate that compounds selectively activating the GABA B receptor might be efficient at limiting absence seizures, hopefully with a decrease in adverse effects. Further, the authors note that identification of the defective lh/lh gene should increase understanding of the neuropathology of absence seizures in both the lh/lh mouse model of absence seizures, and by extension, the neuropathology of the human disorder.

Another paper (Cortez et al. 2001) examines various parameters in a rat model of atypical absence seizures. In this study, a cholesterol synthesis blocker, AY 9944, was administered to newborn rats from postnatal day 2 to postnatal day 33. Electrodes were placed on day 50 after birth; two frontal and two parietal electrodes, epidural, were implanted under anesthesia. ECoG recordings were made from day 55 to 75 days of age. This allowed for assessment of which Long Evans rats were most susceptible to the induction of seizures. Following this, male and female rats were subjected to prolonged video EEG recordings from the depth electrodes during sleep and awake cycles in both developing and adult rats. In addition, animals were administered drugs known either to enhance or to block regular absence seizures.

Results showed the ECoG and behavior of AY 9944 administered Long Evans rats did not depend on body weight. Treatment during postnatal development resulted in spontaneous recurring bilateral 5–6 Hz spike wave discharges, and the Long Evans rats were most susceptible. The spike wave discharges were twice as long as were those seen in control animals. The onset and cessation of spike wave discharges were abrupt in AY 9944-treated animals. In contrast, the ictal behavioral features were gradual in onset and cessation. Complete immobility and frozen stare which is characteristic of the other rat models of absence seizures are not seen in the AY9944 Long Evans model. Video EEG recordings at 14 months of age were similar, suggesting the atypical absence seizures were life long. Spike wave discharges were recorded from cortex, thalamus, and hippocampus.

Treatment of the atypical Long Evans rat model with proabsence and anti-absence drugs was predictable in that baclofen and pentylenetetrazole resulted in a prolongation of the AY 9944-induced spike wave discharge, whereas ethosuximide and CGP35348 abolished the AY 9944-induced spike wave discharge activity.

The authors note that the atypical absence model responds to anti- and proseizure drugs very much like animals with typical absence seizures. The idea that the AY 9944 model was a typical absence seizure was disproved based on the less exact EEG findings, and the slower onset and offset of behavior correlates. The authors state that the AY 9944 model is clinically due to the correlation between it and human atypical absence seizures. The authors state that the AY 9944 model represents an atypical absence model similar to the atypical absence seizures seen in the Lennox-Gastaut syndrome and myoclonic static epilepsy (Kaminska et al. 1999). The authors further state that a key difference between atypical and typical seizures may be the involvement of the hippocampus. The authors note that they have unpublished data suggesting a significant decrease in long-term potentiation in the hippocampus in this model of atypical absence seizures.

A review article by one of the above paper's authors examines basic mechanisms of generalized absence seizures (Snead 2004). The author states that absence seizures are frequently encountered clinically, and are unique as regards physiology, pharmacology, and developmentally. Once again, the availability of excellent animal models of seizures, and absence epilepsy in particular, has facilitated increased knowledge of this disorder.

The basic underlying mechanism of absence seizures seems to be related to thalamocortical circuitry and abnormal oscillatory rhythms thereof. The cellular mechanism of discordance between excitation and inhibition may be related to T type calcium current. Cholinergic, dopaminergic, and noradrenergic mechanisms are features of the thalamus, the disruption of which could easily modulate/disrupt the balance of excitation and/or inhibition. The disruption of the balance of these thalamocortical pathways might secondarily alter the generation of the bilateral synchronous spike wave discharges so characteristic of generalized absence seizures. The author finally states that knowledge of mechanisms of pathogenesis could lead to the "design" of drugs customized to the actual nature of the mechanisms of the development of absence seizures in humans. This is another example of translation.

Another study (Tenney et al. 2003) looks at the thalamocortical modulations in an absence seizure rat model. The purpose of the study was to look at the ability of fMRI to evaluate generalized absence seizures in a rodent model. The authors chose to study the gamma hydroxybutyric acid (GHB) model because of its more predictable time course than other models of absence seizures. A precursor molecule to GHB is gamma butyralactone (BBL), which was actually used because it has a quicker onset than GHB.

For imaging, cerebral blood oxygenation level dependent (BOLD) fMRI was measured using a T2-weighted echo planer imager at 4.7 tesla. This was used before, during, and after absence seizures produced by GBL.

Results from eight animals showed GBL-induced bilateral spike wave discharges which were characteristic of absence seizures. GBL caused significant change in BOLD fMRI in the thalamocortical pathway. The thalamus has an increase in the voxels with BOLD signal and a few thalamic regions had negative BOLD signals. In contrast, the cortical regions of interest (ROI) were mostly negative as regards BOLD signal. The significant negative BOLD signal was present in all cortical areas examined.

The authors note that their fMRI data lend support to previous EEG data that the EEG changes (correlated with BOLD data) show increases in activity following activation with GBL. The increase in thalamic and cortical BOLD signal reflects the initiation of spike wave discharges. This shows the ability of fMRI to provide meaningful data. The positive BOLD thalamic signal reflects an increase in neuronal activity. The cortex shows both positive and negative BOLD response have been reported to occur in patients with photo stimulated seizures (Hill et al. 1999).

The authors point out that their study represents the first reported paper looking at BOLD fMRI on absence seizures. Previous studies using 14C-deoxyglucose autoradiography (Wolfson et al. 1977) showed decreases in glucose use in animals given GBL. Studies using PET (Ochs et al. 1987) have been inconclusive due, in part, to spatial resolution issues. The BOLD fMRI method showed brain areas involved in an absence seizure model which was in agreement with previous studies using less sophisticated methodology. This represents a noninvasive method to show changes globally in awake absence seizure models.

The same group (Tenney et al. 2004a, b) published a similar paper looking at a genetic rat model of absence seizures. The purpose of this study was to use EEG-triggered fMRI to identify brain areas activated by spike wave discharges in a genetic epileptic rat model. Part of the rationale for this study includes the idea that fMRI imaging has been identified as a powerful tool for noninvasive brain imaging. Electrophysiological recordings from genetic models of seizures have been instrumental in identifying neural regions involved in seizures and provide a basis for continued investigations. Further, the ability to use BOLD fMRI as an imaging technique when EEG signals spike wave discharges is possible.

In this study, male WAG/Rij rats weighing 250–350 g were used. For imaging, rats were anesthetized, and EEG electrodes were fixed to the skull. The head was immobilized, as was the body in order to minimize motion artifact. Images were acquired using a 4.7 T magnet. High-resolution images were taken using fast spin echo sequence. Image acquisition was initiated when EEG revealed spike wave discharges twofold higher than the background lasting more than 1 s. ROIs (regions of interest) included sensory, parietal, and temporal corticies, as well as several thalamic nuclei. EEGs were taken continuously while the rat was in the magnet. During seizure activity, there were increases in the sensory cortex, parietal cortex, and the temporal cortex. BOLD signal intensity was increased in ROIs from 4 to 10%. Thalamic areas showing increase included mediodorsal thalamic nuclei, nucleus reticularis thalami, posterior thalamic nuclear group, and the ventral posteromedial/posterolateral thalamic nucleus.

The authors point out that spontaneous spike wave discharges in WAG/Rij rats correlate as regards regions affected, with BOLD fMRI. The BOLD fMRI activation areas have been previously implicated as involved in WAG/Rij seizing rats. Of interest is that the hippocampus showed no activation in this study, as was expected. The WAG/Rij model studied in this experiment showed diffuse positive BOLD activation, and no significant negative cortical BOLD results. This is in contrast to the previous report in which the authors noted that the GBL model of absence seizures not only produced positive BOLD activation in cortical and thalamic areas, but also showed some negative cortical areas. The WAG/Rij model had no negative cortical areas. The patterns of BOLD fMRI in each model (WAG/Rij and GBL) were similar in that cortical and thalamic areas had positive BOLD results. These changes and spike wave discharges characterize the absence seizures.

The mutant mouse stargazer is not only a suitable model for absence epilepsy, but also has abnormal motor features plus ataxia. A paper examining mechanisms of motor behavior in stargazer mice has appeared (Khan et al. 2004). Previous studies on motor abnormalities in mice have focused on vestibular function. The ataxia seen in stargazer mice may be associated with a decrease in cerebellar size (Noebels et al. 1990), and a loss of external granule cells (Qiao et al. 1998). The motor dysfunction was one of the first behavioral abnormalities described in stargazer mice, and the mechanism has not yet been well described. The present paper describes studies examining the abnormal motor behavior in stargazer mice in order to describe mechanisms of the defect.

In this study, motor function, dyskinesia, vestibular function, and auditory function were all tested. EEG recordings were also taken using an implanted telemetry transmitter in the peritoneal space. Electrodes were guided to the occipital region of the skull. Three weeks of postoperative recovery was allowed until actual recordings were taken.

Results showed that stargazer mice were hyperactive, with activity being three to four times greater than normal mice. A set of mice with toxin-damaged vestibular systems were ten times more active than controls. Another function test called the "cling" test, measuring ability to cling to a wire grid showed the stargazer mice could only cling to the grid for 1/3 the time as normal controls. The dyskinesia test showed stargazers were hyperactive, displayed circling activities, and head movements were side to side. Toxin-treated mice with damaged vestibular systems displayed the same behavior. A vestibular function test consisted of dropping mice upside down from a height of 50 cm. Normal mice landed on their feet most of the time, whereas stargazer mice landed on their feet about 1/2 of the time. Results from EEG studies showed no correlation between spike wave discharges and dyskinetic behavior. Auditory brainstem responses were also normal.

When examined with light microscopy, differences were noted between stargazer and normal mice as regards the vestibular apparatus. These consisted of a reduction of hair cells in the cristae, vacuole defects in epithelia, decreased mitochondria in afferent nerve calyces associated with hair cells, and finally disorganized utricle membranes were noted. Electron microscopy results showed some dislocation of basement membranes, and vacuole defects surrounding small mitochondria.

These kinds of alterations are associated with decreases in energy metabolites in other examples of metabolic encephalopathies (McCandless 2011).

The authors state that their efforts to characterize the abnormal motor behavior, including ataxia, tremor, hyperactivity, circling, head bobbing, etc. did not correlate with the spike wave discharges seen on EEG, therefore are not part of the absence epilepsy. They note that, therefore, this complex motor behavior abnormality probably emanates from a dysfunction of several motor control systems including the cerebellum and vestibular system. Changes in mitochondrial structures seen in both light and electron microscopy suggest changes in energy metabolism. Dyskinesia may involve several compensatory mechanisms. Circling, head wagging and bobbing act to increase proprioception and somatosensory input which act to augment that provided by the damaged vestibular system.

Another paper related to absence seizures looks at fMRI and EEG in WAG/Rij rats (Nersesyan et al. 2004). Thalamocortical pathways are key to forelimb function. Previous studies link neuronal activity to regional cerebellar blood flow and metabolic demands (Mintun et al. 2001). Absence seizures and tonic-clonic seizures are generalized seizures which involve the thalamocortical networks. Electrophysiological recordings in WAG/Rij rats demonstrate that the spike wave discharges are most intense in the perioral region of the somatosensory cortex and the thalamus. This paper examines the effects of spike wave discharges in WAG/Rij rats on oxygen and metabolism needs in affected brain regions using BOLD fMRI.

In this study, MRI and fMRI data were obtained using a 7T spectrometer. High-resolution MRI data were obtained with gradient echo image contrast. WAG/Rij rats weighing about 200 g were anesthetized, and the femoral artery and vein cannulated. Body temperature was maintained at 37 degrees. EEG electrodes were placed, and movement artifacts were minimized.

Results obtained on fMRI from a total of 256 spike wave discharges had a mean seizure duration of 3.3 s. During spike wave discharges, BOLD fMRI showed increased signal intensity in the somatosensory cortex, motor cortex, thalamus, basal ganglia, and brain stem. These changes were generally bilateral. The changes in fMRI generally correlated with spike wave discharges, but lagged a few seconds behind. In order to compare the response of brain regions in spike wave discharges with more severe seizures, tonic-clonic seizures were induced in WAG/Rij rats with bicuculline. These generalized seizures had much more intense widespread areas as seen with fMRI. The fMRI signal changes in tonic-clonic seizures were increased in amplitude and in distribution.

The authors state that their study shows that oxygen delivery exceeds metabolic demand in cerebral areas known to have increased electrophysiological firing during both spike wave discharges and generalized tonic-clonic seizures. Results from this study confirm increased BOLD fMRI in the somatosensory cortex, correlating with spike wave discharges localized to the peri-oral area of the somatosensory cortex. In tonic-clonic seizure activity, the authors suggest that increased oxygen delivery is sufficient to meet metabolic demands. The authors further note that even in "generalized" seizures, some brain areas are spared such that the seizures are not strictly generalized. Not all studies show increases in oxygen and metabolism; some PET studies indicate decreases (Salek-Haddadi et al. 2002). In conclusion, the authors state they have shown that BOLD fMRI and EEG recordings in WAG/Rij rats correspond to spike wave discharges in regions known to be involved in the seizures.

Another paper (Steriade 2006) looks, in part, at spike wave seizures which arise in the neocortex, and their behavior as a result of the influence of the thalamic reticular nuclear complex. These thalamic reticular neurons may lead to an inhibition of thalamic cortical neurons. The Lennox-Gastaut syndrome is similar to spike wave discharge absence seizures, except that the frequency of discharges is lower. The thalamic cortical neurons tend to be inhibited by GABAergic thalamic reticular neurons. This inhibitor may explain the loss of consciousness seen in absence seizures.

A good and interesting review paper has been published on the stargazer mouse model of absence seizures (Letts 2005). This review lists over 60 references and addresses important issues such as AMPA-receptor trafficking, other mutant mice used to study stargazin function, knockout mice, etc. The reader is referred to this review.

Another fine review has been published (Nicoll et al. 2006), reviewing many of the papers which have made progress in our understanding of stargazer mice, stargazin, AMPA receptors, NMDA receptors, and small trans-membrane AMPA receptor regulatory proteins (TARPs). This excellent review appropriately raises more questions than it answers.

A study examining hippocampal involvement in absence seizures in WAG/Rij rats has been published (Tolmacheva and van Luijtelaar 2007). The authors note that previous theories regarding mechanisms of spike wave seizures do not involve the hippocampus, although some evidence suggests that the limbic system might be involved. The idea that a hyperexcitable cortex is sufficient for spike wave discharges has been shown in WAG/Rij rats (Tolmacheva et al. 2004). However, non-epileptic inbred rats of the ACO strain have high cortical excitability, but no spike wave discharges. The WAG/Rij rats also show a low threshold for the spread of epileptic activity into the limbic system. Another study suggested that steroid hormones facilitate GABA A receptor activity, and that progesterone and corticosterone may regulate absence seizures (Schridde and van Luijtelaar 2004). The present study looked at whether progesterone or tiagabine, both GABA A stimulators when injected into the hippocampus, affect absence seizures.

In this study, 6-month-old WAG/Rij rats were anesthetized and tripolar EEG electrodes were implanted. Cannulas were implanted in the CA3 region of the hippocampus, or in the cortex above the hippocampus. The cannulas were placed in order to inject either progesterone or tiagabine intracerebrally. Hippocampal and cortical injections were made through a 31-gauge needle, and the volume injected was 1 μl over a 45-s time frame. Proper cannula placement was assured by injecting 2 μl of cresyl violet to verify location just before sacrifice.

Results showed that progesterone and tiagbine acted to reduce spike wave discharges for a period of 60 min after injection. There was no change in behavior or electroencephalographic side effects. Injection of the steroids into the overlying cortex did not have any effect on spike wave discharges. The authors suggest that in the WAG/Rij absence model, the hippocampus may have a regulatory effect on spike wave discharges. The authors note that the rostral pole of the reticular thalamic nucleus, considered part of the limbic system, may be exerting an effect. It is certain that the activation of GABA A neurotransmitter by the two steroids progesterone and tiagabine in the hippocampus acted to inhibit corticothalamic circuits, thereby reducing spike wave discharges. This in turn might be a target for new treatment AEDs.

A recent paper (Beyer et al. 2008) reports a research which has examined the inbred mouse strain C3H/HeJ, which is prone to absence seizures. The authors found that a mutation of the Gria 4 gene is involved. This gene encodes one of the 4 amino-3 hydroxyl-5 methyl-4 isoxazolepropionic acid (AMPA), brain receptor subunits Gria 4 knockout mice also have spike wave discharges, characteristic for absence epilepsy. The authors note that considerable evidence supports the concept that AMPA receptors have a role in synaptic plasticity. The paper cited here provides evidence, both genetic and physiologic, that AMPA receptors are very important for normal thalamocortical function. The paper demonstrates that altering the balance between the various AMPA subunits may result in absence seizure activity.

Another recent paper (Menuz and Nicoll 2008) examines AMPA receptors and ataxia and seizures in the stargazer mouse model of absence epilepsy. In their study, thalamic AMPA receptors (AMPARs) were assessed in stargazer mice. Absence seizures involve thalamic circuits. These consist of the reticular thalamic nucleus (inhibitory) and excitatory thalamocortical relay neurons (McCormick and Contreras 2001). Both areas express AMPARs and stargazing. The authors of this study aimed to determine whether altered AMPAR could influence the stargazer mouse model of absence seizures.

Methods involved the use of electrophysiological recordings to probe the regulation of AMPAR in critical cerebral regions. Results showed that miniature EPSC's frequency and amplitude were reduced in stargazer mice. The AMPR/NMDA ratio was also reduced, indicating a postsynaptic loss of synaptic AMPARs. The decrease in EPSC's frequency could indicate fewer total synaptic contacts. Since seizures appear in stargazers about day 17 or 18, older mice (day 21–23) also showed miniature EPSC amplitude was decreased.

Stargazin is a unique protein because of its purported ancillary subunit for both AMPARs and calcium channels. The authors note that this study has demonstrated that the AMPA receptor loss from inhibitory neurons may contribute to stargazer's overt behavior. Reticular thalamic nucleus and cerebellar Purkinje cells have significantly reduced synaptic AMPAR function in stargazer mice. Taken together, these data strongly indicate that altered AMPAR regulation (cerebellar Purkinje cells and reticular thalamic nucleus) probably contributes to both the absence seizures, and the ataxia seen in stargazer mice. This loss of AMPAR is most probably due to decreased levels of stargazin, since stargazin is intimately involved in AMPAR (maturation, surface trafficking, and synaptic targeting of AMPARs). These data relate to translation from both the stargazer and WAG/Rij rat.

Another paper recently published examines electrocorticographic (ECoG) and intracellular recordings in the genetic absence epilepsy rat from Strasbourg (GAERS) model of absence seizures (Polak et al. 2009). While thalamocortical circuits have been implicated, the exact contribution of each is not yet unequivocally determined.

Recent data implicates the cerebral cortex as an important site of initiation of abnormal electrical activity (D'Arcangelo et al. 2006; Gurbanova et al. 2006). Indeed, clinical behavioral evidence and animal data indicate that spike wave discharges may originate from a highly specific area of the cerebral cortex, the peri-oral region of the somatosensory cortex. This study used ECoG plus cellular recording to examine this hypothesis.

In this study, adult male and female GAERS were anesthetized, and silver electrodes were placed on the dura above one or both facial somatosensory cortices. Intracellular recordings were obtained using glass micropipettes. Recordings from cortical cells in the peri-orbital (facial) somatosensory cortex were taken.

Results showed that in the GAERS model of absence seizures, a functional inactivation of spike wave discharges was produced by systemic administration or local application of the sodium channel blockers, lidocaine or phenytoin. Furthermore, blocking neuronal discharges in somatosensory cortex by tetrodotoxin stopped local and distant surface spike wave discharges.

By contrast, the application of tetrodotoxin to motor cortex had little effect. Further, this study showed that the thalamocortical neurons could not endogenously generate paroxysmal oscillations, nor were cortical spike wave discharges correlated with neuronal oscillations in the thalamus. Blocking thalamic seizure oscillations did not prevent cortical spike wave discharges. The authors note that this novel demonstration of the ability of a small cortical area (facial somatosensory) to initiate spike wave discharges, as well as their role in the generation of distant thalamic and cortical paroxysmal discharges, provides a direct demonstration of the hypotheses of a cortical focus for absence seizures instead of functional disturbances in the intrathalamic neural networks.

Another paper (Aker et al. 2009) examines features of spike wave discharges in GAERS. In this study, ethosuximide was injected intracerebrally at the location of the peri-oral region of the somatosensory cortex. This was achieved by recording electrodes in the cortex, and by an amygdaloid stimulating electrode, and by injection of cannulae in the cerebral cortical perioral region. The result of ethosuximide injection to the peri-oral cortical region was to suppress spike wave discharges in the GAERS.

The results were examined in normal Wistar rats and in saline-injected GAERS rats which received 36 electrical stimulations until seizures reached Racine stage 5. In contrast, ethosuximide-treated GAERS rats displayed suppressed spike wave discharges.

As a way of summarizing, examination of genetic animal models of absence seizures has been productive from the basic neuroscience aspect, as well as from the clinical side. As stated above in this chapter, a basic concept of seizures is that the phenotype represents an imbalance between cerebral excitation and inhibition. In intracerebral imbalance, the phenotype is epileptic seizures. In seizures, the excitation aspects seem increased, and AEDs are designed to either suppress excitation (blocking glutamate receptors) or increasing inhibition (facilitating GABA synaptic transmission).

The stargazer mouse model displays both seizures and ataxic behavior. The seizures resemble those seen in human absence seizures. Soon the genetic defect was identified as a mutation of a Ca^{2+} channel subunit calles stargazin. It was also discovered that cerebellar granular cells of stargazer mice were effectively deafferented by a lack of AMPA receptor currents. At this point, the mechanisms of absence spike wave discharges appeared to result from abnormal thalocortical synchronization.

Additional studies relating to the effects of somatosensory initiation of absence seizures both in terms of phenotypes and thalamic involvement are highly interesting. It has been shown that stimulation of somatosensory cortex can induce spike wave discharges. Furthermore, ethosuxamide applied to the peri-oral region of the somatosensory area of the parietal cortex can block the generation of absence spike wave discharges, and resultant absence seizures. This argues strongly in favor of a cortical site of generation, not thalamic. This would seem to be an area in which further studies on long-term results on other similar AEDs seem warranted. It certainly seems that in certain epilepsies there is a clear predisposition for certain anatomical sites to be selectively vulnerable. Why this selectivity occurs is not known. It is known that in other metabolic encephalopathies, such as kernicterus, there is a remarkable predisposition for certain highly focal areas/cells to be vulnerable. As in seizure states, this includes the hippocampus and cerebellar neurons such as Purkinje cells (McCandless 2011). The anatomical sites of absence seizures in animals correlate well with the sites seen in human cases, as described in the next chapter.

Chapter 5
Absence Seizures in Humans

In general terms, absence seizures in humans comprise about 5% of all epilepsies regardless of age. This type of epilepsy usually manifests in the young child age (5–7 years old) (Sato 1983). In some few cases, age of onset is in infancy, and other cases have an age of onset later in childhood or teens. There may be a relation between absence seizures and febrile seizures.

Clinical features of a typical absence seizure consist of a rapid onset consisting of a blank facial stare, and generally an unresponsive state. The patient usually does not have any behavioral or automatisms associated with the typical form of absence seizures. Duration is just a few seconds, or possibly a maximum of 20–30 s. This type of absence seizure can be missed because it may appear as "daydreaming" to parents or teachers.

Atypical absence epilepsy is more common than typical absence epilepsy, and is most always associated with behavioral (motor) changes and automatisms. The motor involvement is usually quite minor and could be lip smacking or tongue movements. These perioral movements may represent a somatosensory cortex origin of the seizure activity, as has been shown in animal studies (see Chap. 4). Clonic movement may even consist of rapid jerking of the arms. There may be a relaxation of the body and head bobbing (Sato 1983). Autonomic features of behavior include sweating, salivation, etc., and even incontinence.

Atypical epileptic absence seizures can last somewhat longer than typical absence seizures, but there certainly is overlap in duration. From an EEG standpoint, typical absence seizure's EEGs are described as a 3-Hz spike-wave discharge. The EEG of atypical absence seizure patients is slightly lower (1.5–2.5 Hz), although all absence seizure EEG results are usually described as 3-Hz. Background recordings may be normal or abnormal, with slowing and/or paroxysmal spikes (see Fig. 5.1).

As regards etiology, the relation to febrile seizures notwithstanding, there does appear to be a genetic etiological factor (Metrakos and Metrakos 1961). This is based on the frequency of spike-wave discharges in siblings of absence epilepsy patients; the exact nature of the inherited seizures as regards genetic transmission is not straightforward. Finally, the underlying pathology of human absence seizures is not clear. Theories based on animal studies have associated absence

D.W. McCandless, *Epilepsy: Animal and Human Correlations*,
DOI 10.1007/978-1-4614-0361-6_5,

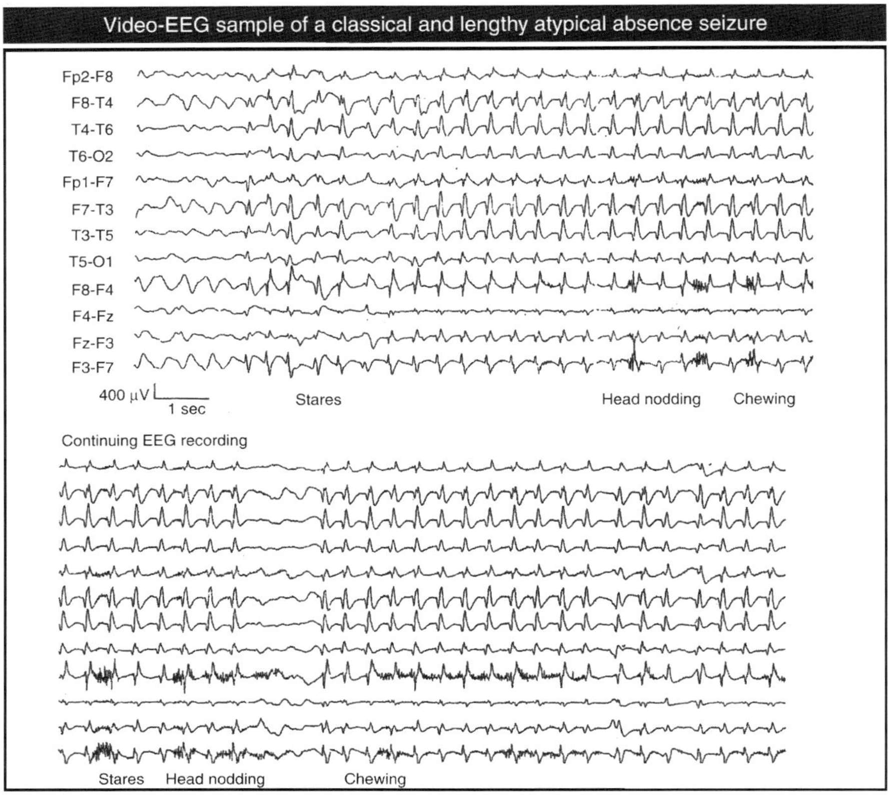

Fig. 5.1 Video EEG from a patient with atypical absence seizures. With kind permission from Springer Science + business Media: Epileptic Syndromes and their Treatment, 2010, p. 291, Panayiotopoulos, C. Fig. 10.6

seizures with thalamocortical networks, then later with a somatosensory origin. These animal studies in some ways predict human studies, as well as theories of pathology in patients. Contributing data, such as effective AEDs and physiological channel data will be examined later.

The differential diagnosis of absence seizures is sometimes difficult as regards complex partial seizures. Complex partial seizures are more common than absence seizures, but both have many similar symptoms: staring, change in levels of consciousness, automatisms, and even incontinence. Complex partial seizures may be longer lasting, but the range overlaps that of absence seizures. Absence seizures are the hallmark of the Lennox–Gastaut syndrome, and may also occur in epileptic encephalopathy (epileptic encephalopathy is defined by the ILAE as EEG abnormalities which themselves contribute to a progressive disturbance in brain function). One reason for accurate diagnosis is that the AED treatment for various etiologies of absence seizures is not the same (Dulac and Kaminska 1997) (see Table 5.1).

Table 5.1 Characteristics of atypical absence seizures

	Typical absence seizures	Complex partial seizures
Clinical criteria		
Duration less than 30 s	Common	Rare
Nonconvulsive status	Common	Rare
Daily in frequency	Common	Rare
Reproduced by hyperventilation	Common	Rare
EEG criteria		
Ictal generalized 3–4 Hz spike and wave	Always	Never
Normal EEG in untreated state	Rare	Common

Adapted from Panaylotopoulos, C. Fig. 10.6, p. 291 Springer, NY 2010

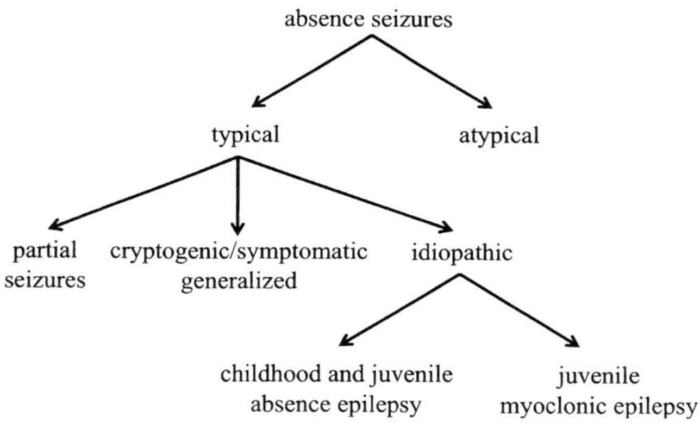

Fig. 5.2 Schematic diagram of possible directions of absence seizures. Adapted from Wirrell, E. Can. J. Neurol. Sci. 30: 184–188, 2003

A couple of review papers have dealt with absence seizures and their treatment (Panayiotopoulos 1999, 2001). Descriptions of symptoms are essentially as above. The author states that there is an inevitable loss of consciousness (absence) and intractable generalized 3–4-Hz spike and slow wave discharges. The discharges are regular, with an abrupt onset and cessation. Clinical features are abrupt stopping of activity, slight movement of eyelids, slight relaxation of the body, and then sudden return to consciousness. The patient may resume previous activity, such as speaking as if nothing happened. Clinical descriptions as above have been published over 100 years ago (Gowers 1881). Absence seizures may occur with only an impairment of consciousness. Absence seizures may occur with clonic components such as eyelid or eyebrow repetitive movements, or repetitive movements of the lips or jaw. Atonic components may occur with drooping of the head and a general relaxation. Other features such as automatisms and autonomic symptoms are described earlier (see Fig. 5.2).

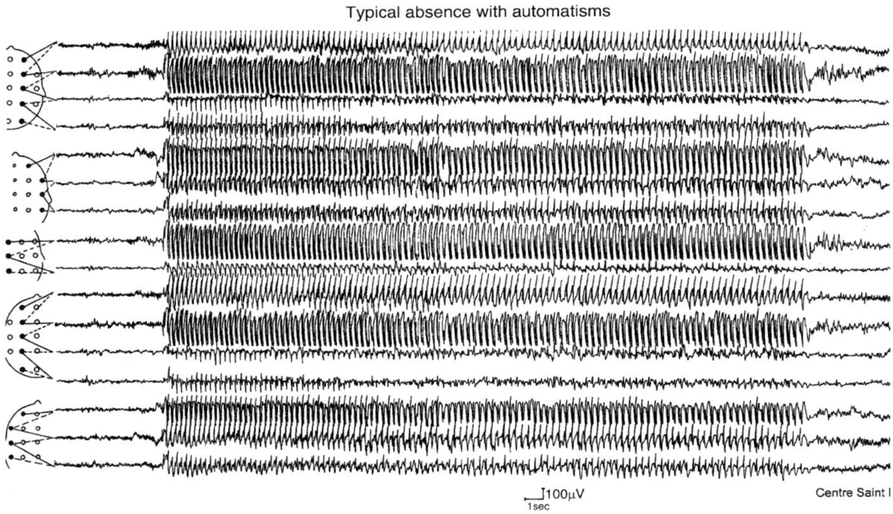

Fig. 5.3 EEG of typical juvenile absence seizures with automatisms. Published in Crespel, A., et al. Atlas of electroencephalography Volume 2: The epilepsies, EEG and Epileptic Syndromes. With permission of John Libbey Eurotext

From a diagnostic view, the authors state that the video EEF is the single most important diagnostic tool for the proper classification of a patient's seizure. Another feature of typical absence seizures is the easy induction by hyperventilation. These occur in over 90% of patients, and can be used clinically when conducting a video EEG in order to assure results. Having the patient count his/her breathes is a convenient way to observe consciousness during the procedure. Sleep EEGs are unremarkable and clinical manifestations rare.

Absence seizures may occur alone or in combination with other seizure types. Thus, childhood and juvenile absence seizures represent two types of absence seizures occurring with no other confounding seizures. Juvenile myoclonic epilepsy and myoclonic absence epilepsy are two separate entities in which absence seizures are key (Commission 1989). Other types, less well defined, may also exist. The accepted criteria for absence seizures are detailed earlier and later in this chapter. Obviously, other seizure variations such as short duration (less than 5 s), mild impairment of consciousness, etc., represent something else (see Fig. 5.3).

One advantage of absence seizures is that childhood absence epilepsy is usually easy to manage with monotherapy, and usually remits in only a few years. The therapeutic AEDs of choice for absence seizures are ethosuximide, lamotrigine, or valproic acid. Most patients will be successfully treated with monotherapy, and that should be a therapeutic goal. Even in the case of absence epilepsy in combination with other types of seizures, monotherapy is the goal, with sodium valproate being a very effective AED. In terms of drug withdrawal, doses may be slowly reduced only after 3 years of seizure free time. One other comment is that the relative efficacy of sodium valproate vs. lamotrigine is not clear. Even though

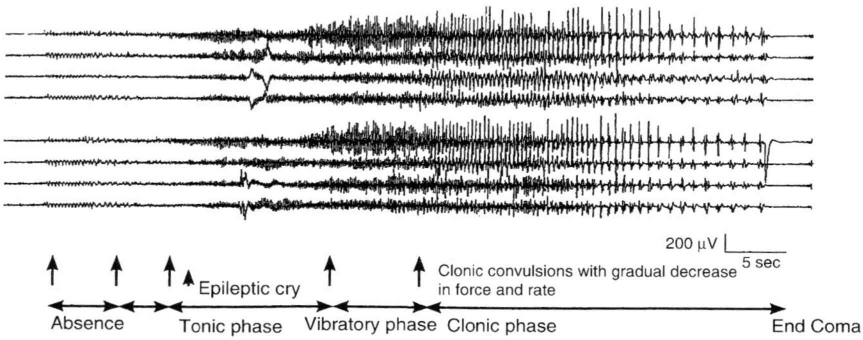

Fig. 5.4 Video EEG of a case of I.G.E. With kind permission from Springer Science+Business Media: Epileptic Syndromes and their Treatment, 2010, p. 346, Panaylotopouslos, C. Fig. 2.2

lamotrigine may be effective in at least 50% of new patients (Frank et al. 1999), there are "ramp up" problems, and sodium valproate has a success rate of about 80% without initial problems.

One interesting study (Kohsaka et al. 1999) examined brain stem auditory evoked responses in human absence epilepsy and related them to EEGs. The studies were conducted both preseizure and during absence seizures. Changes in brainstem auditory-evoked responses consisted of two types related to wave III amplitude and area features. First, was a long range biphasic fluctuation before the seizure onset, and second was an arrhythmic oscillation, synchronized with 3-Hz spike and wave discharges.

The authors state that this method is a valuable tool for the possible role/involvement of brainstem sites in the generation or participation of these sites in seizures. In the case of absence seizures, the authors note that these data argue against the so-called centrencephalic concept regarding absence seizure initiation and propagation.

Another interesting paper describes studies aimed at correlating EEG and fMRI in patients (Salek-Haddadi et al. 2003). In this case study, a 36-year-old female patient with intractable juvenile absence seizures was examined. She had onset of absence seizures at age 14, with 1–4 seizures per day, and with occasional tonic-clonic episodes. She had several EEGs which displayed 3-Hz spike-wave discharges. These were stimulated by hyperventilation and associated with loss of consciousness. She had been treated with sodium valproate, clonazepam, and lamotrigine, all with success (see Fig. 5.4).

Imaging was done using a 1.5 T echo speed system (G.E.). 700 Bold (blood oxygen level dependent) sensitive scans were acquired with a T2-weighted echo-plain imaging sequence. Four episodes of 3-Hz spike-wave discharges were recorded by EEG. No other seizure episodes were recorded. Results showed two distinct patterns of BOLD changes which corresponded with spike-wave discharges. These were activations seen in bilateral thalami, and pronounced negative deactivation in cortical areas, reflecting a reduced activity level. Methods such as PET during spike-wave discharges are inconclusive due to the necessity of obtaining data in preseizure, seizure, and postseizure states, which acts to "average" the results (Engel et al. 1985; Ochs et al. 1987).

The authors note that opposite results were obtained in cortex and thalamus by the BOLD method during spike-wave discharges. This implies that the demands of spike-wave discharges are different (opposite) between cortex and thalamus. Also, the relative participation of inhibitory/excitatory electrical activity during spike-wave discharges is unclear (Heeger and Rees 2002). The authors state that their study is important because it demonstrates a possible role in spike-wave discharges of the thalamus. It also indicates a thalamocortical mechanism in absence seizures, shown directly in a human patient.

In contrast to the above paper, several studies in rodents have shown an activation of both cortex and thalamus (see Chap. 4). Another study (Tenney et al. 2004a, b) looked at this in a primate, *Callithrix jacchus* (marmoset) monkeys. The primates had 3-Hz spike-wave discharges characteristic of absence seizures induced by gamma-butyrolactone. These primates showed similar corticothalamic activation seen in rodents (Tenney et al. 2004a, b). Ethosuxamide prevented the gamma-butyrolactone-induced seizures. This study demonstrated that the marmoset monkey model of absence seizures represents an excellent primate model for studying human absence seizures.

The above paper also points out that there is a difference between childhood absence seizures and juvenile absence seizures. That difference relates to the outcome. Childhood absence seizures tend to spontaneously remit in about two thirds of patients, whereas in juvenile onset seizures, seizures may be successfully treated, but lifelong AED treatment is usually required (Wirrell 2003). Other forms of absence epilepsy, such as that of myoclonic absence epilepsy, generalized tonic-clonic seizures with absence seizures, and atonic epilepsy with absence seizures, are associated with a much poorer prognosis. Cognitive impairment is frequent, and mental deterioration and psychosocial outcomes are not good (Olsson and Campenhausen 1993). Thus, it would seem that the presence of two divergent seizure types in the same patient is not a good prognostic feature.

One interesting issue regarding human absence seizures is the possible variation in cognitive function from case to case during the actual spike-wave discharges (Stafstrom 2004). The basis of this is that the "loss of consciousness" associated with absence seizures and spike-wave discharges may be quite variable from patient to patient. Backing this up are experimental rat studies in which animals were conditioned to exhibit licking movements in response to transient whisker deflections (Wiest and Nicolelis 2003). These were then recorded during spike and wave discharges (oscillations). It was found that this whisker response still occurred even during the spike and wave discharges. This is experimental evidence supporting the concept that human absence epilepsy patients have varying levels of loss of consciousness during clinical absence epilepsy.

Another clinical study of a 7-year-old girl who had typical absence seizure onset in which she looked up, had eyelid fluttering, and was unconscious. Routine EEG recording showed frequent 3-s spike wave bilateral symmetrical discharges.

In this study, fMRI was performed in the usual manner as previously described (Federico et al. 2005). EEG was recorded during the entire examination period. Results showed a significant and pronounced BOLD response. The thalamus was activated bilaterally, and little signal change occurred in cortical areas. Using a more lenient threshold false discovery rate, a more obvious cortical activation rate was seen. Motor cortex bilaterally and some myelinated tracts were activated.

The authors note that this study basically confirms that of Salek-Haddadi (see above), who saw bilateral thalamic activation in an adult absence patient. The present paper also confirms a study by Aghakhani et al. (2004). Activation of the thalamus during absence type spike and wave discharges seems to be in agreement with current knowledge of generalized spike-wave discharges. The present findings indicate involvement of both cortex and thalamus in absence seizures. A negative BOLD result from any cerebral area may well reflect blood flow and/or metabolic decreases in turn associated with cellular inhibition. Activation of the motor cortex in absence seizures could be a reflection of motor activity, and activation of myelinated tracts may reflect propagation of electrical activity.

Over the years, at least four central anatomical theories have evolved in order to explain the pathophysiology of absence epilepsy. These were each developed based on both animal and human studies and clinical observations. The first, described in 1954 (Penfield and Jasper 1954), carries the name "centrocephalic" theory. This theory explaining absence seizures proposes that discharges have an origination from subcortical neurons, which have a nonfocused projection network which serves to dissipate the discharges to distant anatomical sites.

The second idea was proposed in 1968, and is called the "corticoreticular" theory (Bancaud 1969). In this concept, spike-wave discharges originate from thalamic rhythmic spindle oscillations, and become transformed into spike-wave discharges in excitable cortical regions. The third theory, described in 1991 (Buzsaki 1991), carries the name "thalamic clock" theory. This concept was that the reticular nucleus of the thalamus contained some sort of pacemaker cells which in turn controlled rhythm to the cerebral cortex.

Finally, the "cortical focus" theory has been developed (Meeren et al. 2002). This study and theory states that an intact thalamocortical network is required in order for generation of spike-wave discharges. This theory postulates an origin in the perioral region of the somatosensory cortex, which then quickly spreads over the cortex. Initially, the cortex "drives" the thalamus and then the two regions stimulate each other, amplifying the discharges. The cortical focus theory in a sense unites various aspects of earlier theories. For an excellent review, see Meeren et al. (2005).

It should be noted that the exact role of the cerebral cortex in absence epileptic patients is not completely understood (Rodin and Ancheta 1987). Not surprisingly, some cortical areas have stronger spike-wave discharges than do other cortical areas. Some studies (Konishi et al. 1999) show in patients with absence seizures that focal discharges originating from the frontal cortex consistently preceded 3-Hz spike-wave discharges by just a few seconds. These and even PET studies in humans suggest a possible role for the frontal cortex in the generation of spike-wave discharges of absence epilepsy.

As a brief aside, the idea has surfaced that seizure discharges may be a reflection of an intrinsic protective mechanism designed to "reset" the brain from an abnormal state to a more normal condition (Sackellares et al. 2000; Iasemidis et al. 2004). The idea seems to be that epileptic brain makes an abrupt transition into and out of a more ordered state. These concepts further state that failure to return to the ordered state increases the chance and susceptibility for the brain to have a seizure. The interested reader is guided to a review of these ideas (Nair et al. 2008).

An important paper examining the efficacy of the AEDs ethosuximide, valproic acid, and lamotrigine has very recently been published (Glauser et al. 2010). This was a coordinated trial involving 32 locations in the USA, and directed from the Department of Pediatrics, University of Cincinnati. Children (age 2.5–13 years old) were selected for inclusion if they had absence epilepsy as defined by the ILAE., bilateral synchronous symmetrical spike-wave discharges of 2.7–5.0, had normal blood workup, and weighed at least 32 lbs. Excluded were children who had been AEDs for more than 7 days before randomization, had a history of seizures other than absence, had a history of juvenile epilepsy, or a history of any other significant medical history.

The study was a double-blind, randomized, controlled clinical trial consisting of 453 qualifying children. Treatment groups were three, consisting of those treated with ethosuximide (156), lamotrigine (149), or valproic acid (148). The treatment assignments were centrally controlled and computer generated. Baseline neuropsychological testing was performed before or within 7 days following initiation of the protocol. For each AED, doses were gradually increased each week or two until cessation of seizures, or side effects became a limiting factor. When it appeared that the seizures had stopped, bedside hyperventilation was employed to confirm cessation. If a seizure occurred, the dose was increased.

At 16 weeks, a video EEG, hyperventilation, as well as the clinical report were used to determine the seizure status. If seizures were continuing, and AED dose was at maximum, a failure was noted. If the dose was not at maximum, it was increased, and reevaluation was repeated at 20 weeks (4 weeks later). Additional criteria for failure included tonic-clonic seizures, drug toxicity as judged by blood workup, moderate to severe rash, jaundice, etc.

Results showed that at the time of enrollment, the cohort mean age was 7.5 years. A total of 453 were enrolled, and 446 completed the final efficacy analysis. There were no statistically significant differences in age or demographic criteria between groups. Two hundred and nine (47%) of participants were judged to be free from failure at either 16 or 20 weeks of trial. The group treated with valproic acid or ethosuximide enjoyed a higher rate of freedom (58% and 54%, respectively) as those taking lamotrigine (29%). The failures were most often those with no seizure control, and those with intolerable side effects.

More serious adverse events were reported in eight patients who required hospitalization. These included onset of tonic-clonic seizures in three patients, as well as nonepileptic events such as pneumonia, salmonella enteritis, etc. Results from confidence scores of the Connors Continuous Performance Test showed that patients on valproic acid scored higher than those taking either ethosuximide or lamotrigine.

The authors note that their patients diagnosed with childhood absence epilepsy, those receiving valproic acid, or ethosuximide had a greater success rate than those with lamotrigine. There were no differences between groups as regards discontinuation of AEDs due to adverse effects. The key number one criteria for success was the efficacy and tolerability of the AED for producing seizure freedom – this because of the importance of a correct initial AED selection. In addition, ethosuximide resulted in fewer attentional deficits in children in this group, as compared to the other two groups. This is also important because attentional deficits may be associated with poorer school performance. The lack of efficacy of lamotrigine in the current study, the authors note, was somewhat less than has been reported by others (Holmes et al. 2008).

In balance, the authors state that valproic acid, although slightly higher in freedom from seizures than ethosuximide, did produce a negative effect on attention to a higher degree than either ethosuximide or lamotrigine. Taken together, the authors suggest that ethosuximide may well be the initial AED of choice in hopes of achieving monotherapy, freedom from seizures, and least likely to produce a negative cognitive effect. The authors suggest that that long-term follow-up (months to years) is essential in determining further sequelae such as generalized tonic-clonic seizures as the children mature. Ethosuximide is possibly not particularly effective in preventing those later occurring tonic-clonic epilepsies (Glauser and Morita 2005). Double-blind, randomized, controlled studies of this magnitude are infrequent, and are welcomed, and the reader is referred to this excellent study (Glauser et al. 2010).

In the same issue of the New Engl J Med, is a cogent commentary (Vining 2010) about the above study. In the commentary, it is noted that determining AED efficacy is not so easy. In other areas of medicine, progress can be conveniently followed by blood tests, urinalysis, CT scans, MRIs, biopsy, etc. However, in seizure disorders, success takes observation over long periods, and even then there is doubt. For example, even after the "last" seizure, most wait 2–3 years before thinking about slowing withdrawing AED treatment.

A huge advantage of looking at the potential efficacy of AEDs is that in absence seizures, a simple and quick test is to try to induce a seizure with hyperventilation. Having the patient count breaths is a handy way to also assess the level of consciousness. The reviewing commentator noted the importance of this study in that it assessed effects of the three AEDs on an attentiveness as well as efficacy of seizure prevention. This study met both critical clinical criteria as well as stringent statistical analysis. This study is welcomed because, for example, the ILAE states there is an insufficient amount of data to inform clinicians as to which AED is best for childhood absence seizures.

The reviewer notes, as did the authors, that longer term follow-up is the next step, and always very helpful in the evaluation of AEDs. The reviewer correctly notes the difficulty in doing controlled, randomized studies in a pediatric setting. There is always significant concern on the part of parents that the drug their child takes will work, or certainly not make the disorder worse. This takes much time in parental education and reassurance. The reviewer notes that the percentage of children still having seizures (about 40–70% depending on the AED) was troubling, and casts some doubt on the oft quoted statement that absence seizures are benign and easy to treat.

Finally, is the comment that the belief that "newer is better" may not always be true. The frustration on the part of parents and physicians regarding the difficulties of treating epilepsy tends to stimulate the quest for something better. This study showed that ethosuximide, one of the older of AEDs, is still the drug of choice in terms of overall performance. Also is the consideration that new, less tested AEDs are usually very much more expensive. Once again it is noted that efficacy and mechanisms of action are derived from data from animal studies, on both rodents and primates. It is the translation of these data to human studies which is an essential process. This has been done in many instances in the study of absence seizures.

Chapter 6
Tonic Seizures in Animals

The classification of tonic seizures falls under the "generalized" seizure category, implying an alteration of consciousness, and originating bilaterally. Tonic seizures are usually short duration (less than 1 min), and characterized by rapid onset, and increased muscle tone in extensor muscles. This causes a fall if the patient is not seated. Postictal depression is common, with associated headache and confusion. Tonic seizures are a frequent feature of the Lennox–Gastaut syndrome, along with other seizure types. The Lennox–Gastaut syndrome represents as many as 4% of all childhood epilepsies (Kramer et al. 1998). Tonic seizures are the most prevalent seizure type in Lennox–Gastaut syndrome, accounting for more than one-half of seizures (Aicardi 1988).

Tonic seizures can be divided into four groups based on clinical presentation. The first is an "axial" type of tonic seizure which starts with tonic contractions of the neck, jaw clenching, and widely opened eyes. There may be a contraction of the thorax and abdomen, resulting in a cry and apnea. This lasts less than a minute. The second type of tonic seizure is similar to the first, but also has tonic contraction of the proximal upper limbs. Global tonic seizures involve tonic contraction of the entire upper limbs. Fists are clenched, and the lower extremities are involved. Finally, asymmetric tonic seizures affect only one side.

In some tonic seizure patients, gestural automatisms may occur after the tonic phase has passed. The duration of the automatisms can be from a few seconds to a minute. In mild tonic seizures, the clinical feature may only be an upward deviation of the eyes, and slowing of respiration.

Infantile spasms are a frequent seizure disorder in which tonic seizure activity can occur. Infantile spasms usually occur in newborn children less than 1 year old. The mean age of onset is 4–6 months. The seizures last only a few seconds, and tuberous sclerosis is a common cause. Spasms may start with a sudden tonic contraction of the trunk and extremities. These muscular extensions can occur alone or with clonic seizures. Infantile spasms, because they occur early in infancy when many developmental activities are occurring, frequently present with mental retardation, continues to worsen.

D.W. McCandless, *Epilepsy: Animal and Human Correlations*,
DOI 10.1007/978-1-4614-0361-6_6,

Fig. 6.1 Schematic representation of a recording of a tonic seizure

The EEG in infantile spasms shows a typical interictal pattern of hypsarrhythmia. It is thought that in cases of infantile spasms, there are abnormal interactions between cerebral cortex and brainstem structures. Such problems early in life can lead to seizure activity in several other brain sites resulting in hypsarrhythmia.

Tonic seizure EEGs are usually characterized by patterns of diffuse rapid (10–13 Hz) low amplitude activity. This tends to decrease in frequency and increases in amplitude. An EEG from a patient with upward eye deviation as described above may reflect the occipital lobes' involvement. Following a brief generalized discharge, there may be diffuse slow waves and slow spike waves. There is no postictal flattening of the tonic seizure. EEGs are as seen in tonic clonic seizures. Tonic seizures also occur during non-REM sleep. Clinical seizures usually follow by 1 s or so the onset of the EEG manifestations, and last a few seconds after the clinical seizure ends (see Fig. 6.1).

Animal models of seizures have been used for many years providing important data regarding human seizures. Genetically epilepsy-prone rats (Dailey et al. 1989) which were evolved from an audiogenetic seizure strain of rats were used to produce two other types of genetically epilepsy-prone rats. One of these exhibited absence-like seizures; in the other, tonic seizures were more pronounced. Another rat model representing absence seizures in humans has also been developed and is known as WAG/Rij (see Chap. 5) (Van Luijtelaar and Coenen 1988).

Another rat model developed also is an absence seizure animal. This absence model (tremor) and another model of greater severity (zitter) were used (chen et al. 2004) in order to produce a rat model called spontaneously epileptic rats (SER). The symptoms of this rat model include both absence seizures and a tonic seizure component. The absence seizures resemble those seen in tremor rats; the tonic component includes whole body tonic extension (with or without the Straub tail), as well as wild running and wild jumping.

The present study describes various features of the SER as regards the ontogeny of the absence/tonic seizure components. This knowledge is essential to evaluating AED efficacy and treatment paradigms.

A total of 43 SER (21 males, 22 females) were used to evaluate both behavior and EEG. Twenty more were used for weight gain and for survival times, while 20 Wistar rats were used as controls. For EEG recording, electrodes were permanently implanted over the left frontal cortex and hippocampus. After a 1-week equilibration period, rats were placed in a box with a window, and observations plus EEGs were recorded. Frequencies and duration of absence and tonic seizures were quantified using the EEG recordings. Body tremor and gait were quantified using the EEG recordings. Body tremor and gait were estimated on a 0–3 scale where 0 equals normal and 3 equals markedly abnormal.

Results showed that the weight gain of the SER was lower than controls at 5 weeks of age. There were no sex differences. The first mortalities were at 10 and 13 weeks, and all female SERs had died by 18 weeks while the males by 10 weeks. All control Wistar rats lived over 20 weeks.

No differences in male versus female rats occurred in terms of tremor/gait and so the data were pooled. Head tremor was first seen at 2 weeks of age and whole body tremor noticed at 3 weeks was present in all rats. By 12 weeks of age, all tremors had stopped. The staggering gait appeared by 7 weeks, and by 12 weeks all SERs were staggering.

Absence seizures showed 5–7 Hz spike wave complexes in the cerebral cortex and hippocampus. Behavior during the seizure activity included cessation of activity, with just starring. The absence seizures started at 5 weeks of age in three animals, and by 7 weeks all were having absence seizures. The frequency was 1.8 per minute and remained constant, whereas the duration was 16 s per minute, and slowly increased over several weeks.

Tonic seizures occurred spontaneously and could also be elicited by mild stimuli such as hand clapping or an air burst. Phenotypically the SER exhibited body extension, sometimes with tail arching (Straub tail). Tonic seizures were first seen at 6 weeks of age, and by 12 weeks all had tonic seizures. At 12 weeks, most tonic seizures occurred following a stimulus and had a duration of around 20 s. This remained constant.

The authors note that the tonic seizures in the SER are independent of the absence seizures. This is based in part on the different responses of each type of AEDs. When compared to other epileptic rats, such as WAG/Rij rats, cycles of spike wave discharges are not the same (5.7 Hz vs. 7–10 Hz). Onset was 2 months later in appearance in WAG/Rij rats. Another difference was that face twitching was not seen in SER, but is a feature in other rat models.

The authors note that the presence of both absence and tonic seizures in the same SER models will aid in testing efficacy of potential AEDs. The age spread between absence and tonic extension seizures facilitates that process. This model was derived by mating tremor and zitter rat strains. This unique feature promises to provide future data of significant interest.

Another study of tonic seizures in rats involved inducing tonic seizures with pentylenetetrazole, then intra cerebral injecting three different AEDs into the globus pallidus and examining the efficacy of seizure modification (Chen et al. 2004). The idea examined was that the globus pallidus plays a role in "gating" the spread of seizure activity (Depaulis et al. 1994). Other basal ganglia are most likely involved in the spread (Sabatino et al. 1986). Further evidence for the role of the globus pallidus in seizure activity comes from the observation that lesions in the globus pallidus prevent seizure spread, whereas stimulation of the globus pallidus increases neocortical seizures (Makulkin et al. 1992).

In the present study, the AEDs tiagabine, zolpidem, and baclofen were injected directly into the globus pallidus 15 min before pentylenetetrazole injection. Following this, the efficacy of the AEDs and the postsynaptic effects were examined.

Table 6.1 Effect of intrapallidal injection of tiagabine, zolpidem, and baclofen on pentylenetetrazole-induced tonic seizures

Group	Treatment	Number of rats	Incidence of tonic seizures	Seizure rate (%)	Incidence of mortality
I	Control	10	10	100	10
	Tiagabine	6	4	66.7	1
II	Control (DMSO)	8	8	100	8
	Zolpidem	5	5	100	5
III	Control (saline)	7	7	100	7
	Baclofen	9	0	0	0
	CGP55845 + baclofen	6	6	100	5

Adapted from Chen, L., J. Biomed. Sci. 11:457 2004

In this study, 240–270 g Sprague-Dawley rats were anesthetized, and using stereotaxic coordinates, cannulae were implanted above the globus pallidus. The cannulae were fixed in the skull with screws. Three days or more after surgery, the AEDs (tiagabine, zolpidem, or baclofen, or vehicle) were infused into the globus pallidi bilaterally. Fifteen minutes later pentylenetetrazole was infused. The incidence and latency of tonic seizures were observed and mortality noted. The site of injection was checked by injecting dye.

Results showed that 51 rats were used, divided into three groups, each with an AED plus vehicle control. Pentylenetetrazole induced tonic seizures in all 25 globus pallidus-injected rats. Seizures were seen to show rearing, loss of righting reflex, and tonic extension of all extremities. Asphxia and death occurred within an hour in all rats.

Results from the group pretreated with tiagabine showed efficacy in suppressing the pentylenetetrazole-induced tonic seizures. The efficacy was noted in the frequency of animals affected (only 67%), and the onset latency was significantly longer than in vehicle controls. The mortality rate was only 16%. Interestingly, when tiagabine was injected into anatomical structures immediately adjacent to the globus pallidus, no antiseizure effect was seen.

By contrast, zolidem, when injected into the globus pallidus had no antiseizure effect. Mean delay of onset of seizures was not significantly different than control rats. Baclofen pretreatment, in contrast, protected the rats from tonic seizures in all who received the AED (nine rats). The mortality rate was zero in baclofen, and the anticonvulsant effect was zero when adjacent areas were pretreated with baclofen. In another trial, the GABA B receptor antagonist CGP55845 prevented the antiseizure effect of baclofen. This argues in favor of a GABA effect in the globus pallidus (see Table 6.1).

Other studies including cell attached and whole cell current clamp recordings showed that baclofen stopped spontaneous action potentials in 50% of cells, and reduced activity in the other 50% of cell-attached neurons. Results in whole cells current clamp recordings were similar.

The authors note that the goal of the study was to look at the role of the globus pallidus in tonic seizures induced by pentylenetetrazole. The AEDs are three which affect the GABA system when injected into the globus pallidus (Chen et al. 2002). The present data show that activation of GABA B receptor in the globus pallidus may gate pentylenetetrazole-induced tonic seizures. Furthermore, that the GABA B system in the globus pallidus is involved in tonic seizures, and is shown by the baclofen suppression of pentylenetetrazole-induced tonic seizures.

The authors state that their baclofen/pentylenetetrazole data support the concept that GABA B receptors in the globus pallidus play a key role in tonic seizure expression (Schuler et al. 2001). The authors further state that GABA B receptor activation serves to inhibit globus pallidus activity, which "closes the gate," thereby explaining the antiseizure effect on tonic seizure activity. These data on mechanisms might act to suggest other treatment modalities.

Another study (Shehab et al. 2006) has examined the subthalamic nucleus of the brain to see if it is involved in seizure control, especially in terms of tonic seizures produced by the electroshock model of seizures. The importance of possible endogenous antiseizure mechanisms could be critical in designing new AEDs given that about one-third of epilepsy patients are refractory to drug treatment. The idea that there might be an endogenous antiseizure system within the brain which could be manipulated to induce antiseizure outcomes is highly attractive (Depaulis et al. 1994). The goal of the present study was to examine if the subthalamic nucleus might play a role in generalized seizures evoked in the hind brain by electroshock.

In this study, adult Wistar rats were anesthetized and cannulae were implanted 5 mm above the subthalamic nucleus bilaterally. In another group of Wistar rats, electrodes were stereotaxically implanted bilaterally into the subthalamic nuclei. In all cases, both cannulae and electrodes were attached to the skull with screws and dental cement.

Three days after surgery, either muscimol or saline was injected into the subthalamic nuclei. Each rat received a 3-day pretreatment, then electroshock was delivered for 1 s (40 mA, 50 Hz AC) through ear clip electrodes. The duration of tonic hindlimb extension was observed. After 3 days, continuous trains of rectangular constant current pulses (intensity 175–300 μA; 60 μs width) were applied to the subthalamic nuclei via the electrodes. When all testing was finished, animals were sacrificed, perfused for histology in order to localize sites of injection, brains removed, mounted, and sectioned.

Results showed that muscimol did not suppress electroshock-induced tonic hindlimb extension. The muscimol did however suppress neuronyl activity in the subthalamic nuclei. Another test of the suppression of neuronyl activity by muscimol involved administration of ether. This resulted in a suppression of the induction of c-fos in the subthalamic nuclei. This proves that muscimol exerted a powerful subthalamic nuclear inhibition. Finally, results from high-frequency stimulation of the subthalamic nuclei also showed a lack of attenuation of the tonic hindlimb extension. Eleven rats also underwent excitotoxic lesioning, and again, no reliable suppression occurred as regards tonic hindlimb extension (see Table 6.2).

The authors note that unlike results from forebrain seizures (Vercueil et al. 1998), none of the experimental attempts in the current study resulted in any change in hindbrain tonic seizures. The authors further note that these negative results regarding the

Table 6.2 Effect of saline and muscimol injection into the subthalamic nucleus

	Saline	Muscimol (200 pmol)	Muscimol (400 pmol)
Tonic hindlimb extension (seconds)	8.4 ± 0.4	8.2 ± 0.5	7.7 ± 0.6

Data are the time of tonic extension
Adapted from Shehab, S. Exp. Br. Res. 173:274 2006

subthalamic nucleus' involvement in altering tonic hindlimb extension are likely to be correct. This conclusion is based on the muscimol data. These data, using doses previously shown to be anticonvulsant (Dybdal and Gale 2000) in a different seizure model, did not show an effect in the present study.

Similarly, data from the high-frequency stimulation experiments and exitotoxic lesion experiments showing a negative result in terms of the subthalamic nuclei seem convincing. No less than three independent assessing techniques were used looking for involvement of the subthalamic nuclei in its ability to protect against tonic hindlimb extension seizures. They were unable to demonstrate a change in seizure characteristics. This study is in a sense a part of a continuous effort to map the target areas capable of suppressing tonic motor seizures. Drug modification of such a system could lead to new approaches to epilepsy treatment.

In a recent experimental study of a new AED-carisbamate, efficacy against absence seizures and tonic seizures was evaluated (Francois et al. 2008). Carisbamate is a neuromodulator which has completed clinical trials, and is considered to be an adjunctive AED for partial onset seizures (Novak et al. 2007). The drug has broad spectrum efficacy in a variety of animal models of seizures, including electroshock, pentylenetetrazole, audiogenic seizures, picrotoxin, etc. It is also effective in kindled rat models, and has prevented status epilepticus in a pilocarpine seizure model.

The present study uses the genetic model of generalized absence epilepsy in Wistar rats (GAERS) and the Wistar audiogenic sensitive (AS) rat in order to examine carisbamate effects. In the GAERS model, the rats' EEGs show spontaneous spike and wave discharges with simultaneous behavioral arrest. The AS rats respond to an intense audio stimulus with wild running and jumping, followed by a tonic phase with hindlimb extension. A catonic phase follows the tonic extension. The EEG in the AS rats is flattened during the tonic phase, then shows regular and fast theta rhythms.

The GAERS rats were anesthetized and electrodes were placed over the frontoparietal cortex bilaterally. After equilibration, EEG recordings were made on freely moving rodents. Carisbamate was used in three doses (10, 30, and 60 mg/kg). Spike wave discharge duration was recorded for each dose. Behavior was also followed closely since absence seizures seem to occur during a quiet state.

For the Wistar AS rats stimulated at 120 dB, a single audiogenetic seizure with wild running followed by tonic seizures was the result. The tonic extension phase consisted of both forelimbs and hindlimbs extended, head and tail extended, and mouth open. After carisbamate treatment, latency, number and duration of wild runs, and the tonic phase were rated. The EEG was not utilized in the Wistar AS animals due to many artifacts due to the wild running phase of the seizure.

Results showed that the degree of spike wave discharges were inversely related to carisbamate dose. Thus, while the low dose had little effect, the middle dose decreased the duration of the spike wave discharges by 70–90%, and the high dose abolished the spike wave discharges. None of the carisbamate doses used had any effect on the behavior. At a very high dose (120 mg/kg) carisbamate did induce sedation and ataxia.

Results in the Wistar AS rats showed the latency after the lowest dose for the wild running episode, and for the tonic seizures rose 266% and 327% over control rats, respectively. After the two other doses (20 and 30 mg/kg), no animals exhibited either wild running or absence seizures.

The authors comment that these animal models have been bred for many generations assuring pure phenotypes with only one seizure type per strain of rat. The present study confirms that carisbamate has high efficacy for audiogenetic-induced tonic seizures. Additional efficacy against absence seizures also indicates the broad spectrum capability of carisbamate.

The effect was especially impressive in the audiogenetic model, where the middle and high doses of carisbamate essentially eliminated seizure activity. This agrees with another study in a different audiogenetic model of epilepsy (White et al. 2006). These data on audiogenetic rats extend the previous data showing efficacy for electroshock, pentylenetetrazole, bicuculline seizures, etc.

In the GAERS rats, carisbamate decreased the duration of spike wave discharges, and the decrease was dose dependent. There was no effect at the low dose, but the middle and the upper doses both dramatically lowered the spike wave discharge duration. The dose for the effect in the GAERS rats was higher than the audiogenetic animals as regards the AED. There was no sedation or ataxia in this group of rats.

The exact mechanism of action of carisbamate is largely unknown. The authors note that valproate also has a broad spectrum of efficacy as regards seizure types. The exact action(s) of valproate are also not completely clear, but increased GABA may play a role in the regions involved in seizure propagation and control (Loscher 2002). The authors propose carisbamate which shows promise as an (adjunctive) AED, and that more data are necessary as to mechanisms of action.

Another interesting recent paper (Hanaya et al. 2010) examines aspects of hippocampal cell loss and propagation of discharges in the SER. As discussed earlier, SERs are a double mutant derived from two "parents", tremor and zitter (see above). At about 8 weeks of age, SER shows absence seizures, and tonic seizures later. The two seizure states are accompanied by 5–7 Hz spike and wave complexes on EEG examination in the cortex and hippocampus. There is a CA3 located depolarization shift which is accompanied by repetitive neuronal firing. This is thought to be related to calcium channels.

Aberrant sprouting may be a feature of mossy fibers in the hippocampus, frequently found associated with temporal lobe epilepsy and tonic seizures. It also seems that the hippocampus is not involved in absence seizures (Kandel et al. 1996). The present study was undertaken to clarify the actual relationship between the seizures of SER and the abnormality seen in SER hippocampi.

In this study, mature and immature SERs, age-matched Wistar rats, and mature tremor rats were used. For intracellular hippocampal slice recording, brains were removed from decapitated animals and placed in a calcium-free medium. A stimulating electrode was placed in the granule cell layer of the dentate gyrus, and stimuli delivered to the mossy fibers on Schaffer collaterals every 5 s. Recordings were made from CA3 pyramidal neurons.

Quantification of cell density was performed in perfused brains which had been prepared for light microscopy. For immunohistochemistry, immunostaining was performed with rabbit polyclonal antibox antibody. Optical densities of areas of interest were captured and quantified using imaging software. Immunoreactivity was expressed as optical density. Statistical analysis used the analysis of variance and Scheffe's test.

Results from intracellular recording showed abnormal hippocampal firing in 7/10 CA3 neurons, which reconfirms earlier data (Ishihara et al. 1993). In CA3 neurons from mature SER, higher voltage stimulation induced long depolarization shift and repetitive firing. In contrast, mature Wistar rats did not show these changes. Additionally, abnormal firing was not induced by mossy fiber stimulation in any CA3 neurons of tremor rats.

The results from histology showed that in immature SER, the number of hippocampal neurons was equal to controls in the hippocampal areas, whereas in the mature animals, the CA3 cell count was lower than both in tremor rats and in control Wister rats. Other hippocampal areas showed similar cell counts. Mossy fiber sprouting was seen in the dentate gyrus in mature SER, and was not seen in CA1 or CA3 regions. The mossy fiber sprouting was greater in the SER (as seen by Timm's stain) than in mature tremor rats.

Immunohistochemical analysis of BDNF immunoreactivity showed higher levels in the hilus, pyramidal layers, and stria orience of CA3 and dentate in SER, as compared to Wistar rats. Neither Bax nor caspase immunoreactive neurons were found in SER.

The authors comment that hippocampal sclerosis is often found in specimens from temporal lobe resection in cases of refractory epilepsy (Mathern et al. 1996). The sclerosis shows neuronal loss and glial proliferation. Prolonged hyperexcitability in the hippocampus induces mossy fiber sprouting and associated reorganization, and BNDF modulates this and synaptic transmission.

In this study, the authors state that tremor rats (absence model) did not show hippocampal neuronal cell loss. This suggests the conclusion that absence seizures do not enhance the development of hippocampal sclerosis. Other models of absence also do not show hippocampal sclerosis (Sirvanci et al. 2003). Neuronal changes in the hippocampus of SER show findings of hippocampal sclerosis.

The authors state that the semiology of the seizures in SER were not the same as in temporal lobe seizures, and such symptoms did not appear in the lifetime of SER. A combination of repetitive tonic seizures coupled with the vulnerable CA3 neurons of SER could explain the development of hippocampal sclerosis. The SER showed hippocampal sclerosis without the induction of epileptogenesis. Further investigation is needed to clarify this interesting phenomenon. Animal models of tonic seizures are excellent for discerning trends seen in human tonic epilepsy (see Chap. 7).

Chapter 7
Human Tonic Epilepsy

Epilepsy is a relatively common disorder, which affects 0.5% of people in the USA, and 50 million or more worldwide. The impact of these figures in terms of medical needs and costs in terms of human suffering from the ravages of epilepsy is inestimatable. The etiology of many epilepsies remains unclear, and in spite of monumental research efforts, the treatments are sometimes lacking in efficacy, with some 30% of patients refractory to AEDs. As many as one half of this group are poor candidates for surgery. Of the various classifications of seizures, tonic seizures are an important subgroup of generalized seizures.

Tonic seizures are a common epileptic phenomenon in humans. For example, at least one half of children with Lennox–Gastaut syndrome are reported to have tonic seizures as a component of the disorder. These can occur both during awake periods and during sleep, and so may go unnoticed. Tonic seizures may be either focal or generalized. Focal tonic seizures are characterized by sustained posturing of single limbs, sustained asymmetric trunk posturing, and sustained eye deviation. Focal tonic seizures cannot be initiated by stimulation. Generalized tonic seizures' features include sustained symmetrical posturing of limbs, trunk, and neck. The phenotype may be flexor, extensor, or mixed. Generalized tonic seizures can be provoked by stimulation and can be suppressed by restraint.

Clinical features may include tonic contraction of neck muscles, widely opened eyes, and clenching of the jaw. Tonic contraction can extend to the distal portion of the limbs, with fists clenched. Depending on the patient's position when the seizures start, the patient may fall. EEG usually shows bilateral synchronous spikes of 10–25 Hz. Infantile spasms are frequently associated with tonic seizures, which begin with a rapid and sudden contraction of the trunk and limbs. This form of tonic seizures carries a poor prognosis for newborns, with premature death rates running from 5 to 30% depending on the study.

Neonatal seizures that are recognizable clinically are serious occurrences as regards the potential for serious outcomes such as neurodevelopmental sequelae, and death. A study (Scher et al. 1993) has reported investigations of preterm and term infants using criteria such as etiology and outcomes, and with EEGs as definitive confirmation of seizures.

D.W. McCandless, *Epilepsy: Animal and Human Correlations*,
DOI 10.1007/978-1-4614-0361-6_7, © Springer Science+Business Media, LLC 2012

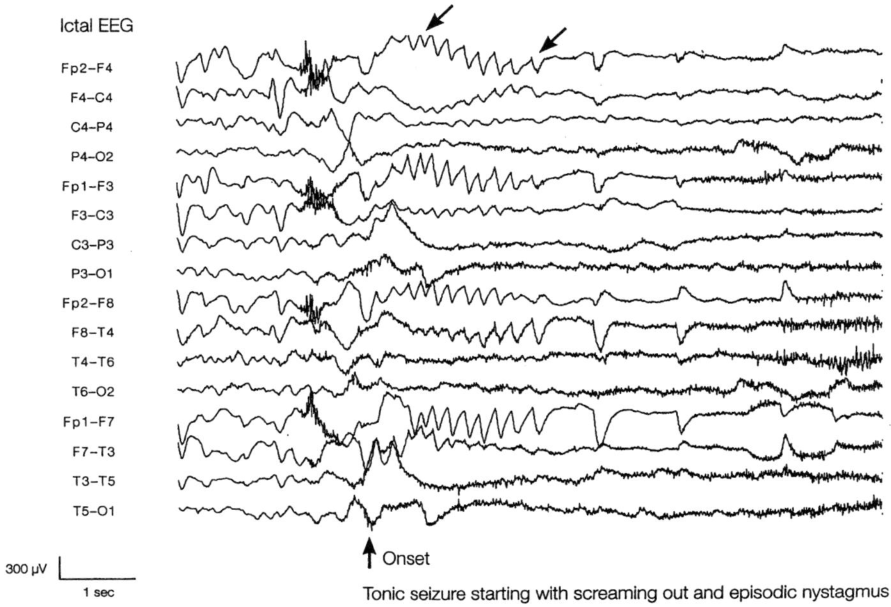

Fig. 7.1 Tonic seizure starting with a scream. With kind permission of Springer Science + Business Media: Epileptic Syndromes and their Treatment. 2010, p. 289, Panaylotopoulos, C. Fig. 2.2, Fig. 10.3

This study involved 92 neonates, of which 62 were preterm and 30 were term. All had EEG-confirmed seizures. Clinical seizures if present were classified as subtle, focal clonic, multifocal clonic, and focal or generalized tonic or myoclonic. Careful observations recorded features such as facial movements. Autonomic changes were recorded. All EEGs were interpreted by the same person. Features including frequency, structure, and amplitude were all taken into account. A duration of 10 s or more was considered sufficient for a seizure to be recorded. If more than one abnormal movement type was seen, the most frequent was noted.

Of the 92 patients in the study, 42 were females and 50 were males. The mean gestational age for the full-term infants was 40.8 weeks, and for the preterm group, the gestational age was 29.7 weeks. EEG results showed that 48% of patients had neonatal seizures. Clinical diagnosis revealed that 25 neonates had asphyxia, 27 had pulmonary hypertension, 17 had hyaline membrane disease, 7 had CNS infection, and several others had disorders such as hyperbilirubinemia, and CNS malformations.

EEG recordings revealed that 48% had clinical seizures. Subtle seizures were the most frequent seizure type identified, and 61% of these had subtle seizures only. About 10% had tonic seizures, 20% had myoclonic seizures, and 41% had clonic seizures. Over one-third of the preterm neonates had an autonomic component. Ninty percent of the entire group showed CNS structural lesions. The difference in malformation percentages between preterm and term numbers was not statistically different (see Fig. 7.1, and Table 7.1).

Table 7.1 Difference between tonic seizures and the tonic phase of generalized tonic–clonic seizures.

	Tonic seizures	Tonic phase of GTCS
Generalized seizure EEGs	10–25 Hz	8–10 Hz
Postictal background attenuation	Brief	Prolonged
Tongue bite	Not seen	Commonly present
Postictal confusion	Brief	Prolonged

Adapted from Patel, H., et al. Child Neurol. 21:813, 2006

In terms of outcomes, 30% of full-term and 58% of preterm infants did not survive. Survivors were followed for 6.5 years, and normal outcomes were noted in 25% of preterm and 60% of full-term infants. Forty-two percent of preterm and 35% of full-term infants showed developmental delay/cerebral palsy. Epilepsy was present in childhood in 17% of premature and 30% of full-term neonates.

The authors note that their results confirm others' data (Keen and Lee 1973; Radvanyi-Bouvet et al. 1985) showing that a relatively low EEG confirmed incidence of seizures, based on hospital-born neonatal population. Tonic seizures were certainly the seizure type that was frequently seen, but there was no discussion as to how tonic versus clonic and other seizures might have evolved in these similar cases. AEDs were tried in some of the patients in this study, but with mixed results. The authors state that this retrospective study did not have a consistent protocol for AED treatment.

Generalized tonic seizures are rarely associated with some other brain disorder, and are then termed reactive epilepsies. Such a case in which generalized tonic seizures occurred in a patient with ganglioma has been published (Lee et al. 2001).

This was a case of an 18-year-old, right-handed female patient who had refractory epilepsy since 1 year of age. She initially had epileptic spasms, which progressed into generalized tonic epilepsy, with about three seizures per week. The patient's prenatal and postnatal histories were unremarkable.

Neurological examination showed a mentally retarded person (I.Q. <45), with no focal deficits. Video EEG showed 10–20-s episodes of generalized tonic seizures. They materialized as sudden extensions of upper and lower extremities, and flexion of the head and trunk. The EEGs showed a 7-Hz background which was interrupted by diffuse attenuations and/or 5–10-s fast rhythm, and then slow spike waves. MRI showed a mesial temporal cystic tumor and left hippocampal atrophy. The temporal cystic tumor appeared to be a ganglioma. Seizure frequency dropped after lobectomy, and finally stopped. This is an example of a reactive epilepsy.

The authors note that the differentiation between Lennox–Gastaut syndrome and partial epilepsy with secondary bilateral synchrony should be made since the surgical approaches are not the same for the two syndromes. The authors state that the diagnosis of partial epilepsy with secondary bilateral synchrony was confirmed by the success of the surgical resection. Since a misdiagnosis can occur causing an inappropriate surgical procedure (Benbadis 1999a, b), caution is greatly needed before proceeding with surgery on such patients.

Automotor and tonic seizures are the most common seizure types with an uncommon occurrence when epileptiform spike discharges occur in the midline.

This paper attempts to describe signs, symptoms, and MRI findings in 35 patients with midline spikes shown on EEG (Kutluay et al. 2001). In a retrospective study, 35 patients with midline spikes were identified, and were the focus of the study.

These 35 patients had midline spikes as the only epileptiform abnormality. They were defined as those with localized spikes at vertex scalp electrodes F2, C2, and/or P2. EEGs were performed using 21 channel recordings and the 10–20 system. MRIs were evaluated by a neuroradiologist blinded to the clinical diagnosis. Seizure semiologies were confirmed by communicating with family members. Seizure classification was made according to Luders (Luders et al. 1998).

Results showed that 18/35 were males. The mean age at the time of the EEG was 11.8 years. Mean age of seizure onset was 5.1 years. Mental cognitive testing showed that 63% were normal, but the rest were mild to moderately retarded. Of these patients, 57% had EEG results which were localized to 1–2 midline electrodes and 43% had midline localization in which the recordings were of maximum amplitude. Of these, 57% had the highest response at the C2 electrode; response at the other two electrodes was of about equal intensities.

According to the semiology seizure classification scheme, automotor and tonic seizure types were the most frequent. About half of the patients had a partial complex onset with or without generalization. Eight (27%) patients had tonic seizures as the major seizure component. Automotor seizures had complex movements which varied in different patients.

MRI was performed on 29/35 patients, and results showed that 45% were abnormal. Results showed that one-fourth had focal seizures lateralized to a frontal lobe, and six had diffuse cerebral abnormalities. In patients with midline spikes, there was a positive correlation between partial onset or generalized tonic–clonic seizures and the midline spikes. There was no significant correlation between midline spikes and age, seizure type, or MRI results.

The authors note that few large studies have addressed the question of seizure semiology, and imaging results in patients with midline spikes. The authors state that their results are in keeping with another rigorous study showing similar results (Bagdorf and Lee 1993). Bagdorf also reported that complex partial seizures are the most frequent seizure type associated with midline spikes. While nearly all seizure types can be seen in patients with midline spikes, most will have partial complex onset, generalized tonic–clonic seizures, and simple complex seizures with a motor component.

Another interesting paper examines the so-called tonic-absence seizures as a heretofore underrecognized seizure disorder (Shih and Hirsch 2003). Generalized paroxysmal fast activity (GPFA) greater than 13 Hz and associated brief tonic seizure have been described (Gastaut et al. 1963). There is also considerable literature on generalized slow spike and wave (less than 3 Hz) absence seizures. The present paper looks at patients with GPFA immediately followed by absence seizures.

In this study, eight patients were identified with symptomatic generalized epilepsy, and 6/8 had multiple seizure types. Twenty-nine seizures were recorded with video EEG. Of the 29, 26 showed evidence of GPFA followed by absence seizures which lasted for 3–88 s. The predominant clinical correlate was bilateral tonic activity.

The duration of the GPFA ranged between 2 and 30 s, with frequency between 14 and 35 Hz. The EEG was complicated by diffuse muscle artifact. The absence seizures were between 1 and 3.5 Hz, with an amplitude of 70–500 μV.

The clinical phenotype consisted of head or trunk flexion, tonic extension, and elevation of the proximal upper extremity. Tonic deviation of the eyes upward was also seen. The absence seizures that followed was characterized by relaxation of the neck, trunk, and upper extremity. The patient was generally unresponsive. Automatisms and urinary incontinence were also seen. The tonic-absence seizure was not associated with other seizure types. For example, some patients had tonic seizures independent of the tonic-absence episodes, which could easily be distinguished.

In terms of patient data, most had childhood seizure onset; etiology consisted of two patients with intrauterine insult, and two had poorly defined encephalitides. Six patients were mentally retarded. The incidence in this study was 8 out of 1,500, or a 0.5% rate. The authors state that this is probably an underestimate.

The authors note that this is an infrequent diagnosis which is most often seen in patients with the Lennox–Gastaut syndrome. The absence seizures of Lennox–Gastaut syndrome are prolonged and cannot always be distinguished based on onset. The authors state that the advent of continuous video EEG monitoring has improved the ability to correlate EEG recording with semiology. The authors say that their data show a tonic seizure followed by an absence seizure, clearly distinct from a tonic seizure. The reorganization of their electroclinical pattern is important. The authors believe that this is a variation of tonic seizure, albeit an independent "free-standing" seizure type.

Another paper looked at clinical factors in the differential diagnosis of frontal lobe versus posterior cortex epilepsy. Tonic seizures are a feature of both, but occur significantly more frequently in frontal lobe epilepsy (Fogarasi et al. 2005).

The study involved 35 patients, all under 12 years of age, who had MRI-proven evidence of either frontal lobe epilepsy or posterior cortex epilepsy. There were 20 frontal lobe epilepsy patients and 15 with posterior cortex epilepsy. Of the posterior cortex epilepsy patients, nine had pure occipital lobe epilepsy, and six more had a slight involvement of the occipital border of the parietal lobe.

A total of 177 seizures from 35 patients were analyzed. The author followed typical classification schemes, and the data included tonic seizures, versive seizures, myoclonic seizures, clonic seizures, atonic seizures, etc. Inter-observer differences in classification of seizures occurred in 20% of cases, and were reviewed a second time in order to gain consensus. Both parametric and nonparametric statistics were utilized to evaluate data.

Results showed a total of 19 males and 16 females with AED refractory epilepsy. Range of onset of seizures was from 3 days old to 9 years of age. Following surgical resection, all patients became seizure free. Histologic examination of resected tissue showed malformations of cortical development in 25 cases, dysgenic tumor in 3 cases, and stroke in 3 cases. Age of onset, lateralization, gender, etc., did not show any significance between groups.

In terms of seizure components, tonic seizures were the most frequent seizure type. Tonic seizures were seen in 17 of 35 children and were seen more frequently in the frontal lobe epilepsy group than in the posterior cortex epilepsy group (0.01), (14 vs. 3). In the tonic seizure group, all patients had upper extremity, and trunk and lower extremity signs. Of the tonic seizure patients, seven of the frontal lobe epilepsy group and two of the posterior cortex epilepsy patients had unilateral tonic components.

Two seizure components occurred exclusively in each seizure group: hypermotor in the frontal lobe epilepsy group and versive in the posterior cortex epilepsy group. Eye deviation and automatisms were seen in about one half of all children, but did not aid in localization. Nystagmus was seen exclusively and often in posterior cortical epilepsy, but mention of this sign was not made as a diagnostic feature.

The authors do state that the most important feature in differentiating frontal lobe from posterior cortex epilepsies was the aura. Somatosensory auras occurred exclusively in frontal lobe epilepsy patients, and visual auras occurred exclusively in posterior cortex epilepsy patients. Neither of these groups had auras which had a frequency of over about 25% of patients, however.

The authors comment that tonic seizures were the most frequent seizure type occurring in about one half of all subjects. They were most frequent in the frontal lobe group. There were also differences between groups as regards types of tonic seizures. The five frontal lobe epileptic children had a tonic elevation of the arms, whereas none of the posterior cortex epilepsy children had that particular phenotype.

The authors summarize by saying that ictal features only helped a little in distinguishing frontal lobe epilepsy from posterior cortical epilepsies. More pronounced differentiating features of these two seizure types in adults were not seen in children, especially younger children. The semiology of childhood extra-temporal epilepsy has a wide range of symptoms. The nocturnal prevalence of seizures associated with the frontal lobe do present in childhood, and along with somatosensory auras can help differentiate frontal lobe epilepsy from posterior cortical epilepsy.

A case report of a 14-year-old boy who had a dysembryoplastic neuroepithelial tumor accompanied by refractory tonic seizures was published (Patel et al. 2006). This patient was mildly retarded and was seen due to frequent seizures. The seizures were reported to be preceded by an aura which was described as a feeling of his stomach "dropping." He was being treated with valproate and lamotrigine without any relief. Onset of seizures was at 2 years of age. These early episodes consisted of starring, drooling, and mouth twitching. Several AEDs failed to ameliorate the seizures. Development, history, etc., were normal, except for the occurrence of febrile seizures in a younger brother.

Video electroencephalography recorded seven episodes, which were confirmed by his parents as his typical seizure. Examination revealed six tonic seizures and one focal seizure. The tonic seizures had an abrupt onset, with sudden flexion of head/trunk, tonic extension of upper and lower extremities, and jaw clenching. This was followed by several mild clonic jerks of the extremities.

EEG results showed that the onset of the seizure was associated with a paroxysm of generalized bilateral synchronized rhythmic fast activity (20–25 Hz) with some muscle artifact. This lasted about 12 s and was followed by slowing at 1.5–2.5 Hz with spike wave discharges. Postictal EEG showed a generalized decrease in amplitude, associated with the patient appearing lethargic. Five of six of these tonic episodes occurred during non-REM sleep and one occurred while the patient was awake.

MRI showed a focus of increased signal activity in the right posterior parietal lobe. There was no evidence of mass effect, edema, or contrast enhancement. Magnetic resonance spectroscopy was normal. PET showed a hypometabolism focus in the right parietal lobe on the same location as the MRI lesion. The patient subsequently had a total resection of the right parietal lobe lesion. Neuropathological examination revealed a dysembryoplastic neuroepithelial tumor. The patient is now seizure free, except for a couple of breakthroughs.

The authors note that tonic seizures are a common type of seizure and were first described in 1902 (Jackson and Singer 1902). Furthermore, early studies showed tonic seizures to usually be nocturnal and difficult to treat. The EEG is usually characterized by high-frequency (12–35 Hz) rhythmic activity. It is important to note that tonic seizures are not the same as the tonic component of tonic–clonic seizures in that they are of a higher frequency. The ictal pattern of tonic seizures is called "paroxysmal fast activity" because the activity is not always generalized (Markland 2003).

The authors state that although tonic seizures are most often associated with Lennox–Gastaut syndrome, they do occur in some patients with highly focal lesions. This patient had tonic seizures associated with focal seizures. Right parietal lobectomy revealed the dysembryoplastic neuroepithelial tumor, and rendered the patient seizure free.

Another interesting paper reports investigations of startle-provoked epileptic seizures (SPES) in which the most common semiologic findings were generalized tonic seizures in 50% of patients examined (Tibussek et al. 2006). In this study, clinical, neuroradiologic, and neurophysiologic data from 22 patients diagnosed with proven SPES were evaluated. This was a retrospective study, and patients' records were reviewed regarding etiology, neurologic comorbidities, response to AEDs, etc. A total of 89 SPES were documented by video EEGs, and seizure semiology and EEG results were recorded.

Results showed that 15 boys and 7 girls were diagnosed with SPES. Mean age was 5 years, 8 months. Many patients had comorbidities including spastic tetraparesis, mental retardation, and delayed communicative skills. Two children had normal development. The SPES were clearly seen clinically following an acoustic or somatosensory stimulation. These were accompanied by a clear EEG pattern (Rosenow and Luders 2000).

In 50% of children, the seizures were predominantly tonic in nature. The seizures consisted of generalized tonic, myoclonic tonic, or tonic myoclonic seizures. Generalized tonic seizures showed bilateral upper extremity extension and neck and trunk flexion. Some patients (5) had myoclonic jerks which preceded the tonic phase. Myoclonic jerks after the tonic phase were less frequent and represent the tonic myoclonic group.

Provoking the seizure by unexpected sound ($n=14$) or touch ($n=3$), or both ($n=4$), was of interest. In some cases, following a quiet period, just speaking at a normal volume would initiate a seizure. Loud hand clapping was also used to initiate a seizure, but not always after a clinical response. Other seizure types seen during SPES included gelastic and hypermotor seizures.

Diffuse electrodecremental pattern (DEP) was the most frequently seen ictal finding on EEG. DEP was seen in three patients with myoclonic seizures and in nine children with tonic seizures. Others showed generalized sharp wave EEG patterns, polyspike waves, and irregular spike waves. Four patients showed EEGs of focal seizures. Interictal patterns were abnormal in 14 children and normal in 8. The 14 abnormal interictal EEGs were either in complete disorganization or showed generalized slowing.

The authors note that the 22 patients actually showed a wide variety of results both clinically and on EEGs. Thus, startle epilepsy seems much more heterogeneous than previously reported (Aguglia et al. 1984; Arzimanoglou et al. 2004).

The patients had a diverse list of etiologic possibilities in their histories. Frequently occurring features were tetraparesis and mental retardation. Somatosensory evoked potentials were a frequent cortical abnormality. Other workers have found focal cerebral pathologies (Manford et al. 1996). The authors of the current paper believe that the key is the severity of cerebral pathology, rather than any specific etiology.

As regards semiology, the authors state that all patients except two had daily seizures, and one with a high of ten seizures per day. The main seizure type was tonic extensions. Others agree with Rosenow and Luders (op cit) stating that the frequency of tonic seizures in SPES patients was 90% or more. Variations in these numbers may reflect selection criteria. The authors conclude that the high proportion of severely brain damaged patients in their study resulted in the incidence of tonic seizures.

The authors state that since SPES are easy to provoke, ictal video EEG is easy to obtain and should be used in diagnosis. Focal origin seizures were seen in three patients with bilateral tonic seizures and versive head movements. This suggested a contralateral onset zone in the temporal or frontal lobes.

Ictal EEG results were quite variable. Most frequently seen were DEP. Fast activity frequently was 15–20 Hz. The DEP was related to tonic or myoclonic tonic seizures. DEP is seen in Lennox–Gastaut syndrome patients. Nineteen of the children in this study were refractory to AEDs, a reflection of the poor prognosis. During an average of over 4 years, none of the drug-refractory patients showed cessation of SPES.

The authors conclude that their study showed that startle epilepsy is not a consistent, uniform seizure form. There were many patterns of semiology and EEG findings. They speculate that it is unlikely that there is a common underlying neuropathologic mechanism of startle seizures.

In a brief review article (Scantlebury et al. 2007), the authors discuss the severe epilepsies seen in newborn infants, and how to handle these potentially devastating cases. The authors note that in early infantile epileptic encephalopathy (EIEE), tonic seizures are the characteristic feature associated with ictal burst suppression discharges as seen in EEGs. In early myoclonic encephalopathy (EME), the main seizure phenotype is myoclonus, although tonic seizures may become predominant as the EME becomes advanced.

In EIEE, the neuropathology is that of diffuse changes in the cerebral cortex, subcortex, and brain stem regions. The changes in the brain stem may be present and help to explain what tonic seizures are a major ictal phenomenon (Djukic et al. 2006). In EME, the brain stem damage is progressive, and metabolic/genetic alterations are probably key. The EME tonic seizures may evolve through kindling of brain stem structures. An autosomal recessive form of EME has been described (Molinari et al. 2005) with an alteration in glutamate metabolism.

Infantile spasms represent another category of very serious newborn seizures. Infantile spasms are characteristically associated with brief spasms that can be of the flexion (clonic), extension (tonic), or a mixed type. They are generally refractory to treatment, but some success has been reported using ACTH or vigabatrin. The prognosis is poor, in some measure due to lack of any satisfactory animal models.

One animal model consists of the intracerebroventricular injection of corticotrophin releasing hormone. This model, however, does not adequately reproduce the signs/symptoms of human infantile spasms. Another model attempt involves producing damage to the cortex, subcortex, and brain stem in rapid succession. Newborn rats were injected on day 3 with doxorubicin and lipopolysaccharide. On day 5, *p*-chlorophenylalanine was injected.

By days 7–12, rat newborns were exhibiting recurrent seizures which resembled human flexion/extension spasms. EEG results from this rat model showed ictal discharges that were similar to the EEGs of human infantile spasm patients. The behavioral abnormalities in the rat infantile spasm model and similar results in humans increase the possibilities that this triple injection model may prove to be worth further investigation, possibility leading to new effective treatment paradigms.

In another study, three stages in the evolution of seizure encephalopathy associated with infantile spasms are examined (Bahi-Buisson et al. 2008). Mutations in the x-linked cyclin-dependent kinase-like five (CDKL5) gene are responsible for the severe encephalopathy. This disorder usually starts by 2 months of age and displays developmental delay in addition to spasms. Mental retardation is severe. About 20 mutations of CDKL5 are known, and no clear relationship seems to exist between the mutation location and the phenotype.

This paper presents a retrospective study in which the clinical and EEG aspects of 12 patients with CDKL5 mutations were studied. The authors considered the cases to be epileptic encephalopathy when the epileptiform abnormalities contributed directly to progression of the disorder. MRIs were performed on all patients, and ten channel EEGs of more than 2 h captured ictal/interictal recordings in each case.

Results showed that the electroclinical descriptions fit into three phases. The first (early epilepsy) consisted of epilepsy as the presenting symptom, with an average age of onset of 4 weeks. Seizures were that of generalized tonic epilepsy with face flushing, or tonic followed by clonic seizures. Seizures were brief, lasting less than 1 min, but frequent. Patients were treated with valproate, vigabatrin, or combinations thereof. Early ictal recordings showed a general flattening and fast activity discharge followed by spike and wave discharges in frontal and central regions. Interictal recordings were normal in nine cases.

Stage 2 (epileptic encephalopathy) was characterized by nine progressively relapsing cases. This phase occurred between 6 months and 3 years. Typical or modified hypsarrhythmia and slow EEG activity were seen. Neurological examination during the second period showed profound mental retardation, hypotonia, no language or visual interaction, and no developmental progress. Eight cases showed encephalopathy consisting of brief tonic seizures intermixed with infantile spasms. At this stage, treatment was with corticosteroids, which allowed some partial and transient improvement in behavior. In three cases, seizures stopped a month after corticosteroid treatment, and there was improvement in behavior.

In the third stage (late multifocal and myoclonic epilepsy), seven patients aged 2.5–19 years were seizure free and five patients were refractory to their seizures. This group had previously failed to respond to corticosteroids. Seizure types included tonic seizures and spasms in four patients, myoclonic seizures in three cases, and atypical absences in two patients.

Brain MRI showed hyperintense areas in posterior white matter in six cases, which were associated with hyperintensities in the dentate nuclei in three cases. Delayed myelination was seen in five patients, and cerebellar atrophy was noted in one case. A total of 12 possible genetic mutations were identified. Patients with early truncation mutations or missense mutation had the worst disease progression.

The authors note that the various CDKL5 mutations are each responsible for specific phenotypes. The authors further state that other workers (Scala et al. 2005) believe that early epilepsy (stage 1) is very important in predicting which patient will have CDKL5 mutations. The patterns in stage one of convulsive seizures and neurological delay, poor contact, and axial hypotonia are a bad combination. In stage 3, two groups can be distinguished – one with a relatively favorable outcome and the other with a poor outcome.

The authors conclude saying that the results contribute to determine the epileptic phenotype of the CDKL5 mutations. The data also relate to the site of mutation and phenotype. They state that mutation analysis will optimally be tested in patients with early epilepsy and developmental delay.

In another paper (Besag 2001), the risks of drowning are briefly discussed in terms of tonic seizures. While it is accepted that epileptic people should not swim, or if so, only when accompanied and supervised, particular problems exist in those with tonic seizures.

The case of a 14-year-old epileptic boy who drowned is presented. His seizures were all similar and consisted of an upward deviation of head and eyes, a stiffening or flexion of the trunk, facial twitching, and twitching or jerking of the extremities. He was being treated with valproate plus carbamazepine, and had stopped carbamazepine 3 months before his drowning, with only two seizures since.

The patient had gone to a lake to swim with other students and several teachers. He was playing in the water when he disappeared. He was swimming when the group was about to leave, divers were called, and he was found in 5 ft of water. His arms were crossed over his chest as often occurred, according to his parents, whenever he had a tonic seizure.

Previous studies have stressed the risk of drowning for patients with epilepsy, and the importance of not swimming alone if epileptic. Supervision should be close. In one study (Kemp and Sibert 1993), results showed that out of 306 drowned children, none who were supervised while swimming were in the group. Of several published studies examining drowning and swimming, none emphasize the added risk of drowning associated with specific seizure types.

The boy in this case had tonic seizures. The author notes that during a tonic seizure, the chest wall muscles contract, and air is quickly expelled almost completely. When this occurs while swimming, the body density becomes greater than water density, and the person quickly submerges. Then, when the chest muscles relax and inspiration occurs, the patient is underwater and having a seizure. The outcome is clear. The answer is that especially in certain types of epilepsy (tonic), supervision must be physically close and the accompanying person should be trained in what to do.

Chapter 8
Atonic Seizures

Atonic seizures are not infrequent, and are one of the more dramatic seizure types. As the name implies, they are an almost instantaneous loss of muscle tone, which results in a rapid fall if the patient is standing. Less dramatic phenotypes are a dropping head, or slumping forward of the body. The loss of muscle tone can be without any preceding or concomitant myotonic or tonic events. Falling seizures, or drop attacks are a heterogeneous group of seizures with atonic seizures as a feature, but also with other phenotypic expression. The Lennox-Gastaut syndrome is a seizure disorder in which atonic seizures are a feature.

Seizures characterized by sudden atonic episodes have been in the literature for at least 200 years. Tissot is credited with the first published description of a patient who had daily episodes of drop attacks (Gastaut 1982). Hunt (1922) has been credited with the first publication containing a detailed description of atonic seizures. Controversies as regards terminology and classification result because of the variety of seizure disorders which have atonic seizures (and therefore drop attacks) as a key feature.

Two types of atonic seizures have been described. The first is a brief atonic seizure in which only the head may droop, or the seizure can involve all postural muscles resulting in a fall to the ground which happens rapidly, and the patient sits on his/her buttocks, then topples over. The second type of atonic seizure is a prolonged episode in which loss of consciousness and atonia may last for from 1 min to several minutes. After falling, the patient remains motionless. The variety of seizures which display drop attacks as a feature suggest a variety of mechanisms can cause atonia.

The clinical features of generalized atonic seizures consist of falling to the ground in less than 1 s. The sequence involves a short head drop followed by trunk and leg drop. Patients may resume standing in a second or two (Gastaut et al. 1966). In partial atonic seizures, diminution of muscle tone is mainly in the axial muscle groups. This fall is relatively slow, taking 2–4 s. Focal atonic seizures can have a 30 min or longer duration. Thereby being classed as status epilepticus.

The Lennox-Gastaut syndrome is an electroclinical syndrome of seizures which have drop attacks as a main feature. The syndrome is also referred to as childhood epileptic encephalopathy with diffuse spike and wave discharges. In addition to the

D.W. McCandless, *Epilepsy: Animal and Human Correlations*,
DOI 10.1007/978-1-4614-0361-6_8,

above, features of Lennox-Gastaut syndrome are mental retardation, and seizures including myoclonic jerks, atypical absence seizures, and atonic seizures.

The Lennox-Gastaut syndrome is responsible for 1–4% of childhood epilepsies, and 10% start in the first 5 years of life (Hauser 1994; Kramer et al. 1998). The incidence is relatively similar from world region to world region, and the annual incidence is about 2 per 100,000 children (Rantala and Putkonen 1999). The proportion of Lennox-Gastaut syndrome with mental retardation in institutions is obviously much higher.

The mean age at onset is 26–28 months (Chevrie and Aicardi 1972).

Lennox-Gastaut syndrome is classified as either idiopathic or symptomatic. The ratio is about 25–75%, respectively. Idiopathic Lennox-Gastaut syndrome is classed as such if there is normal development up to diagnosis, and there are no underlying potential causes such as encephalitis, brain malformations, birth injury, frontal lobe lesions, etc.

Clinical manifestations of the Lennox-Gastaut syndrome include mental retardation, psychiatric symptoms, and various seizure types including atonic, tonic, myoclonic, and atypical absence seizures. IQ examination has shown mental retardation in from 2/3 to 3/4 of Lennox-Gastaut syndrome patients. Patients with a later unset have better cognitive results. Psychiatric symptoms consist of mood changes, and disturbances of personality. Psychoses with aggressiveness are frequent psychiatric features.

EEG features of patients with Lennox-Gastaut syndrome are characterized by a slow background. Permanent background slowing is associated with poor cognitive prognosis. Diffuse slow spike and waves pattern with a frequency of 1.5–2.5 Hz predominates (Markand 1977). The amplitude is often highest in the frontal brain areas.

The neuropathophysiology of Lennox-Gastaut syndrome is not clear. This is complicated by the absence of any animal models. PET imaging has not shown any reproducible obvious metabolic pattern, although a focal abnormality is suggested (Gur et al. 1982). Areas of hypometabolism have also not been found (Theodore et al. 1987). Other investigators (Chugani et al. 1987) have had more positive results, showing unilateral focal hypometabolism, unilateral diffuse hypometabolism, bilateral diffuse hypometabolism, and normal metabolism. Focal sites include the inferior frontal gyrus. Another investigator found hypometabolism in the frontal to temporal regions (Miyauchi et al. 1988).

MRI is the preferred imaging strategy for patients with Lennox-Gastaut syndrome. Abnormalities stated above such as brain malformations, hypoxia–ischemia injury, and frontal lobe lesions, when present, suggest treatment modalities, and may predict outcomes. The EEG is critical to the correct diagnosis of Lennox-Gastaut syndrome. A prolonged EEG serves to minimize missing crucial findings. EEG recordings from seizures in Lennox-Gastaut syndrome patients, while asleep are also important in the evaluation.

The AED felbamate was a drug tried early to assess efficacy in Lennox-Gastaut syndrome. Previously, valproate and benzodiazepines were shown to be beneficial in the Lennox-Gastaut syndrome (Jeavons et al. 1977; Farrell 1986). Preclinical trials

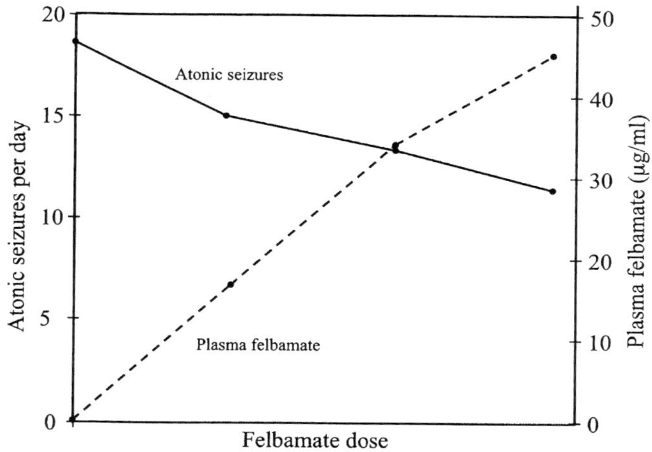

Fig. 8.1 Relation of the felbamate dose to plasma concentration and frequency of atonic seizures. Adapted from Ritter, F., et al. New Eng J. Med. 328: p. 29, 1993

of felbamate were effective in animal models, and the present double blind controlled study was designed to look at the efficacy of felbamate as an add on AED (Ritter et al. 1993) (see Fig. 8.1).

In this study, 73 patients with Lennox-Gastaut syndrome were studied. Criteria for inclusion included multiple seizure types, and at least 90 atonic seizure attacks per month.

The study had a two phase course. Two days before the first 28-day base line phase, valproate or phenytoin was reduced by20%. Patients were carefully monitored and blood samples drawn. After the 28-day initial phase, a 14-day felbamate/placebo titration phase was incorporated, followed by a 56-day maintenance period. At the start of the 56-day period, patients were randomly assigned to felbamate or placebo.

Results showed that the felbamate and placebo groups were similar as regards demographic and pretreatment characteristics. As assessed by closed circuit T.V. and EEG, the group treated with felbamate had an 11% decrease in seizure frequency, whereas the placebo group had a 1% increase in seizure frequency. Side effects are similar between the two groups. Most side effects were mild to moderate, and self-limiting. Neither vital signs nor laboratory values were different between placebo and felbamate groups. During the treatment phase, valproate and phenytoin concentrations did not vary more than 22% from base line values.

The authors note that as compared to placebo, patients with the Lennox-Gastaut syndrome were significantly improved as regards atonic seizure frequency, and the felbamate patients had improved life quality due to increased alertness. The improvement of quality of life, and better seizure control, makes felbamate an efficacious drug. The authors choose atonic seizures as a measure of efficacy because these seizures are always identifiable, and reliably documented. The authors state that the efficacy of felbamate as an AED, improvement of the quality of life, and safety, render felbamate an important add on drug in the treatment of epilepsy.

A patient (6-year-old girl) with atonic seizures and cortical dysplasia was studied using ictal EEG and SPECT (Aihara et al. 1997). The purpose of the study was to try to determine neurophysiologic mechanisms of atonic seizures.

The right handed patient was born of an uneventful pregnancy. At age 6 months, the patient was noted to have a limited rotation of her left arm, and had a neurological work up at 8 months. She presented with hypertonia and increased deep tendon reflexes. C.T. showed right cerebral hemisphere atrophy and the EEG was unremarkable. At 2 years, the patient showed MRI results of an unfolding of cortical gray matter, and a thickening of the right fronto-parietal lobes. Interictal EEG showed isolated spikes. Prescribed AED treatment stopped seizures until age 5, when frequent head nodding occurred 50–100 times per day. Ictal SPECT showed marked bilateral mesial frontal and right frontal hyperperfusion as compared to the contralateral region. Another interictal SPECT study showed that 24 h after a seizure, there was bilateral frontal hypoperfusion, especially in the bilateral mesial frontal lobes.

The authors state that their findings showed interictal epileptic activity was seen in the right central region, and also localized in the frontocentral region, as well as localized in the frontocentral area. Ictal polygraphic recording showed that interruption of EMC discharges were associated with bilaterally synchronous spike and wave complexes (BSSW). The mechanisms in this case may be unique, and the presence of cortical dysplasia confounding matters. The authors note, however, that there may have been a greater energy requirement in the right fronto-parietal lobe. Also, the long period of slow waves in the right hemisphere may have played a role. It is thought that atonic seizures are an ominous sign in patients with intractable epilepsy because of the high daily frequency of these seizures.

The features of atonic epilepsy drop attacks and associated generalized spike and slow wave complexes have been carefully studied using video-polygraphic methods in two patients (Oguni et al. 1997). In this study, video-polygraphic examinations were performed during 51 atonic seizures in one patient, and 18 atonic seizures in a second patient. The phenotypic range was from head nodding to complete falls.

Results showed the first manifestation was a flexion of the knees and waist, followed by a straight down fall such that the patient lands on the buttocks. Ictal polygraphs of the atonic seizure drop attacks showed sudden interruption in ongoing EMG potentials. They were in trunk muscles in mild attacks, and were in lower anti-gravity muscle groups in intense seizures. The authors conclude that features of atonic seizure drop attacks are characteristic, and different from those of tonic drop seizures.

Most atonic attacks occur in patients with generalized epilepsy; however, they also occur in complex partial seizure states, and a recent paper examines two such patients (Satow et al. 2002). One of the two patients was diagnosed as having partial epilepsy with atonic seizures emminating from the frontal lobe. The other had

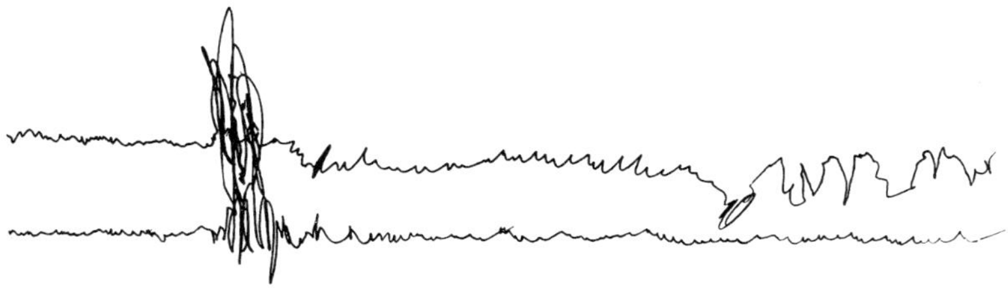

Fig. 8.2 Schematic representation of atonic seizure with falling. Adapted from Satow, E. Epilepsia 43: p. 1425, 2002

parietal lobe epilepsy with atonic attacks. Long term video EEG monitoring, MRI, and interictal fMRI (FDG-PET) were used to evaluate patients' seizures.

Results showed an awake EEG in the first patient had 8.5–9.0 Hz posterior dominant rhythm without asymmetry or epileptiform discharges. While undergoing video-EEG monitoring, the patient had a focal epileptiform discharge in the bifrontal area every 5 min during a sleep period. Patient 2 showed slow posterior dominant rhythms and frequent multifocal spikes in the left parieto-occipital area, and right temporal region. The atonic phenomena started with head droop, then atonia in axial/limb muscles on the right side. The fall took about 3–5 s, whereas the fall in the first patient took 2–3 s. FDG-PET showed glucose hypometabolism in bilateral frontal regions. The glucose hypometabolism in patient 2 was in the bilateral parietal areas.

The authors note that their two patients had slower falls when compared to atonic attacks seen in the Lennox-Gastaut syndrome. The authors point out that two frontal lobe regions cause the inability to initiate or maintain voluntary movements. The primary negative motor area rests in the inferior frontal gyrus. In patients with Lennox-Gastaut syndrome, atonic seizures are likely caused by seizure activity spreading to the brainstem (Egli et al. 1985) (see Fig. 8.2).

The authors summarize saying slow fall atonic seizures in complex partial epilepsy could be explained by long lasting atonic seizures resembling negative motor response from the negative motor area and/or sustained atonia caused by successive EMG silent periods. This in turn could be due to epileptiform activity involving inhibitory areas.

A retrospective study was taken to examine the efficacy of the AED topiramate as an adjunctive treatment for patients with myoclonic astatic epilepsy (Jayawant and Libretto 2003). Myoclonic astatic epilepsy is difficult to treat seizure type which starts during the first 5 years of life. Myoclonic jerks usually are the first seizures seen, but later, atonic seizures appear and are prominent.

Twenty seven children diagnosed with myoclonic astatic epilepsy were followed in an outpatient clinic. Children's ages ranged from 3 to 17 years, and seizure types included myoclonic, atonic, as well as non-convulsive status, generalized tonic clonic seizures, tonic attacks, and febrile seizures. There was fast spike wave or polyspike wave activity seen with EEG. Children with myoclonic astatic epilepsy who were adjunctively treated with topiramate were selected. Features such as seizure frequency, AED dosage, effects of AEDs were all carefully documented.

Results from 16 boys and 11 girls showed that complex partial seizures and atonic seizures were the most frequently seen seizure types. Two had Lennox-Gastaut syndrome, and nocturnal epilepsy. Generalized tonic clonic seizures and absence seizures were also seen. AED results showed that the patients in the study had had an average of 4 (range 1–8) AEDs before adding topiramate. Topiramate was not used as a first line AED in any of the patients, but was used as a second, etc. adjunctive AED.

Results of topiramate therapy showed all except one patient had seizure control improvement. Four patients had improvement of over 50% in frequency of seizures. One had an elimination of atonic seizures, whereas the two Lennox-Gastaut syndrome patients did not show improvement.

The authors note that open label retrospective studies can provide important information. While valproate is the most commonly used AED for myoclonic astatic epilepsy, many other AEDs are either ineffective or aggravate the seizures (Genton et al. 2000). Topiramate, however, is a potent AED with more than one mechanism of action. Its use as an adjunctive therapy has included several seizure types (Glauser 1997; Sachdeo et al. 1999).

The author's study extends and supports positive effects of topiramate to myoclonic astatic epilepsy, and encourages its use in the hard to successfully treat seizure type.

The occurrence of myoclonic atonic seizures as a first symptom of subacute sclerosing panencephalitis is rare (Dimova and Bojinova 2004). This paper describes a case in which myoclonic atonic epilepsy was the first symptom in a case of subacute sclerosing panencephalitis.

Subacute sclerosing panencephalitis is a CNS slow viral infection which usually affects children and adolescents. The initial symptoms are usually those of intellectual and behavioral decay. Scant few cases of early (first) presentation of myoclonus astatic seizures in subacute sclerosing panencephalitis have been previously reported (Lahl and Dorer 1972).

This is a case of an 11.5-year-old boy who at age 4 months had a face and body rash and slight fever. This cleared in about a week, and development progressed normally. At age 10, with no other preceding illness, the patient had myoclonic seizures, some with accompanying atonic attacks at a daily rate of 50–70. EEG showed general slowing with high polymorphic delta waves at the frontal and temporal regions. Spontaneous generalized bursts of sharp and slow waves and irregular spike wave complexes were associated with head and shoulder myoclonic jerks.

A diagnosis of myoclonic astatic seizures was followed by successful AED treatment with valproate and clonazepam. Following 3 months of freedom from seizures, they began to reappear. Soon there was an intellectual decline, manifesting in school work, and visual problems. The seizures continued. A neurological exam showed a severe intellectual deficit including behavioral and cognitive disturbances. Speech was impaired. CT scan and MRI showed mild bitemporal atrophy. A variety of seizure types were evident, including myotonic and atonic. Aggressive AED treatment including valproate, clonazepam, and ethosuximide resulted in seizure control. There were, however, residual myoclonic spasm.

The intellectual/cognitive decay continued, and finally subacute sclerosing panencephalitis was confirmed by CSF findings. An acute febrile illness resulted in a rapid worsening with severe extrapyramidal and pyramidal hypertonia. A state of decortication and autonomic crises developed. The patient has fixed decorticated rigidity without any life threatening situations, and remains in a vegetative state.

The authors note that although there has been a decrease in subacute sclerosing panencephalitis due to measles vaccination, the disease is a serious CNS infection, and a difficult diagnostic/treatment problem. The initial presentation of subacute sclerosing panencephalitis can be variable, even to the point of the rare initial presenting feature being myoclonic astatic epilepsy as seen in this case. Epilepsy occurring in the case of subacute sclerosing panencephalitis is reported in at least 40% of cases (Kissani et al. 2001).

The authors state this case consisted of a 6 months history of myoclonic atonic seizures plus a typical EEG recording. AED efficacy and normal neuropsychologic findings supported the misdiagnosis. The authors speculate that the early (4 months) exanthemia, or a measles vaccination could be the cause of the subacute sclerosing panencephalitis in this case. This case represents an extremely rare incident of myoclonic atonic seizures and absences as the first initial symptom of subacute sclerosing panencephalitis.

The authors note the extremely rapid progression of this disorder may have an anatomic basis. Seizures in myotonic atonic epilepsy originate in the thalamus (Bonnani et al. 2002). Spread of thalamic discharges to the brainstem and frontal/temporal lobes could explain both dementia and a terminal vegetative state. The authors note that subacute sclerosing panencephalitis, presenting as myoclonic atonic epilepsy, although rare, should be considered as a possibility.

A more common neurological problem in children with other medical conditions is the association of seizures (atonic) with leukemia and brain tumors (Kahn et al. 2003). This paper reports a retrospective study in which 93 survivors of childhood cancer with uncontrolled seizures had an atonic seizure rate of 11%. Early recognition of atonic seizures is important because AED treatment depends on an accurate diagnosis, and because atonic seizures are associated with accidental falling injuries.

Atonic seizures were defined as seizures of short duration (3–6 s), loss of muscle tone, and lack of post-ictal depression. Results showed that of 185 post-cancer patients, 93 (50%) had uncontrolled seizures, and ten had atonic seizures. The median age at cancer diagnosis was 9 months, and the first seizure occurred at a mean age of 40 months. Of the ten patients with atonic seizures, all also had other seizure types. EEGs showed multifocal slow waves and spikes in nine of ten children with atonic seizures, and slow diffuse waves in one patient.

All patients had significant impairments of intellectual and functional development. Neuropsychologic testing showed six of the ten to be moderately deficient, and four were too impaired to be tested. At the last follow-up, three children had no bladder/bowel control, and two could not walk. Six of ten have some seizure control in part due to the atonic diagnosis and appropriate AED treatment.

The authors note that atonic seizures have been seen and reported before in patients with brain tumors and leukemia (Mitsufuji et al. 1996). Risk factors for developing Lennox-Gastaut syndrome with atonia would seem to be young age at cancer diagnosis and brain irradiation. Mean age of cancer diagnosis was low at 9 months in this study. Brain development continues for many years after birth, and a key developmental feature in early months/years is myelination. Cranial irradiation is associated with neurocognitive deficits (Reimers et al. 2003).

The authors note that with ever increasing cure rates for childhood cancers, an increasing number will be having seizures. Atonic seizures might be more common than previously thought because they may not be readily recognized. Recognition is important due to possible head trauma from falling. While Lennox-Gastaux syndrome has not previously been seen in post-cancer patients, except for a single case, there should be an increased awareness of this possibility.

Chapter 9
Tonic-Clonic Epilepsy in Animals

The tonic-clonic seizure (grand mal) is one of the most dramatic seizures, lasts several minutes, and is what is commonly associated with a "seizure." It is a generalized seizure in that consciousness is lost, and if standing, the patient falls to the ground (the falling sickness). The seizure originates, or is immediately a bilateral phenomenon. There is frequently, but not always, an aura, and the patient might emit a moan, or scream just as the seizure begins. Following the seizure (5 min or more) there is an obvious and significant post ictal depression period, in which the patient may sleep for over an hour. Incontinence is common, and the patient should only be prevented from hurting himself, never restrained. Moving the patient onto his/her side helps prevent aspiration of vomit.

The tonic phase usually consists of a sudden tensing of muscles. The arms may be drawn to the body. This phase is the shortest, and may only last several seconds. The clonic phase usually represents the remainder of the seizure. This phase consists of a rapid contraction/relaxation of extremities, and sometimes to a lesser extent, trunk, head, and neck. This movement can be exaggerated "twitching" of extremities to violent extremity movement. In this case, the patient must be moved free of furniture and room walls in order to prevent broken bones and soft tissue damage. The patient's eyes frequently roll back, and the tongue may be bitten. Following the seizure and sleep, there usually is amnesia of the episode.

The aura, which is usually thought of as a simple partial seizure, may alert the patient of the impending seizure a few seconds, or even minutes before onset. This allows for a clearing of impediments, or moving to a safe environment. Not all patients experience auras in tonic-clonic seizures, or in other types of seizures. In patients having auras, they can lie down and await the onset.

Most tonic-clonic seizures are idiopathic. Some start as a partial seizure, then spread very quickly to both hemispheres, and this process is termed a secondary generalization. The threshold for seizure generation is frequently altered (lowered) by stress, fatigue, hypertension, flashes of light, rapid motion, noises, etc. In many cases, especially when symptoms have been present for years, MRI and other imaging techniques show scarring and loss of neurons. The lesions themselves may represent the focus for the initiation of generalized tonic-clonic seizure episodes.

D.W. McCandless, *Epilepsy: Animal and Human Correlations*,
DOI 10.1007/978-1-4614-0361-6_9, © Springer Science+Business Media, LLC 2012

A variety of animal models of tonic-clonic seizures are in existence, and these have provided solid data, some of which is translatable to human patients. Two of the most utilized are the maximal electroshock model, and the pentylenetetrazole model. Generally speaking, one model of tonic-clonic seizures suggests brainstem regions such as the lateral geniculate body, ascending pathways passing through the thalamus, and the substantia nigra and locus ceruleus are intimately involved in seizure propagation. The spread of excitation through this route results in the tonic phase. Subsequently, an inhibitory phase, originating in the thalamus, disrupts the tonic phase, resulting in the alternating phase of muscle contraction and relaxation called clonus. Generalized tonic-clonic seizures are the most common type seizure. The usual sequence is an initial tonic phase, followed by the rhythmic clonic phase which gradually decreases in frequency. Sometimes the seizure episode begins with clonic or myoclonic jerks, thereby being an idiopathic generalized epilepsy/clonic-tonic-clonic seizure.

EEG tracings from patients with tonic-clonic seizures show generalized epileptiform discharges with a normal background. The discharge may be spikes, spike-wave complexes, sharp waves, or polyspikes. There is of course a variety of EEG combinations seen in different patients. EEGs from tonic-clonic seizure patients usually have bilateral spike wave complexes.

Tonic-clonic seizures tend to respond favorably to AEDs, making the correct diagnosis very important. Too many such patients are initially erroneously treated with an inappropriate AED, which, in some cases, can worsen seizures (Benbadis et al. 2003). The AED of choice for tonic-clonic seizures is valproic acid. Ethosuximide is also widely used.

Many tonic-clonic seizures in patients occur in the morning, and may be photosensitive, and/or precipitated in part, by fatigue, etc. There may be a family history of similar seizures. There are variations in the actual phenotype of various patients with idiopathic generalized epilepsy/tonic clonic seizures. Not all patients, for example, have seizures upon awakening. Others may have a mixed event, with an absence seizure occurring prior to the tonic-clonic episode. In addition to the clinical manifestations, EEGs are also of a "mixed" variety. It is very important to make the correct diagnosis of idiopathic generalized tonic-clonic seizures. This is what determines the initial (and hopefully only) treatment AED. Many idiopathetic generalized epilepsies start after age 18 (Gastaut 1981), so it should not be assumed that late onset absence like seizures are complex partial seizures. Similarly, it should not be assumed that tonic-clonic seizures are secondarily generalized (Marini et al. 2003). The key is the correct diagnosis and correct treatment, first time.

In terms of animal models, it should be immediately apparent that maximal electroshock, and partly pentylenetetrazole are not the best models of tonic-clonic seizures. Chemically induced models of generalized seizures have a distinct advantage over the maximal electroshock model. The advantage is that there is a preseizure period in chemical models which permits analyses to be made, then compare them to those in the seizing and recovery period. Chemical seizures of tonic-clonic are more easily regulated. A disadvantage of chemically induced seizures is that they rarely occur in humans, so exact applicability of these models to human seizure types is always a question.

Since the mechanisms of action of convulsants vary, it is best to try any proposed AEDs on more that one chemical convulsant model. As compared to in vitro models, in vivo models have a highly significant advantage in that they more closely resemble conditions seen in human seizure states. This is essential in examining, for example, tolerability, behavioral and neurological sequelae, etc. (McCandless and FineSmith 1992).

Chemical models of epilepsy can be roughly classed into two groups: topically administered and systemically administered convulsants. In some ways, the questions being asked help to determine the choice of animal. For example, questions on EEG activity can be conveniently answered in primates. Conversely, if energy metabolism is the focus, small animals such as mice are best since they permit rapid sacrifice in liquid nitrogen, which helps to prevent decal in labile metabolites. In the final analysis, there is no optimal model for generalized tonic-clonic epilepsy. Another excellent review has been written on this subject (Fisher 1989).

Topically applied convulsant models of tonic-clonic seizures consist of the following: penicillin, tetnus toxin, strychnine, and alumina cream, among others. Problems to be dealt with as regards topical administration include choice of anesthetic. It is well established that various anesthetics are also potent anticonvulsants, so frequently two are used, and results compared. Also, inhalation anesthesia can be used, as well as a local anesthetic.

Drug administration is also of concern. If injected into brain, pH must be adjusted. The volume needs to be small to prevent forced spread to adjacent areas. Convulsants layered onto the cortical surface are difficult to control. A soaked pladget can be used, but the amount absorbed by blood and CSF is hard to quantify. Methods of animal surgery and post surgery are federally regulated, and are in constant flux; so the investigator must always be concerned about protocol.

The topically applied convulsant penicillin has been used widely as a seizure inducer. Early studies on paroxysmal depolarization were performed on animal models using penicillin (Prince 1968). The mode of action of penicillin appears to be on GABA metabolism (Wong and Prince 1979). The usual method of administration of penicillin is by pledget, which is soaked with an appropriate volume of convulsant and layered on the pial surface of the cerebral cortex. It can also be administered subcortically (Gloor et al. 1977).

Tetanus toxin (from Clostridium tetni) is another widely used convulsant. It can be injected into the hippocampus and produce a limbic seizure which is long lasting (Mellanby et al. 1977). Status epilepticus can be produced with this model, in which animals have up to 100 seizures per day. This method has value because of the chronicity, comparing relatively well with the human counterpart.

Strychnine is another long-studied model, and a GABA-altering anticonvulsant. It can be administered topically, or systemically. Sometimes multibarreled pipettes are used to both administer strychnine plus record from the site. When topically applied, a pledget is used, and placed onto the cerebral cortex. Systemic administration is usually I.P. at a dose of about 1.75–2.50 mg/kg.

Alumina cream application is achieved by placing the compound in a thin cup, then inverting, and placing it on the exposed cortex. This method has widespread use. This compound shows significant effect on the sensorimotor cortex and temporal lobes.

Alumina cream can also be injected. The dose of alumina cream, as is the case with other topically applied convulsants, must be established in each laboratory.

Systemic convulsants have the clear advantage over topically applied convulsants in that they do not require surgical preparation and potentially confounding anesthetics. A disadvantage concerns lack of precise drug placement. EEG recording can minimize this disadvantage, as the site of onset can be determined. Systemic convulsants widely used include kainic acid, bicuculline, and pentyltetrazole.

Kainic acid, used extensively, acts at least in the hippocampi, and is used as a kindling model of epilepsy. Use of this convulsant produces a pathologic lesion also seen in human cases of epilepsy, hippocampal sclerosis. Kainic acid is an excitatory convulsant, having an effect on the excitatory neurotransmitter glutamate. The usual dose is about 4 mg/kg.

Bicuculline also has seen widespread use as a model of tonic-clonic seizures. Bicuculline is a GABA antagonist, thereby producing its effects. At a dose of about 1 mg/kg, a short preseizure state (90 s) is produced, followed by generalized tonic-clonic seizures. This model has been used in mice and in primates (McCandless et al. 1986). Bicuculline is also used as an agent to produce status epilepticus.

Pentylenetetrazole is used as a convulsant, producing tonic-clonic experimental seizures. The dose varies depending on the animal used. Pentylenetetrazole is also a GABA antagonist, acting at the synapse. A low dose of pentylenetetrazole can produce a longer preseizure state and a reduced strength seizure. The convulsant can be administered I.P. or I.V. It is important in all these models for the investigator to determine the proper dose in their own laboratory.

A nonchemical method for producing tonic-clonic seizures is the maximal electroshock technique. In this method, clips are applied to the ears of mice or rats, or are touches to the corneas, and an electrical pulse is administered. This produces a rapid and dramatic convulsion. The disadvantage is that there is no preseizure state, however, the resultant seizure is consistent. The maximal electroshock method has been used many years ago to evaluate several anticonvulsant drugs to reduce the recovery time from tonic-clonic seizures in rats and mice (Tedeschi and Swinyard 1958). In this study, anticonvulsants tested in terms of recovery time were diphenylhydantoin, phenobarbitol, mephobarbitol, trimethadione, paramethadione, and phenacemide. Recovery time from maximal electroshock, effects of anticonvulsants on tonic hind limb extension, and terminal clonus were all determined.

Results showed that all anticonvulsants decreased the time of hind leg tonic extension, the most significant result being produced by trimethadione and paramethadione. The hind limb tonic flexion was increased in each case. Clonus was also increased in time by the anticonvulsants, with the greatest effect produced by phenocemide. The authors note that there was a prolongation of post ictal recovery by all anticonvulsant drugs tested. The conclusions from this early study are that it is the severity of the seizure which is important in determining post ictal recovery. With anticonvulsants present, both the severity of the seizure and the recovery time are affected. The authors conclude that the recovery time is most significant.

An excellent paper (Collins 1976) examines the metabolic response to focal penicillin-induced seizures in the rat. In the study, a variety of modalities were examined in control and experimental animals. In this study, 10 units of penicillin injected into the motor cortex of rats was the threshold for producing spike discharges. Spikes with afterdischarges required 300 units. The rats were studied as regards location of the penicillin focus using a tracer dose of 14-C-penicillin, and dissected samples were counted in a scintillation spectrometer. The focus was further defined using 14C-penicillin and autoradiography. 14-C-deoxyglucose quantitative autoradiography was also used to assess glucose metabolism in the brains of animals with focal seizures. EEG activity was measured in order to assess the spread of electrical activity. Tissue was harvested for histology and for measuring energy metabolites.

Results showed that 10 units of penicillin produced spike discharges, each of which correlated with a synchronous contralateral motor jerk. Twenty-two of 24 animals which received 300 units or more of penicillin had afterdischarges and/or contralateral tonic-clonic seizures. The manifestation was contralateral tonic extension of head, neck, and tail, with flexion of the limbs. The tonic phase lasted about 15 s, with an additional 15 s of rapid repetitive clonic jerks. Twenty percent of clonic events included a similar ipsilateral response. In some cases there was a 1–3 min period of post ictal depression.

EEG showed maximal discharge in the area of the penicillin injection. Areas adjacent to the injection site showed synchronous discharge. With a 300-unit injection of penicillin, afterdischarges were present at 60 min, and were greatest at the injection site. These occurred bilaterally. The injection was monitored histologically and showed a thin needle track, surrounded by a shell of rarefaction and a few dark neurons. There were no other histological changes.

The half life of 14-C-penicillin was about 15 min. At the time of the first spike, the extent of 14-C-penicillin was 2.6 cubic mm. Some animals injected with 300–450 units showed that penicillin had infiltrated a volume of tissue equal to 32 cubic mm. This was after the first tonic-clonic seizure. Using 14-C-deoxyglucose, resulted in an increase in glucose utilization in the seizure focus, and cortical columns in the contralateral homotypic cortex. When the high dose (500 units) was given, 14-C-deoxyglucose reflected that after tonic-clonic seizures began, autoradiography revealed a large focus, and dense bilateral involvement of the medial frontal cortex and surrounding columns.

In terms of high-energy metabolism, results showed similar results between the low and high doses of penicillin. Both phosphocreatine and ATP were depleted. Contralaterally, tissue showed an increase in metabolites in the spike with after discharge animals, and a slight decrease in energy metabolites in spike only animals.

The author states the major finding was that the penicillin seizure focus in both dose animals were quantitatively and qualitatively different between spike only rats and those with spike/discharges. Animals with afterdischarges needed many more units of penicillin and a greater tissue exposure volume. The mechanism of transition to spikes with after discharges is not understood.

The dose of penicillin may have a bearing in that the actual tonic-clonic seizure may affect membrane pumps. There may be interplay between concentrations of penicillin and the numbers of neurons exposed. The speed of onset of afterdischarges was not, if present, increased with doses of penicillin over 300 units. The volume of cortex involved could be important to the appearance of the afterdischarges. The volume as determined by 14-C-deoxyglucose for after discharges was 4.6 times that of the spike only rats. The metabolic focus was not completely homogeneous, but had distinct boundaries. Thus, the focus was sharply contained by inhibitory surround. Metabolite values in this nonaffected adjacent tissue would have been interesting.

Metabolite levels in the penicillin rats showed both the 10 unit animals and the 300 unit rats had about the same overall decrease in high-energy phosphates. The alteration in ATP was relatively greater than the phosphocreatine change in the spike only group of rats. The elevation of energy metabolites in the contralateral homotypic cortex were opposite to those seen in the focus. An elevation of energy metabolites is not a usual occurrence, but can be associated with decreased utilization.

The author concludes saying that the modest changes in high-energy phosphates are important in light of the other significant changes such as those shown by 14-C-deoxyglucose. This suggests as proposed earlier (Folbergrová et al. 1969) that seizures may start, proceed, and end independent of major energy metabolite changes. This is without knowledge of levels of metabolites in highly discrete regions, or of turnover rates. Also not known are the effects of high-energy phosphate depletion on other cellular energy requiring processes.

Newborns are often exposed to various levels of hypoxia, and a study looked at the effects of hypoxia plus ischemia on the effect on seizure susceptibility in rats (Cataltepe et al. 1995). Seven-day-old rats were subjected to unilateral carotid artery occlusion, and to 8% O_2 for 2 h. Subsequently, the hypoxic/ischemic rats were administered bicuculline subcutaneously at doses of 4–6 mg/kg at 2 and 24 h after the hypoxic/ischemic insult. Results showed that rat newborn pups subjected to hypoxia/ischemia had a decreased seizure susceptibility to bicuculline at 2 h post insult, but an increased susceptibility at 24 h. Tonic seizures were the predominant type of response to the bicuculline at both times. Controls receiving hypoxia/ischemia, but not bicuculline showed only lesion sided circling behavior, not any actual seizures. The results suggested that the hypoxic insult serves to render the newborn rats more susceptible to seizures by 24 h after insult. The lack of an effect at 2 h could be related to decreased energy metabolism at such an early time point after the hypoxic/ischemic insult.

An interesting study looked at the inosital 1,4,5-triphosphate (InsP3) receptor, which acts as a gated calcium release channel in brain (Matsumoto et al. 1996). The type 1 InsP3 receptor (IP3R1) is found in cerebellar Purkinje cells, cerebellar cortex, hippocampal CA1 areas, and in the caudate-putamen. Gene targeting is a mechanism by which IP3R1 mice can be produced, most of which die in utero. Those born, show severe ataxia and tonic-clonic seizures. The rats have EEG results consistent with tonic-clonic seizures, however, histological results looking for localized lesions are negative. These data show that in spite of significant insult of the gene targeting and resultant depletion of IP3R1, and a functional exhibition of tonic-clonic seizures, the effects were not yet associated with structural lesions. This again points out the early biochemical changes only, and indicate the importance of IP3R1 for normal brain function. It also indicates IP3R1 decrease may lead to tonic-clonic epilepsy.

Another interesting paper looks at the influence of the Jnk3 gene on seizure activity and apoptosis (Yang et al. 1997). Earlier studies had shown that a member of the mitogen-activated protein kinases could be required for stress-induced neuronyl apoptosis. This study shows that disruption of the encoding gene Jnk3 causes mice to be resistant to kainic acid-induced seizure activity and apoptosis.

The study was accomplished by the designing of a targeting vector which replaced the 4-kitlo base Msch-Spel Jnk3 genomic fragment with a PGKneo cassette. The deleted region encompasses 1.5 exons. The generated Jnk3-deficient mice had hippocampi which were deficient in Jnk3 activity as reflected by protein kinase activity measurement. The Jnk3-deficient mice showed no abnormality in any cerebral tissue examined using immunocytochemistry, nissl stain, or light microscopy. The number of neurons in the facial nucleus were similar between the deficient mice and wild-type controls.

The altered phenotype was obvious when animals were injected with sufficient kainic acid to produce a tonic-clonic seizure. Kainic acid was chosen since it acts on the hippocampi. Kainic acid injection into wild-type mice and heterozygous Jnk3 mice produced dramatic seizures. The seizures consisted of shaking, forelimb tremor, rearing, falling, and continuous tonic-clonic seizures. Homozygous Jnk3 mice when injected with kainic acid had only staring spells and mild myoclonic tremors. They recovered much more quickly than did wild types or heterozygous Jnk3 mice. A higher dose of kainic acid did induce seizures in the homozygote Jnk3 mice, but the same high dose resulted in a 60% mortality rate in the heterozygous mice.

Five days after kainic acid administration (30 mg/Kg) to wild-type mice, heterozygous mice, and homozygous Jnk3-deficient mice were examined for hippocampal damage. Results showed significant cell destruction in the hippocampus of both control animals (75%), and the most damage was in the CA3 subfield. Interestingly, none of 18 homozygous Jnk3 mice showed any CA3 cell damage. The high lethal dose (45 mg/Kg) of kainic acid which produced seizures in the homozygous rats only had cell damage in 2 of 15 mice. Apoptosis was documented by electron microscopy in the wild-type mice, and prevented in the homozygous Jnk3 mice.

The authors note that their study demonstrated that the absence of the Jnk3 protected against kainic acid induced seizures and apoptosis. Selective expression of Jnk3 in the brain renders neurons especially vulnerable to stress. Data indicate that neuronyl protection may be due to the extinction of the Jnk3-mediated signaling pathway. Previous results show the Jnk3 amino acid sequence is highly consumed in rodents and in humans (Mohit et al. 1995). Further, expression of Jnk3 in rodents and humans is also similar (Gupta et al. 1996). This indicates similar functions between the rodent/human Jnk3, rendering results of this mouse study applicable to human tonic-clonic seizures, and to neuronyl apoptosis. This, the authors state, makes the Jnk3 gene a potential subcellular target for future therapeutic attempts.

Fig. 9.1 Schematic representation of tonic clonic seizure in rats. Adapted from Sansig, G. J. Neuroscience 21: p. 8734, 2001

The effects of the anticonvulsants nifedipine and nimodipine as anticonvulsants have been examined in rats and mice (Khanna et al. 2000). The convulsants were maximal electroshock and pentylenetetrazole. Maximal electroshock and pentylenetetrazole represent an acute and chronic, respectively, method of induction of tonic-clonic seizures. The calcium channel blockers were administered 30 min before electrical or chemical seizure induction. In the case of pentylenetetrazole seizures, a subthreshold dose was used to establish kindling. An acute dose of pentyenetetrazole was also used.

Results showed in acute studies, both of the anticonvulsants nifedipine and nimodipine were capable of reducing the duration of the tonic hind limb extension, as well as decreasing the duration of clonic seizures. Nifedipine alone acted to increase the latent period. In chronic (kindled) animals, nifedipine given prophylactically over the kindling development period significantly decreased the intensity of the tonic-clonic seizure, but nimodipine had no significant effect on kindling. The authors state that their study shows that both channel blockers nimodipine and nifedipine had anticonvulsant effects, but of the two, nifedipine had a better anticonvulsant response.

An interesting paper was published examining seizure susceptibility in mice which do not have metabotropic glutamate receptor 7 (Sansig et al. 2001). In this study, mice were created by homologous recombination which completely lacked metabotropic glutamate receptor 7 (mGluR7). Aging and health of these mGluR7 mice were similar to heterozygous littermate controls except for a slight weight difference. There were no discernable histological differences in brains of the littermates. The morbidity in the mGluR7 mice consisted of seizures at age 10–12 weeks, and some abnormal fear responses at age 6–8 weeks.

Spontaneous seizures occurred frequently in homozygous mGluR7 mice, starting at 10 weeks, and were precipitated by handling/cage transfer. The seizures were mostly clonic, but tonic-clonic seizures also occurred. The seizure susceptibility was thought to be olfactory induced from the bedding material. Usually the response only occurred again after a lag of three days between cage transfers. Controls had no such response (see Fig. 9.1).

In order to further examine seizure susceptibility in these animals, two convulsants were examined for their ability to induce seizures. Both pentylenetetrazole and bicuculine were utilized. Results showed subthreshold doses of both convulsants were able to induce seizures. Pentylenetetrazole produced tonic-clonic seizures in 3/4 of susceptible mice, and of these, 50% progressed to death. Bicuculine produced only clonic seizures.

Anticonvulsants ethosuximide, clonazepam, and valproate were tested for seizure protection in the susceptible pentylenetetrazole mGluR7 animals. All three anticonvulsants provided protection from seizures by pentylenetetrazole. The authors conclude that since receiving anticonvulsants protected from pentylenetetrazole-induced seizure, the mGluR7 receptors do not contribute to mechanisms involved in the efficacy of these three anticonvulsants. The authors conclude that there is likely reduced GABAergic synaptic inhibition in the depleted mGlutR7 animals. The authors believe that further defining this interesting model of tonic-clonic seizures could indicate potential treatment paradigms.

New descriptions of epileptic animals occasionally are published, including one recently observed in Wistar rats (Tsubota et al. 2003). This mutant, termed the Wakayama epileptic rat (WER) is a spontaneously occurring model which exhibits absence seizures and tonic-clonic seizures. The authors studied this new mutant strain both genetically and with EEGs.

Results showed that genetically, the seizure propensity was inherited as an autosomal recessive trait with an 86% incidence. The seizure component of the WER animals had two parts: absence-like seizures and tonic-clonic seizures. The absence-like seizures consisted of a cessation of motor activity and a head droop. The absence seizures began after the F2 generation. The tonic-clonic seizures consisted of neck and forelimb clonus, wild running, opisthotonus, and tonic-clonic and clonic seizures. The seizures occurred in 87% of progeny from crosses between epileptic rats.

Results from EEGs showed a 4–6 Hz spike and wave complexes in absence seizures, and low voltage fast waves in tonic-clonic seizure episodes. The authors state that this new seizure model in which seizure onset usually occurred between 25 and 70 days of age, may represent an interesting new seizure model, study of which could be a useful model of human inherited seizures, perhaps providing translatable data.

The different susceptibilities for seizures between a sensitive mouse (DBA/2j) and a more resistant mouse strain [C57Bl/6j have been studied (Ferraro et al. 1999)]. The idea has long existed that in human epilepsies there is a genetic component which has been hard to characterize (Ottman et al. 1989). Difficulties are based on the highly variable nature of both the clinical and genetic features. Even though seizure disorders are complex in humans, some mutated genes have been identified (Biervert et al. 1998). The idea of the present study was to identify and characterize strain differences in two mouse models with varying seizure susceptibility in order to better understand human features of epilepsy.

In this study, the response to pentylenetetrazole in the two mouse strains, and their F1, F2 progeny was studied. A single subcutaneous injection of the convulsant was administered to the mice at age 8–12 weeks of age. The response to pentylenetetrazole was defined as having four stages: (1) hypoactivity; (2) partial clonus affecting face, head, and forelimbs. This lasted 1–2 s; (3) generalized clonus in which there was loss of posture, and whole body clonus, including all four limbs, rearing, and autonomic signs; and (4) tonic-clonic seizures. These seizures usually ended in death.

Fig. 9.2 Schematic representation of tonic clonic seizure in a rat model of epilepsy. Adapted from Singh, N., et al. J. Physiol. 586: p. 3405, 1998

Further results showed that all doses of pentylenetetrazole used, DBA/2j mice had tonic-clonic seizures, whereas C57Bl/6j only had the generalized seizures at the highest dose. Analysis of seizures in F1 and F2 intercross progeny for heritability indicated a genetic component for latencies. Multipoint interval mapping indicated significant and/or suggestive evidence for linkage on three chromosomes, 1, 4, and 5.

The authors note that their study showed that results of a full genome scan show a locus on chromosome 1 near the D1 mit 16, and 17, which have a prominent effect. This site has also been shown to be similar to that seen in kainic acid seizure models (Ferraro et al. 1997). The finding is perhaps more important since pentylenetetrazole and kainic acid have diverse modes of action, yet affect the same chromosome locus. This site may be involved in a "final pathway" in neuronyl excitability. The authors conclude saying that the genetics of this seizure model are complicated, but the chromosome 1 locus may be involved in several seizure models.

An autosomal dominent form of epilepsy in humans called benign familial neonatal convulsions (BFNC) causes daily partial or generalized seizures (Singh et al. 2008). Mutations of the homologous potassium channel genes KCNQ2 and KCNQ3 exist in patients with the BFNC (Lucarini et al. 2007). BFNC is a highly interesting seizure state in that in only a few weeks of onset, humans undergo remission of clinical features, and have a favorable prognosis. The present paper describes studies in which missence mutations were introduced into mice. These mice were examined as regards altered thresholds of seizures, key seizure features, EEG characteristics, and hippocampal histology.

Results showed that many KCNQ2 and KCNQ3 homozygotes died prematurely between postnatal days 16–30. Mice were implanted with electrodes and video EEG results displayed frequent generalized interictal discharges. Both mutant mouse groups showed behavioral signs of seizures before the discharge. Duration of discharges ranged from 1 to 3 min. Clustering was observed, consisting of multiple mid-severity seizures which occurred 1–5 min apart. Seizure thresholds were shown to be lower in older animals (see Fig. 9.2).

Histologically, KCNQ2 mice following one seizure had normal hippocampal histology. Nissl-stained hippocampal sections demonstrated no pyknotic cells in pyramidal or granule cell layers. Sections from KCNQ3 mice were examined looking for evidence of chronic excitability such as neuropeptide Y upregulating, neuronyl loss, and mossy fiber sprouting. Evidence showed an upregulation in all mice after postnatal day 29, while there was no evidence of upregulation during postnatal days 8–14. Glial fibrillary acidic protein (GFAP) was seen in mice from day 30–120. Special stains showed no hippocampal neuronyl cell loss. There was no associated mossy fiber sprouting into the inner molecular layers of the dentate gyrus.

The authors note that their study details experiments in which they have successfully knocked in to mice the disease producing mutation, which in humans produces BFNC. The possibility that seizures were present soon after birth was not demonstratable because the pups were not very visible during nursing. This research demonstrates that the spontaneous recurrent seizures continue into adulthood, with substantial changes in Ik(m) amplitude, current density, and activation kinetics. However, there was no discernable hippocampal structural lesions as shown in light microscopy. The study emphasizes the important roles that KCNQ2 and KCNQ3 play in regulating neuronyl excitability. Importantly, this mutation suggests these malfunctions can cause seizures in human patients. The ability therefore is present for the KCNQ3 homozygous model especially, to provide a basis for the evaluation of potential AEDs. Such studies, using an ideal seizure model could have important translation to the human counterpart.

In terms of models, there is a double knockout mouse model in which synapsin I/II activity is inactivated (Etholm and Hegglund 2009). Synapsin I and synapsin II are synaptic terminal proteins involved in neurotransmission. The double knockout mice (Syn-DKO) show no behavioral abnormalities above and beyond seizures, and no pathological lesions have been reported. The mutation of synapsin I in human patients is related to the development of epilepsy.

In the mouse model Stn-DKO, seizures can be classified into one of three possible behavioral outcomes: (1) truncus dominated behavior, which includes tonic-clonic seizures (flexion/extension) of the trunk, (2) myoclonic seizures, and (3) wild running and tonic-clonic behavior. Tonic-clonic seizures involving the trunk, and seizures largely consisting of myoclonic episodes, are the most frequent components of the Syn-DKO epileptic component. The evolution of the symptoms associated with the first two seizure sequences are different, reflecting possible different neurobiological mechanisms. The Syn-DKO mouse model may have importance in the development of new AEDs for human seizures since the underlying mechanism in each may be similar.

Another study was performed looking at the vesicular glutamate transporters (vGluts) and seizures (Schallier et al. 2009). Alterations in transport of the excitatory neurotransmitters glutamate could increase synaptic presence of glutamate, with the effect of increasing excitation, and initiating seizure activity. Of the three vGLUTs, vGLUT2 was studied. Pentylenetetrazole was used as a convulsant in vGLUT2 knockout mice and seizure thresholds were compared to wild-type mice.

Results showed that pentylenetetrazole acted to induce seizures, as reported many times previously. When compared to knockout vGLUT2 heterozygous mice, the seizure threshold was lowered. Thus, a lower dose of pentylenetetrazole was needed to induce epilepsy in the knockout mice. This was judged by the time needed until the onset of the first myoclonic jerk. The threshold for onset of clonic seizure activity was also lower than wild-type controls. The authors state that this is the first report suggesting that the glutamate transporter vGLUT2 is probably involved in the genesis of generalized seizures.

The relation to pentylenetetrazole-induced seizures and nitric oxide has been examined (Itoh and Watanabe 2009). This study utilized mice which lacked the neuronyl nitric oxide synthetase (nNOS) gene. The convulsant used was pentylenetetrazole, and the study showed that a subconvulsant dose induced seizures in mice lacking the nNOS gene. The seizures were such that they were lethal at a slightly higher dose of convulsant. Additional studies showed that seizures induced by pentylenetetrazole are modulated by endogenous nitric oxide and ionotropic glutamate receptor-mediated stimulation.

Succinic semialdehyde dehydrogenase deficiency is a disorder of GABA degradation. Results of this deficiency include increased levels of GABA in brain. The syndrome consists clinically of retardation, hypotonia, ataxia, epilepsy, and sleep and psychiatric disorders. Results from T2-weighted MRI showed an increased signal in the globus pallidus, cerebellar dentate nucleus, and subthalamic nucleus. There was evidence of cortical and cerebellar atrophy. EEG results showed generalized spike-wave activity.

Further results showed a downregulation of GABA A and GABA B receptors in the succinic semialdehyde dehydrogenase-deficient mice. Human studies indicate a similar downregulation of GABAergic activity. Various translational attempts for treatment in which the ketogenic diet is effective in the mouse model of this disorder have been equivocal in patients (see Chap. 30).

A new third-generation AED, pregabalin has been evaluated in combination with phenobarbital using the mouse maximal electroshock model (Luszczki 2009). Pregabalin has recently been approved as an adjunct therapy for simple/complex partial seizures, with or without secondary generalization. In the present study, maximal electroshock using ear electrodes was administered to mice pretreated with phenobarbital and pregabalin.

Results showed that pregabalin and phenobarbital in a 1:1 ratio had an additive effect on the maximal electroshock induced seizures in mice. Previous studies have looked at the interaction of phenobarbital and gabapentin (a second-generation AED with a similar pharmacologic action as pregabalin), and so the results were predicted to be similar (Borowicz et al. 2002). Indeed, pregabalin had an additive effect, whereas gabapentin had a synergistic additive effect. The apparent discrepancies between the effects of the two AEDs (additive vs. super additive) results from differences in methodology in the interaction analysis.

The authors conclude that a 1:1 combination of pregabalin and phenobarbital have no acute adverse effects, and are efficacious in the treating of maximal electroshock models. The results of the two AEDs in combination in mice should translate to an effective combination in patients with refractory seizures.

A sedative drug, zolpidem, which is widely used, acts on GABA A receptors. Some studies show that zolpidem has anticonvulsant properties, so a study examined this in adult and aged mice using pentylenetetrazole (Vlainić and Pericic 2009). The thresholds for myoclonus and tonic-clonic seizures were assessed.

Results showed that zolpidem increased the threshold for pentylenetetrazole-induced seizures and death. The effect against tonic seizures was greatest. The anticonvulsant was also effective against seizures in aged mice. These data provide experimental evidence for a potent anticonvulsant action of zolipidem. This bears trial in human epileptics. Much data from tonic clonic seizure models is applicable to human tonic clonic epilepsy.

Chapter 10
Tonic-Clonic Epilepsy in Humans

Tonic-clonic seizures (grand mal seizures) are those affecting the entire brain. The convulsions involve the entire body and consist of muscle rigidity, violent muscle contractions, and a loss of consciousness. When the seizure activity stops, unconsciousness usually continues for many minutes, then occurs a period of confusion, amnesia, and drowsiness. Generalized tonic-clonic seizures can be roughly classified into two varieties. The first is secondary generalized tonic-clonic seizures, which arise as an aura (simple partial seizure), then become partial complex seizures with a distinct focus. Secondary generalized tonic-clonic seizures then spread bilaterally throughout the brain. The second type (primary generalized tonic-clonic seizures) is bilaterally generalized initially without an initiating focus.

The differences between these two types of tonic-clonic seizures are important in determining which antiepileptic drug (AED) might be effective. Thus, an accurate diagnosis is important. Some types of AEDs may aggravate associated non-convulsive seizures such as absence seizures. Broad-spectrum AEDs are good starting drugs for generalized tonic-clonic seizures. As is always the case, thorough histories, complete diagnostic evaluation and counseling should be achieved. The diagnostic work-up must include blood studies, neurological exam, MRI, genetic evaluation, the ruling out of any ancillary contributing disorder, etc. This thoroughness in the initial evaluation can serve to suggest initial drug trials. The goal, as usual, is for a first-time monotherapy success.

An example of this approach involved the use of topiramate (Montouris et al. 2000). The AED topiramate is a broad-spectrum AED in which tissue culture studies show a mechanism of action which involves inhibition of neuroexcitation via voltage-gated sodium channel blockage (Coulter et al. 1993). Topiramate also acts to enhance GABA inhibition, and also inhibits excitatory amino acid activation. There may be additional mechanisms of action of topiramate (Shank 1991). Animal data suggest that topiramate has efficacy against both primary and secondary epilepsies and kindled seizures (Gardocki et al. 1986; Wauquier and Zhou 1996).

In the present study, 131 patients were included in the study. Inclusion criteria were age 4 or older, EEG consistent with generalized tonic-clonic seizures, seizure rate of 3 plus seizures during an 8-week period, and current AED treatment.

D.W. McCandless, *Epilepsy: Animal and Human Correlations*,
DOI 10.1007/978-1-4614-0361-6_10, © Springer Science+Business Media, LLC 2012

Patients with partial onset seizures/Lennox-Gastaut epilepsy were not included. All patients had a baseline medical and seizure history taken, and medical/neurological exam. Blood, urine, etc. analyses were performed, as well as each patient underwent an EEG. Patients/families were carefully instructed as to correct recording of seizures.

Results showed an age range of 3–59 years, and a median seizure rate of 4.4 per week, and 17.5 per month. The mean duration of treatment was 387 days. Results of therapy showed that 2/3 of patients had a seizure reduction equal to or greater than 50%. Sixteen percent of patients treated with topiramate for 6 months or more were seizure-free. Other seizures (absence, myoclonic, and tonic) had similar results. Only 16% of patients withdrew from the study due to inadequate seizure control, or adverse effects. Adverse effects consisted of headache, somnolence, influenza, etc.

The traditional first-line therapies for generalized tonic-clonic seizures are phenytoin, carbamazepine, or valproic acid. Secondary treatment options are limited to vigabatrin, or gabapentin, among others, but studies show no appreciable improvement with these 2 AEDs when used as adjunctive therapy. Vigabatrin/gabapentin may actually aggravate absence and myoclonic seizures (Perucca 1998). The AED lamotrigine is effective, but has some serious side effects. Topiramate, according to the authors, seems effective for generalized tonic-clonic seizures, and may be beneficial in absence and myoclonic seizure patients. Adverse effects are minimal.

The authors note that long-term follow-up shows that efficacy persists indicating a long-term response in seizure reduction. The authors further note that the relative lack of adverse effects and tolerability are attractive benefits of topiramate.

At the other extreme in treatment modality for intractable generalized epilepsies is corpus callosotomy (Feichtinger et al. 2006). The longitudinal section of the anterior portion of the corpus callosum effectively prevents the propagation of epileptic discharges from one side to the other. The method is effective in tonic seizures, tonic-clonic seizures, absence seizures, etc. (Oguni et al. 1994; Gates et al. 1984; Oguni et al. 1991). This surgical approach is a "last resort" for patients with severe intractable epilepsy and in whom other less dramatic surgery is not an option.

The current study reports experience with a non-invasive alternative – radiosurgical corpus callosotomy – which is gaining acceptance for generalized severe epilepsy. Advantages include decreased morbidity and sequelae from open brain surgery. The use of the gamma knife for callosotomy has been previously described (Pendl et al. 1999).

In the Feichtinger report, eight patients underwent radiosurgical corpus callosotomy. Pretreatment diagnostic work up consisted of MRI, EEG, neurological testing, neuropsychological testing, etc. No risk factors for this method were present in any patients. Follow-up after surgery occurred at 3 months, 6 months, and every year after for 3 years. Seizure frequency reduction was assessed after the radio surgical callosotomy.

Follow-up ranged from 1 to 12 years, mean 4.33 years. Results showed a 100% reduction in tonic epilepsy in three patients, and a 60% reduction in two other patients. Another tonic epilepsy patient required an additional radiosurgical disconnection of the middle third of the corpus callosum. This second surgery produced a 100% reduction in tonic seizures. Two patients with additional generalized tonic-clonic epilepsy enjoyed a 100% cessation of seizures, and 2 others had a 50% and 60% reduction.

Adverse effects of the treatment consisted of headaches, and in some patients nausea with vomiting. The neurologic and neuropsychologic exams were unchanged from presurgery. In three patients in whom IQ testing was done, there was no change following surgery. Post-surgical MRI showed the expected post-radiogenic necrosis within the targeted site of the corpus callosum. These lesions were visible 6–12 months after surgery, and remained the same for the duration of the follow-up period. The dosage of AEDs remained unchanged following radio surgical callosotomy.

The authors note that there are several risks to conventional corpus callosotomy which are not present in radio surgical callosotomy. These include intracerebral hematoma, wound infection, infarction, meningitis, and mutism. In addition, conventional surgery may not be 100% effective. The authors note that in their series of 8, there were no immediate severe complications. The successes in this study were very positive, with three tonic epileptics having a 100% reduction in seizure frequencies, and two more with 50% or better. In addition, generalized tonic-clonic seizures were dramatically reduced in four patients. One point of reservation is that the gamma knife method might be a cause of neoplasms. The use of stereotaxic techniques for gamma knife radio surgery limits the surgery to a well-defined target, thereby reducing exposure of adjacent tissue to radiation. The possible incidence of this is however extremely small.

The authors conclude saying that they believe the radio surgical approach to corpus callosotomy is preferable to the standard surgical approach. The results are comparable to the conventional method, the complications of the surgery are almost nil, and an additional area can be disconnected if need be. The authors suggest increased numbers of patients should be studied with the gamma knife radio surgical callosotomy technique. Special attention should be noted as regards the possible induction of neoplasia.

As a follow-up study to the one above, the outcome of corpus callosotomy in patients with primary idiopathic generalized epilepsy, has been published (Cukiert et al. 2009). These were patients (11) who had extended callosal section through a parasagittal craniotomy. Only the splenium of the corpus callosum remained in place. Pre-operative work up consisted of history, neurological examination, ictal and inter-ictal EEG, MRI, IQ testing, etc.

Results showed the age of onset ranged from 4 to 8 years, and age at surgery was 21–53 years. Ten of the 11 patients had the predominance of seizures upon awaking, and 10 had associated absence seizures. All were receiving high-dose polytherapy. AEDs included: valproate (ten patients), lamotrigine (six), phenobarbitol (eight), clonazepam (three), and ethosuximide (one). EEGs showed spike wave discharges ranging from 2.5 to 3.0 Hz. MRI was normal in seven, and showed mild atrophy in four. IQ ranged from 75 to 100 pre-operatively.

Postoperatively, MRI showed adequate extensive callosal section, with only the splenium spared. There was a 75% plus reduction in seizures after surgery, 40% of the patients had at least a 90% reduction in seizures. IQs were statistically unchanged (range 80–98). All patients exhibited acute postoperative syndrome with urinary incontinence, apathy, low verbal output, etc. These were all resolved by 3 weeks after surgery. At the last follow-up, four were still receiving a high dose of valproate therapy, while the remainder were receiving polytherapy.

The authors state that extensive sections of the corpus callosum are more effective than partial sections. The authors state their study was a large/homogeneous one. The patients all enjoyed a better quality of life, although still on AEDs. The extensive corpus callosal section did not impair global function, or IQ. In fact the authors believe there was a significant cognitive improvement in their patients. While subcortical structures might be involved in seizure spread, bilateral synchrony in these patients was disrupted by extensive corpus callosal section. This implies that corticocortical interactions are critical. Although diffuse electrical activity is restrained in each hemisphere after callosal section, focal areas of seizure activity cannot be documented. The corpus callosal section clearly reduced seizure frequency in these patients. The authors end saying that corpus callosotomy is safe, efficacious, and an underused palliative procedure for selected refractory epilepsy patients.

The EEG is usually a key component of any diagnosis involving seizures. A recent retrospective study in which the EEG was evaluated as regards timing of the diagnosis in idiopathic generalized epilepsies was published (Betting et al. 2006). For this study, idiopathic generalized epilepsies were divided into five types based on the age of onset and symptoms. The types were: childhood and juvenile absence seizures, juvenile myoclonic epilepsy, generalized tonic-clonic seizures, and generalized tonic-clonic seizures upon awakening. The purpose of the study was to examine the initial and subsequent EEGs to see what features might suggest an accurate diagnosis.

This study included all patients (180) who had generalized seizures over a 5-year study period. Focal onset epileptic patients were excluded. A positive family history for epilepsy was noted when at least one first- or second-degree relative had a history of epilepsy. Patients were classified as having tonic-clonic seizures when they occurred at any time during the day, and not associated with awakening or leisure time.

EEGs were classified into three groups: the first had irregular or regular synchronous generalized spike wave discharges, the second group had atypical features with well-defined sharp or slow waves, focalities, and asymmetries. The third group was classified as normal.

Results from the 180 patients showed a mean age at evaluation of 31.1 years (range 8–79). When assessed, 17% were taking valproate, and 3% were taking both valproate and lamotrigine. The remainder were taking a wide variety of AEDs, and 8% were not taking anything. Results from the first EEGs showed 45% were normal, 55% were abnormal. EEGs performed after the first, showed 51% were normal, 30% had typical, and 19% had atypical EEGs. This approximate ratio continued through the study. In the tonic-clonic seizure group, 68% had a family history of epilepsy.

The authors comment that while 40% of epileptic patients have at least one normal inter-ictal EEG (Binnie and Prior 1994), the EEG provides strong supportive evidence for the clinical diagnosis. Of some interest was that 70% of the patients referred to this study from primary care centers were taking AEDs other than valproate and/or lamotrigine, the two first-line therapies for idiopathic generalized epilepsy.

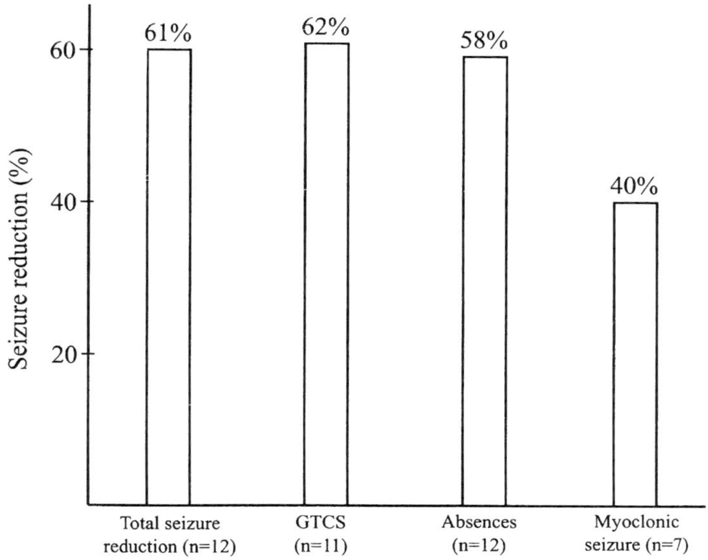

Fig. 10.1 Total seizure reduction after vagus nerve stimulation. Adapted from Kostov, H., et al. Acta Neurol Scand 115: p. 55, 2007

The authors note that the pathophysiology of idiopathic seizures is not clear. There is the possibility of a cortical focus which initiates the seizure. The EEG focalities in this study might indicate mild cortical structural changes, which can be seen in MRI (Leutmezer et al. 2002). The authors state their study supports the notion that AEDs should be chosen in keeping with the clinical diagnosis. Waiting for EEG evidence can result in significant delay in initiation of treatment.

A study of patients with drug-resistant idiopathic generalized epilepsy was evaluated as regards the efficacy of vagal nerve stimulation (Kostov et al. 2007). While complete seizure control is frequently obtained (54–82% of patients), many patients do not have adequate seizure control (Faught 2004). In the present study, 12 patients were treated with vagus nerve stimulation, and the results analyzed. The vagus nerve stimulation began within 2–3 days after implantation.

Results showed an overall reduction in seizure rate of 61% for all types of seizures, and a 62% reduction in tonic-clonic seizure patients. In the primary generalized tonic-clonic seizure patients, 8 out of 11 had a 50% or greater reduction, and of these, three were seizure-free. Three more had a 90% reduction in seizures (Fig. 10.1).

These results show success in the efficacy of vagal nerve stimulation treatment for drug refractory generalized primary tonic-clonic seizure patients. This study is one of the few which looks at efficacy of vagal stimulation in tonic-clonic patients. The ability of the patients/care givers to increase stimulation provides a mechanism to increase seizure control. The authors conclude saying that vagal nerve stimulation is a favorable treatment option for the drug-resistant idiopathic generalized epilepsies.

Seizures are associated with an increased cerebral blood flow, which in turn results in increased blood volume (Hollo et al. 2001). These vascular changes are present in focal seizures as well as generalized seizures. Because of the resistant skull, increased blood flow during seizures can raise pressure, resulting in herniation, coma, and death. The present study (Shah et al. 2007) examines this problem. The application of intracranial EEG monitoring using subdural grid electrodes can produce increased pressure, although usually only temporarily.

In this study, 16 children with intractable seizures were analyzed. The mean age of participants was 10.2 (range 1–18 years). The number of subdural electrodes used was a mean of 79, range 48–116. After placement, the dura was loosely closed and the bone flap replaced. A strain gauge monitor was used to obtain pressure recordings.

Results showed that the baseline intracranial pressure was from 4 to 34 mm Hg. The pressure was positively correlated with age. Nine of 16 children had baseline pressure greater than the age-related upper limit of normal (Tamburrini et al. 2004). A total of 48 seizures were analyzed in the epileptic children. The increase in intracranial pressure was related to the type epilepsy, with secondary generalized tonic-clonic seizures showing the highest level of pressure (35 mm Hg). In addition, there was a positive correlation between pressure and length of time of the seizures.

These results show that older children with tonic-clonic seizures, and long lasting, have the greatest risk for clinically significant seizures. Even baseline pressure was higher than was expected, no doubt due to previous repeated intractable seizures. This suggests that when performing intracranial grid recording of EEGs, loosely suturing the dura, and not firmly placing the bone flap serves to minimize pressure. Other reports of increased intracranial pressure in seizure patients have been published, some showing even higher pressures than this report (Perlman and Volpe 1983).

The authors note that in their study, none of the 16 patients showed any signs of deterioration in neurological status. Even so, there are reports of herniation and death associated with grid placement (Wong and Birkett 2005; present study). The ability to record intracranial pressure adds another important method of evaluation to improve the efficacy and safety of invasive EEG monitoring.

Possibly related to the increased intracranial pressure associated with seizures is the rare occurrence of sudden unexplained death in pediatric epilepsy patients (McGregor and Wheless 2006). There are obvious accidents and risk of injury in epilepsy in children, and sudden death is occasionally noted (Pellock 2004; Berg et al. 2004). Previous studies regarding risk factors have shown a correlation between sudden death with subtherapeutic doses of AEDs in adults (Donner et al. 2001). The present study by McGregor and Wheless was undertaken in order to evaluate risk factors in children who died unexpectedly at a comprehensive epilepsy center.

This was a retrospective study in which the records of children under 18 years old who died unexpectedly were examined. The study consisted of 17 cases, of which seven were deemed definite, nine probable, and one possible as regards the deaths relation to epilepsy. Attempts to find a specific cause via autopsy were not successful.

Results showed the 17 patients consisted of ten females, seven males, with an average age of onset of seizures of 4.3 years, and an average age of death of 12.6

years of age. The types of seizures of the 17 children in the study were: secondary generalized tonic-clonic epilepsy in ten patients, asymptomatic generalized tonic-clonic seizures in six patients, and one patient with febrile generalized tonic-clonic seizures, three of these patients had a corpus callosotomy for their seizures. At the time of death, the average number of AEDs was 1.6. A variety of other treatment modalities had been tried in these patients, including vagal nerve stimulation, ketogenic diet, and corpus callosotomy. The oldest patient at death was 25, and four others were over 18 years.

The authors note that 70% of patients in this series had structural lesions, cognitive delay, or both, and all had secondary, symptomatic, or febrile generalized tonic-clonic seizures. Other studies show patients with tonic-clonic seizures are at a risk of sudden death (Earnest et al. 1992). Given the data in the previous paper regarding intracranial pressures in tonic-clonic seizures, it is not surprising that the seizures per se are an important "proximate" cause of the sudden death (Walczak et al. 2001).

Many possible contributions to the sudden death epilepsy syndrome have not shown any statistical significance. These include: various AEDs as treatment, vagus nerve stimulation, and nongeneralized seizure types – partial seizures. Possible contributors which have merit include pulmonary edema, present in all patients in this study, which has been described in previous studies. Another common finding appears to be that the patients at death were asleep in bed, and prone. This suggests that the reduction of nocturnal seizures might help lower the incidence of sudden unexplained death.

The idea was presented above that increased intracranial blood volume/pressure was deleterious. This high pressure is capable of causing herniation of the brain stem out the foramen magnum, and in this way might contribute to sudden death. This is seen almost exclusively in tonic-clonic seizure patients.

In terms of regional cerebral blood flow and volume, a recent study using single photon emission computed tomography (SPECT) looked at small areas of blood flow in patients with idiopathic generalized epilepsy. The study used 21 patients with idiopathic generalized epilepsy (mean age 21.3 years), and 21 normal matched controls. Two seizure types were present in the 21 patients: 14 with idiopathic generalized seizures, and seven with myoclonic seizures and generalized tonic-clonic seizures. The differences in regional cerebral blood flow (rCBF) between groups were analyzed using statistical parametric mapping of the SPECT images.

Results showed that rCBF of the idiopathic generalized patients was reduced in the anterior/posterior cingulated gyri, cerebellum, bilateral anterior nuclei and right dorsolateral thalamic nucleus, and the right superior colliculus. No idiopathic generalized epilepsy patients had any increases in CBF. The authors note that these studies of inter-ictal rCBF in these patients are reduced in the cingulate gyrus, thalamus, brainstem, and cerebellum, and resultant probable dysfunction in these four regions is related to the seizures of these patients.

Theophylline was a widely used drug for therapy for respiratory disorders such as asthma. Because of adverse effects, theophylline has been curtailed. One serious side effect is the possible production of seizures, which when severe is a neurological emergency. The theophylline seizures can easily progress to status epilepticus, and become intractable. The present study (Yoshikawa 2007) is a retrospective study examining potential first-line treatment in this serious adverse effect. Not all cases of theophylline toxicity result in CNS involvement. Other side effects include arrhythmias, nausea, insomnia, etc.

In this study of 833 patients admitted to a hospital for seizure complaints, about 6% (54 cases) had theophylline-associated seizures. The criteria were that the patients were taking theophylline at the time of their seizures. This group consisted of 26 males and 28 female patients. The treatments of these 54 cases were compared to those of the other 779 seizure patients not taking theophylline. Success was defined as cessation of seizures within 30 min, and freedom from seizures for more than 12 h.

Results showed that 87% of the theophylline patients had a fever at the time of seizure, 36 cases had generalized tonic-clonic seizures, and 18 had partial seizures. Only one patient had toxic levels of theophylline. When arriving at the hospital, 42% of patients from non-theophylline group were seizing, compared to 65% of seizing patients from the theophylline group, and the duration of seizures was greater in the theophylline group. Seizures in the theophylline-associated seizure groups lasted longer than in the non-theophylline patients when treated with diazepam. Those requiring mechanical ventilation were lower in numbers in the non-theophylline group as compared with the theophylline-associated seizure group.

The mechanisms of theophylline toxicity involve a decrease in seizure threshold. Theophylline may also inhibit enzymes involved in GABA production, thereby lowering seizure threshold. Theophylline may also interfere with adenosine metabolism and production and also inhibits the binding of adenosine to receptors. The effect of the alteration of adenosine on energy metabolism has not been examined. Theophylline may not actually cause seizures, but rather acts to potentiate the seizure (Odajima et al. 2003).

Theophylline is an antagonist to benzodiazepines such as diazepam. The authors suggest the use of barbiturates for the treatment of theophylline-associated seizures, but more studies are clearly needed for this neurological emergency.

Secondary generated tonic-clonic seizures are those that begin as a focal seizure which can produce partial symptoms only, or become generalized. The actual features and mechanisms of the secondary generalization have received little attention. The present study utilizes patients with intractable partial seizures and the matching pursuit method and Garbor Atom Density (GAD) to assess these patients' seizures (Jouny et al. 2007).

The patients in the study were selected from over 120 who had intractable partial seizures, and had intracranial grid electrodes in place. From these patients, seven were found for inclusion, who had at least one partial seizure and one generalized tonic-clonic seizure of either mesial temporal or neocortical onset. The generalized seizures were secondary generalized, and were tonic-clonic seizures. The seizure classifications were based on both clinical and EEG findings. The study was based on data obtained from 24 seizures in three mesial temporal seizure patients, and 26 seizures from four neocortical seizure patients. Median ages were 17 from the first group and 23 years of age in the second group. All seven had evidence of lesioned areas.

The GAD measures the signal complexity. The data from the matching pursuit is used to derive GAD measures. The matching pursuit algorithm is a distinct improvement over the usual time-frequency measures of signal complexity used to compare ictal dynamics of partial and secondary generalized tonic-clonic seizures. The localization of the seizure focus, onset, and termination was determined by the intracranial grid.

Results showed that there was a similarity of intracranial ictal onset patterns in focal seizures in the same patient. This had been seen before in intracranial EEG, but not quantified (Franaszczuk et al. 1998). These data improvements justify further use of GAD to study the dynamics of seizure onset. An interesting finding was that as the seizures in mesial temporal lobe onset generalize secondarily, the GAD complexity remains relatively constant at the focus. Thus, the mesial temporal lobe focus is relatively little changed by electrical events in remote areas.

Conversely, neocortical onset seizures have a different pattern during secondary generalization. Partial seizures which originate from neocortical areas have a relatively short duration, and showed a more dramatic effect on GAD maximum at the focus as compared to the same results in mesial partial seizures. This suggests a basic difference in secondary generalized seizures between these two types. The authors conclude by stressing the small sample size of their study, but saying these findings show significant differences between the generalizations of the two types. This concept may have significant implications for intracerebral connections in seizures.

In a somewhat similar vein, the effects of electroconvulsive therapy (ECT) on regional cerebral blood flow were examined using SPECT (Enev et al. 2007). ECT produces a relatively consistent generalized tonic-clonic seizure which closely resembles those seen in spontaneously occurring tonic-clonic seizures in patients. ECT is used to treat psychiatric problems such as depression. ECT permits the study of the spread of seizures from the initial focus to determine what structures are involved. Previous studies have supported the concept that bilateral focal areas are involved in the generalization of seizures (Ackermann et al. 1986).

Patients (8) were studied who were undergoing ECT for significant polar depressive disorder. The eight were divided into two groups, one with SPECT injection at "0" time after ECT, the other group at 30 s after ECT. Standard bitemporal ECT electrodes were placed as previously described (Blumenfeld et al. 2003). SPECT (single photon emission computed tomography) was used to measure regional CBF during ECT. Image analysis was achieved using statistical parametric mapping.

Results showed a different pattern at 0 s after ECT as compared with 30 s after ECT. In the early 0 s group of patients, areas of hyperperfusion were related to the location of the electrodes. The greatest increases in hyperperfusion occurred in the inferior frontal gyri and anterior insula. The areas "activated" in the early group included the putamen and thalamus. Decreases in perfusion were not seen in the early 0 s group. At 30 s after ECT, hyper perfusion was now seen in bilateral parietal and occipital cortex. Intervening regions were spared. Decreases in perfusion were seen in the bilateral cingulate gyrus and left dorsolateral frontal cortex. The authors state the idea that generalized seizures involve the entire brain may not be correct.

Previous work by this group, plus the results of SPECT on ECT generalized seizures support the concept that even this most extreme seizure paradigm does not affect the entire brain. The authors show that rCBF increases in focal areas following ECT. The authors speculate that the region of seizure onset dictates the initial pattern of seizure increases. This information might result in novel seizure reduction modalities.

The authors speculate that the spread of seizure activity is via corticocortical networks or by cortico-thalamo-cortical networks. The superior longitudinal fasciculus could mediate propagation of seizure activity between the frontal and parietal cortex, sparing intervening tissue. Orthodromic conduction could achieve this. Another possible scenario is that cortical seizure activity "recruits" subcortical sites to spread the activity to other cortical sites to which the subcortical (thalamus) area is connected.

The involvement of the thalamus, anterior cingulated and frontoparietal cortex suggest that altered function in these regions contribute to loss of consciousness (McNally and Blumenfeld 2004). The authors note that their study should be cautiously interpreted due to the need for further studies with more patients. Also, ECT seizures may not accurately resemble spontaneous seizures in epilepsy patients. This ECT model is different as compared to spontaneously occurring seizures with regard to chronicity, which may have subtle differences. In spite of these reservations, the authors state that there are significant similarities between the two tonic-clonic seizures. Therefore, the results from this study are relevant, and emphasize the importance of propagation mechanisms, and their association with impaired consciousness.

The use of single photon emission computed tomography (SPECT) as a clinical technique has recently been reviewed for its use in secondary generalized tonic-clonic seizures (Varghese 2009). Complex partial seizures cause an increase in regional cerebral blood flow which is sufficient enough to be detected by SPECT. This serves to confirm the locus. One problem is that when complex partial seizures generalize, the SPECT images are less optimal. The present study examines this problem more thoroughly than before.

In this study, 59 secondary generalized seizures from patients who were candidates for surgery were evaluated. Fifty-three patients' data were used for analysis. Ictal vs. inter-ictal SPECT data were statistically compared. Video EEG permitted the classification of data as: pregeneralization, during generalization, or after generalization.

Results showed that in both pregeneralization and generalization phases, there were significantly more focal areas with increased blood flow than in patients with partial epilepsy not generalized. These results rendered it impossible to accurately predict the initial focus in at least half of cases. SPECT was able to identify correctly the hemisphere of onset in 84% of cases. When a single focus increased in size as shown by SPECT, that area was the site of localization in 80% of cases. Post-ictal SPECT added nothing in being able to identify an initial focus of activity. The authors note that careful interpretation of SPECT data can help identify areas of localization, but in no way is it 100% accurate. It is a useful tool to assist in cases of secondary generalized tonic-clonic seizure patients.

A recent paper (Koutroumanidis et al. 2008) looked at electroclinical characteristics of 33 consecutive patients with generalized tonic-clonic seizures. Of the 33, 18 had generalized tonic-clonic seizures only, whereas 15 also had previously unrecognized mild absence seizures. The purpose of the study was to define these two entities. The authors had previously described a syndrome of idiopathic generalized epilepsy in which there were phantom absence seizures (brief – 2 s) present. These were of such brevity and mildness that they had gone unnoticed (Panayiotopoulos et al. 1997).

Results from this study showed that there were no differences between patients with generalized tonic-clonic seizures, and those with additional previously unrecognized phantom absence seizures as regards gender, age at referral, age at follow-up, duration of seizures, age at seizure onset, or family history. The group with generalized tonic-clonic only seizures had a frequency twofold that of the group with additional absence seizures. Most patients with the additional absence seizures had a more variable time of day onset, as opposed to the generalized tonic-clonic group in whom seizures characteristically occurred upon awakening or immediately thereafter.

Results also showed no other types of seizures such as independent myoclonic or partial seizures. There was a correlation with family history in that both groups had a first- or second-degree relative with seizures. This frequency was approaching 40% in the generalized tonic-clonic seizures only group. Situations thought to precipitate seizures included stress, sleep deprivation, and alcohol consumption. All patients with additional absence seizures had at least one attack during hyperventilation. Generalized spike wave discharges were recorded in 16 patients, while in the phantom absence group, EEG was able to detect phantom absence discharges in eight patients. Thus the total numbers of generalized spike wave discharges was comparable between groups. The phantom absence group was more likely to show poly spikes or polyspike and wave complexes. None of the patients regardless of group had any prolonged trains of poly spikes.

The authors note that of 33 patients whose EEGs showed generalized tonic-clonic seizures, 15 actually, in addition, had brief phantom absence seizures as well. They state that detecting phantom seizures by EEG was easy. The circadian pattern was such that the group without absence seizures occurred mostly upon awaking, while those with additional absence seizures experienced attacks more evenly distributed throughout the day. Polyspikes appeared in the EEGs of both groups, but more in the group with phantom absence seizures. The spike wave events were also longer in the phantom absence groups.

The authors also note that some (Genton et al. 2008) state that phantom absence seizures are subclinical discharges and are therefore less significant, but the present studies show that it is possible to demonstrate a distinction between the two groups, therefore they represent two distinct clinical entities. Lastly, the identity and characterization of these minor phantom absence seizures provides a useful tool for classification of seizures, and contribute to an improved understanding of generalized tonic-clonic epilepsy.

Therapeutic drug monitoring is a process in which the plasma concentrations of AEDs are monitored in order to evaluate potential problems such as compliance, toxicity, drug–drug interactions, treatment success/failure, etc. Overall assessment of this procedure in terms of pharmacologic and economic costs and benefits has been evaluated (Rane et al. 2001).

In this study, two groups were chosen. The first consisted of 25 patients who had undergone therapeutic drug monitoring, while the "control" group had epilepsy patients who had not been monitored. Criteria for inclusion were patients with onset of seizures between ages 18 and 35, seizure duration of 4 years, visits to the clinic for at least 2 years, and at least two monitoring sessions per year for at least 2 years. Extensive questionnaires were used to gather data regarding seizure history, medications, etc.

Results showed a mean onset of epilepsy of about 21 ± 3.5 years for each group, and a mean duration of 9.6 ± 4 years for the first group, and 8 ± 2.6 years for the second group. These differences were not significant. In the first group, 23/25 had uncontrolled epilepsy at the time of the first therapeutic drug-monitoring session. When interviewed, this group showed that 11/25 had complete seizure control, and 10/25 had a 50% reduction in seizure frequency. Four patients had uncontrolled epilepsy. The pre- and postmonitoring data were statistically significant. In the non-monitored group of 25 patients, 25/25 had uncontrolled epilepsy. When interviewed, 12/25 were still uncontrolled, 11/25 had greater than 50% reduction in seizure frequency, and only two had complete seizure control.

Quality of life assessments showed that at the start of monitoring, none of the 25 patients in the first group were married, while at interview, 60% were married with children, whereas those not monitored went from 0% married at the start to 28% married at the interview.

The authors comment that the study shows a significant benefit of therapeutic drug monitoring as compared to those not undergoing the drug monitoring. There was significant reduction of seizure frequency, plus an overall improvement in life quality. The cost effectiveness of this procedure is quite small compared to the clear benefits. This method of AED drug monitoring to assure compliance, lack of toxicity, and benefit, or lack of benefit should be considered as beneficial to patients' well being.

Chapter 11
Juvenile Myoclonic Epilepsy in Animals

Juvenile myoclonic epilepsy is a common form of generalized idiopathic epilepsy which usually has an onset between about 12 and 18 years of age, hence its name. The seizure types which may be seen include myoclonic jerks and seizures, absence seizures, and tonic clonic seizures (Zifkin et al. 2005). The seizures usually occur during daytime, but can occur upon awakening. Clinical examination does not reveal any focal neurological signs, and MRI shows no definitive characteristic anatomical lesions. Psychiatric examination does show a specific personality pattern in about 15% of patients which consists of low self-esteem, emotional instability, lack of discipline, mood changes, etc.

EEG changes consist of inter-ictal epileptiform discharges of 3–6 Hz spike wave, or spike slow wave complexes. Ictal discharges are bilateral. Pathological data support the concept of thalamic and corticothalamic involvement. There is also clear evidence of frontal lobe involvement. The frontal lobe role may be of a modulating nature. Spike wave discharges in animal studies show discharges in the frontal cortex during sleep which indicate involvement of corticothalamic networks during both sleep and spike wave discharges (Steriade and Amzica 2003).

The idea of a generalized epilepsy in juvenile myoclonic epilepsy has become controversial due to increasingly more sensitive imaging techniques. Frontal lobe preponderance has been defined, and there is a possibility of lateralization. Studies have shown evidence for foci in frontal lobes, and occasionally the temporal lobe at the onset and propagation of seizures in absence seizures (Tucker et al. 2007).

There are several excellent animal studies relating to both juvenile myoclonic epilepsy and progressive myoclonic epilepsy. The murine models have genetic components, and EEG and clinical features similar to those seen in human cases. Use of animal models of various types of myoclonic epilepsies helps explain mechanisms of pathophysiology as well as suggests treatment approaches. Use of these models translates data to similar features of the human myoclonic epilepsies.

A mouse model for one form of progressive myoclonic epilepsy has been developed and the characteristics explored (Pennacchio et al. 1998). The progressive myoclonic epilepsies are classed in at least five major forms: myoclonic epilepsy associated with ragged red fibers, neuronal ceroid lipofuscinoses, sialidos, Lafora disease,

D.W. McCandless, *Epilepsy: Animal and Human Correlations*,
DOI 10.1007/978-1-4614-0361-6_11, © Springer Science+Business Media, LLC 2012

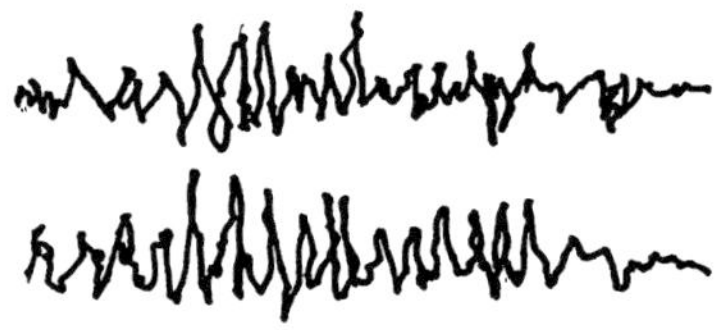

Fig. 11.1 Schematic representation of myoclonic seizures in the cerebellum of cystatin beta-deficient mice. Adapted from Pannacchio, L., et al. Nature 1998

and Unverricht-Lundborg disease. These are progressive myoclonic epilepsies with severe ataxia and dementia. Gene mutations are associated with CLN3, CLN5, NCL5, Cstb, and others. Genes encoding mitochondrial tRNA in myoclonic epilepsy and ragged red fiber disease have been studied. The pathogenesis of this type of progressive myoclonic epilepsy are due to mitochondrial oxidative phosphorylation defects which result in malfunction and destruction of cells dependent on proper ATP levels and function (Bindoff et al. 1991) (Fig. 11.1).

By contrast, Unverricht-Lundborg disease is not associated with cellular defects such as inclusion bodies, or mitochondrial disruption. There is a loss of cystatin B function. This study looks at features of a mouse mutant with a deficiency of cystatin B. Cystatin B is a member of a class of proteins which inhibit cysteine proteases (Turk and Bode 1991). Plasmid vector pBluescript II SK was used. About 10–15 ES cells from three targeted clones were injected into C57Bl/6 mouse blastocysts, and transfected into female recipient mice. All clones produced multiple chimaeras with over 50% agouti coat color marker. About one half had the insertion. Subsequent interbreeding heterozygous mice produced homozygous offspring for the disruption of Cstb.

Results showed that mice which were homozygous for the targeted insertion lacked cystatin B. Only by the narrowing of the palpebral fissures and increased tearing could Cstb deficient mice be identified from littermates. Size, weight, behavior, etc. were indistinguishable between littermates. By 3–7 months, homozygotes developed easily seen corneal opacities in one or both eyes. By 6 months of age, ataxia was discernable. Later, affected mice walked with a wide-based stance and occasionally fell.

The histological appearance of tissues lacking cystatin B was compared to controls at 6 and 9 months of age. Results showed few changes except in the cerebellum, where a reduction in the granule cell layer density was seen. The observation was that granule cell layer nuclei were pyknotic. Specific stains showed the nuclei contained fragmented DNA, and the overall appearance was of apoptotic cell death and granule cell loss in the cerebellum of Cstb-deficient mice.

It was also observed that the Cstb-deficient mice developed myoclonus/seizures by 1 month of age, mostly during sleep. As this progressed, there was twitching of ears and tails, then facial spasms, and shaking of the body and limbs. Shaking increased and myoclonic jerks were seen. At the end of seizure development, there would be a large myoclonic outburst, and the mouse jumped in the air. These outbursts occurred only during sleep. EEG showed abnormal cortical activity.

This period was characterized by 4–6 Hz spike discharges. Normal recordings were observed immediately following the cessation of the epileptic event. The seizure events were not observed in control littermates.

The authors note that the Cstb mouse model has the same genetic mutation found in Unverricht-lundborg disease. Further, the clinical progression, ataxia, and seizure development sequences also mimic the human disorder (Koskiniemi 1990), making this mouse model a good animal model for translation. This should be capable of providing useful information on mechanisms. The authors state that even the EEG recordings are similar between the mouse model and those seen in humans (Kyllerman et al. 1991). Some differences between the human and mouse models exist such as a lack of photosensitivity in the Cstb mice, and no observed tonic clonic seizures. Other genes may be involved to a lesser extent in the two corresponding phenotypes.

The authors conclude that Unverricht-Lundborg disease should be classed as a primary neuropathologhic degenerative syndrome which targets specific cells (cerebellar granular cells). They state that the mouse model of Unverricht-Lundborg syndrome supports the concept of focus on early development events, centered on cerebellar development. These animals should serve as a splendid reagent for further studies on mechanisms and on possible treatment modalities.

Another paper (Shannon et al. 2002) examined cerebral neuropathological alterations in the mouse model of Unverricht-Lundborg syndrome described in the previous paper. These authors confirmed the above studies' findings of a lower density of cerebellar granule layer neurons. In addition, similar apoptotic cells were found in the hippocampi and entorrhinal cortex, neocortex, and striatum. Also seen was white matter gliosis, which could have been a secondary phenomenon. Cellular atrophy was also noted in the cerebral cortex.

The authors note that their study also showed that Cstb mutant mice from a seizure-prone or seizure-resistant genetic background had the same neuropathological alterations in cerebral architecture. The authors say this indicates neuronal atrophy is a key consequence of cystatin B deficiency, independent of the level of seizure phenotype. This in turn may indicate a physiological role for cystatin B.

Other relevant mouse mutant Cstb model findings have observed that, in kindled rats, seizure activity has been shown to induce widespread upregulation of Cstb mRNA and protein in the rat forebrain neurons (D'Amato et al. 2000). Further, in Cstb mice, mRNA profiling shows an increased expression of seven genes involved in apoptosis, proteolysis, and glial activation. Also, there was an increased expression for cathepsin S (Lieuallen et al. 2001). Studies of models of these rare progressive myoclonic seizures should, in addition to providing important specific data, provide insights into molecular pathogenic mechanisms of neuronal function/survival and seizure generation/propagation in general. This will benefit the understanding of more common seizure disorders such as juvenile myoclonic epilepsy.

There is a succinic semialdehyde dehydrogenase-deficient mouse model of juvenile absence epilepsy (Cortez et al. 2004). This mouse model starts as an example of absence seizures, but ultimately develops into myoclonic seizures and status epilepticus. The succinic semialdehyde dehydrogenase-deficient mouse is a reliable model for the human counterpart. It is characterized by elevated levels of both gamma hydroxybutyric acid (GHB) and GABA in brain. GHB can induce absence seizures. This study tested the idea that the phenotype of the mutant mice had seizures due to elevated GHB.

EEG and video EEG recordings were made in homozygous succinic semialdehyde dehydrogenase-deficient mice, and heterozygotes and normal controls. Spontaneous seizures occurred during postnatal days 8–15 of life in the homozygous mutant mouse. The seizures were of a 7-Hz spike and wave discharge recorded from the thalamocortical circuitry. The seizure was initially seen as vibrissal twitching, facial myoclonus, and a "frozen stature."

The absence seizures became worse, and by day 18 evolved into myoclonic generalized seizures, becoming status epilepticus, then death. Ethosuximide and the GABA agonist CGP 35348 abolished the seizures. The authors comment that the homozygous mouse model strongly resembles the human counterpart. Succinate semialdehyde dehydrogenase deficiency was seen in humans. The mouse mutant therefore represents an excellent model for the human metabolic deficiency, as well as a model for the transition from juvenile absence epilepsy to juvenile myoclonic epilepsy.

A mouse model with a defect at Scn1a was produced using stem cells and a vector for delivery. The result was a mutant mouse model which had severe ataxia and died on postnatal day 15. Heterozygous mutants had seizures and sporadic deaths after day 21. This was a murine model for human severe myoclonic epilepsy in infancy, and was used in their study to examine loss of function of voltage-gated sodium channels (Nav) (Yu, F et al.).

Results from this study showed that the gene encoding the Nav1.1 channel was disrupted in two mouse lines, and produced a similar phenotype. The homozygous phenotype displayed was of a mouse which began to show ataxia and seizure activity by day 9. This soon developed into limb tremors, swaying, and loss of righting response. By day 15, the homozygous mice were essentially in status epilepticus, and did not survive. Diazepam treatment acted to reduce frequency of seizures, but mice still died on day 15. Hand feeding extended the life span by 2.5 days.

Heterozygous mice (Scn1a$^{+/-}$) started having seizures and sporadic deaths between ages 21 and 27 days. Heterozygous mice seemed to die in status epilepticus. Some few heterozygotes lived with seizures for over 1 year. EEG recordings of heterozygous mice showed normal periods of baseline cortical EEG activity. Video/EEG recordings showed spontaneous electrographic discharges and seizures in nearly one half of heterozygous mice. Phenytoin results were included an arched tail (Straub tail), myoclonic jerks, forelimb clonus, head bobbing, etc. Post-ictal depression inevitably followed.

Results from studies of sodium currents showed that about half of the sodium current was lost in heterozygous interneurons, and a smaller additional decrease was observed in homozygous interneurons. Additional studies showed that interneuron excitability and inhibitory control of targets was reduced in the hippocampi of mutant Nav1.1 mice. There was an upregulation of Nav1.3 in the hippocampal interneurons in both heterozygous and homozygous mice. Further study showed that there was a compensatory upregulation specific to GABAergic inhibitory interneurons. The upregulation was region specific for the hippocampi, as no other upregulation was seen in brain.

The authors comment that the complete loss of Nav1.1 is incompatible with normal brain function, and results in ataxia, seizures, and death as is seen in the human counterpart of severe myoclonic epilepsy in infancy. This correlation between the mouse model and the human disorder includes phenotypes of significant severity, early age of onset, rapid downhill progression, and a genetic basis (Dravet et al. 1992; Nabbout et al. 2003). The authors note that similar cellular and molecular bases between the two syndromes stress the utility of the mouse model for further investigations of the human disorder.

As pointed out above, sodium channel subtype mutations may affect sodium currents in the hippocampi (Yu et al. 2006). It was not clear in the above paper just how this effect in the hippocampi might have a phenotype of such severity. Mutation of the Nav1.1 sodium channels caused generalized epilepsy with febrile seizures and also Dravet's syndrome (George 2005; Fukuma et al. 2004).

The question remained as to how the depletion of Nav1.1 sodium channel could cause such a severe disorder of ataxia, failure of motor coordination, and spasticity. In this paper (Kalume et al. 2007), the previously described mouse model of severe myoclonic epilepsy of infancy (SMEI) was used to further explore mechanisms of neuropathology. Specifically, the authors focused on alterations in function in cerebellar Purkinje cells.

In this study, Nav1.1 mutant mice were produced by deletion of the last exon encoding domain IV from 53 to 56 and the C terminal tail of Nav1.1 channel. Breeding strategies resulted in Nav1.1 (++), Nav1.1 (+–), and Nav1.1 (––). The distribution of voltage-gated sodium channels in cerebellar Purkinje cells was investigated in 13–14 day old mutant mice and controls. Behavioral tests were performed to evaluate ataxia, muscular hypotonia, etc. Electrophysiological characteristics were determined on dissociated Purkinje cells.

Results showed that mutant mice were readily generated through the disruption of the Scn1a gene, which encodes Nav1.1 channels. Results from immunohistochemistry compared expression of brain sodium channels in wild-type mice and experimental mutants. Data showed that Nav1.1 and Nav1.6 were the predominant sodium channels in Purkinje cells. In Nav1.1 (+–) heterozygotes, there were no detectable staining of either Purkinje cell bodies or dendrites in Nav1.1 knockout mice. Anti-Nav1.6 showed strong staining of Purkinje cell bodies and dendrites.

Ataxia results showed that the Nav1.1 mice are distinguishable by days 9–10 as signs of ataxia appear as compared to controls. The symptoms progress to death by day 15. Ataxia characteristics were such that 100% of mice showed limb tremors and loss of balance. The Nav1.1 mice showed uncontrolled movements when walking. The reflex responses were also abnormal in the Nav1.1 heterozygous mice. Similarly, human patients with SMEI are ataxic at a comparable time frame as the mice (1–2 years of age).

As regards sodium current in Purkinje cells, studies showed that deletion of Nav1.1 channels reduced the amplitude of the sodium current, but did not affect the sodium current of cerebellar Purkinje cells. The reduced expression of Nav1.1 channels did, however, reduce persistent sodium current in Purkinje cells. This suggests Nav1.1 substantially contributes to persistent current. In addition, the reduced expression of Nav1.1 reduces the amount of resurgent current in Purkinje cells. Finally, the actual firing rate of cerebellar Purkinje cells was lower in Nav1.1-depleted mice than in appropriate control mice.

The authors conclude saying that there are at least two regional defects – the hippocampi and the cerebellum. These two defects can produce the epileptic phenotype of the Nav1.1 mutant mice. Sodium currents are significantly reduced in the hippocampus, sodium currents and excitability are reduced in Purkinje cells. The Purkinje cell alteration leads to decreased cerebellar nuclei output and ataxia. This model is excellent for exploring mechanisms of altered function in human severe myoclonic epilepsy in infants, and the results translate accordingly.

Yet another paper from the same group (Oakley et al. 2009a, b) looks at temperature and age-dependent seizures in the same Nav1.1 mouse model of severe myoclonic epilepsy in infancy (SMEI). The authors note that disorders such as SMEI are quite rare, and therefore an available mouse model such as Nov1.1, which contribute to action potential generation (Gong et al. 1999), can contribute in many ways to the definition of SMEI. With this caveat in mind, the authors sought answers to several questions, including those associated with aspects of temperatures and SMEI, age and seizure onset, and EEG features in the Nav1.1 model of SMEI.

Results showed that wild-type controls showed seizure activity only at 42.5°C, whereas the mouse model of SMEI had temperature-induced seizures at 39.5°C. This implies a temperature-based cause of seizures, not an inflammatory mechanism in the febrile seizures of SMEI. Human infants with SMEI do not usually have seizures until about 6 months of age. The authors tested temperature-related seizures at different ages of the mutant mice. Results showed the temperature-induced seizure in the mouse model occurred at a lower temperature in the older mice (30–46 days of age – 39°C) as compared to the younger mice (20–22 days of age – 41°C).

One distinctive feature of the human disorder of SMEI is the progressively increasing severity of seizures with age. In the mouse model, the investigators found a similar phenotype. Thus, the average seizure duration at 30–46 days in the mutant mice was 25.3 ± 1.5 s, whereas at 20–22 days of age the average duration was 13.3 ± 3.1 s ($P = 0.002$). These results demonstrate a significant increase in severity of seizures in the mouse model, mimicking the human phenotype. Collectively these data imply that mechanisms underlying both temperature- and age-dependent features are similar between the Nav1.1 mouse mutation model and the human SMEI counterpart. This certainly facilitates translation between the two.

The authors clearly state that the mouse model recapitulates the human SMEI disorder with surprising consistency. Such close phenotypes mean not only qualitative correlations can be made, but also quantitative comparisons are possible. This important paper, already defining important phenotypic features, should stimulate further studies on this particularly severe form of epilepsy in an excellent mouse model.

Another paper describes studies on a new animal model of seizures. This seems to result from a single genetic defect of the gene encoding RNA-binding protein Bruno-like 4 (Brunol4), which results in limbic and tonic clonic seizures, as well as having features of absence seizures, and myoclonic seizures (Yang et al. 2007). The model has been named "frequent flyer" because of a phenotypic behavior consisting of running and bounding as though the mouse is trying to fly.

The Brunol4 mice were derived from C57Bl/6J transgenic mice at the Jackson Laboratory. Phenotypic behavior was produced by routine handling such as transferring the mice to a clean cage. The most mild form of seizure behavior in the heterozygotes consisted of facial muscle twitching and forelimb clonus. A more severe stage consisted of tail arching (Straub), back arching, myoclonic jerks, and falling. The most severe phenotype included wild running and bouncing, and tonic extension of the hind limbs.

The symptoms did not begin until the third postnatal month of age, but by 7 months the heterozygous mice showed decreased electroconvulsant seizure thresholds. The homozygous Brunol4 mice were much more severely affected in that most died in the first day after birth. Only 1% survived until age 4 weeks. The homozygous did not display seizure activity and had some alternative mutation which was lethal.

Some mutants showed spike wave discharges, thought to represent electrical evidence of absence seizures. Homozygotes tested on an F2 hybrid background did show frequent spike wave discharges. The discharges were synchronous, rhythmic, and generalized as compared to other absence mouse models such as stargazer mice. Although the duration was short (1.5 s), the mice were motionless, suggesting absence seizures.

Further studies showed that the mutation was represented on chromosome 18. Localization studies showed that Brunol4 is brain specific as had been shown previously (Kislinger et al. 2006). Brunol4 showed a predominant neuronal expression brain pattern which included cerebral cortex, olfactory bulbs, granular layers of the cerebellum, and the hippocampi. It was in the hippocampi where the expression was highest. The authors suggest the Brunol4 is associated with nRNA encoding events which in turn are involved in synaptic function. This yields an imbalance in neuronal excitability.

The authors note that this model, affecting chromosome 18, seems to render it an ideal model for adolescent onset idiopathic generalized epilepsy such as juvenile myoclonic epilepsy. This model, once further investigated, could serve to suggest effective AEDs, or other efficacious treatment in human cases of juvenile myoclonic epilepsy.

Chapter 12
Juvenile Myoclonic Epilepsy in Humans

Juvenile myoclonic epilepsy is a common epilepsy, most often seen in the adolescent population. There are two distinct forms of adolescent myoclonic epilepsy: juvenile myoclonic epilepsy and a more severe type called progressive myoclonic epilepsy. Both these forms will be considered in this chapter.

The first, juvenile myoclonic epilepsy is sometimes classified as one of three syndromes included in the broader class of idiopathic generalized epilepsy with an adolescent onset. The idiopathic generalized epilepsies are considered to be a genetic disorder characterized by a low threshold for seizures as opposed to the disorders with a specific pathologic structural defect. Close microscopic examination has yielded inconclusive results as regards lesions (Opeskin et al. 2000). The idiopathic generalized epilepsies all represent similar genetic defects, most of which involve problems with ion channels, and have come to be called ion channelopathies (Hirose et al. 2005).

Juvenile myoclonic epilepsy was first described 150 years ago. Juvenile myoclonic epilepsy characteristically begins in the teen years, and contributes from 4 to 10% of all epilepsies. This incidence reflects U.S. data and may be much higher in some ethnic groups worldwide (Nair and Thomas 2004). The age of onset is between 12 and 18 years of age (Panayiotopoulos et al. 1994). Forty percent or more juvenile myoclonic epilepsy patients have a family history of seizures. This number is higher than that of most other forms of seizures. Genetic mutations in patients with juvenile myoclonic epilepsy occur most frequently on chromosome 6, and are involved with ion channels. The actual mechanisms associated with these mutations remain unclear.

Clinically, seizures occur mostly in the morning after awakening and may be precipitated by sleep deprivation. When the myoclonic seizures occur in clusters, a generalized tonic-clonic seizure almost always follows. Absence seizures are found less often. The three seizure types occur in a ratio of 100%, 95%, and 20%, respectively (Dhanuka et al. 2001). The myoclonic and generalized tonic-clonic seizures are usually the types seen in the morning.

These early AM seizures may progress to myoclonic status, but that is infrequent; these seizures usually do not progress. Despite being generalized, loss of

D.W. McCandless, *Epilepsy: Animal and Human Correlations*,
DOI 10.1007/978-1-4614-0361-6_12, © Springer Science+Business Media, LLC 2012

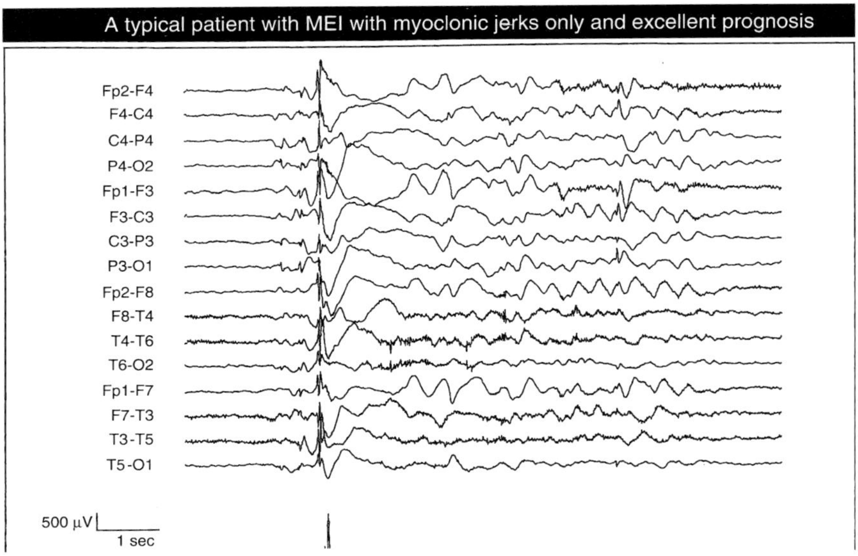

Fig. 12.1 EEG from a patient with MEI. Patient had frequent jerks and grunting sounds. With kind permission from Springer Science+Business Media: Epileptic Syndromes and their Treatment, 2010 p. 270, Panayiotopoulos, C. Fig. 9.2

consciousness usually does not occur. The seizures are notably brief. The associated absence seizures are the usual episodes which resemble, for example, day dreaming. The associated absence seizures may represent a separate subclass of absence seizures because of clinical boundaries (Obeid 1994). Other precipitating factors besides sleep deprivation include alcohol consumption, and in women, menstruation. Photosensitivity is another common feature that may affect seizure rates (see Fig. 12.1).

The diagnosis can be problematic. It has been noted that from 25 to 90% of patients diagnosed with juvenile myoclonic epilepsy were originally misdiagnosed prior to the correct diagnosis. Diagnosis errors can be connected to incomplete histories, mistaking myoclonic seizures for complex partial seizures, and lack of familiarity with juvenile myoclonic epilepsy. Patients, family members, and caregivers should all be specifically asked about seizure history.

While neurological and cognitive examinations are usually unremarkable, interictal EEGs are abnormal in half to 2/3 of patients. The EEG consists of 4–6 Hz generalized synchronous, polyspike and wave complexes. Because of the photosensitivity, flashes of light may change EEG patterns, or precipitate a myoclonic seizure. Photostimulation should be a required part of an EEG in suspected cases of juvenile myoclonic epilepsy (see Fig. 12.2).

From the treatment standpoint, juvenile myoclonic epilepsy is considered a lifelong disease, with careful withdrawal from AEDs unsuccessful. Therefore, the AED treatments should be made carefully as even optimal AED treatment may not stop

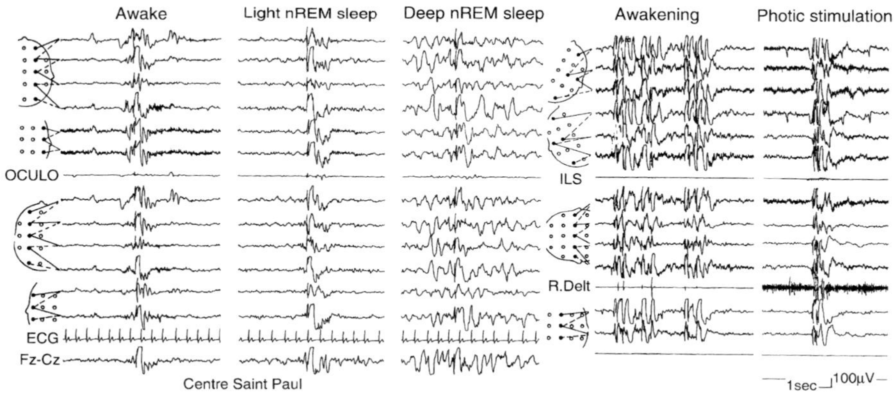

Fig. 12.2 EEG from a patient with juvenile myoclonic epilepsy. Published in Crespel A. et al. Atlas of electroencephalography Volume 2; The Epilepsies, EEG and Epileptic Syndromes. With permission from John Libbey Eurotext

occasional "triggered" attacks. The AED of choice for juvenile myoclonic epilepsy is valproic acid. It has been judged successful in 85–90% of cases (Bourgeois et al. 1987). If side effects are significant, lamotrigine is a candidate for this disorder.

The disorders called progressive myoclonus epilepsy are quite a different seizure disorder. They consist of myoclonic seizures, tonic-clonic seizures, and a progressive neurological deterioration. The neurological deterioration usually consists of locomotor dysfunction (ataxia) and dementia. Age of onset is in adolescence, but it can develop anytime. The myoclonus epilepsy can be precipitated by various stimuli including flashing lights, or other external stimuli. Massive bilateral myoclonic jerks can occur (Berkovic et al. 1986). The disease is relentless and in later stages, the diagnosis is obvious. The diagnosis may be more difficult in early stages, and is sometimes misdiagnosed in the early stages. Fortunately, progressive myoclonic epilepsy is rare.

There are a number of subtypes: Unverricht–Lundborg disease, myoclonus epilepsy with ragged red fibers, Lafora disease, neuronyl ceroid lipofuscinoses, and sialidoses. The first of the above, Unverricht–Lundborg disease is the archetypal pattern for progressive myoclonus epilepsies. In this type, there are no storage particles, but neuronal loss and reactive gliosis are present in the spinal cord, medial thalamus, and cerebellum. An excellent review of the progressive myoclonus epilepsies has been published (Berkovic 2008).

The question of the genetic relation in terms of mutations between juvenile myoclonic epilepsy and progressive myoclonus epilepsy has been addressed (Rees et al. 1994). The locus for the Unverricht–Lundborg disease was mapped to the chromosome 21q22.3. The locus for juvenile myoclonic epilepsy has also been localized, in this case to chromosome 6p. Segregation analysis suggests that there are two chromosomal loci for juvenile myoclonic epilepsy. This translates to genetic phenotype heterogeneity for juvenile myoclonic epilepsy. This led to an investigation of the locus of the severe form of the disorder, progressive myoclonus epilepsy.

Results were obtained using linkage analysis and three microsatellite DNA markers in the EPM 1 gene region. This was carried out on 25 families. Multipoint linkage analysis provided definitive exclusion for 20 cM around PFKL, the closest marker to EPM 1 in 75% of models tested. The authors conclude that these results show that the EPM 1 gene is not linked in any way to the phenotypes which were expressed in the families. This unequivocally demonstrates that Unverricht–Lundborg and the mild form, juvenile myoclonic epilepsy, are not allelic variants.

Familial myoclonic epilepsy has more than one gene locus. The disorder is an autosomal dominant epileptic syndrome which frequently starts in adolescent patients, but can have an adult onset. The clinical features, noted by the authors, of adult onset include tremulous finger movements and extremity myoclonus, and those after adolescence include infrequent seizures, abnormal polyspike and wave EEG data, and increased photosensitivity. Other phenotypic features include a non-progressive course without ataxia or dementia. The present study was undertaken in a large Japanese family in order to attempt to localize the locus of this autosomal dominant seizure syndrome.

In this study, 27 patients, which included 17 with seizures, were studied. The average age of onset of symptoms was 30.5 years (range 18–45 years). Tremulous finger movements were initial symptoms. Myoclonus occurred in half of the cases in upper extremities only, and in both upper and lower extremities in the other half. Tonic-clonic seizures were rare, occurring only a few times in the entire course of the disease. None had any evidence of ataxia or dementia, hallmarks of the much more severe progressive myoclonus epilepsy. There was photosensitivity, and "photomyoclonus" could be induced in some patients.

Results also showed significant evidence to indicate a benign adult familial myoclonic epilepsy gene at chromosome 8q23.3–24.11. Several other epilepsy syndromes have been linked to chromosome 8q. The authors note that additional studies of different families have been done to determine possible allele and locus heterogeneity. The authors state that future studies might show a relationship between benign adult-onset myoclonic epilepsy and the CHRNA4-nAChr subunit, as has been shown in neonatal frontal lobe epilepsy, but that awaits further genetic research.

Abnormalities in genes in juvenile myoclonic epilepsies have been identified as also occurring on chromosome 6p21. This was found in the human lymphocyte antigen (HLA) region of chromosome 6p21, and is called EJM3, and its association with BRD2 has been described (Pal et al. 2003). Another paper has described a gene responsible for juvenile myoclonic epilepsy in affected families which have a defect mapped to the EJM1 locus (Suzuki et al. 2004).

The EMJ1 locus contains 18 genes, and there can be multiple mutations present, which all result in a single amino acid substitution. The EFHC1 gene was the mutation target gene. It has been shown that at least two separate families had a similar mutation, thereby showing a common haploid surrounding the EFHC1 gene. This implies that the two mutations started on a founder chromosome common to both families.

The authors comment that overexpression of EFHC1 in mouse hippocampus cultured neurons induced apoptosis, which was lowered by the mutation. This would in turn increase the number of neurons present over the optimal number that should have been present, resulting in an increased neuronal density. Indeed, quantitative MRI shows increased gray matter in juvenile myoclonic epilepsy patients (Woermann et al. 1999). This increased density might lead to hyperexcitability circuits, or increased sensitivity.

While juvenile myoclonic epilepsy is a more benign form of myoclonic seizures, the progressive myoclonus epilepsy, nevertheless, may develop into status epilepticus, although this rarely occurs. An excellent review paper has been published looking at various aspects of status, including those associated with progressive juvenile epilepsy (Shorvon and Walker 2005). In this review, absence seizures, myoclonic epilepsy, tonic-clonic epilepsy, and autonomic seizures (Panayiotopoulos syndrome) are all examined.

In terms of juvenile myotonic epilepsy, status epilepticus is considered a very rare event and has a frequency even less than that of absence epilepsy status epilepticus. When myoclonic status epilepticus actually occurs, it usually manifests as a so-called myoclonic storm. This is described as a slowly developing, but ever-increasing series of myoclonic jerks which increase in severity. The frequency increases, and the intensity is reached in minutes to hours depending on the patient. Consciousness may be impaired, and the final episode is frequently a generalized tonic-clonic seizure. Again, the frequency of progressive myoclonic epilepsy is rare.

The actual issues of classification are subject to review as improved imaging and other diagnostic methodologies are improved or developed. In the case of juvenile myoclonic epilepsy, increasing sensitivity of neuroimaging such as quantitative MRI and 1H-magnetic resonance spectroscopy (MRS) has shed additional light on this issue, including showing multiple foci, including foci in the frontal lobe (Koepp 2005). Although juvenile myoclonic epilepsy is ordinarily associated with normal intelligence (Delgado-Escuetta and Enrile-Bacsal 1984), studies have shown a personality peculiar to juvenile myoclonic epilepsy, and some neuropsychological studies suggest a frontal lobe involvement (Savic et al. 2000). The present paper aims to examine this concept using improved imaging techniques.

Results from PET using the uptake of 18F-fluoro-2-deoxyglucose demonstrated that patients with juvenile myoclonic epilepsy had frontal lobe alterations different from control which might affect cognitive function and epileptogenic potential (Schwartz et al. 1998). Another study using quantitative MRI has shown subtle widespread cerebral cortical anatomic changes in 8 of 20 juvenile myoclonic patients (Woermann et al. 1998). Statistical analysis of the data showed an increase in cortical gray matter in the mesial frontal lobes of the juvenile myoclonic patients. The use of proton magnetic resonance spectroscopy was able to demonstrate normal thalamic volume, but a reduction in *N*-acetyl aspartate in the thalamus in idiopathic generalized patients.

The authors conclude that more sensitive neuroimaging methodologies are capable of increasing neuropathological damage, which serves to correlate and explain neurobehavioral alterations in juvenile myoclonic epilepsy. This additional evidence showing multifocal anatomical regional sites is an important finding. It is important in that the frontal lobes have been shown to be involved for the first time.

This suggests that juvenile myoclonic epilepsy has a frontal lobe involvement, and is a multiregional thalamocortical network form of epilepsy.

The idea that juvenile myoclonic epilepsy is associated with a unique neuropsychologic personality dates back 55 years (Janz and Christian 1957). Several personality characteristics are expressed including lack of discipline, poor self-confidence, poor affect, a child-like attitude, etc. There is also a tendency to sleep late in the A.M. and appear drowsy, and then go to bed late in the P.M. The present paper (Pung and Schmitz 2006) examines some of these features in light of very recent studies using more sophisticated imaging techniques. These show involvement of the frontal cortex in juvenile myoclonic epilepsy patients (see above).

The authors used a variety of psychological tests and questionnaires aimed at examining diurnal rhythms and personality traits. This study included 20 juvenile myoclonic patients and 20 temporal lobe epileptic patients as controls. The subjects were matched as regards age, sex, gender, etc. Results from a "morning/evening" test showed that juvenile myoclonic epileptic patients could be classed as "evening" people, while those with temporal lobe epilepsy as morning subjects.

Furthermore, the temporal lobe epilepsy patients chose work which was theoretical and did not directly involve others, whereas the group with juvenile myoclonic epilepsy chose jobs with direct personal contact with other people. This represented a slightly more extroverted display by the group with juvenile myoclonic epilepsy.

The authors note a significant sleep/wake diurnal rhythmic difference between the two groups. Thus, the juvenile myoclonic epilepsy group was reluctant to arise, and seemed not ready to "face the day" until afternoon. This is believed to fit a feature of subclinical epileptic activity in the juvenile myoclonic epilepsy patients. This is thought to be a frontal lobe involvement, in some ways consistent with features of nocturnal epilepsy.

In another study (Fittipaldi et al. 2001), looking at 1,000 patients' diurnal epilepsy features revealed that only 4.6% had epileptic activity during awakening, and those were predominantly patients diagnosed as having juvenile myoclonic epilepsy, or generalized tonic-clonic epilepsy. The authors speculate that patients with "distorted" lifestyles have this feature due to juvenile myoclonic epilepsy. An alternative explanation is that the altered diurnal features are an independent feature of juvenile epilepsy. Further studies in which testing and imaging are done in the same patients are warranted.

Features of a rather unique neuropsychological personality implying frontal lobe involvement have been described by others (Pascalicchio et al. 2007). These authors listed personality traits such as impaired attention, language, verbal memory, visual perception, etc., in patients with juvenile myoclonic epilepsy. The authors state that their findings strongly suggest frontal lobe involvement in juvenile myoclonic epilepsy patients based on neuropsychological clinical examination.

Another paper examines the anatomical location of lesions associated with juvenile myoclonic epilepsy (Deppe et al. 2008). This study was performed on ten patients using diffusion tensor imaging as an imaging strategy. Results showed a cohort with a mean age of 28 years. Age of seizure onset was 13.7 years, and duration of seizures was 14.2 years. Five of the ten patients were on valproic acid monotherapy. Results from imaging studies showed normal images. Diffusion tensor imaging showed a significantly reduced fractional anisotrophy in the thalamocortical fibers, and fibers of the basal ganglia, parahippocampal, and frontal basilar region. The degree of changes in the frontal cortex correlated with age, duration, and number of tonic-clonic seizures.

The authors conclude that these structural alterations have reduced the integrity of the fibers in the parahippocampal, basal ganglia, frontal, and thalamocortical regions. Further studies will determine if the localized structural lesions are related to the cause or result of seizures, or possibly are related to treatment modalities.

By 2007, a paper appeared listing 15 separate chromosomal mutations associated with people with juvenile myoclonic epilepsy (Delgado-Escueta 2007). In fact, juvenile myoclonic epilepsies are considered to be primarily genetic in origin. It is possible that in some cases, multiple mutant genes may act together producing various epileptic phenotypes. Various gene mutations producing a variety of neurochemical alterations have been identified. Juvenile myoclonic epilepsies are the most common form of idiopathic generalized seizure states, accounting for 12–30% of epilepsy patients in hospitals.

The authors describe two major types of juvenile myoclonic epilepsies, including the "standard" form and the childhood absence seizures, which evolve into juvenile myoclonic epilepsy. The second form involves both awakening seizures and an evolution to tonic-clonic seizures. Minor subtypes include those which start late (after 18 years of age), and a form which includes astatic drop seizures (Martinez-Juarez et al. 2006). Juvenile myoclonic epilepsy is transmitted as a Mendelian dominant or recessive trait, or as a complex oligogenic trait. The authors note that improvement in treatment may result from an understanding of the major genes which cause juvenile myoclonic epilepsy.

An example of successful therapy for one form of progressive myoclonus epilepsy, called severe myoclonic epilepsy of infancy (Dravet syndrome), shows the possible ability of an AED to control a serious epilepsy syndrome (Striano et al. 2007). In the same issue of Neurology, the study was critiqued (Krauss and Morrison 2007).

Dravet syndrome is a rare early childhood syndrome characterized by generalized seizures often stimulated by fever. The disorder is usually poorly controlled by drugs and may develop into status within the first year of life. The syndrome in these cases was treated with high doses of levetiracetam, and in a trial of the AED, 2/3 of 28 patients had a 50% or greater reduction in seizures.

This is a syndrome associated with specific genetic mutation in the sodium channel alpha 1 subunit gene (SCN1A). The defect seems to produce a phenotype in which there is a decrease in normal GABA-mediated inhibitory tone in the dentate gyrus and an upregulation of sodium channels in hippocampal neurons. Once again, the idea emerges that with this knowledge, drugs might be developed and/or administered which are efficacious in treatment attempts.

Levetiracetam has been shown to act positively in cases of Dravet syndrome. The precise mechanism of action is unclear. Levetiracetam does bind to the synaptic vesicle protein 2A (SV2A) (Meldrum and Rogawski 2007). Mouse mutants that have no SV2A present with abnormal GABAergic neurotransmission in the hippocampus. This is an example of the concept that when a gene mutation is identified

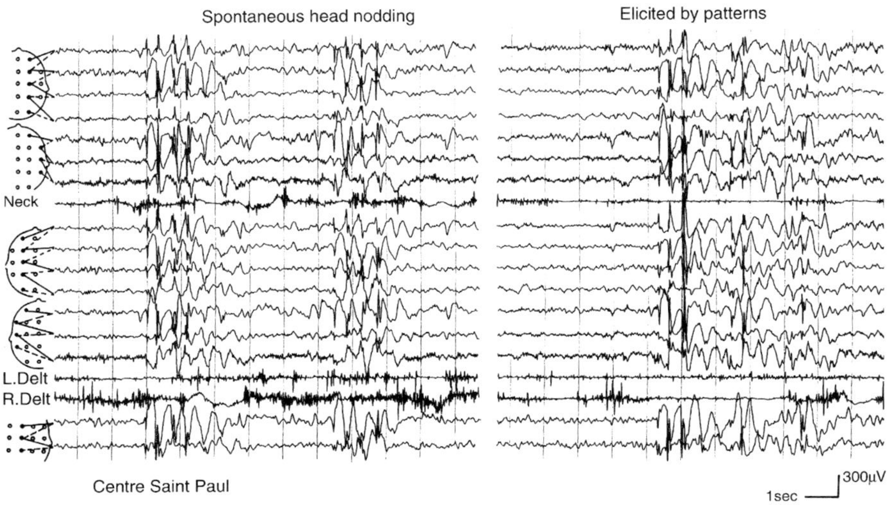

Fig. 12.3 Dravets syndrome – severe juvenile myoclonic epilepsy. Published in Crespel A. et al. Atlas of electroencephalopathy Volume 2; The Epilepsies, EEG and Epileptic Syndromes. With permission from John Libbey Eurotext

in an epilepsy disorder, proper AED therapy may become more obvious. The SCN1A gene mutation is associated with more than one phenotype, which can lead to misdiagnosis. Many infants diagnosed with mutations of the SCN1A gene were diagnosed with severe myoclonic epilepsy of infants – borderland. The "borderland" patients had severe seizures/encephalopathy, but not generalized spike wave activity. Not all SCN1A gene mutation patients, however, have Dravet syndrome. Further investigations are needed (see Fig. 12.3).

Another study sought to examine the risk factor of the BRD2 gene for juvenile myoclonic epilepsy. The study examined the association between the promoter variant rs3918149 and juvenile myoclonic epilepsy. The patient population included 531 with juvenile myoclonic epilepsy, plus 1,390 normal controls. Extensive clinical data were collected, including those pertaining to etiology, symptoms, etc. The classification scheme corresponded to the scheme of the International League Against Epilepsy. The 531 patients were from 5 cohorts originating from England, Germany, Ireland, Australia, and India.

Results looking at the risk of the promoter variant and juvenile myoclonic epilepsy were mixed. There was a significant effect (association) in the British cohort, and a borderline significant effect in the Irish population. Surprisingly, there was no association in patients from Indian populations, or those from Australian or German populations.

The authors state that four possibilities may explain the discrepant results. One explanation would assume that the positive British/Irish results were false positives. Another possibility states that the association of rs 3918149 is real, but the causal variant is outside BRD2. Yet another possible explanation is that the effect of BRD2 is not limited to rs 3918149 or to juvenile myoclonic epilepsy. This explanation would assume that different alleles contribute in different populations. The final explanation is the notion that there is phenotypic heterogeneity within the classification of juvenile myoclonic epilepsy. This would call for juvenile myoclonic epilepsy subgroups with BRD2 limited to one or several but not all subgroups. One recent paper (Martinez-Juarez et al. 2006) provides data supporting this concept. Overall, the present data do not support a conclusive role for BRD2 in juvenile myoclonic epilepsy.

A comparison was made of cognitive function between patients with juvenile myoclonic epilepsy, frontal lobe epilepsy, temporal lobe epilepsy, and control patients (Piazzini et al. 2008). There were 40 patients in each group, with the exception of 50 in the juvenile myoclonic epilepsy group. The study is of importance because although juvenile myoclonic epilepsy is a generalized epilepsy with EEG discharges generalized over the entire cranium, evidence supports the idea that there are focal frontal lobe sites of discharge from specific anatomical structural sites (Kim et al. 2007).

Earlier studies suggested that cognition in juvenile myoclonic epilepsy was normal, but more recent studies find evidence of focal brain dysfunction, as well as possible psychiatric symptoms. In the present study, cognitive tests such as the word fluency test, and the Wisconsin card sorting test were administered. Results showed that patients with juvenile myoclonic epilepsy had the worst performance as regards cognitive tests, followed closely by patients with frontal lobe epilepsy. Temporal lobe epilepsy patients faired better, followed by controls.

The clinical significance is not clear, except for the significance of the involvement of the frontal lobe. It could be that the epileptiform activity in turn modifies the frontal lobe functionally and probably structurally such that the frontal lobe develops overt seizure activity. Another hypothesis would be that the "fundamental" neuropathology might arise from a variety of foci, which result in the juvenile myoclonic epilepsy phenotype.

A multicenter prospective, long-term, open label study was performed in order to examine the effects of the AED levetiracetam on the EEG in juvenile myoclonic epileptic patients (Specchio et al. 2008). Ictal and interictal EEG patterns of juvenile myoclonic epilepsy show generalized 3–6-Hz spike/polyspike slow wave discharges. Thirty-three percent of patients show photosensitivity, and if onset of juvenile myoclonic epilepsy occurs below the age of 12 years, over 80% of patients have photosensitivity. Although valproate is a first-line treatment for juvenile myoclonic epilepsy, it has a significant adverse effects record. Valproate is also efficacious in cases involving photosensitivity. The drug levetiracetam, a new AED, is effective in juvenile myoclonic epilepsy, and the adverse effects are lower. This study generated much data on EEG and levetiracetam in cases of juvenile myoclonic epilepsy (see Table 12.1).

After an initial 8-week baseline period, patients were given a starting dose of 250 mg b.i.d., which was increased to 500 mg b.i.d. after 2 weeks. Visits to a clinic were scheduled about every 3 months, and patients were followed for up to 36 months.

Table 12.1 Antiepileptic drug treatment in cases of juvenile myoclonic epilepsy

Drug	Common adverse effects	Important concerns
Valproate	Sedation Tremor Weight gain Menstrual irregularities	Teratogenicity Polycystic ovarian syndrome
Lamotrigine	Rash Insomnia Headache	Reduced clearance with valproic acid
Levetiracetam	Sedation Irritability	
Topiramate	Paresthesia Dysnomia Impaired cognition Weight loss Oligohydrosis Renal calculi	Metabolic acidosis in children Rare acute glaucoma
Zonisamide	Weight loss Renal calculi Rash Oligohydrosis	Possible cross-reactivity with sulfur allergies

Adapted from Welty, T., Pediatr Drugs 8: p. 303, 2006

EEG assessment was performed at each visit. Bipolar/monopolar 10–20 electrode placement system was used. EEGs were performed at least 3 h after awakening. This reduces the chances of interpreting paroxysmal activity (Labate et al. 2007). Reviewing examiners were blinded. Photoparoxysmal responses were separately examined. A total of 48 patients were evaluated.

Results showed a female/male ratio of 38/10. The mean age was 27.4 years, age of onset was 14.1 years, and family history was 21 yes and 27 negative. Of the 48 patients, 38 were resistant to previous AEDs, and ten were newly diagnosed. Photoparoxysmal sensitivity was seen in 17 of the 48 patients. After the initiation of levetiracetam treatment, 27 of 48 (56%) had normal EEGs during the study. Of the patients, 76% showed a suppression of photoparoxysmal responses following levetiracetam. The frequency of days with myoclonic and generalized tonic-clonic seizures was decreased with levetiracetam treatment. Of the patients, 37.5% had no further myoclonic seizures after initiation of levetiracetam, and 73% showed no more generalized tonic-clonic seizures.

The authors state that this study shows the efficacy of levetriacetam treatment both in terms of EEG and clinical improvement in seizures. While valproate treatment for juvenile myoclonic epilepsy may be effective in up to 80% of patients (Gelisse et al. 2001), levetiracetam could be an excellent option for patients not responding well to valproate, or those experiencing adverse effects. Both the EEG and clinical outcomes show favorable results from levetiracetam treatment. The authors point out that their study was not a controlled one; nevertheless, the results strongly support the use of levetiracetam as indicated above.

A randomized, double-blind, controlled (placebo) study was performed in order to evaluate the new AED levetiracetam in patients with juvenile myoclonic epilepsy (Noachtar et al. 2008). The efficacy and tolerability at a dose of 3,000 mg/day in both adolescents and adults were excellent. The study was performed after an 8-week baseline. There was a 4-week increase in dose and 8 weeks of full dose, followed by a 6-week downregulation of levetiracetam.

Results showed that of 60 epileptic patients (56 with juvenile myoclonic epilepsy, 4 with absence seizures), a 50% reduction or more occurred in 58.3% of patients taking levetiracetam, and 23% of patients taking placebo. Additionally, the levetiracetam-treated patients had higher freedom from myoclonic seizure episodes than did the control placebo group. The authors comment that levetiracetam is an effective adjunctive treatment that has few adverse effects. It is efficacious for the treatment of generalized epilepsy with myoclonic seizures.

Myoclonic seizures can occur not only in juvenile myoclonic epilepsy, but also in severe myoclonic epilepsy in infants, in benign myoclonic epilepsy in infants, and in idiopathic epilepsy with myoclonic astatic seizures. In the study described (Hirano et al. 2009), videopolygraphic methods were used to evaluate 550 seizures in 26 epileptic children having myoclonic seizures. The nature of the muscle involvement – neck, trunk, and proximal/distal upper extremities – postural changes, frequency, and duration of seizures were all evaluated and documented.

Results showed that the median age of onset of seizures was 30.5 months, and the age of video polygraphic studies was 1–2 years after onset of myoclonic seizures. he age of onset for myoclonic seizures was the lowest in the severe myoclonic epilepsy group, and the highest in the juvenile myoclonic group. The other two groups fell midway as regards age of onset of seizures. Seizures were seen exclusively during the daytime in juvenile and severe myoclonic epilepsy groups. The other two drugs showed seizures during sleep and during wakefulness. The most predominant body regions involved in seizures were trunk and proximal upper extremities in three groups, but the distal upper extremity was most frequently involved in juvenile myoclonic epilepsy. Generalized spike wave discharges were highest in severe and juvenile myoclonic epilepsies. The duration of electromyography potentials was similarly longer in the two groups of severe and juvenile myoclonic epilepsy.

This study was undertaken to define how to make an early diagnosis between these four types of myotonic epilepsy effectively in order to choose an effective AED in the first attempt. These descriptions are extensive, involving 550 seizures from patients of each of the four types of seizures. It is noted that benign myoclonic epilepsy and idiopathic epilepsy with myoclonic astatic seizures may be a difference only in degree. The difference between atonic drop attacks and myoclonic astatic seizures may be difficult. Atonic drop attacks usually involve the patient dropping straight down and landing on their buttocks (Oguni et al. 1992, 1997). When patients fell, the fall was sometimes backward because of strong trunk extension.

A difference between severe and juvenile myoclonic epilepsies was the involvement of proximal muscles as opposed to distal muscles in the severe myoclonic epilepsy group. Borderline severe myoclonic epilepsy has all the clinical features as the severe form, including SCN1A mutations, except for absence seizures.

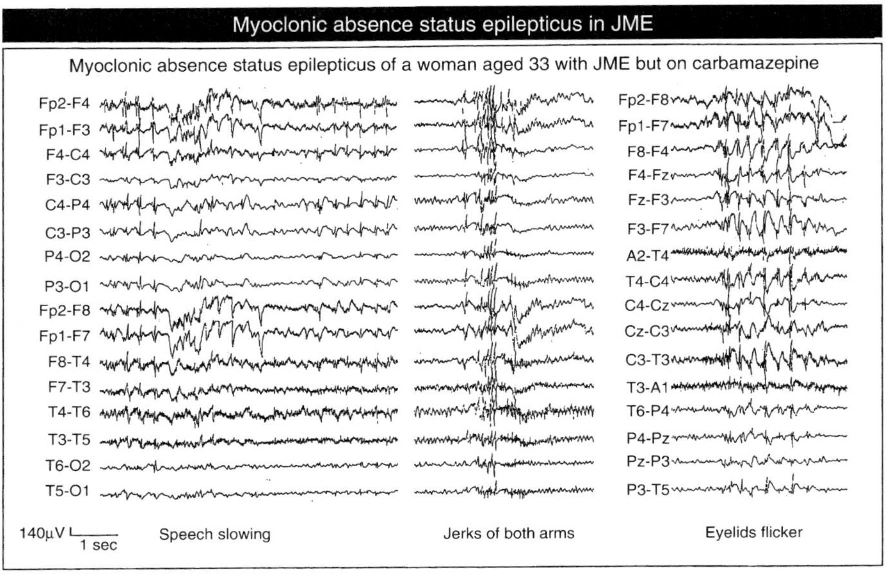

Fig. 12.4 Video EEG myoclonic absence status in juvenile myoclonic epilepsy. The patient was mildly confused and taking carbamazepine. With kind permission from Springer Science + Business Media; Epileptic Syndromes and their Treatment, 2010 p. 407 Panaylotopoulos, C. Fig. 13.9

The authors conclude by saying that there are distinct differences between the four groups and many similarities. Although the patient numbers are low, this report should assist significantly in diagnosis.

Another paper (Lin et al. 2009) looked at brain metabolic differences between patients with juvenile myoclonic epilepsy and normal controls. The patients were 60 long-term juvenile myoclonic epilepsy patients, plus 30 healthy volunteer controls.

In this study, multi-voxel proton spectroscopy data were collected from prefrontal cortex, primary motor cortex, parietal and occipital cortices, thalamus, striatum, cingulate gyrus, and the insula. Results showed a reduction in *N*-acetyl aspartate/creatine-phosphocreatine (NAA/Cr) ratios in the primary motor cortex, medial prefrontal cortex, and thalamus in juvenile myoclonic epilepsy compared to that in controls. Glutamate-glutamine (GLX)/Cr ratios differed from controls in the primary motor cortex, medial prefrontal cortex, insula, striatum, and cingulum. The difference was an increase in the ratio in the insula, and a decrease in the other areas.

The authors state that initially, juvenile myoclonic epilepsy was thought to be a generalized seizure, but improved neuroimaging studies show a multifocal seizure syndrome. The authors note that a network of neurochemical alterations in juvenile myoclonic epilepsy was found especially in thalamocortical related structures. This strongly suggests that the cortical hyperactivity in juvenile myoclonic epilepsy patients is not necessarily diffuse (generalized). Furthermore, differences between patients who had different AED treatments were not shown. The authors note that this body of data needs further study (see Fig. 12.4).

The issue of altered personality traits (emotional instability, immaturity, mood changes, etc.) has been re-examined psychiatrically and also examined using quantitative multi-voxel MRS (de Araujo-Filho et al. 2009). The multi-voxel MRS was used to examine areas previously examined, including the thalamus, insula, cingulate gyrus, striatum, and frontal, parietal, and occipital cortices. One hundred juvenile myoclonic epileptic patients were evaluated for personality disorders, and 16 patients who met the criteria were selected.

Results showed a significant decrease in the NAA/Cr ratios localized mainly to the left frontal lobe in the juvenile myoclonic epilepsy and personality group. The GLX/Cr ratio was increased in the same brain region in the same group. These data confirm data previously published (Lin et al. 2009), and extended them to a psychiatrically defined personality disorder group. These behavioral manifestations most likely represent frontal lobe involvement in juvenile myoclonic epilepsy.

The authors conclude that the coexistence of significant consistent personality disorders in juvenile myoclonic epilepsy may represent a more severe form of this seizure disorder. In this study, this form represented 16% of the cases. This form is seen as one involving the thalamic cortical network and also the frontal lobe cortex. The authors encourage more studies in which psychiatric evaluation and neuroimaging are correlated.

Yet another paper has examined the relationship between juvenile myoclonic epilepsy and suspected frontal-temporal corticothalamic networks (Holmes et al. 2010). This study used dense array electroencephalographic (dEEG) epileptiform discharges in patients with juvenile myoclonic epilepsy.

Results from ten patients showed some variability in epileptiform discharges from the same structures and between patients. Unique discrete unilateral regions in orbital frontal/frontopolar cortex localized epileptiform components. The authors note that thalamic circuits may be involved in normal and abnormal cerebral rhythmicity (McCormick 2002). Furthermore, the circuits which modulate normal sleep patterns may also be involved in seizure spike wave discharges (Steriade 2003). It should be remembered that although the subcortex plays an important part in the spread of seizures, the cerebral cortex plays a key role in the generation of seizure activity.

The authors note that their study provides evidence of specific regional involvement in seizure initiation. It appears that restricted networks of cortical sites are involved (medial orbital, frontal, and anterior basal medial temporal lobe). This precludes a generalized definition of juvenile myoclonic epilepsy. A decrease in frontal and temporal lobe thickness is consistent with this being an area of epileptogenesis (Tae et al. 2008). The authors note that neuroimaging and recording which improve resolution will increase important knowledge which may improve treatment.

A paper looking at differences in cortical excitability as predictors of progressive vs. juvenile myoclonic epilepsies has been published (Badawy et al. 2010). In this study, transcranial magnetic stimulation was used to evaluate cortical excitability characteristics. Six patients with progressive myoclonic epilepsy were compared to nine patients with refractory juvenile myoclonic epilepsy, and ten with chronic, but well-controlled juvenile myoclonic epilepsy.

Results showed a significant increase in cortical excitability compared to that in refractory juvenile myoclonic epilepsy, and also to well-controlled juvenile myoclonic epilepsy patients. There were no significant differences in motor thresholds in the progressive myoclonic epileptic patients compared to the other two groups. The authors state that their findings demonstrate clear-cut differences in cortical excitability between progressive myoclonic epileptics and either of the two juvenile myoclonic epileptic groups. The data indicated a role for GABA B-mediated networks. Clearly, more studies are suggested by these interesting results.

A recent study (Manganotti et al. 2011) looked at the effect of afferent input on cerebellar electrical activity in a patient with progressive myoclonic epilepsy. This was assessed by examining somatosensory evoked potentials. Functional MRI (BOLD) was used to observe somatosensory output. High-amplitude evoked potentials were noted in association with highly focal BOLD activation contralaterally. There was no diffuse activation of frontal or parietal cortical areas in control patients. These results suggest that controlling hyperexcitability is limited to the sensorimotor cortex in myoclonic epilepsy patients.

Part III
Partial Epilepsies

Chapter 13
Simple Partial Seizures

Simple and complex partial seizures have in common by definition a focal or localized area of onset. A complex partial seizure involves impaired consciousness, whereas in patients exhibiting simple partial seizures do not have impaired consciousness. They are able to answer questions, respond, etc. during the seizure. In contrast, impaired consciousness involves altered awareness/responsiveness. The patient cannot respond to questions or follow directions during the seizure (Dreifuss 1981). Awareness refers to the patient's contact during the seizure, with the environment, and to recall events.

The etiology of partial seizures is manifold and can usually be identified. Causes include hypoxia/ischemia, infections such as hemolytic uremic syndrome, syphilis, viral infections, etc. Some infectious agents seem to favor the temporal lobe, a site of many partial epilepsies (Ounsted et al. 1966). The pathological finding of mesial temporal sclerosis is associated with over one half of patients with temporal lobe epilepsy (Brown and Babb 1987).

The phenotype of partial seizures coupled with EEG monitoring can usually pinpoint the location of the seizure focus. For example, foci in the precentral gyri may result in contralateral motor signs in simple partial seizures. In fact, simple partial seizures showing motor signs are the most common type. This is reflective of the most common foci of simple partial seizures.

Simple partial seizures may have somatosensory or special sensory signs. These are frequently the initial signs of seizures initiated in the postcentral area. These simple partial seizures are also frequently the "aura," announcing a more involved impending seizure such as a complex partial seizure, or a generalized seizure (bilateral). The "linkage" of the sensory/motor phenomenon is due to the extensive connections between the two areas.

Various types of simple partial seizures manifest depending on the site of activity, For example, if the auditory cortex is the site of initiation of the simple partial seizure, the seizure may be associated with subtle whistling, or buzzing sounds. They may also be less than subtle, and involve music or roaring noises. Simple partial seizures presenting with a bad smell or bad taste perception, originate from the anterior lobe or the insula.

D.W. McCandless, *Epilepsy: Animal and Human Correlations*,
DOI 10.1007/978-1-4614-0361-6_13,

Simple partial seizures with visual features are those which involve the calcarine fissure region of the visual cortex. Initiating foci in the visual type of simple partial seizures may manifest as odd color and/or light flashes in the contralateral visual field. The visual phenotype may take the form of hallucinations. As in the case in all types of simple partial seizures, the initiating type may spread producing complex partial seizures. The phenomenon of vertigo (posterior temporal lobe) may be the simple partial seizure proceeding a complex partial seizure, and is the aura for the more involved seizure.

Simple partial seizures may also materialize with an "autonomic" presentation. In these cases, autonomic involvement includes flushing, sweating, nausea and vomiting, and epigastric sensations, among others. Autonomic features occur in complex partial seizures, incontinence being one example. Ictal G.I. phenomena most frequently occur in children, and pallor and the cold sweats may occur simultaneously with G.I. symptoms in simple partial seizures. Tachycardia is a frequent cardiac autonomic symptom of simple partial seizures. This plus other cardiac autonomic occurrences may be the basis for the sudden unexplained death associated with partial epilepsies.

Symptoms such as déjà vu, dysphasia, fear, anger, dreamy states, hallucinations, etc. may all represent psychic symptoms in cases of simple partial seizures. This type of simple partial seizure is nearly always followed by a complex partial seizure. The "spread" of a simple partial seizure to a complex partial seizure and/or spread to a generalized tonic clonic seizure is always a possibility, but occurs infrequently. The corpus callosum represents a site for generalization, and is a surgical candidate in order to stop generalization (see Chap. 35).

The simple partial seizure is considered to be an aura of complex partial epilepsies. At least 50% of patients with complex partial epilepsy report the presence of an aura just before the complex seizure starts. Aura reporting is less common in children, probably due to lack of familiarity with sometimes subtle changes in sensations, and possibility because of post-ictal amnesia. The aura is a simple partial seizure, which can proceed to become a complex partial seizure. Complex partial seizures which involve the mesial temporal cortex often have auras consisting of G.I. sensations. The complex partial seizures have varying levels of loss of awareness/consciousness, which may impair memory of the aura.

Simple partial seizures are idiopathic and include benign childhood epilepsy (rolandic epilepsy). Benign rolandic epilepsy has an excellent prognosis without AED treatment (Astradsson et al. 1998) and is probably the most common childhood seizure disorder (the name "rolandic" is derived from the affected brain region). It is frequently misdiagnosed. The disorder surfaces between age 3 and 12 and often occurs at night.

Symptoms include awakening from sleep, and blank staring. There may be facial twitching and drooling, and seizures last 1–2 min. They are simple partial seizures. The epileptic patient can comprehend and is responsive. The EEG shows blunted high-voltage spikes, which are not seizures but rather a marker of rolandic seizures.

Simple partial seizures can originate from the occipital lobe and are also benign in that they may not need treatment if mild and infrequent. The seizures can be treated, especially if they generalize. They tend to undergo spontaneous remission.

The EEG shows frequent occipital discharges with eyelid closure. This seizure may be associated with migraine headaches.

In another paper (Holmes 1986), a group of patients were monitored using telemetry EEG with videotape recording (TEEG-VR). All patients were followed during recording by an experienced EEG technician. Simple partial seizures were classed by the absence of an altered state of consciousness. When consciousness was altered, the seizure was classed as a complex partial seizure.

Results showed that in the study, 15 patients had a total of 62 simple partial seizures, and 38 patients had a total of 133 complex partial seizures. Three patients had epilepsia partialis continua. The average age of the simple partial seizure group was 11.3 years (range 4–18 years of age). Age of onset was 2.6 years old. The average duration of simple partial seizures was 18 s; the duration of seizures in the patients with epilepsia partialis continua (3) was 60 min. Eighty percent of patients with simple partial seizure had motor manifestations including some unilateral clonic/tonic movements.

Clonic movements involved a range of anatomical structures from whole extremities, face, and neck, to one finger. Versive (eye movement) and head turning were present in 42% of patients. One patient had autonomic symptoms consisting of flushing and sweating. Three patients had somatosensory symptoms including a bad taste in the mouth, and chest discomfort. The three epilepsia partialis continua all showed motor symptoms involving face and arms, and in one patient, only one finger. The movements were irregular and occurred from every 2–3 s to continually.

As regards the EEG, all patients with simple partial seizure had abnormal interictal EEG findings except one. Spikes, sharp waves, or theta activity occurred unilaterally or bilaterally, and multifocal activity was also seen in one-third of patients. Initial EEG alterations were localized in 80% of patients. In the three patients with epilepsia partialis continua, no obvious sharp waves or spikes were seen.

The author notes that the most common symptom in patients with simple partial seizure was motor activity. Movements were usually asynchronous. Psychic symptoms were not seen in this series, although they are a feature of simple partial seizures (Dreifuss 1981). The clinical features of epilepsia partialis continua were as previously reported (Thomas et al. 1977). In complex partial seizures, as many as one half experience an aura of motor, sensory, autonomic, or psychic symptoms. The auras precede the complex partial seizure, and are considered to be a simple partial seizure. Many patients, especially children do not remember the aura. The author notes that in his series of patients with complex partial seizure, less than one-third of patients could verify an aura.

Another study (Deonna et al. 1986) was a clinical and EEG study on 107 children with partial seizures. Sixty-three patients had simple partial seizures, and the rest had complex partial seizures. Of the 63 with simple partial seizures, 38 cases had benign partial epilepsy of children with rolandic spikes, and 25 had simple partial seizures not associated with rolandic spike seizures. . The patients with the rolandic features were easily diagnosed, and the final benign outcome was confirmed. Some of the patients with rolandic seizure had less controlled seizures during the course of the disorder.

In cases of non-rolandic simple partial seizure in patients, no clear distinguishing group(s) could be identified. Some children with non-rolandic simple partial seizures had as few as two seizures. The authors note that most benign course simple partial seizures occur within the framework of the rolandic subclass, but may occur in the subclass of non-rolandic simple partial seizures.

Another interesting and excellent paper (Sharma et al. 1984) has detailed various aspects of mesial temporal lobe epilepsy. The pathology and especially features of the related animal models of mesial temporal lobe epilepsy have been emphasized.

Of interest to the study of simple partial seizures is the reminder that simple partial seizures may be associated with sensory, somatosensory, autonomic, and/or psychic symptoms. The key differentiation from complex partial seizures is that the simple variety does not cause loss of consciousness. These seizures are almost always of shorter duration than complex partial seizures. In both types, the initial focus is single and unilateral.

Both simple partial seizures and complex partial seizures can develop into secondary generalized seizures. Simple partial seizures represent an aura if complex partial seizures follow. While the partial seizures may become secondarily generalized, they may not as well. Some patients with simple partial seizure never have but a few secondary seizures. Unlike the partial seizures, primary generalized seizures are a result of paroxysmal discharges from both cerebral hemispheres.

The authors note that status epilepticus can be either partial or generalized. The authors note that status epilepticus is a state of continuous seizures lasting 5 min or more (Chap. 23). Obviously, any prolonged seizure activity has the potential to produce or be associated with significant damage to the normal functional/structural integrity of the brain.

The action of ketamine on an animal model of simple partial seizures has been examined (Velisek et al. 1993). In this study, rats were treated with pentylenetetrazole by topical application at a dose which produced simple partial seizures. The action of ketamine was examined in this model, as well as a similar model of absence seizures.

Results showed that in the simple partial seizure model, ketamine exerted biphasic effects. In the simple partial seizure model, a dose of 20 mg/kg of ketamine acted to suppress ictal neocortical discharges, whereas at a dose of 40 mg/kg, ketamine accentuated the onset of seizures, and increased inter-ictal spikes. Ketamine showed similar results in the absence model, but those results were not significantly different from control results.

These results, state the authors, show a bimodal result in the case of ketamine and the simple partial seizure rats. The low dose served to suppress seizures, whereas the higher dose accentuated seizures. Additionally, ketamine efficacy or lack thereof depends on the seizure model employed. This model/dose represents a new model of simple partial seizures which occur in freely moving rats.

Auras, as examples of simple partial seizures, have been examined in a cohort of adult patients having refractory complex partial epilepsy (Manchanda et al. 2000). The focus of the study was to determine the aura characteristics, and its association with the seizure, and relation to behavior. Patients underwent a detailed seizure observation protocol of their seizures as well as auras. The seizure focus was determined using standardized EEG, as well as subdural electrode recording when indicated. Auras were divided into nine categories.

Results showed that of the 144 patients studied, 77% had one or more symptoms during the auras. Twenty-three of the complex partial seizure patients exhibited no aura. Eighty percent of the patients had a temporal lobe seizure focus. The three most frequent reported auras were those labeled viscerosensory (32%), experimental (30.6%), and cephalic (14.6%). Experimental auras are those of, for example, fear and déjà vu. A psychiatric (DSM-III-R) diagnosis was made in slightly over one half of patients. Patients with two or more auras were most likely to have a concomitant psychiatric diagnosis.

The authors note that the nature of auras in complex partial seizures has not been well studied. Data presented indicates that auras did occur in the majority of patients with refractory to treatment partial seizures. The occurrence of two or more auras was associated with psychiatric disorders. At least one half of the patients had demonstrable psychiatric disorders in addition to partial epilepsy.

Another paper has described studies relating to auras (Kohler et al. 2001). In this study, 22 patients with fear auras and with intractable seizures were compared to groups with other types of auras, or no auras. All were about to undergo surgery for intractable epilepsies.

Auras are simple partial seizures which may precede complex partial seizures, are secondarily generalized seizures. They may also occur independently, reflecting a close anatomical relationship to the main focus. Auras related to mesial temporal lobe epilepsy include both visceral auras and experimental auras such as fear and déjà vu (Taylor and Lochery 1987). The amygdala has been noted to be associated with fear and déjà vu auras in temporal lobe epilepsy (Gloor et al. 1982). Aggressive behavior has also been associated with the amygdala, and is a feature of patients with mesial temporal lobe epilepsy.

Results showed that 68% of patients in the fear group, 55% of patients in the non-fear group aura group, and 57% of patients in the no aura group experienced mood and anxiety symptoms. One to three months after temporal lobectomy, the three groups showed 86%, 41%, and 27% of patients, respectively, had mood and anxiety disorders.

The authors note that mood disorders are more common in temporal lobe epilepsy than other seizure disorders. This signals a hippocampal/amygdala limbic system involvement. In the fear aura group, anxiety and mood disorders continued after surgery, and can be attributed to amygdala involvement, as the structure is involved in the expression of fear (Gray 1989). Another possible explanation states that kindling has taken place as regards fear/anxiety, and persists after surgery.

The authors further note that the psychiatric diagnoses were made in part by the history and assessment gleaned from previous evaluations. Direct assessment would have been better. The authors also state that mood and anxiety assessment was unblended. Future studies should include a more quantitative analysis of the amygdala complex.

Another study reports studies on the nature of olfactory auras in patients with temporal lobe epilepsy (Chen et al. 2003). In this study, 217 patients' records were reviewed, all of whom had undergone temporal lobectomy for intractable epilepsy. Of these, 12 were identified who had olfactory auras, and 11/12 had structural lesions in the mesial temporal lobe as shown by MRI. Patients with olfactory auras described an awful smell just preceding a complex partial seizure. Eleven of the 12 patients also had other auras concurrently. Lesions in the amygdala were also noted.

The authors note that in their study, the incidence of 5.5% of olfactory auras was higher than that in other studies (Penfield and Perot 1963), possibly due to the thoroughness of presurgical work up. Consistent with most other studies, the quality of the odor was described as highly unpleasant. Most patients in this study had multiple auras. The authors state that the site of lesion was not a factor in determining olfactory auras. The present findings stress the mesial temporal structures, and especially that the amygdala plays an important role in olfactory auras.

Autonomic auras in complex partial epilepsies associated with goose bumps and cold shivers are not common, but were the focus of a descriptive retrospective study (Stefan et al. 2002). In this study, out of 420 patients with refractory temporal lobe epilepsy, 16 were identified with cold shivers and goose bumps-autonomic auras.

Results showed that lateralization of mesial temporal lobe lesions was predominantly on the left side. In fact only three were right side lateralized. As a "symptom" breakdown, goose bumps alone were least frequent. Goose bumps were seen on the face, upper body, and both upper extremities.

The authors note that previous studies have shown a localizing effect between viscero-sensory auras and temporal lobe epilepsies (Gupta et al. 1983). Epigastric auras were often associated with the right temporal lobe. Déjà vu auras tended to be associated with the right temporal lobe, and cephalic auras originate from the frontal lobe (Palmini and Gloor 1992).

The authors note their results of statistical assessment of the autonomic auras conclusively show a relation of shivers/goose bumps to the left mesial temporal cortex. The authors conclude saying that the close anatomic relation of the amygdala to the temporal lobe was most important for localization, rather than seizure etiology. In addition to the amygdala, the cingulum, and hypothalamus may be at least partially involved (Walker 1940; Smith 1945). The exact mechanisms are still unclear, but may involve neurochemical and/or physiologic mechanisms.

Elucidation of mechanisms will provide valuable information prior to the neurological treatment of patients with AED refractory partial seizure.

Status epilepticus is described in several nonconvulsive seizure types which include both complex partial and simple partial seizures. Complex partial status epilepticus consists of a variation of periods of motionless staring and unresponsiveness and of partial responsiveness. Simple partial status consists of focal seizures which can last for days, and this condition is termed epilepsia partialis continua. Nonconvulsive status may be difficult to diagnosis in the absence of EEG recordings.

The classification of aura continua is generally restricted to continual subjective feelings in the absence of visible motor phenomena (Wieser 1997). Thus, aura continua is seen as distinct from simple partial status epilepticus. Autonomic only aura continua would be included here if there were no motor correlates. This would, for example, include chills and goose bumps. Auras continua also would include involvement only of special senses.

The issue of whether epilepsia partialis continua can originate in the cerebellum needs further investigation, and such a study was undertaken (Vander et al. 2003). Although previous data had implicated the cerebellum in focal seizures, there were no cases of epilepsia partialis continua directly caused by a cerebellar lesion. This paper is a case report of a patient who had a cerebellar hemorrhage and then developed epilepsia partialis continua.

The patient was a 58-year-old female who 2 months earlier had a right lobar cerebellar hemorrhage. She was diabetic and presented with rhythmic clonic jerks of her right hand and lower right lip. These movements had a frequency of 1–2 Hz. The movement continued in sleep and was exacerbated by voluntary movement of the affected areas.

EEG recording showed epileptiform discharges over the left frontotemporal region. An MRI showed a hypointense lesion in the right cerebellar hemisphere, appearing as post-hemorrhagic gliosis. 18F-FDG showed hypometabolism in the cerebellar lesion. AED treatment was not successful in controlling the seizures (Fig. 13.1).

The authors note that the diagnosis made in this patient was epilepsia partialis continua with motor involvement and evidence of a cerebellar lesion on MRI and PET imaging. The authors state that the cerebellum has extensive connections including the cerebral cortex, which in turn connects to the thalamus. There are also direct connections between the cerebellum and thalamus (Norden and Blumenfeld 2002). The patient described had a strong temporal association with the cerebellar hemorrhage, and in the absence of any cortical alteration is highly suggestive of a cause/effect.

In another case report, the connection between epilepsia partialis continua on multiple sclerosis (MS) was made (Striano et al. 2003). There is a correlation between MS and seizures, the incidence of seizures being between 1 and 10% of cases (Sokic et al. 2001). Seizures may also be the first symptom in about 10% of patients with MS (Moreau et al. 1998). The current paper is a case report of such a patient.

The patient, a 21-year-old right-handed male, suddenly presented at age 19 with long-lasting left hand jerks, sometimes spreading to the entire left upper extremity. There were at least two episodes of generalized tonic clonic seizures. Myoclonic jerks also occurred at night. On admission, brain MRI showed small T2-weighted hyperintensive lesions in supra tentorial and cerebellar white matter. The right frontal region showed two larger lesions. Carbamazepine was started for seizures, which disappeared in a week.

The diagnosis of MS was made based on clinical, laboratory, and MRI findings. In this case, epilepsia partialis continua was a presenting feature. The authors state that this case provides detailed documentation of an MS case which first presented as epilepsia partialis continua. Cortical plaques noted in the right hemisphere could have been directly involved in the seizure of this patient.

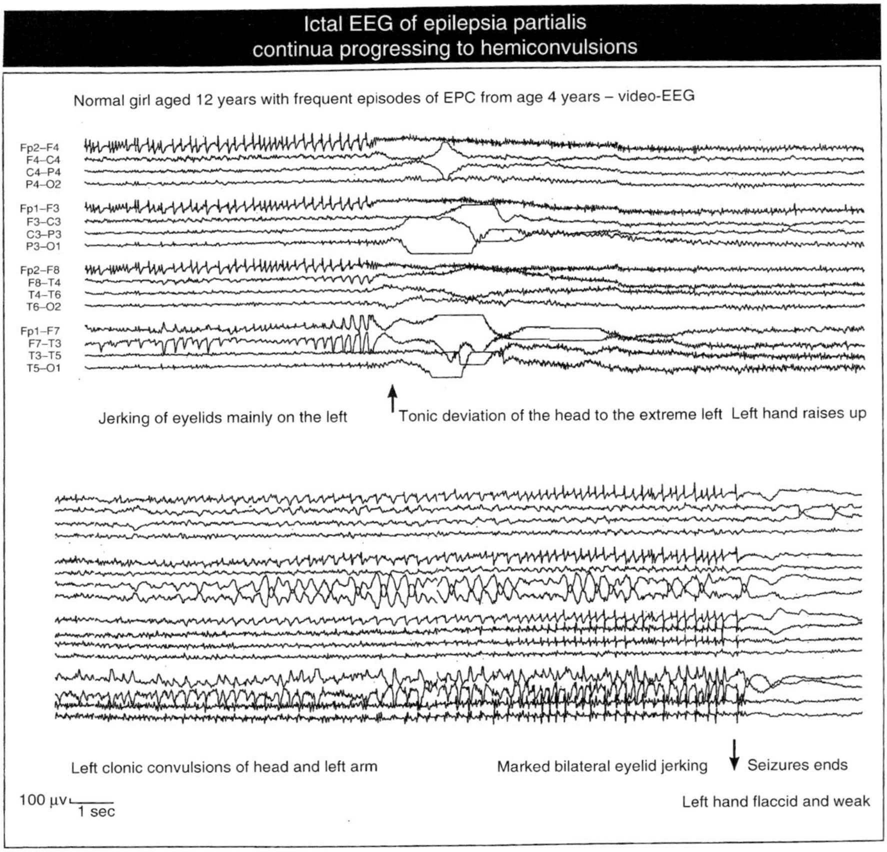

Fig. 13.1 Ictal EEG from a patient with epilepsia partialis continua extending to hemiconvulsions. With kind permission from Springer Sciences+Business Media; The Epilepsies, EEG, and Epileptic Syndromes and their Treatment, 2010, p. 466, Panaylotopoulos, C. Fig. 15.11

Another case report links epilepsia partialis continua with Creutzfeldt-Jakob disease (Lee et al. 2000). This case involves a 42-year-old male with diabetes and colon cancer who presented with forgetfulness, ataxia, and involuntary movements of his right arm. An EEG showed sharp and slow wave complexes in the left hemisphere. Multiple AED treatments were unsuccessful, and AEDs were terminated.

A thorough work up showed impairments in immediate memory, and three step commands. Motor examination showed semirhythmic right arm clonic movements and mild right hemiparesis. There was a significant impairment of right hand fine motor control was seen. The patient had moderate ataxia.

The initial MRI was normal, but another MRI 1 month later showed a small focus of signal increase in the left occipital cortex on fluid attenuated inversion recovery (FLAIR). Continuous video EEG showed complex partial seizures presenting as decreased responsiveness plus head turning. The final diagnosis as regards the seizures was epilepsia partialis continua as well as complex partial seizures. The patient continued to deteriorate, and a final diagnosis of Creutzfeld-Jakob disease was supported by brain biopsy

The authors note that epilepsia partialis continua often coexists with other seizure types such as complex partial seizures (Cockerell et al. 1996). Spikes and sharp waves are frequently seen in cases of epilepsia partialis continua. The authors suggest the differential diagnosis includes infections, tuberculosis, AIDS, etc. Many other conditions may coexist such as diabetes, MS, trauma, neoplasms, etc. A similarity exists between myoclonus and epilepsia partialis continua, but the twitch of myoclonus is considered to be more rapid (Juul-Jensen and Denny-Brown 1966).

The authors state that this patient presented with semirhythmic jerks which suggested a diagnosis of epilepsia partialis continua. The EEG sharp wave complexes were also suggestive. The rapidly progressive dementia accompanying the seizure phenotype was suggestive of Creutzfeld-Jakob disease.

Anti-glutamic acid decarboxylase antibodies are involved in autoimmune mechanisms which result in type 1 diabetes. These antibodies may be involved in epilepsia partialis continua, and a case report has been recently published (Baglietto et al. 2009). The case was of a patient who actually developed epilepsia partialis continua only 5 months after being diagnosed with type 1 diabetes.

Analysis of cerebral spinal fluid showed that anti-glutamic acid decarboxylase was present. There was evidence of oligoclonal bands. The epilepsia partialis continua worsened and was refractory to AEDs. EEG showed continuous spike wave activity and severe behavioral abnormalities. The authors note there seemed to be an imbalance between metabolism and GAD autoimmunity. This is a relatively rare occurrence, but more studies are needed.

An interesting study (Brodie 2004) has looked at the efficacy of a relatively new AED (pregabalin) for partial epilepsies including simple partial seizures. The patient population consisted of 1,052 highly refractive subjects. Three studies collectively represented the population base of this report (French et al. 2003; Arroyo et al. 2004; Beydoun 2000).

Results on the cohort of patients with AED refractor partial seizure showed a significant positive effect of pregabalin on patients with simple partial seizure, complex partial seizure, and also those with secondarily generalized tonic clonic seizures. Results were similar in all of the above classifications. The frequency of seizures was reduced in about 50% of all groups including those with simple partial seizures. The authors note that this result was especially encouraging given the drug refractory nature of the seizures.

A paper has been presented in which the first documented case of a simple partial epilepsy patient has presented with autonomic cardiopulmonary symptoms (Cohen-Gadol et al. 2004a and 2004b). In this case, the patient had experienced stereotypical episodes of dyspnea. There was no associated loss of consciousness. A thorough work up showed conclusive intermittent arterial oxygen desaturization was associated with the patient's dyspnea.

After failure to demonstrate pulmonary embolism, cerebral MRI showed a right mesial temporal lobe lesion. As the suggestion was made that the dyspnea may have been related to the simple partial seizures, phenytoin as a final AED was administered. This treatment resulted in the disappearance of both dyspnea and oxygen desaturization. The authors state that this case preventing autonomic cardiopulmonary symptoms caused by simple partial symptoms illustrates an example of limbic and limbic autonomic network.

Another interesting case involving simple partial seizures relates to a patient with a right temporoinsular glioma, which was removed (Rossetti et al. 2005). Following surgery, refractory simple partial seizures developed. These consisted of somatosensory sensations in the left leg, followed by an odd mothball taste. Intracranial recording showed as many as 50 electrographic seizures per day. The ictal discharges were confirmed to the postero-superior insula.

The authors note that the initial ictal manifestations were consistent with involvement of the insula, stimulation of which produces symptoms seen in this case. Thus, disagreeable taste can be elicited by insular stimulation, and the insula thought of as being involved with visceral, motor, sensory, and gustatory sensation (Isnard et al. 2004). This patient's results suggest that insular seizures might exist with simple partial seizures of a gustatory nature, and may lack scalp EEG correlates. The diffuse nature of dysgeusia suggested to the author that these central pathways are not exclusively crossed.

Chapter 14
Models of Complex Partial Seizures: Animal Studies

In the most general way, seizures can be classified as partial or generalized. Partial seizures have a regional (focal) cerebral localization, whereas generalized (tonic–clonic) seizures occur throughout the brain. Partial complex seizures are common, occurring in about one-third of epilepsy patients. Complex partial seizure activity is also quite common in laboratory rats and mice, as well as many other animals. Complex partial seizures are characterized by: impairment of consciousness, an immediately preceding event called an aura (simple partial seizure), automatisms, and usually arise from the temporal lobe, hence are also called temporal lobe seizures. The nature of the impairment of consciousness is variable and may range from an inability of the patient to respond to simple commands, to an inability to respond at all.

Automatisms are a common feature of complex partial seizures. These usually occur during the period of impaired consciousness, and take the form of chewing movements, lip smacking, fumbling/grasping movements, blinking, leg movement suggestive of running, etc. The automatisms are inappropriate for the situation, and tend to be similar in succeeding seizures. They may represent initial EEG activity from the focus. This has been shown in cases of oral facial movements which originate from the oral facial somatosensory cortex.

At least 60% of complex partial seizures originate in the temporal lobe. The unique features of the automatisms in complex partial seizures are highly suggestive of the seizure focus. Of the 40% of extratemporal lobe complex partial seizures, 30% originate from the frontal lobe (Manford et al. 1992). Features of frontal lobe complex partial seizures include an early loss of consciousness, head and eye movements, clonic jerks, falling, etc. More specific location within the frontal cortex is difficult.

The etiology of complex partial seizures in humans may consist of several causative features. These factors include birth defects, such as intrauterine developmental disorders such as aberrant neural migration (neural crest cells), cortical dysplasia, lissencephalopathy, etc., trauma associated with birth such as hypoxia, and asphyxia, infections such as rubella or syphilis, postnatal infections such as

D.W. McCandless, *Epilepsy: Animal and Human Correlations*,
DOI 10.1007/978-1-4614-0361-6_14,

hemolytic uremic syndrome, parasites, etc. The majority of these can be identified using MRI (Kuzniecky et al. 1993). In adolescents and adults, mesial temporal sclerosis is thought to be causative in over half of patients (Brown 1973; Brown and Babb 1987).

The aura associated with complex partial epilepsy is in fact a simple partial seizure. In children, the aura is either less expressed or less recognized. As the patient ages, the aura is more easily recognized. The aura leads into the complex partial seizure. Auras may include gastrointestinal sensations, fear, a particular smell, or any repetitive sensation foretelling an epileptic seizure. As many as one half of patients may have a complex partial seizure progress to a generalized tonic–clonic seizure. This occurs most frequently during sleep. Postictal depression is a usual occurrence which may last several hours (Delgado-Escueta et al. 1981). Confusion and lethargy are common during this period as is dysphasia. Rarely, complex partial status epilepticus can occur. This may last for many days, and the patient is totally dysfunctional. A continuous focal ictal discharge seen on an EEG recording in company with clinical features is almost always diagnostic.

A variety of animal models of complex partial seizures have been used to gather experimental data. The use of chemical models has advantages, for example, in allowing preseizure and postseizure signs/symptoms, and data collection from these time points, as well as during the actual seizure episode. In the use of chemicals, the dose is critical since too much convulsant may lead to tonic–clonic seizures instead of complex partial seizures (McCandless and FineSmith 1992).

One major advantage of systemic convulsants is ease of administration. The subsequent behavioral state is easily monitored, without concern about anesthetic effects which occur with topical convulsants. A disadvantage of systemic convulsant administration is that exposure of brain is general, as opposed to highly specific, as is the case of topical convulsants in which the drug is placed in a localized site.

Kainic acid is a frequently used convulsant in experimental animals to resemble complex partial seizures. Administration can be by either systemic delivery or by intracerebral injection. The mode of action of kainic acid is to inhibit inhibitory interneurons. This in turn offsets the balance between excitatory and inhibitory neurotransmission. Kainic acid has a structure similar to glutamate, which may explain its mode of action. The site of action of kainic acid is invariably the hippocampus, irrespective of route of administration.

Kindling is another animal model of complex partial seizures (Abel and McCandless 1992). The term kindling refers to the ability of some convulsants to induce seizures in animals at a dose at which earlier repeated doses failed to elicit seizures. Some kindling doses are administered electrically and some chemically. The site usually selected or involved is the hippocampus/amygdala. In electrically produced kindling, a low electrical pulse is administered through implanted electrodes. After many low doses, the animals (usually rats) begin to have seizures. Electrode placement is frequently in the hippocampus.

The sequence of seizures displayed by the experimental animals includes: mouth and facial (oral) movements, head nodding, forearm clonus, rearing, and rearing with falling. The sequence of seizures is consistent, but the timing can be altered by dosage.

In electrical kindling (Goddard 1967), electrodes are surgically placed, and up to 2 weeks elapses before kindling is initiated. The most common site for electrode placement is the hippocampus/amygdala because of the involvement of these sites in several types of seizures including complex partial. With chemical kindling models, the dose is important. The difference between overt tonic–clonic seizures and the gradual onset of kindled seizures is dose-related.

There also exists a genetic mouse model of complex partial seizures, the EL mouse (Imaizumi et al. 1959). In this mouse model, 14C-deoxyglucose has been used to examine functional activity of brain structures (Suzuki 1976). Results showed that the hippocampus and cerebral cortex had a higher level of activity. Subcortical areas such as the thalamus, locus ceruleus, and central gray had decreased metabolic activity. This shows that this seizure type had discrete anatomic foci with increased metabolism. This is in contrast to maximal electroshock, in which all cerebral areas have increased metabolism. The increases are not however uniform throughout the brain (McCandless et al. 1979). Histological examination of brain showed poorly stained neurons, irregular shapes, and synaptic changes (Suzuki et al. 1983; Hochi et al. 1987). A priming event using vestibular stimulation is required for maximal seizure production in EL mice.

Animal models are an ideal starting point for examining anticonvulsant properties of potential seizure therapies. One such study looked at the potential anticonvulsant characteristics of D-23129 (2-amino-4-4-fluorobenzylaminp phenyl) carbamic acid ethyl ester in an amygdala kindled model of complex partial seizures (Tober et al. 1996). D-23129 is a potent anticonvulsant drug, which when given to kindled rats, exerted a dose-dependent increase in the threshold for production of after discharges. High doses of D-23129 affected other characteristics of amygdala kindled rats such as severity and duration of seizures. D-23129 showed no adverse behavioral effects in the open field, or in the Rotaval tests. The authors note that D-23129 is an effective drug for amygdala kindled models of complex partial seizures, and shows no neurotoxic effects in normal treatment dose ranges. They state that this is a potential drug for antiepileptic treatment in humans.

The issue of behavioral changes in seizure states is important in complex partial seizures since production of seizures by stimulating the amygdala/hippocampus seems to produce adverse behavior. This appears as hyper-defensiveness in the cat and anxiogenic effects in rats (for review, see Depaulis et al. 1997). The speculation is that neuroplasticity produced changes which are associated with epileptogenesis such as changes in neurotransmitter activity, may be involved in fear-promoted behavioral reactions in these animal models of complex partial seizures.

Refractory epilepsy is epilepsy in patients in whom the result of pharmacological drug therapy is not deemed satisfactory by the patient. The number of these patients may be as high as 30%. For these patients, surgery is an option, and is detailed in another chapter in this book. Cell therapy is another potential treatment. Most experimental animal studies of the potential benefit of cell therapy have been performed in complex partial seizure animal models. Cell therapy has been used in some other neurological disorders (Raedt and Boon 2005). Two variations of cell therapy exist. The first uses transplanted cells to replace damaged neurons, whereas in the second, endogenous cells can be modified to ameliorate the seizures. Temporal lobe epileptic humans are those with the worst prognosis as regards refractory seizures.

Both human patients with complex partial epilepsy, and animal models of same, frequently have hippocampal sclerosis. This pathological lesion has neuronyl cell loss throughout the hippocampus, and especially in the dentate gyrus (Thom et al. 2002). Other areas such as the amygdala may also be involved (Salmenpera et al. 2001). While microscopically, neuronyl cell loss includes both excitatory and inhibitory (GABAergic) neurons, the loss of inhibitory neurons is considered to be key in this type of seizures. The inhibitory neurons which are decreased express neuropeptide Y (see chapter on alternative treatments/gene therapy). Mossy fiber growth/sprouting may also serve to enhance the excitatory output of the hippocampus. A correlation exists between mossy fiber proliferation and the excitatory output of the hippocampus (Cavazos et al. 1991).

In terms of cell therapy, the hippocampus is an obvious target. The goal of therapy is to re-establish the excitatory/inhibitory balance. This can be difficult since the ability of grafted neurons to integrate properly may not always occur. In this case, the nonendogenous neurons may become cells which increase excitability instead of having inhibitory action (Scharfman et al. 2000).

Fetal hippocampal neurons have been successfully transplanted in experimental kainic acid-induced epileptic rats (Shetty and Turner 2000; Zaman and Shetty 2001). Results show a survival rate of transplanted cells of over 75% in some animals. The transplanted cells were able to extend projections to endogenous cells and vice versa. When CA3 neurons were transplanted homotopically, loss of inhibitory cells which occurred before treatment was reversed. While this technique appears to be valuable, several fetal brains are needed to treat one patient. Also, the viability of transplanted tissue is variable, resulting in uncertain results.

The other method of cell therapy is that of trying to induce endogenous cells to function more normally. For example, the intraventricular infusion of growth factor in experimental models of hippocampal ischemic damage resulted in a regeneration of functional cells (Nakatomi et al. 2002). Thus, the goal of this type of therapy in complex partial seizures would be to find a method in which functional endogenous cells could be regenerated in damaged brain areas. Further work on this potentially important area is required.

One perplexing issue regarding complex partial seizures has to do with the mechanism of loss of consciousness in this disorder. One hypothesis is that lateralization of seizure activity to the contralateral temporal lobe may result in loss of consciousness. Data from rats with partial limbic seizures show that unconsciousness is associated with frontal cortical slow waves, and a slowing of frequency, and hypometabolism (Englot and Blumenfeld 2009). The large amplitude slow EEG activity and imaging signal decreases were shown in both the frontal and parietal cortex association areas. The slow wave discharges in neocortex areas is contrasted by fast spike activity in limbic and subcortical structures.

The authors state that a network inhibition hypothesis supports the concept that the spreading of seizure activity to subcortical regions permits the cortex to progress into an inhibited state. Rat studies have shown that loss of consciousness is associated with slow waves in the frontal cortex.

Another study examined the effects of partial kindling on dendrite spine morphology, neurogenesis, and astrogliosis (Kraev et al. 2009). Results demonstrated that mild focal seizure activity induced an increase in postmitotic bromo deoxyuridine-labeled cells; there was also an increase in double cortin-labeled cells; both increases were in the dentate gyrus. Using electron microscopy, it was noted that the rats showed movement of mitochondria to the base of dendritic spine stalks, and large spinules formed at the head of hippocampal dentate gyrus neurons.

The authors state that these changes are occurring during early stages of kindling, before generalized seizures have occurred. They interpret this as an indication that the development of neurogenesis, gliosis, and dendritic spine changes in the dentate gyrus are important changes which result in the development of seizure susceptibility in the hippocampus.

Another recent study (Chuang et al. 2009) reports the examination of cellular changes associated with temporal lobe status epilepticus. In this study, Sprague–Dawley rats were injected with kainic acid unilaterally into the hippocampal CA3 subfield, and various cellular functions assessed. Results showed that after eliciting a sustained seizure (status epilepticus), bilateral CA3 hippocampal subfields showed an increase in nitric oxide and peroxynitrite, and a decrease in mitochondrial electron chain enzyme activity. DNA fragmentation was also noted. These changes were seen bilaterally. Injection of inhibitors of these changes into the bilateral hippocampal CA3 subfield attenuated these alterations. The authors note that the changes in cell markers resulted in apoptotic cell death in hippocampal CA3 subfields after induction of status epilepticus in the rat temporal cortex.

Recent studies have examined various aspects of anticonvulsant activity in experimental models of partial complex epilepsy. In one study (Sardo et al. 2009), levetiracetam (keppra) was used as an anticonvulsant in the partial complex seizure model called maximal dentate gyrus activation model. Rats were administered either levetiracetam alone, or in combination with 7-nitroindazole. Results showed that the anticonvulsant effect of levetiracetam was enhanced by the co-administration of 7-nitroindazole, which is a potent inhibitor of neuronyl nitric oxide synthetase activity. This is a significant result in that levetiracetom alone is already a highly effective anticonvulsant.

In a similar study by the same investigators (Ferraro and Sardo 2009), the combined effects of vigabatrin and cholecstokinin sulfate were evaluated in the maximal dentate gyrus activation rat model of partial complex epilepsy. In this study, rats were anesthetized and pretreated with vigabatrin alone, or in combination with cholecystokinin sulfate, and results were analyzed. Repetitive electrical stimulation of the dentate gyrus led to dentate gyrus epileptic activity. Latency, duration, and poststimulus after discharge duration was associated in these rats. Vigabatrin alone was able to reduce the duration of seizures, and also of after discharge duration. When administered in combination with cholecystokinin, the anticonvulsant effects were enhanced. The authors speculate that the effect of cholecystokinin sulfate is mediated by an increase in GABA levels.

The occurrence of febrile seizures and their relation to temporal lobe epilepsy is not well investigated. In one study looking at this question (Dube et al. 2009), a febrile model of epilepsy was established in immature mice and rats.

Hyperthermia-induced limbic seizures were produced in experimental animals, and were shown to stimulate the secretion of endogenous fever mediators such as interleukin-1-beta. As regards mechanisms of action, results showed an expression of specific ion channels. The authors speculate that these changes may be key in febrile-induced temporal lobe epilepsy, and may suggest new pharmacologic treatment of these types of complex partial seizures.

In the class of metabolic encephalopathies, a subclass exists, mitochondrial encephalopathies in which there are distinct mitochondrial alterations. The observation that seizures are an important feature in the phenotypic expression of these mitochondrial discharges suggested that mitochondrial changes may be generally associated with epilepsy. An example of seizures and mitochondrial genetic changes are seen in thiamine-related disorders such as inherited ataxias and subacute necrotizing encephalomyopathy (SNE) (McCandless 2010).

Several animal models of complex partial seizures have been used to examine structural alterations in the hippocampus. One such model involves inducing status epilepticus in rats with pilocarpine. Within a couple of weeks after pilocarpine treatment, animals begin to have spontaneous seizures. The clinical similarities of this rat model resemble complex partial epilepsies seen in humans (Turski et al. 1983). In human patients, the hippocampus, with evidence of sclerosis, showed hypometabolism in the CA3 region using 18F-fluorodeoxyglucose PET (Vielhaber et al. 2003). Microscopically, the hippocampus shows a loss of pyramidal neurons in the CA1, CA3, and CA4 regions, and gliosis and mossy fiber sprouting.

Oxygen radicals are produced by the mitochondrial respiratory chain, and these radicals may cause neuronyl mitochondrial DNA damage (Kudin et al. 2002) when levels are high. Inhibition of the function of the respiratory chain may lead to an increase in oxygen radicals which exceeds the capacity of the protective enzymes such as glutathione peroxidase. There is a decreased function of the mitochondrial respiratory chain in the CA1 and CA3 regions of the hippocampus in pilocarpine-induced seizures in the rat model. These data indicate that oxygen radicals play an important role in the pathogenesis of selective hippocampal seizure activity.

Further supportive evidence is derived from mouse studies showing that inhibition of respiratory chain enzymes such as cytochrome c oxidase is capable of producing seizures (Yamamoto and Tang 1996). Impairment of other respiratory chain enzymes also has been shown to produce seizures (Urbanska et al. 1998). Mechanisms which might be involved in these observations include a decrease in intracellular ATP, and/or changes in calcium balance, which permits mitochondria to modulate neuronyl excitability and synaptic transmission (Bindokas et al. 1998). It is therefore evident that a decrease in energy metabolism in the affected hippocampal regions may be a key primary mechanism which is permissive to seizure generation in mouse and rat models of complex partial seizures, and perhaps in human patients with this form of epilepsy.

CO_2 has a potential adverse effect on seizures, as well as affecting several other cerebral features such as pH. In a recent study (Dulla et al. 2005), hippocampal slice techniques were used to evaluate the effect of CO_2 on endogenous neuromodulators such as adenosine. Results showed that manipulation of CO_2 levels resulted in an endogenase hippocampal response which included ATP receptors, ATPase, and adenosine receptors.

Increased levels of CO_2 served to inhibit excitatory glutamate neurotransmission, and decrease seizure activity in the hippocampal slice preparations. The authors showed that the effect was due to pH since increasing CO_2 while blocking its effects on pH had no effect. Lowering CO_2 levels had an opposite effect in that adenosine levels dropped, glutamate neurotransmission increased thereby increasing seizure activity (excitatory postsynaptic potentials). The authors note that these data provide a link between CO_2 and neuronyl electrical activity, and a link between hyperventilation and the reduction of seizure threshold, commonly seen in complex partial seizure patients.

Levels of intracellular ATP are also considered to be directly involved. In hypocapnia, lower CO_2 levels result in an increase in ATP. This facilitates the excitatory state in that ATP is available. By contrast, increasing levels of CO_2 has an opposite effect by decreasing ATP levels. In addition, the ratio of adenosine and ATP also has a significant role. These data support the concept that both adenosine and ATP serve as neuromodulators (Pascual et al. 2005). The authors state that their data on mechanisms of hippocampal excitability and its regulation may be important future predictors of effective seizure control in patients. The mechanisms of the effect of hyperventilation and lowering of the seizure threshold in complex partial seizure patients are especially important.

The role of altered energy metabolism (ATP and phosphocreatine) as a feature of seizures in experimental models is certainly not new (King et al. 1967). One problem with some animal models examining electrical or chemically induced seizures is that increased muscular activity or decreased breathing (hypoxia) may have their own influence on energy metabolism independent of the seizure per se. In order to circumvent these potential problems, an early study used paralyzed and ventilated mice as models of seizures (Collins et al. 1970). In these studies, the paralyzed/ventilated mice were electrically stimulated to seize using electrodes attached to the bases of the animal's ears. At various times after induction of seizures, the mice were quickly frozen in liquid Freon, whole brains were dissected, and metabolites such as ATP, phosphocreatine, glucose, lactate, etc. were measured.

Results showed that in unparalyzed and unventilated mice, electroshock was associated with significant decreases in glucose, ATP, and phosphocreatine. Cerebral lactate levels were increased in these same mice. By contrast, when ventilated on 100% O_2 and paralyzed mice were given electroshock, energy metabolites were unchanged except for minor decreases in glucose at a couple of time points. During the postictal period in both groups, brain metabolite levels were elevated and use of energy metabolites decreased. These data show that the brain had the energy capacity to sustain the seizure process if the confounding muscular activity and hypoxia of the seizure were controlled. Changes in brain pH were considered to reflect an increase in H ions due to increased anaerobic metabolism.

C14-2-deoxyglucose and quantitative autoradiography have been used to evaluate and map seizure spread (Collins et al. 1976). In these studies, varying amounts of penicillin were injected intracerebrally in order to produce a model of complex partial seizures. The use of tracer doses of C14-2-deoxyglucose quantitative autoradiography is a well established method to provide a visual record of increases and decreases in energy metabolism in discrete brain regions (Sokoloff et al. 1977).

Results showed that rats injected with a low dose of penicillin into the right motor cortex immediately had electrical spike waves. Five minutes after removal of anesthesia, synchronous, contralateral motor movements of the left paw occurred. Autoradiography showed increased metabolic activity in the ipsilateral extra pyramidal areas and in the contralateral cerebellum. A large dose of penicillin produced a more significant seizure, consisting of contralateral jerks of the face, paw, and hind limbs, as well as head and tail extension. Autoradiography showed a larger focus in the ipsilateral cortex and thalamus than in the low dose rats, and activation of cortical columns were defined on both sides.

A high dose of penicillin resulted in after discharges and contralateral tonic–clonic seizures. Autoradiographic images in rats with bilateral motor seizures showed an increase in metabolic activity in the cortex and thalamic regions on both sides. Areas affected included the globus pallidus, thalamus, putamen, and small areas of the cerebellum, among others.

Penicillin has long been known as a strong convulsant (Walker et al. 1945). The use of C14-2-deoxyglucose as an imager of altered metabolism has one major advantage. The advantage is that discrete anatomic areas can be identified, as compared to measuring metabolites in whole brain. Data from this study showed that there was a seizure-induced increase in glucose utilization in highly specific regions. This method cannot reveal the sequence of events because they occur in milliseconds. The authors speculate that their data support the concept that the penicillin-induced focal seizures become generalized by a "march" from the initial focus to other specific areas through existing pathways. Seizures remain localized and unilateral until they reach a level in which the frontal cortex is involved, then spreads to the other side.

Studies such as the above by Collins emphasize the highly focal nature of the initial onset of seizures, as well as most metabolic encephalopathies. Both the cerebral cortex and cerebellum are layered structures, which permits metabolic analysis in even more discrete regions. Some studies show that in a maximal electroshock (MES) model of seizures, mice have a decrease in metabolites in both cortex and cerebellum, but when pretreated with phenytoin, the alteration is prevalent in the cerebellum only (Lust et al. 1978).

A subsequent study (McCandless et al. 1979a, b, c) examined the effects of MES and phenytoin pretreatment on energy metabolites in cerebellar layers. Results demonstrated that the actual levels of ATP phosphocreatine, glucose, glycogen, lactate, GABA, and cyclic nucleotides were quite similar between cerebellar layers. As expected, MES significantly decreased phosphocreatine, ATP, and glucose in all cerebellar layers only 10 s after MES administration. Phenytoin attenuates most of that change. Lactate, increased with MES, was unaffected when mice were pretreated with phenytoin. Cyclic nucleotides, increased in MES, were hardly changed when the mice were pretreated. GABA levels were unaltered during MES. These data suggest that that cerebellar cyclic nucleotides may play a key role in seizures. Decreased cyclic AMP would favor the suppression of seizures, and the prevention of a cyclic GMP increase in MES by phenytoin would also act to suppress seizures.

Another similar study examined the effects of isoniazid and sodium valproate on both cerebral cortical and cerebellar layer energy metabolism (McCandless et al. 1979a, b, c). Results demonstrated a similar response across cortical and cerebellar layers. Thirty minutes after isoniazid administration in the preseizure state, both GABA and cyclic nucleotides were changed. GABA was decreased in cerebellar layers, and cyclic nucleotides were increased in most layers in both cortex and cerebellum. When seizures started, these changes were even more dramatic. Valproic acid alone raised GABA in all layers of both areas, and decreased cyclic nucleotides. Valproate and isoniazid together showed a valproate attenuation of the isoniazid effect. The decrease in cyclic GMP in the cerebellum caused by valproic acid is conducive to an increased Purkinje cell output. The convulsant isoniazid increases cyclic AMP, increasing seizure sensitivity. The combined effects on the cerebellum, whose output is inhibitory, suggest this structure may be important in the cessation of seizures.

Changes in energy metabolites were also analyzed in the motor cortex of the primate Cynamologus fascicularis with bicuculline-induced status epilepticus (McCandless et al. 1986a, b). While this was not a true complex partial epilepsy model, the results are of interest since complex partial seizures may develop into status epilepticus.

In this study, the monkeys were immobilized, then anesthetized with ether and a craniotomy was performed. Continuous recordings were obtained through EEG electrodes over the occipital cortex. After a period of equilibration and EEG recording, bicuculline was infused through an indwelling catheter. Samples of motor cortex were removed for metabolite analysis just prior to bicuculline infusion, 20 min after onset of seizures, and 2 h later. Samples were also taken and prepared for electron microscopy.

Results showed that bicuculline produced a dramatic neurophysiologic response in which the EEG showed prolonged epileptic activity. Overt behavioral manifestations (tonic–clonic seizures) accompanied these EEG results. Samples for electron microscopy were unremarkable 20 min after seizure onset, but by 2 h after the 20 min samples, samples from the motor cortex showed changes in both neurons and oligodendrites which consisted of dilated endoplasmic reticulum and swollen mitochondria.

Results from metabolite analysis demonstrated that phosphocreatine levels were decreased in pyramidal cell layers 20 min after bicuculline infusion. ATP levels at 20 min were unchanged, whereas 2 h later, ATP levels were over 50% decreased. At that time, phosphocreatine levels were returning toward normal. These results are consistent with the concept that at 20 min after bicuculline infusion, ATP levels were maintained at the expense of phosphocreatine. The decreases of ATP at 2 h and 20 min after bicuculline infusion were in the two pyramidal cell layers, the exact location of electron microscopic changes. It is easy to speculate that the changes in endoplasmic reticulum and mitochondria, which correlate with the decrease in ATP levels, might indeed be the result of the metabolic stress. This in turn would likely lead to cell death. The morphological changes, as expected, followed the changes in energy metabolites.

The previous few papers stress highly regional changes in animal models of complex partial seizures, yet these samples for analysis contained multiple cell types. In yet another study using maximal electroshock in mice rapidly frozen in liquid nitrogen, single cortical pyramidal cells and single Purkinje cells were dissected from freeze–dried sections (McCandless et al. 1979a, b, c). Adjacent neuropil was also analyzed. Samples weighed from 1 to 10 ng on a quartz fiber fish pole balance. Glucose, ATP, and phosphocreatine were analyzed using previously described techniques (Passonneau and Lowry 1993; Passonneau et al. 1980).

Results showed that glucose in the four cells (Purkinje, pyramidal, cerebellar neuropil, and cortical neuropil) was dramatically lowered (by 75%) equally in all four cell types. Both ATP and phosphocreatine were also significantly decreased at all time points except that ATP in the Purkinje cells was only about 20% lower at the 30 s interval after maximal electroshock.

The result of maximal electroshock is a rapid depolarization of cerebral nerve cells, and a need to restore ionic gradients. This represents a condition in which there is a high energy demand. There may be a switch, in part, to anaerobic glycolysis due to the apnea associated with the tonic phase of the seizure.

The difference between the four areas analyzed was that the Purkinje cells were spared the same degree of metabolic stress exhibited by the other three layers in terms of ATP values. Purkinje cells were also slightly less affected as regards phosphocreatine. There are no obvious reasons to predict this level of sparing of Purkinje cells in maximal electroshock-induced seizures. The speculation is therefore that either the maximal electroshock stimulus does not reach the Purkinje cells as readily as other areas, or that the cellular response is dampened or inhibited. There exists the concept that the circuitry in the cerebellum is such that an excitable state cannot be sustained.

The Purkinje cell is the primary source of cerebellar output, which is inhibitory. The changes in energy metabolites reflect the output of the cells in question. The reduced metabolic response in Purkinje cells may reflect a decreased inhibitory output, which could be viewed as conducive to the seizure state. A more pronounced Purkinje cell output could possibly modulate the seizure. This sparing effect could be seen as a deleterious characteristic of cerebellar circuitry.

These data also emphasize clearly the problem in analyzing tissue with many cell types, or whole brain with many regions, each with different functions. The attempt should always be made to use as small a sample as possible in such a heterogeneous organ as the brain.

The convulsant pentylenetetrazole is chemical convulsant which has been used extensively for the study of experimental seizures. By varying the dose, seizures range from mild with low dosage to tonic–clonic seizures with high doses. Intermediate seizures resembling complex partial seizures can be achieved with moderate doses. Pentylenetetrazole has been utilized to examine effects of seizures on energy metabolites in the lower primate Tupaia glis (McCandless et al. 1987).

This study used an intermediate dose of pentylenetetrazole (125 mg/kg) to induce seizures. Previous studies used a low dose (80 mg/kg) to produce seizures, with only minimal changes in energy metabolites (King et al. 1967) or high doses (150 mg/kg) which produced a major seizure and large decreases in both ATP and phosphocreatine (Duffy et al. 1975). In the present study, advantage was taken of microassay techniques to analyze cerebral cortical and cerebellar layers. Sacrifice was by a 25 kW microwave instrument. This certainly represents one of the best sacrifice methods with which to preserve labile metabolites. The cerebral cortical area sampled was the motor cortex, and the cerebellar area was the vermis.

The pentylenetetrazole dose used (125 mg/kg) after injection produced a 60–90 s period of quiescence, followed by a tonic–clonic seizure. Metabolite analysis showed a drop in ATP only in the outer small pyramidal cell layer of the cerebral cortex, and no changes in cerebellar layers. Phosphocreatine demonstrated a decrease in the outer small pyramidal cell layer and also in the white matter. Phosphocreatine was also unchanged in the cerebellar vermis. Glucose was decreased in all cortical layers and in the granular cerebellar layer. Phosphocreatine was depleted about one-third in the outer small cortical layer, whereas ATP was depleted in the same layer by 30%.

These results show once again the highly regionalized nature of changes in energy metabolites in experimental seizures in a primate model of complex partial seizures. This selective effect has also been demonstrated repeatedly by C14-2-deoxyglucose methodology (Lothman and Collins 1981; Collins et al. 1983). The only problem with imaging with C14-2-deoxyglucose is that it only measures uptake and phosphorylation of deoxyglucose. That might not reflect overall energy metabolism in a perturbed state (McCandless et al. 1986a, b).

The present data are of interest in terms of the cerebellum. The cerebellar output is largely inhibitory, and the lack of, or attenuated response of the cerebellum in experimental seizures has been previously noted (see above). The cerebellar response is seen as conducive to the seizure state. The finding of changes in ATP and phosphocreatine in specific motor cortex layers again emphasizes the possibility that studies examining whole cortex could have missed more localized changes by inclusion in the sample of nonaffected tissue. Similarly, even more regional studies (single cells) might be expected to show more distinct changes. This is why the materials and methods are critical in order to assess results. Similar results as above have been demonstrated in bicuculline-induced seizures in mice in which changes in energy metabolites were localized to layers of motor cortex, and the cerebellum was spared (Dworsky and McCandless 1987).

The anticonvulsant lacosamide has recently been tested as to efficacy in modulating various seizure models of partial complex seizures (Chung 2009). Lacosamide proved effective in inhibiting tonic extension in a maximal electroshock model of seizures in both mice and rats. Furthermore, lacosamide proved effective in reducing the frequency of a kainic kindling model of hippocampal seizures by about 40%. Lacosamide was able to decrease after discharge duration by 85%. This result was superior to that produced by other complex partial seizure anticonvulsants such as phenytoin, valproic acid, ethosuximide, or carbamazepine. Locasomide was also effective in protecting from status epilepticus in a cobalt/homocysteine rat model. Lacosamide was effective in protecting and increasing seizure threshold in amygdala kindling in rats, but had no effect in the WAG/Rij genetic rat model of absence seizures.

Another animal study revealed that lacosamide was effective in complex partial seizure models including maximal electroshock in mice and rats, and in a rat model of kindling, as well as in audiogenic seizures. Notably, the AED displayed a good margin of safety as regards side effects (Ohr et al. 2007). Yet another positive feature of lacosamide is that it apparently has no effect on the autonomic nervous system. These studies were done in rats and dogs (Beyreuther et al. 2007). It also seems to have no teratogenic effects or adverse effects on organ systems such as G.I., renal, etc.

Another recent study examined the relationship between limbic lesions and temporal lobe epilepsy in an animal model (Parekh et al. 2010). In this study, electrodes were implanted in the ventral hippocampus and status epilepticus was produced. In vivo diffusion tensor and T2 MRI images were evaluated at intervals between 3 and 60 days after seizure onset. Seizures were documented using time-locked video-monitoring.

The self-sustaining rat model of status is a good one because it shows both spontaneous recurring limbic seizures, and a latency between the acute injury and the onset of chronic seizures. This sequence is frequently seen in human cases. A period of 2–8 weeks follows status epilepticus before the spontaneous seizure onset. This model also has pathological lesions and EEG activity similar to human cases of temporal lobe epilepsy. Thus, this particular animal model is an excellent one for making comparisons with patients and data translation.

Results showed that self-sustaining status included 30–45 min seizures during and after electrical stimulation. Wet dog shakes, head bobbing, and seizures continued after stimulation for 2–8 h poststimulation. Subsequently, spontaneous seizures occurred in 73% of animals, a ratio similar to other studies (Lothman et al. 1990). Both diffusion tensor imaging and T2 MRI were performed in vivo. During the acute phase, diffusion and T2 changes occurred only in those animals which exhibited spontaneous limbic seizures. Rats developed moderate to severe seizures when average diffusivity and T2 were seen in the parahippocampal gyrus in the acute phase. Rats also had increased diffusivity and fractional anisotropy changes in bilateral hippocampal subfields and the parahippocampal gyrus. There were also changes in white matter (fimbria and mossy fibers) as seen with diffusion tensor imaging.

In excised brains, bilateral increases in anisotropy occurred in the dentate gyrus. This corresponded to mossy fiber sprouting. The authors state that the results do indicate that early changes in the amygdala and piriform cortex in the status epilepticus phase are related to irreversible damage which causes a sequence of structural alterations during the latent period, which then results in spontaneous seizures. Importantly, the authors speculate that these hippocampal and parahippocampal alterations could represent an anatomical biomarker suitable to serve as an indicator of early epileptogenesis in complex partial epilepsy.

The authors further state future studies are necessary in order to make a correlation between histological changes and those seen in MRI. Further, it is necessary to examine the temporal MRI changes with the temporal neuropathological alterations. These also need correlation with seizure progression. This is an important paper since it correctly looks at highly focal structural changes in an excellent animal model of temporal lobe epilepsy. The model is somewhat complicated to set up, but is closely correlated to human complex partial seizures, rendering meaningful interpretation possible.

There appears to be a correlation between febrile seizures and complex partial seizures (Annegers et al. 1987), although not all studies agree. Examination of possible features related to mechanisms, and association of the two seizure types require animal studies in order to look at early correlations. Studies described in the paper used an animal model of febrile seizures, then examined features (Koyama and Matsuki 2010).

Several models of febrile seizures have been developed, and the so-called "hair dryer" model has seen much use. In this model, young postnatal rodents (postnatal day 10–14) are placed in a glass jar, which is then heated with a hair dryer. Hyperthermia is then maintained for 30 min at a temperature of 40–42°C. The age of newborn mice and rats is chosen because at that time the limbic system is developing. In the model, the hippocampus displays long-term hyperexcitibility (Dube et al. 2000). With a low dose of kainic acid, EEG hippocampal seizures are induced, which tend to progress to status epilepticus. There is a tendency for mature mice and rats to develop limbic seizures at 3 months of age (Dube et al. 2006).

From a mechanistic standpoint, it has been shown that interleukin-1 beta is likely activated in hyperthermia, which leads to febrile seizures (Dube et al. 2005). In addition, intracortical pH increases in hypothermia to a level at which seizures occur using other pH increasing methods (Schuchmann et al. 2009). It has also been shown that hyperthermia has a suppressing effect on GABA release from presynaptic terminals as well as a decrease in function in GABA receptors.

As regards structural changes in the hippocampus, it has been shown (Toth et al. 1998) that febrile-induced seizures do result in neurons with altered structure in the amygdala and hippocampus within 24 h, and these changes last for over 2 weeks. Some of these disrupted neurons could survive and produce long-lasting problems even including hippocampal sclerosis. Other studies have shown that hyperthermia at this critical period could alter the ability of certain cells to remove glutamate from the synaptic cleft. This suggests an influence of hyperthermia on later lack of ability to control excitability (Lemmens et al. 2008). Thus, while excellent animal models of febrile seizures have not revealed the precise mechanisms of neural damage and epileptogenesis, the correlation between febrile seizures and complex partial seizures is strong.

In another paper (Gill et al. 2010), the effect of the glutamate agonist domoic acid on developing rat brain was examined. Results showed that domoic acid treatment in the second postnatal week decreased glutamic acid dehydrogenase in the hippocampal dentate gyrus and CA3 areas. Additionally, these changes were noted to occur more frequently in male rats. Non-GABAergic cells were not affected.

Some of the changes seen in this study resembled those seen in other studies, some did not. The authors note that careful experiments are important since subtle changes in neurochemistry may have pronounced effects later.

As regards new AEDs, lamotrigine and topiramate have been experimentally evaluated in terms of efficacy and cognitive effects (Chen et al. 2010). In this study, a lithium pilocarpine model of complex partial epilepsy was employed. Five hours after the onset of status epilepticus, lamotrigine or topiramate were administered to seizing rats, two times/day throughout the time of the study. Results showed that both AEDs significantly inhibited the seizure-induced proliferation of hippocampal neuronyl progenitors. Neural differentiation was not affected. Both AEDs decreased the frequency of recurrent seizures.

A difference was observed between the two AEDs in that lamotrigine acted to reduce the number of ectopic hilar neurons after seizures, whereas topiramate increased the number of dentate granule cell layer neurons after seizure activity. Hilar basal dendrites formed in spite of either AED in the hippocampus. The authors state that topiramate facilitates altered neuronyl hippocampal regeneration after status epilepticus. This is viewed as possibly conducive to changes in cognitive function.

Chapter 15
Complex Partial Epilepsy in Humans

About two million people in the USA have epilepsy. As many as 70% of cases begin during childhood, with the majority appearing in the first postnatal year. World wide of course, the numbers are staggering, with the vast majority of cases going undiagnosed as well as untreated (Chap. 1).

In the most general way, seizures could be classified as partial or generalized. Partial seizures have a regional focal cerebral localization, whereas generalized (tonic clonic) seizures occur throughout the brain. Partial complex seizures are common, occurring in about one-third of epilepsy patients. Complex partial seizures are characterized by impairment of consciousness, an immediately preceding event called an aura (simple partial seizure), automatisms, and all usually arises from the temporal cortex, hence the sometimes used name temporal lobe epilepsy. The nature of the impairment of consciousness is variable and may range from an inability of the patient to respond to simple commands to an inability to respond at all.

The term complex partial seizure acknowledges a possible frontal lobe origin, thus the term temporal lobe seizure is used less frequently. It is not unknown for a complex partial seizure to actually originate in the parietal or occipital lobes, and spread to the temporal lobe. Thirty percent of extra temporal lobe seizures originate from the frontal lobe (Manford et al. 1992). Features of frontal lobe complex partial seizures include an early loss of consciousness, head and eye movements, clonic jerks, falling, etc. The exact location within the frontal cortex is difficult to determine (Figs. 15.1 and 15.2).

In complex partial seizures, there is an associated initial period in which there is altered behavior/consciousness (aura). The patient may not remember these sensations. Psychic experiences such as sensory illusions are frequent. These may also involve altered perceptions of space/distance, and changed perception of people's size. Hallucinations may occur, and they can be visual or auditory. Emotional experiences may range from happiness to sadness, and even strong anger. Auras can involve gastrointestinal sensations, fear, a particular smell, etc. In children, the aura is either less experienced or less recognized. As the patient ages, the aura is more easily recognized.

D.W. McCandless, *Epilepsy: Animal and Human Correlations*,
DOI 10.1007/978-1-4614-0361-6_15,

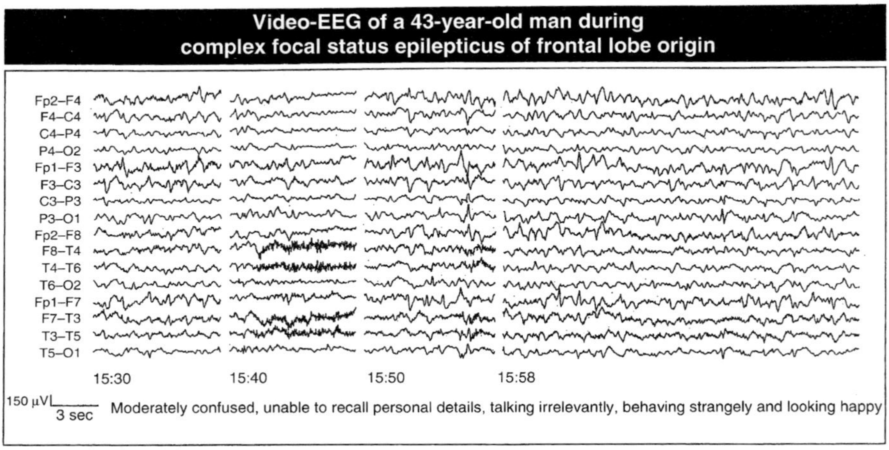

Fig. 15.1 Video EEG of a case of complete focal status epilepticus from the frontal lobe. With kind permission from Springer Sciences+Business Media; The Epilepsies, EEG, and Epileptic syndromes and their Treatment. 2010, p. 80, Panaylotopoulos, C. Fig. 3.2

Automatisms are a common feature of complex partial seizures. These usually occur during the period of impaired consciousness, and take the form of chewing movements, lip smacking, fumbling/grasping movements, blinking, leg movements resembling running, etc. The automatisms are inappropriate for the situation and tend to be similar in succeeding seizures. They may represent initial EEG activity from the focus. This has been shown in cases of oral facial movements to originate from the oral facial somatosensory cortex.

The etiology of complex partial seizures may be of several features. These etiological factors include birth defects, trauma associated with birth such as hypoxia, asphyxia, infections such as rubella, postnatal infections such as hemolytic uremic syndrome, parasites, etc. The majority can be identified with MRI (Kuzniecky et al. 1993). In adolescents and adults, mesial temporal sclerosis is thought to be causative in half of patients (Brown 1973; Brown and Babb 1987). Intrauterine developmental disorders – aberrant neural migration (neural crest cells), cortical dysplasia, lissencephaly, etc. – may be present.

The duration of complex partial seizures can be highly variable, ranging from a few seconds to as much as 2 min. Brief complex partial seizures may closely resemble absence seizures, making clinical diagnosis difficult. One differentiating factor is that complex partial seizures almost always have a post-ictal depression stage, whereas absence seizures do not. In terms of EEGs, complex partial seizures show a post-ictal slowing of wave forms. Extended time/place disorientation may suggest a right side seizure source. The post-ictal period in complex partial seizures may be associated with a prolonged period of paranoia or psychosis which can last several weeks. Post-ictal behavior/personalities have been described as hyposexuality, hyperreligiosity, and hypergraphia (Geshwind 1983).

As many as half of patients may have a complex partial seizure, progressing to a generalized tonic clonic seizure. This occurs most frequently during sleep. Post-ictal

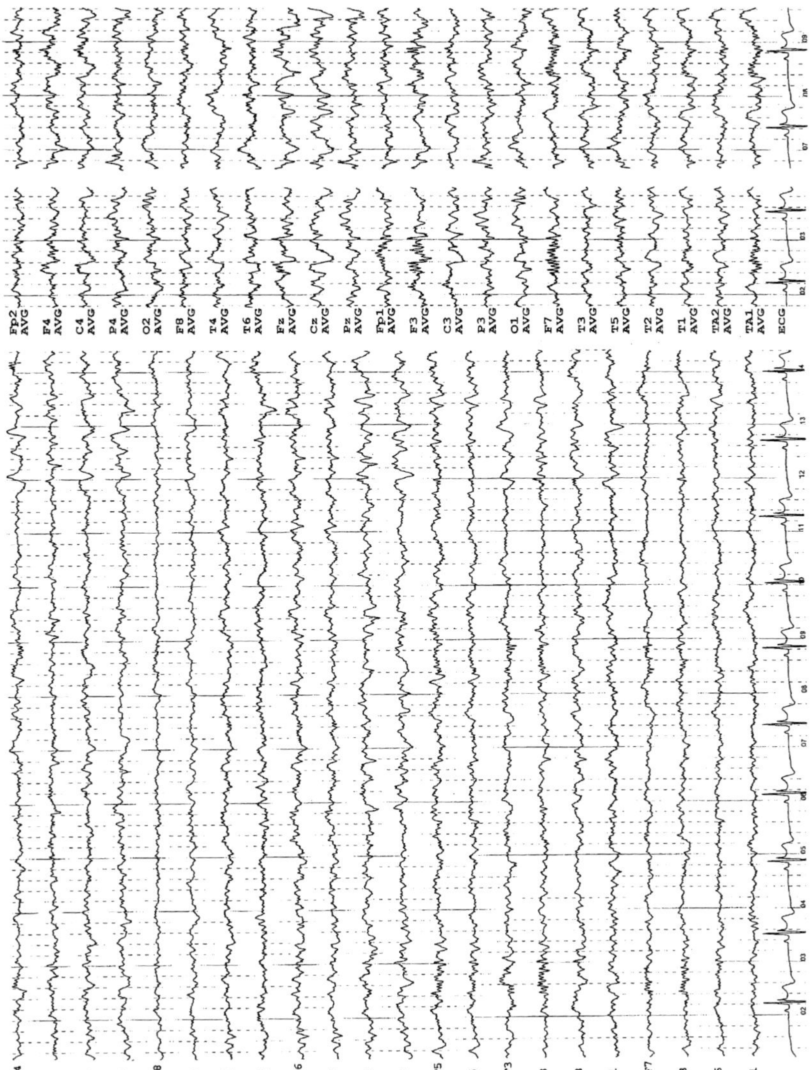

Fig. 15.2 Example of frontal lobe epilepsy EEG Pattern. Published in Crespel, A., et al. Atlas of electroencephalography Volume 2; The Epilepsies, EEG, and Epileptic Syndromes and their treatment. With permission from John Libbey Eurotext

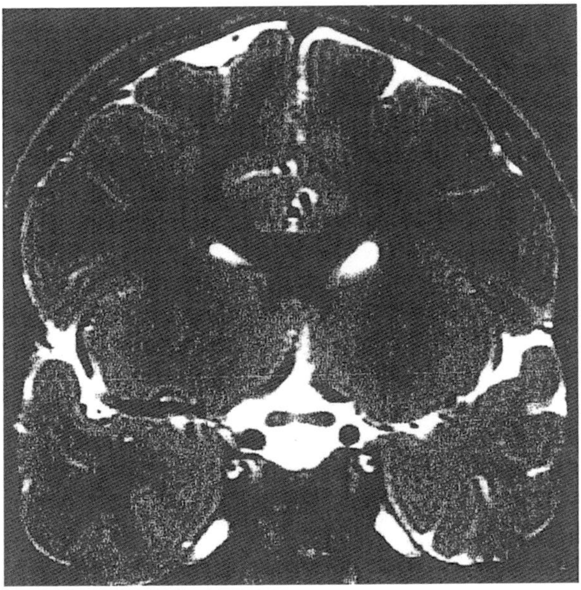

Fig. 15.3 MRI of a 10-year-old with temporal lobe epilepsy and left-sided hippocampal sclerosis. With permission from Elsevier-Kuzniecky, R., and Jackson, G. Magnetic Resonance in Epilepsy, 2nd ed

depression is a usual occurrence, which may last several hours (Delgado-Escueta et al. 1981). Confusion and lethargy are common during this period as is dysphagia. Rarely, complex partial status epilepticus can occur. This may last many days, and the patient is dysfunctional. A continuous focal ictal discharge seen on EEG in company with clinical features is diagnostic.

Pathologic changes in complex partial epilepsies are common. These consist of structural changes such as hippocampal sclerosis and associated amygdala alterations are seen in removed temporal lobectomy, and can be visualized with MRI. Some patients (the minority) do not demonstrate cerebral lesions. Of all the pathological changes such as fibrosis, vascularization, gliosis, etc., it is difficult to identify that which is responsible for the seizures. A common finding histologically is a neuronal cell loss in the hippocampal CA1 region. This may involve the dentate gyrus as well. Whether the cell loss is primary or secondary is not known. Trauma and cerebral infection may cause mesial temporal sclerosis (Figs. 15.3 and 15.4).

The diagnosis of complex partial epilepsy is perhaps the most difficult seizure diagnosis to make. This is due to the highly variable nature of these seizures and to associated behavioral and psychic alterations. Seizures rarely cause mental retardation (Ellenberg et al. 1986).

Complex partial seizures often start somewhere in later childhood or adolescence and may be idiopathic. Sometimes there has been an earlier seizure which was unnoticed, and the patient is diagnosed with absence seizures. Infections such as hemolytic uremic syndrome, which is increasing in frequency, will have a cerebral component which can lead to complex partial epilepsy.

Some of the difficulties in diagnosing complex partial seizures have been discussed in a brief review of the subject (Restak 1995). This author emphasizes the

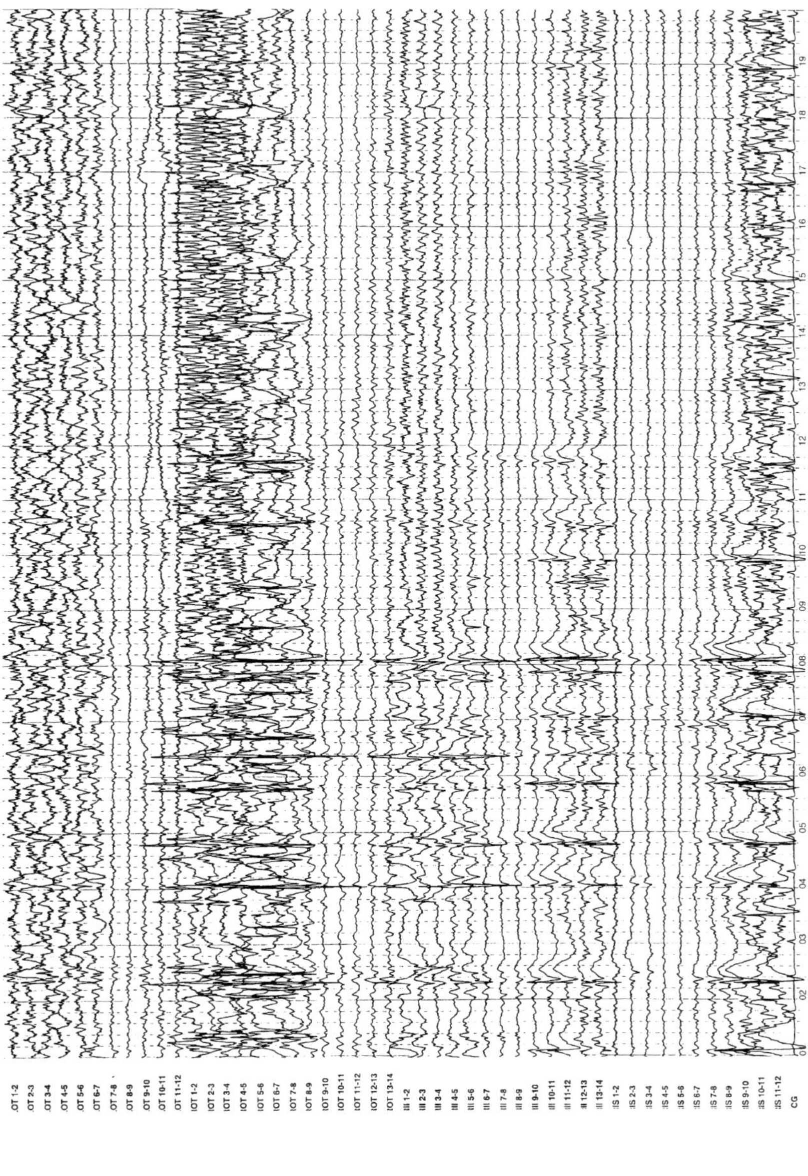

Fig. 15.4 Medial temporal lobe epilepsy. Published in Crespel, A. et al. Atlas of electroencephalopathy Volume 2; The epilepsies, EEG and Epileptic Syndromes. With permission from John Libbey Eurotext

importance of a thorough history. For example, questions regarding migraine headache should be asked as many symptoms of migraine mimic those of complex partial seizures. The patient should be questioned about déjà vu, depersonalization, sudden mood swings with visceral or oral sensations, memory disturbances, etc. All these may suggest complex partial seizures.

Overt psychosis can also be a presenting sign of complex partial epilepsy. The author relates a story of a businessman who after two all night marathon work, decided he needed to see the President. After arrest at the White House and transfer to the psychiatric ward, an EEG ordered because the history included childhood seizures, revealed complex partial seizures. Complex partial epilepsy is also associated with chronic psychoses, the most common being paranoia, with the EEG again being a key in determining the presence of a seizure state. The author notes that the complex partial seizure diagnosis is a clinical one, and that even several negative EEGs do not necessarily rule out a diagnosis of epilepsy. MRI, fMRI, PET, and SPECT are all helpful, if available.

Another aspect reviewed in this paper is the common occurrence that effective AEDs for complex partial seizure control may have significant psychiatric consequences. Schizophrenia-like psychoses may accompany successful or partially successful improvement in seizures. The psychosis can easily become more troublesome than the epilepsy. Treatment of these problems is sometimes difficult because neurologists may not be familiar with psychiatric aspects, and psychiatrists not familiar with aspects of seizure treatment. The ideal solution may be a team approach.

Unusual cases and diagnosis are emphasized by the next report (DiRocco et al. 2001). This was a case study of a 4-year-old boy who had staring episodes, déjà vu sensations/feelings, and automatisms, all features of partial complex epilepsy. These signs increased in frequency, and finally he had a tonic-clonic seizure and was referred for imaging studies. His only medication was carbamazepine, started earlier when the complex partial seizures were increasing in frequency.

Results showed that the axial T1-weighted post-gadolinium demonstrated a large tumor in the right frontal lobe. The axial T2 images enabled visualization of mild edema in adjacent white matter. The edema seemed due to compression by the tumor. Some bone remodeling was seen, indicative of some degree of chronicity of the tumor.

The authors note that about 25–50% of similar cases present with seizures. In the pediatric population, the tumors associated with epilepsy are almost always benign. Conversely, before the development of modern imaging, only 0.2–0.3% of epileptic children had their seizures attributed to tumor growth. Subsequently, the number rose by tenfold. Even though the authors note that there is an increased awareness of the association of tumors and epilepsies, there is reluctance among pediatricians to request imaging studies on their patients, believing that the diagnosis can be made on the basis of clinical features.

The authors voice concern about these cases in that in the absence of imaging tests, early epileptic signs/symptoms may respond to AED treatment, delaying recognition of the presence of a growing tumor. In fact, the mean interval between first seizures and the diagnosis of a tumor when present is about 5–8 years. Such a delay is not good.

This case represents the difficulty in making this distinction without imaging. What presented as, and was complex partial epilepsy, was successfully treated by removal of a large temporal lobe tumor.

While on the subject of surgery, a study examining the effects of surgery on memory in complex partial epileptic patients offered interesting results (Martin et al. 2002). The reader is also referred to the chapter on surgery. In this study, pre- and postoperative evaluation of 55 patients with drug-resistant temporal lobe seizures was performed. Thirty of the patients had right-sided localization of the complex partial seizures, 25 patients had left-sided localization.

Criteria for inclusion in the study was an IQ over 75, generalized changes in cognitive function, age of 10–15 years at seizure onset, and right handedness. Drug resistance was of course also a criteria; all patients were on multiple drug therapy. All patients were evaluated neuro-psychologically before surgical intervention and after intervention. The usual intelligence tests were used in evaluations. The Wechsler Clinical Memory Scale-R was used to evaluate memory. Surgical resection was performed using the Spencer technique (Spencer 1991). The area of cortex and the extent of the hippocampus and parahippocampal gyrus were defined by electrocartigraphical means (Penfield and Jasper 1954).

Changes in memory are a significant feature in those with complex partial seizures (Selwa et al. 1994; Sawrie et al. 2001). Overall, results of examination of post-surgical memory deficits are conflicting. Some show improvement, whereas others show little changes, or even a decrease in performance.

The results of this study demonstrated an improvement in IQ, plus an improvement in memory in those patients who had right-sided resection. Those with left-sided resection had a deterioration in memory. This was in keeping with other results, such as those of Helmstaedter (Helmstaedter et al. 2000). The authors note that their results are of prognostic interest in that they may predict the presence or lack of risk of further deterioration or improvement.

Another similar report (Alsaadi et al. 2003) looked at the laterality of unilateral hand posturing. Unilateral hand automatisms, nonforced head turning, and postictal dysphasia signs in treatment intractable complex partial seizure patients prior to surgery. Follow-up lasted at least 2 years. Results showed that the sidedness (laterality) and consistency of these signs did not have a statistically significant relation with postsurgical outcomes.

In a report on long-term chronic epilepsy and intelligence (Jokeit and Ebner 2002) various features of complex partial epilepsy were examined for impact on mental performance. Features such as age of seizure onset were analyzed, as well as the duration of seizures. Results of examination of over 275 patients showed that the most important characteristic in predicting decreased intellectual function in temporal lobe epileptic patients is duration of seizures. Patients with the presence of seizures for over 30 years were most likely to show measurable intellectual deterioration. This is believed to be due to the areas remote from the temporal lobe focus (prefrontal cortex) are associated with a deterioration in mental function. Fluorodeoxyglucose-PET showed areas of hypometabolism in unilateral temporal cortex as well as the prefrontal cortex (Figs. 15.5 and 15.6).

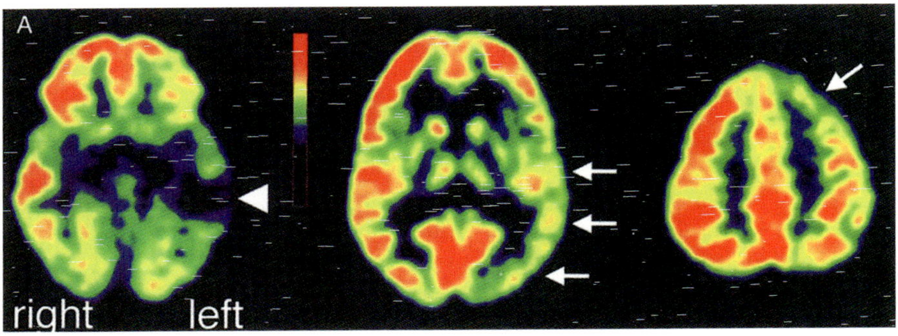

Fig. 15.5 FDG-PET from a patient with left temporal lobe epilepsy showing severe hypometabolism in the left temporal lobe. With permission from Elsevier-Kuzniecky, R., and Jackson, G. Magnetic Resonance in Epilepsy, 2nd ed, 2005

Fig. 15.6 Left frontal/temporal hypometabolism in a patient with cryptogenic infantile spasms (West syndrome). With permission from Elsevier-Kuzniecky, R., and Jackson, G. Magnetic Resonance in Epilepsy, 2nd ed, 2005

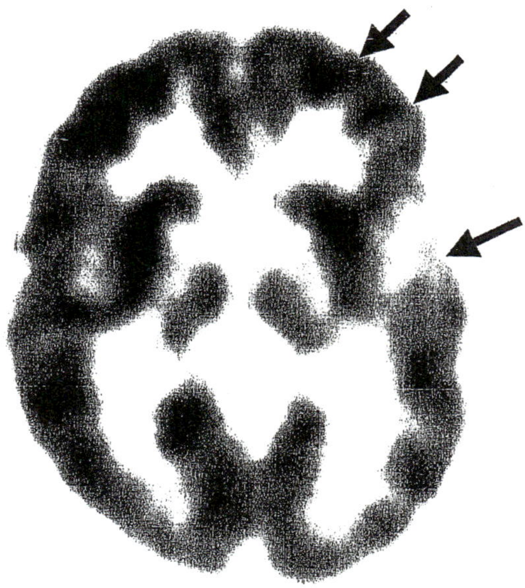

These studies not only demonstrated a correlation between length of time with seizures, but also that patients with a high level of education demonstrated a longer period of stability of mental function than those with less education. The authors speculate that the brain may acquire a "functional reserve" in terms of intelligence based on efficiency, plasticity, re-organizational capacity, etc. (Stern et al. 1996). These studies may help to recognize potential adverse factors, and identify those patients with increased risks.

Many patients with complex partial epilepsy are cognizant of memory deficits, and a recent paper looks at ideas involved in cognitive rehabilitation of memory problems (Ponds and Hendriks 2006). Of all, cognitive problems in epileptics such as attention deficits, language problems, concentration defects, the loss of memory abilities are the most obvious to the patient. Memory impairment problems are most commonly complained about by patients, who seek help to improve memory. In this study, epilepsy features such as age of onset, etiology, cerebral site of involvement, or AEDs did not correlate with the pattern of memory deficits. It has been shown that lateralization of the focus in complex partial seizures is an important factor. Thus, patients with a unilateral left temporal lobe focus are more prone to show memory deficits than patients with a right-sided focus. As mentioned above, the duration of seizures (over 30 years) is a key characteristic in determining memory problems (Jokeit and Ebner 1999). It is also possible that this deficit over time could on part be a normal memory loss seen in normal aging (Elger et al. 2004).

In terms of memory rehabilitation, most reports are anecdotal, or in small groups not well controlled. Patients with memory loss may have only a mild loss of memory, but perceive the problem to be greater than it is. So one area for improvement is in memory self-confidence. Otherwise, help in training memory improvement should be centered on a tailor-made goal, such as learning to remember friends' names, major events, where things are placed, remembering plans, etc. As everyone knows from school, repetition improves memory and learning. Mnemonic training techniques are highly efficient for many patients. In fact, the use of mnemonics is an efficient and effective method to improve memory. Additionally, epilepsy patients and families often attribute memory and cognitive deficits to AEDs.

Another interesting treatment modality involves the use of electrical stimulation of the hippocampal seizure focus as a method of seizure control (Velasco et al. 2007). The authors note that not all patients are surgical candidates for temporal lobe epilepsy, and therefore alternative non-resective treatments are needed. Over the years, several cerebral targets for electrical stimulation have been tried including the cerebellum (Velasco et al. 2005). These targets have had less than optimal outcomes as regards complex partial seizures. Most trials were on single patients, so the results were anecdotal. In the present study, nine patients were selected for bilateral hippocampal electrode stimulation from a group with intractable complex partial seizures. Eight contact electrodes were implanted in the anterior hippocampus and amygdala. MRI verified the placement of electrodes. These electrodes were used to verify seizure foci. Later, four permanent contact electrodes were placed in the epileptic site. Placement was again confirmed by MRI. Stimulation consisted of 1-min square pulse with a 130 Hz frequency, and an amplitude of 300 μA.

Results demonstrated that seizure reduction occurred in all nine patients, dropping from an average of about 30 per month to just over 5. Patients with normal MRI findings became seizure free, whereas patients with demonstrable hippocampal sclerosis had a partial/slower response to the electrical stimulation. A major criterion for inclusion in this study was the risk of memory loss associated with temporal lobe resection. The nine patients showed a preservation of memory following hippocampal electrical stimulation. Earlier studies (Velasco et al. 2006) suggest that the efficacy of the treatment is related to an inhibitory mechanism. The procedure seems entirely safe except that some skin erosion around the implanted electrodes occurred after 24 months of implantation of electrodes. One patient required plastic surgery and antibiotic treatment. All three required explanation due to this complication.

AED treatment remains the most common treatment modality available for partial complex epileptic patients. The relatively new AED lacosamide has been examined as to its safety and efficacy as an adjunctive AED in patients with complex partial seizures. The study was a double blind, randomized, placebo-controlled trial (Ben-Menachem et al. 2007). Placebo and three doses of lacosamide (200, 400, and 600 mg/day) were tried. As many as 312 patients completed the trial.

Results showed that for the three lacosamide doses stated above, there was a 26%, 39%, and 40% reduction in seizure frequency. Placebo reduction was 10%. This represents a dose-dependent effect and a highly statistically significant reduction in seizure activity. The mean reduction over placebo of the 400 mg/day lacosamide treatment was significant at the $p=0.002$ level of confidence. Adverse effects included diplopia, ataxia, dizziness, nystagmus, fatigue, etc. Lacosamide did not have any negative interactions with other AEDs.

The authors of this multicenter study conclude that adjunctive lacosamide reduced significantly the frequency of complex partial seizures in the participating cohort. Tolerability results indicate the efficacy, and further studies are warranted on lacosamide as an important AED.

Another paper looking at several features such as efficacy and safety of lacosamide has been published (Beydoun et al. 2009). As noted earlier, lacosamide is a relatively new AED, recently approved in the USA and Europe as an add-on treatment for partial complex epilepsy. The mode of action is thought to be on voltage-gated sodium channels. It may also interact with collapsing response mediator protein-2. As stated above, 400 mg/day seems the ideal dose. The overall effect is a stabilization of hyper excitable neuronal membranes, as well as inhibition of repetitive neuron firing. The dual mechanism of action of lacosamide is somewhat novel, and the slow inactivation of Na channels is unique. Yet to be thoroughly evaluated are the effects in pediatric patients and efficacy as a momotherapy. Lacosamode has a favorable safety profile.

A recent review article summarizes most of the information available on lacosamide (Halford and Lapointe 2009). Lacosamide has been shown to be effective in numerous animal models of seizures, including maximal electroshock and pentylene tetrazole, among others (Stohr et al. 2007). Animal trials, as noted above, have shown that 400 mg/day dose was statistically significantly effective in reducing seizure frequency. The reduction was not however greater than pregabalin or topiramate.

Pharmacokinetic studies showed that lacosamide is quickly absorbed (within 4 h) and not affected by food. Lacosamide is excreted by the kidneysand has a half life of 13 h. Studies show no significant differences between adult and pediatric patients as regards pharmacokinetics. From a safety standpoint, lacosamide does not have adverse effects on GI, renal, or respiratory systems (see above). The autonomic nervous is also unaffected by lacosamide. From the safety standpoint lacosamide seems to have no real significant side effects except for dizziness and nausea and vomiting.

To conclude, lacosamide seems an effective new add-on AED, with few side effects, and no drug interactions. As much as 50% or more of patients on lacosamide had at least a 40% reduction in seizure frequency. Efficacy is good, but more studies in children are warranted.

The goal of AED treatment in complex partial epilepsy patients is to identify a highly effective mono therapy. Studies have been undertaken to determine the efficacy of various AEDs when used alone. One such study looked at the relative presence or lack, of merits of oxcarbazepine vs. carbamazepine (Koch and Polman 2009). This paper is a review, examining results of three actual studies having a total of 723 patients (Dizdarer et al. 2000; Donati et al. 2007; Marson et al. 2007). Carbamazepine is the current most commonly used AED for complex partial seizure patients, while oxcarbazepine is a newer AED.

Results from these studies show that there were no statistically significant differences between the two drugs regarding efficacy in any criteria such as the length of time necessary for a 12-month remission from seizures. Also examined were length of time to cessation of drug use due to ineffectiveness and overall withdrawal time.

There were no statistically significant differences in the overall number of adverse effects between oxcarbazepine and carbamazepine. Adverse effects included headache, dizziness/vertigo, allergic rash, fatigue/drowsiness, and nausia and vomiting. Patients receiving carbamazepine had fewer episodes of nausia/vomiting as compared with patients receiving oxcarbazepine. This could be an important difference as regards side effects.

The authors conclude that little evidence exists to suggest that either drug is better than the other. They seem equally well tolerated and equally well able to control complex partial seizures. The only real difference related to the side effect of nausia/vomiting. The authors state that the confidence levels in the data were large, so statistically, some differences could have been missed. Further studies of these two AEDs as candidates for monotherapy for complex partial epilepsy should be conducted.

In another similar review of two important AEDs, phenytoin and valproate, comparisons were made as regards monotherapy for complex partial epilepsy (Tudur Smith et al. 2009). The authors note that phenytoin is generally thought of as being more effective than valproate in the treatment of complex partial seizures. Phenytoin is also usually the first AED tried due to its low cost. The data analyzed were obtained from 669 patients in five separate trials (Craig and Tallis 1994; de Silva et al. 1996; Heller et al. 1995; Ramsay et al. 1992; Turnbull et al. 1985a, b).

This comparison was for complex partial epilepsy patients and for those with generalized tonic clonic seizures. Results are presented here for patients diagnosed with complex partial epilepsy. Overall results show that in terms of complex partial epilepsy patients, the time to withdrawal of allocated treatment suggested an advantage for valproate. The time to achieve a 12-month remission (244 patients) showed no clear advantage for one or the other AED. In terms of time to achieve a 6-month remission (244 patients), again there was no clear advantage between the two AEDs. As regards time to first seizure postrandomization, patients with complex partial seizures and who were taking phenytoin had an advantage.

The authors note that they were unable to detect a statistical difference between phenytoin and valproate as regards time to withdrawal of treatment. The outcome is balanced between efficacy and safety. These two criteria tend to balance the data. Once again, confidence levels are large. In spite of the clinical impression that valproate is more effective than phenytoin, this was not borne out by the data. The authors certainly found evidence for the clinical misclassification of seizures (misdiagnosis), which can further confound data analysis. The authors conclude saying that there is insufficient data to alter the practice of using valproate for patients with generalized onset tonic-clonic seizures, and phenytoin for those with complex partial epilepsy. Future studies should take into account the "smoothing" of data from highly heterogeneous populations of patients. The importance of well-defined seizure classification and diagnosis and elimination of misdiagnosis of seizures is essential to meaningful data interpretation and treatment. One problem is of course that not all epilepsy patients can be easily classified, and how to treat such cases, is not straightforward.

Another recent paper reports a study looking at the efficacy of other add-on AEDs (Sun et al. 2009). In this study, patients with complex partial seizures who had experienced a lack of response to the attempted monotherapy with carbamazepin were randomized into two groups: one group receiving valproate as an add-on AED and the other receiving primidone. The rational for this study was that when the initial monotherapy fails, frequently other monotherapy drugs are tried instead of proven add-on drugs.

In this study, carbamazepine-unresponsive patients were given, in addition, the add-on drug valproate or primidone and the combination efficacy determined. This was a 2.5-year study of 136 patients, over half of whom were randomized to each add-on AED group. Not all finished the trial due to various reasons such as adverse effects, etc., and 120 patients completed the trial.

Results showed that both add-on AEDs acted to lower seizure frequencies as compared to carbamazepine therapy alone. The valproate add-on group had better results than the primidone gtoup. In the valproate group, 35 patients experienced a greater than 50% reduction in seizures compared with 23 patients in the primidone group. These data are similar to that of other investigators (Dean and Penry 1988).

The authors comment that there were no significant differences between the two add-on AEDs as regards tolerability. The reasons for the difference between add-on groups is not clear as each has a similar mode of action on the GABAergic system, and a blockage of sodium channels. If cost is an issue, phenobarbital may be a low-cost alternative to valproate or primidone. Otherwise, valproate is the recommended add-on AED of choice based on this study.

A possible genetic feature of complex partial seizures has been described in epileptic patients (Li et al. 2010). In this study, the glutamate receptor 5 was investigated given its cerebral role as an excitatory modulator of synaptic transmission. Resected temporal lobe and hippocampal tissue were examined from patients undergoing surgery for intractable complex partial epilepsy. Results showed that glutamate receptor 5 was up regulated in the hippocampus, but not the temporal lobe in complex partial epileptics as compared with control subjects.

Patients also showed a hippocampal mossy fiber sprouting which seemed correlated with glutamate receptor 5.

Of interest was that a macaque model of the epilepsy-inducing agent coriaria lactone did not show an up regulation of hippocampal glutamate receptor 5. These data indicate that glutamate receptor 5 is associated with hippocampal mossy fiber sprouting.

Another recent paper discusses the neurophysiological features of complex partial seizures (Cavazos and Spitz 2009). These authors note that the focal inter-ictal epileptiform spike wave is almost pathognomic of complex partial seizures. These are associated with a paroxysmal depolarization shift. Calcium-dependent potassium channels mediate the post-hyperpolarization phase. Scalp EEG recordings are sufficient if the underlying discharge is about 6 square cm or more. Various mechanisms may be responsible, singly or in combination, to induce an overt epileptic seizure following inter-ictal spikes. When these mechanisms become permanent, intractable epilepsy will occur.

Mechanisms of complex partial epileptogenesis fall into two categories: those involving a decrease in inhibition, and those involving an increase in excitation. Mechanisms which involve a decrease in inhibition involve the inhibitory neurotransmitter gamma aminobutyric acid (GABA).

One mechanism of altered GABA function relates to the GABA receptors, GABA-A and GABA-B. The GABA-A and GABA-B receptors are linked to chloride channels. Thus, these receptors represent a target of AEDs such as phenybarbitol, diazepam, clonazepam, topiramate, etc. These anticonvulsants act to modulate the calcium channel characteristics. Another mechanism involves a failure in activation of GABA neurons. This can be the result of faulty feed forward or feedback projections by excitatory neurons. The mechanism of feedback for example permits GABAergic cells to exert control in repetitive firing of neurons such as pyramidal cells. Recurrent collaterals from pyramidal neurons stimulate inhibitory GABAergic neurons after the firing of an action potential (Sloviter 1999).

Mechanisms of increased excitation are largely mediated through the release of the excitatory neurotransmitter glutamate. Mossy fiber sprouting may be one mechanism of increased excitation (Sutula et al. 1989). In this scenario, when neurons of the hilar region are lost because of continual seizures, their feedback mechanism is lost. The dentate granule cell degeneration induces sprouting of the neighboring mossy fibers. This results in the formation of excitatory collaterals, thereby increasing the overall excitation of the dentate neuronal output. Animal models of status epilepticus and those kindled to seizures show neuropathological evidence of hippocampal sclerosis.

The authors state that these mechanisms listed above, plus others, may not act alone, but in concert to produce complex partial seizures. These mechanisms probably vary in importance in different patients depending on numerous variables such as age of onset, duration of seizures, etiology, etc. When the net increased excitability becomes a permanent feature, then one expects intractable to AEDs complex partial epilepsy. The authors correctly note that most AEDs were developed in order to stop and block acute seizures. They were not developed with long-term treatment in mind.

The hope is that when causative mechanisms of seizure initiation and long-term excitability are better understood, then design and development of AEDs that are more effective in long-term treatment of seizure disorders will evolve.

Cell loss is certainly a feature of complex partial epilepsies, frequently in the hippocampus. The hippocampal cell loss is usually of pyramidal cells in CA1, CA3, and CA4. The low seizure threshold of the hippocampus may be the reason for the cell loss, as well as the resistance of complex partial epileptic patients having recurrent seizures often AED resistant, along with progressive memory loss as well as poor behavioral and social outcomes (Kudin et al. 2009).

Limbic seizures are associated with hippocampal necrosis of both neurons and glial cells (Bengzon et al. 1977). The similarity of ischemic lesions and those seen in the hippocampus of complex partial seizures have been commented on for many years by many investigators (Meldrum 1993). In addition, mitochondrial structural alterations are seen in electron microscopy which are completely consistent with the idea that energy metabolism is impaired. Further, the changes are seen in inhibitory GABAergic neurons. These changes are frequently seen in, or in the immediate vicinity of the epileptic focus (Kunz 2000).

Further evidence of mitochondrial dysfunction and decreased oxidative phosphorylation in patients with complex partial epilepsy is derived from 18F fluorodeoxyglucose PET images of glucose hypometabolism in the hippocampal CA3 region (Vielhaber et al. 2003). Further, an increase in lactate and succinate were seen with NMR spectroscopy in the hippocampal CA3 region (Vielhaber et al. 2008). Also observed in patients was a decrease in mitochondrial DNA. In addition, patients with complex partial epilepsy showed a decrease in nucleotides associated with the respiratory chain. All of the above data support the concept that, in human complex partial epilepsy, altered energy metabolism is a key feature of epileptogenesis. This holds true for the feature of the hippocampal sclerosis.

As mentioned above, complex febrile seizures (15-min duration, plus more than one episode/24 h) have been associated with adult temporal lobe epilepsy in 40% of patients (Theodore et al. 1999; Cendes et al. 1993). Neuropathological changes include hippocampal sclerosis (Fisher et al. 1998). While these patients described alterations in complex partial epilepsy are certainly indicative of a direct influence of early febrile seizures on subsequent hippocampal seizure activity, this is another area where animal studies have played a critical role in the elucidation of mechanisms of toxicity. These mechanisms can not be controlled as well in human studies as they can in animal models of febrile seizures and resultant temporal lobe epilepsy.

Post-ictal sequelae include the condition post-ictal psychosis. This disorder is said to represent at least 1/4 of all psychoses in epileptic patients (Schmitz and Wolf 1991). Post-ictal psychosis occurs in several forms of epilepsy, but is often a feature of complex partial epilepsy (Gibbs 1951). The post-ictal psychosis frequently follows a cluster of partial complex seizures. The study described here (Falip et al. 2009) was a retrospective one in which 55 patients with complex partial epilepsy were enrolled. Of the 55, 5 had post-ictal psychoses.

Results showed there were no differences between groups as regards auras, seizure semiology, sex, neuropathological lesions, etiologies, age, age of onset, duration, etc. There were statistically significant differences as regards seizure type and

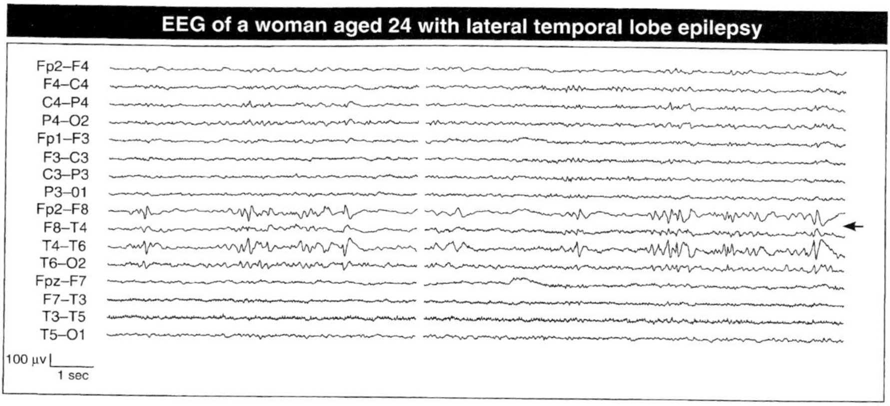

Fig. 15.7 From a 24-year-old patient with lateral temporal lobe epilepsy and marked focal abnormalities around the right anterior midtemporal region. With kind permission from Springer Science+Business media, Epileptic syndromes and their treatment 2010, p. 438, Panayiotopoulos, C. Fig. 15.2

presence or absence of previous status epilepticus episodes. Regarding seizure characteristics, the authors found that there was a higher incidence of post-ical psychoses in patients with dialeptic or automotor seizures which then become overt secondary generalized seizures than were seen in controls. In addition, patients with post-ictal psychosis were more likely ($p=0.019$) to have had some antecedent incidence of status epilepticus. Also, bilateral dysfunction and secondary generalized seizures were associated with post-ictal psychoses. Each of these, especially status epilepticus might be considered as a precursor to structural brain damage.

The authors note that in two of their patients with post-ictal psychosis, seizures only had a 5-year duration. Good predictors of outcome seemed to be a unilateral localization of the seizure, and features suggesting unilateralization of the focus, for example, febrile seizures. The idea that mesial temporal lobe sclerosis is a causative factor in post-ictal psychoses is still uncertain. Patients with post-ictal psychoses do have bilateral cerebral dysfunction, which has been known for a long time (Taylor 1972) (Fig. 15.7).

The authors state that while the underlying pathophysiology of post-ictal psychosis is not determined yet, complex partial seizure patients who have ictal clusters subsequently have significant post-ictal depression. This takes the form of drowsiness and lassitude, and could portend later structural changes leading to post-ictal psychoses. The authors correctly state that their study has certain limitations including the retrospective nature, which would have influenced history taking and the selection of epileptic patients, since other epilepsy types are associated with post-ictal psychosis.

The authors further think that another perhaps more significant limitation of their study was the small number of patients, especially in the post-ictal psychosis group. Further studies correcting these deficiencies are warranted. This is another area where animal studies might suggest further studies on patients with complex partial epilepsy.

Chapter 16
Nocturnal Epilepsy Animal Models

Animal models of nocturnal frontal lobe epilepsy are not as easy to produce as other animal models of human epilepsy. This is in part due to the main feature of these seizures – they occur during sleep, so are not readily observable. They are of course, complex partial seizures, so have many features of those originating from temporal lobe structures. One reason the classification of "temporal lobe epilepsy" has fallen from widespread use is that these seizures can originate from foci adjacent to, but removed from the temporal cortex.

In humans, phenotypic expression of nocturnal frontal lobe seizures include abrupt awakening, frequent loud vocalizations, and meaningless stereotypic motions such as "bicycling" movements of the lower extremities. The seizures are classed as complex partial seizures. The nocturnal seizures can occur many times per night, and have a significant negative effect on proper rest. One longtime problem with nocturnal seizures, animal and humans, is that they occur under circumstances which render observation unlikely. An important finding in this disorder is that genetic mutations have been identified and defined in the human form of some of the nocturnal frontal lobe epilepsies. These are inherited as autosomal dominant, and large families have been studied since 1995.

While these seizures are complex partial seizures, a good animal model should show epileptic activity during sleep in order to be representative. For example in humans, nocturnal seizures do not occur during REM sleep. That needs animal model study. The murine models have been somewhat restricted until recently, when genetic models of nocturnal epilepsy have been developed.

A human mutation of CHRNB2 is the second acetylcholine receptor subunit to be associated with autosomal dominant nocturnal frontal lobe epilepsy. Mutational analysis of nicotinic acetylcholine receptor beta 2 subunit was identified in a cohort of affected people. Mutations in subunits of CHRNA4 or CHRNB of the neuronyl nicotine acetylcholine receptor may lead to nocturnal frontal lobe epilepsy, and provide a basis for studying murine models of partial complex seizures.

A paper with some possible relevance examines a genetic animal model of human neocortex heterotopia which is associated with seizures (Lee et al. 1997). This study

D.W. McCandless, *Epilepsy: Animal and Human Correlations*,
DOI 10.1007/978-1-4614-0361-6_16, © Springer Science+Business Media, LLC 2012

may have some bearing on nocturnal frontal lobe epilepsy because the defect involves the frontal lobe (as well as the temporal lobe).

The proper development of the brain in utero requires many steps which are sequentially timed in order for a fully functional brain to develop. In the case of the cerebral neocortex, neurons migrate, establish themselves in proper sites, and make accurate connections. Some level of cerebral dysgenesis is thought to occur in 1% of all newborns (Meencke and Veith 1992), and it is estimated that 14% of all epilepsy patients have some degree of cortical malformation.

The dearth of animal models for nocturnal epilepsy is also a problem as regards cortical malformations. Obviously, such animal models are conducive to defining sequential events as they happen, not possible in human subjects. Early mutants like the reeler mouse were important in defining altered cerebral cortical development (Caviness and Rakic 1978). The present paper describes a mutant rat termed "tish" due to the mutation which shows as a tencephalic internal structure heterotopia.

The cerebral cortex of the mutant shows that under the neocortex are large areas of heterotopic gray matter. This area is separated by a thin layer of white matter from the overlying cortex. This bilateral layer extends from the frontal lobe to the occipital lobe. The heterotopic region is obvious in the frontal and parietal lobes, but may be reduced or absent under the temporal lobes. Proton density MRI images of the animal's brain showed a bilateral iso density below the neocortex. Breeding studies demonstrated an autosomal recessive pattern of inheritance, consistent with a mutation in a single gene.

The histologic appearance of the overlying cortex is laminar, whereas the heterotopia lacks lamination. There is disarray in the heterotopic region of neurons, some of which were inverted with apical dendrites oriented 180° from normal. Unaffected areas such as the cerebellar cortex and the hippocampus are structurally normal. Results from anatomical studies suggest the heterotopic region of the tish brain is not fully established. Tract-tracing studies demonstrate that the heterotopic neurons have restricted afferents to sub-cortical structures in spite of improper orientation and lack of proper laminae.

Spontaneous seizures resulting from this cortical dysgenesis have been observed in many animals. MRI and/or postmortem studies show those seizures have the tish phenotype. Spontaneous seizure activity has been followed for up to 4 months. The frequency is 2.5 episodes per day. Seizure duration is about 45 s.

The authors comment that the structural result of the mutation is most significantly located in the fronto-centro-parietal areas, and less obvious in the medial temporal cortices. They state that the tish mutant represents a unique animal model for human epilepsies associated with cortical dysplasia. It is also a good model for seizure emanating from the frontal lobe. Whether these seizures are in part nocturnal is not clear. EEG changes are characterized as having increased frequency and amplitude. The tish region has an unknown function, but may represent a second/parallel cortex, duplicating in part the role of the normotopic cortex. Xenopus oocytes have been used in order to study mechanisms of action of the two CHRNA4 nicotinic mutations and nocturnal frontal lobe epilepsy (Figl et al. 1998). The purpose of the study was to examine possible mechanisms responsible for the seizures associated with nocturnal seizures.

The study involved the construction of two rat homologues (S252F and +L264) of the human mutations seen in the disorder, and then co-expressed with rat beta 2 subunits in Xenopus oocytes. Following this, the expressed receptor properties were studied.

Results showed that the S252F and +L264 mutations had at least three effects on the acetylcholine response. First, the two mutations potentiate the response to a train of brief acetylcholine pulses. They also delay the response to a low nanomolar concentration of acetylcholine. Third, the two mutations reduce calcium-induced increases in the acetylcholine response. Both mutations had the above three results in common. In addition, the +L264 mutation increases the voltage jump relaxation time constants. The S252F mutation reduces voltage jump relaxation time constants. It also reduces maximum acetylcholine response, raises the level of steady-state desensitization, and decreases single channel conductance. Neither mutation alters the quantity of surface receptors.

The authors comment that there are two possible mechanisms by which the autosomal dominant nocturnal frontal lobe seizures could be produced. The first possibility would be that the mutation of this disorder acts to reduce cortical nicotinic-mediated inhibitory neurotransmitter release. This would be accomplished by decreasing calcium permeability of nicotinic receptors. A second mechanism capable of generating seizures could be that use-dependent potentiation of the mutant response could produce an increase in nicotinic-mediated neurotransmitter release. Use-dependent potentiation of the mutant response might act to increase both inhibitory and excitatory neurotransmission release. Both effects could contribute to cortical synchronous neuron electrical activity. A sudden increase in the firing rate of cholinergic cortical projections might initiate seizure activity. This fits with the observation that autosomal dominant nocturnal frontal lobe epileptic patients usually have no other chronic neurological deficits – just nocturnal seizures.

An interesting paper appeared describing nocturnal frontal lobe seizures in Shetland sheepdogs (Morita et al. 2002). This paper was based on several generations of sheepdogs and a total of 28 puppies. Six dogs were produced for this study and available from birth for clinical examination and EEG.

For EEG examination, a standard 10–20 surface grid was used, recording from stainless steel needles. Cerebral spinal fluid (CSF) was collected 1 week after an observed seizure. CSF was analyzed for aspartate, glutamate, asparagines, glutamine, arginine, taurine, and GABA. Complete autopsies were done on seven dogs. Routine histology and immunohistochemistry were performed.

Results showed that many of the dogs from this single breeder had seizures, with an age of onset between 1.0 and 1.5 years. The average seizure lasted from 1 to 2 min, and the frequency of observed seizures was from one per week, to one per 6 months. When seizures occurred, they were mostly at night. During occasional daytime seizures, the dogs were resting or napping. They were not watched continuously. The seizures consisted of lateral recumbency, tonic–clonic activity, repetitive jaw movements, salivation, and incontinency. There was a loss of consciousness and postictal depression of up to 30 min.

EEG recordings showed paroxysmal discharges localized to the frontal lobe. The recordings consisted of sharp waves ranging from 50 to 150 μV. By 10 months, the spike and sharp waves were recorded from all leads. The discharges from frontal leads were characteristically more pronounced, those from parietal and temporal lobes were less severe.

Neurohistologically, the frontal lobes of all cases showed pathologic changes. These consisted of perineuronal basophilia, swelling of astrocyte nuclei, and mild vascular endothelial layer swelling. There was an acute nerve cell damage consisting of cell body shrinkage, homogeneous cytoplasm, and atrophic nuclei. The damaged neurons were interspersed among normal ones. The changes were most pronounced in the frontal cortex, but were also seen to a lesser extent in the parietal and occipital lobes. The authors comment that this represents the first time the 10–20 electrode placement paradigm has been used on puppies. The dogs had an initial focus in the frontal lobe, then later extended to other cortical regions. While the seizures were initiated in the frontal lobes, their later generalization to secondary seizures has been seen before in baboons (Silva-Barrat et al. 1986) and in mice (Ishida et al. 1993). The finding of increased CSF levels of glutamate may indicate a possible role of increased excitatory neurotransmitters in these dogs' brains. The authors suggest further studies are needed in the dog model of autosomal inherited nocturnal frontal lobe epilepsy in order to elucidate the underlying neurochemical mechanisms, the exact relation to the human syndrome, and the precise mode of inheritance.

In another paper (Dobelis et al. 2003), the pharmacology of nicotine receptor-mediated seizures was studied in C3H mice in part because of possible involvement in autosomal dominant nocturnal frontal lobe epilepsy. In this study, 11 nicotinic agonists and six antagonists were given to the C3H mice. The goal was to see which nicotinic cholinergic receptors modulate brain excitability.

Results showed that all nicotinic agonists induced seizures in a minimum of 90% of the mice, except cytosine. Epibatidine was the most potent, whereas acetylcholine was the least potent. The ability of nicotinic agonists to produce seizure activity says that nicotinic receptors are proconvulsant. All six antagonists were effective in producing seizure activity. The most effective was d-tubocurarine, and mecamylamine was the least effective of the antagonists. The ability of nicotine antagonists to elicit seizures suggests that receptor activation is not necessary for seizure induction, and that decreased receptor function is proconvulsant.

The authors note that alpha 4 beta 2 and alpha 7 receptors are involved in the generation of nicotine-induced seizures. Two pathways are presented which involve activation of alpha 7 receptors, another involves the inactivation of alpha 4 beta 2 receptors. Both are thought to work through the GABAergic system in the hippocampus. The authors state that both models are consistent with the already described mechanisms of seizure activity.

Mutated mouse lines have been developed which express hypersensitive alpha 4 receptors (Fonck et al. 2003). These mutant knock in mice have a mutation at the Leu9′A position in the M2 region of the nicotinic acetylcholine receptors (nAChRs) subunit. This renders the alpha 4 receptors sensitive to low agonist concentrations. It was already known that autosomal dominant nocturnal frontal lobe epilepsy mutations lead to altered ACh sensitivity. The present study examines various aspects of the alpha4 L9″A mutation when expressed in oocytes. The alpha 4 beta 2 receptor was also studied in forebrain regions of L9″A and wild type mice.

Results showed that when expressed in Xenopus oocytes, alpha 4 (L9″A) beta 2 nAChRs were about 30 times more sensitive to ACh and nicotine as compared to controls. The alpha 4 beta 2 receptors were less sensitive than those which contained alpha 4 L9″s. The L9″s mutation served to increase nicotine-induced desensitization in Xenopus oocytes. When thalamic cell culture from L9″A embryos were identified, they showed an increased sensitivity to nicotine as indicated by fura-2 fluorescence ratios. Neurons displayed this increased sensitivity.

The functional expression of mutant nAChRs was measured in the forebrains of adult L9″A mice by measuring quantitating nicotine-stimulated 86Rb efflux from synaptosomes. A very brief (5 s) exposure of synaptosomes to nicotine caused a concentration-dependent efflux of 86Rb in both cortical and thalamic preparations from wild type mice. The results above were more pronounced in the thalamus preparation than from those of the cortex.

All heterozygous/homozygous mice injected with nicotine had seizures, and a much higher dose was epileptogenic in wild type mice. Pretreatment with a small dose of nicotine acted to protect against seizures previously produced by a modest nicotine dose. The AED carbamazepine administered 30 min prior to nicotine had a protective effect. The nicotine antagonists DHbetaE and hexamethonium did not suppress or delay the nicotine-induced seizures. The nicotine antagonist mecamylamine was however capable of suppressing nicotine-induced seizures in all mice tested.

EEG results showed spike wave discharges at a frequency of 1–5 Hz following nicotine injection in wild type mice. Heterozygous/homozygous mice seized, but had normal EEGs. EEG recordings were from electrodes placed over the parietal cortex above the hippocampus. When a mechanotransducer was recording from the cage bottom, the transducer clearly showed large (muscular) movements during seizures.

Results from still and video recordings characterized the phenotype of the seizures. They were somewhat varied, but showed loss of righting response, rearing, jumping, clonic limb movements, tail arching, etc. Results between wild type (10 mg/kg nicotine) and heterozygous/homozygous (2 mg/kg nicotine) were similar. Homozygous mice given 10 mg/kg of nicotine showed a dual result – first forelimb clonus then followed by a second wild type seizure with multiple movements and a spike wave EEG.

Results from studies of activity during sleep show that the homozygous/heterozygous mice had more "brief awakenings" than the wild type mice. The awakenings were of a few seconds duration and the eyes were not opened. The differences were more pronounced during non-REM sleep. This correlates with human results showing nearly all of nocturnal frontal lobe seizures occurring during non-REM sleep (Zucconi and Ferini-Strambi 2000).

The authors note that their studies do not elucidate the pathophysiology of this seizure disorder. The characteristics of a polygenic seizure syndrome are even more difficult to sort out than a monogenic disorder. They state that their results do emphasize the usefulness of iterative experiments looking at biophysical properties of receptors, and phenotypes of corresponding knock in mice. This approach permits the targeting of a precise set of receptors, spot lighting unique behavioral responses. Such an approach provides data which hopefully has a direct application of this animal model to autosomal dominant nocturnal frontal lobe epilepsy. A study (Klaassen et al. 2006) examined two mutations (CHRNA4s252f and CHRNA4 = L264) in mice in order to explore underlying mechanisms of the mouse models of autosomal dominant nocturnal frontal lobe epilepsy. Because of the similarities in phenotypes between human autosomal dominant nocturnal frontal lobe seizures and mouse mutants, the suggestion has been made that these mutations underlie the human disorder. This paper reports results from genetically engineered mouse mutants developed by the authors, and compare them to human counterparts.

The mutant mice CHRNA4s252f and CHRNA4+L264 generated in this study show persistent abnormal cortical EEGs, characterized by an increase in delta wave activity (0.5–4.0 Hz). There was also an increase in theta wave activity (4.0–8.0 Hz). These mutant mice show a tendency for recurrent spontaneous seizure activity. These seizures correspond to high amplitude, low frequency cortical EEG patterns. The mutant mice show an increased sensitivity to the pro-convulsant action of nicotine. Repetitive EEG seizure discharges show paroxysmal onset, sudden termination, and asymmetric rhythmic patterns in both of the CHRNA4 mutants.

Further electrophysiologic study showed that the effect of nicotine on IPSC frequency was significantly different between wild type and mutant mice. In the wild type mice, nicotine did not alter sIPSCa frequency of kinetics, while in CHRNA4s252f mutants the frequency and amplitude of sIPSCa were increased. The authors then calculated the mean inhibitory current (I_{mean}) and measured its value before and after nicotine perfusion. The authors state that they believe the increase observed in I_{mean} results from an enhanced nicotine-evoked GABA release from interneurons synapsing on cortical pyramidal cells. To further delineate mechanisms, the effect of nicotine on miniature IPSCs was examined. Nicotine increased the amplitude and frequency of mIPSCs in CHRNA4s252f mutants but not wild types.

Data indicate that a significant fraction of the autosomal dominant nocturnal frontal lobe epilepsy mutant alpha 4 subunit which contains nAChRs are on the presynaptic terminal of GABAergic interneurons. This serves to elevate presynaptic calcium levels.

The authors point out that this study describes the development and characterization of two murine models for autosomal dominant nocturnal dominant nocturnal frontal lobe epilepsy. Data are presented documenting the similarities between these mutants and the human disorder. The observation is made that in the human version, REM sleep protects from nocturnal frontal lobe seizures. The corresponding situation in the mutant mice needs further study. The data suggest a model which is consistent with a model of epileptogenesis in which ACh significantly enhances GABAergic neurotransmission. This is thought to involve cortical pyramidal layers II and III pyramidal neurons.

These data together suggest that CHRNA4s252f and CHRNA4+L264 are dominant, cause abnormal cortical EEGs, cause interictal spikes, and repeating seizures. Noncortical contributions (hippocampal) await elucidation. These mouse models certainly seem to be ideal ones for the study of the autosomal dominant nocturnal frontal lobe epilepsy in humans. Such ideal models tend to facilitate translation of animal data to the human counterpart. This mouse model is splendid for such translational research.

Another paper reports results of a mouse model for autosomal dominant nocturnal frontal lobe epilepsy (Teper et al. 2007). The mutation was the alpha 4 nicotinic receptor S248F knock in strain. This is a strain in which a relatively low nicotine dose (1–2 mg/kg) will produce a behavior complex consisting of stereotypic head movements, body jerks, forelimb dystonia, and tail arching. This sometimes carries the same dystonic arousal complex (DAC). As is the case in the human counterpart, carbamazepine can partially block DAC in these mice, as can very low doses of nicotine.

Studies of 22 of the S248F knock in mice failed to show any epileptiform spike and wave activity. Spontaneous behavioral seizures were also not seen in the mutants. Initially, 13 heterozygous/homozygous mutants were compared to wild type controls as regards video monitoring during both daytime and night. The mutant mice showed no behavioral or EEG abnormalities suggestive of spontaneous seizures. In an additional attempt to find EEG abnormalities, epidural EEGs with video monitoring were utilized. Eleven mice were tested for 10 h during daylight. EEG spike and slow waves were never seen. Additionally, spontaneous motor seizures were never seen during daytime handling of over 160 mutants over a 2-year period. With nicotine injection, the mutants display the stereotypic behavior detailed above.

The heterozygous mutants seemed to be protected from nicotine-induced seizure activity. This was shown by the observation that the wild type mice actually had a more severe nicotine seizure response than did the heterozygotes. The finding that a very small dose of nicotine acts to protect from subsequent otherwise DAC producing doses of nicotine probably reflects the suggestion that chronic low dose nicotine results in receptor desensitization.

The authors state that the mutation they generated shows a high penetrant and strong phenotype which is not seen in other mutant models of human autosomal dominant nocturnal frontal lobe epilepsy. In human seizures, there is a sudden elevation of the head and neck, followed by dystonic posturing of the upper extremities (Oldani et al. 1998; Bertrand et al. 2005).

The authors state that the stereotypic upper extremity dystonia in the mouse mutants represent the most convincing parallel to human autosomal dominant nocturnal frontal lobe epilepsy. The attenuation of DAC by carbamazepine in the mouse model further aligns the model with the human counterpart, as carbamazepine is efficacious in the human correlate. Also, the authors note low doses of nicotine act to suppress or block seizures in the mouse model and also in humans with this disorder (Brodtkorb and Picard 2006). This phenomenon would appear to be a "reverse" kindling-like effect.

The issue of spontaneous seizures is interesting because in early reports of the mouse mutant L9′, spontaneous seizures were not reported (Fonck et al. 2003). In more recent papers, spontaneous seizures were reported in this mutant (Fonck et al. 2006). The authors did exhaustive studies looking for spontaneous seizures/EEG abnormalities in the S248F mutants with no success.

They further state that there are only two reasons to have not seen spontaneous signs/symptoms, plus EEG correlates. First is the chance that the methodology lacked the sensitivity to detect very subtle changes. The second possibility is that the DAC is not actually a seizure, but rather a paroxysmal movement disorder. Several previous studies have shown simultaneous epilepsy and movement disorders in the same patient (Guerrini et al. 2002; Du et al. 2005).

The possible effect of modifier genes and or genetic background could explain the phenotypic difference between various mutant models of autosomal dominant nocturnal frontal lobe epilepsy. These types of "extraneous" modifiers have been reported in animal models of myoclonic epilepsy (Yu et al. 2006).

Examining these nicotine responses from a different angle are reports of studies looking at expression changes in mouse brains after nicotine-induced seizures (Kedmi and Orr-Urtreger 2007). The aim was to study brain transcriptional response to seizures and identify genes in which there was a measurable alteration following nicotine-induced seizures.

Results found that no less than 62 genes in whole brain were altered when nicotine-induced seizure mice were compared to controls. These gene changes are each detailed in a two page table in the manuscript. Classes of genes altered include among others: transcription regulation, protein binding, GTP binding, nucleic acid binding, ATP binding, cell cycle, etc. Brain expression in each specific genotype group in nicotine-induced seizure mice compared to untreated mice revealed 559 differentially expressed transcripts. Data also showed that the differentially expressed genes share common transcriptional regulatory responses in their promoter sequences. The 62 differentially expressed genes participate in pathways which participate in regulatory and cellular processes.

One unique gene which was upregulated following nicotine-induced seizures was Per1 (period homolog 1). This is a gene for a transcription factor, part of the period gene family. This gene is expressed in a circadian fashion in the supra chiasmatic nucleus. This nucleus is the major circadian rhythm pacemaker in mammalian brain (Reppert and Weaver 2002). The finding that nocturnal frontal lobe seizures rarely occur during daytime suggests this gene (per1) could be affected in the human autosomal dominant nocturnal seizure disorder. Knowledge of the genetic/molecular mechanisms of seizures in models of human epilepsy might be beneficial in suggesting treatments.

Other changes of note are those of Dusp 1 and Dusp 6. These phosphatase-regulating genes inhibit MAP kinase phosphatase activity and have been shown to be involved in other seizure types such as electroshock (Kodama et al. 2005) and kainic acid-induced seizures (Boschert et al. 1998). This paper provides a plethora of new data which pertains to several seizure types, including nocturnal frontal lobe epilepsy. The delineation of genetic and molecular factors, and their correlations as animal models for human epilepsies serves to provide a basis for translational research aimed at defining new treatment paradigms. This paper promotes these future activities.

The mutation of the genes encoding alpha 4, alpha 2, or beta 2 (CHRNA4, CHRNA2, and CHRNA2) of nAChr are associated with autosomal dominant nocturnal frontal lobe epilepsy in human patients. These mutations in humans produce one of three characteristic phenotypes: paroxysmal arousal, episodic wandering, and paroxysmal dystonia, plus other less pronounced semiology. Most but not all nocturnal frontal lobe seizures occur at night. This led to the development of a mutant rat model which has a S284L missense mutation which had been previously identified in CHRNA4 patients with nocturnal frontal lobe epilepsy (Zhu et al. 2008).

The transgenic rats showed no evidence of defective CNS formation sometimes seen in nocturnal epilepsy, nor did they exhibit any behavioral abnormalities. mRNA levels in the mutant rats were not different as compared to wild type controls, and expression in brain was also similar to the wild type rats. Conversely, the mutant rats did show seizure phenotypes which were similar to those seen in human autosomal dominant nocturnal frontal lobe epilepsy. These symptoms were present during EEG recordings which showed slow wave sleep as noted in human patients. The phenotypes of seizures seen in human patients were present in these mutant rats. AED treatment attenuated the mutant rat seizure.

The authors note that the rats showed a decrease in extra synaptic GABA and an increase in glutamate release. The authors also note correctly that previous models of autosomal dominant nocturnal frontal lobe epilepsy have been in mice. The availability of a rat model facilitates certain experimental studies on models of this disorder in which a larger animal and brain are beneficial.

Animal models such as the rat model described here plus mouse mutant models of nocturnal frontal lobe epilepsy play an important role in seizure research. The careful correlation of genetic mutation, semiology, phenotypic characteristics, and electrochemical, neurochemical features, etc. of the model with the human counterpart aids in interpretation of model data. This in turn benefits the translation of results to humans, in this case with autosomal dominant nocturnal frontal lobe epilepsy.

The main problem with this type seizure and animal models thereof is that by definition the seizures occur almost exclusively at night during sleep. This has served to highly complicate diagnosis and made difficult definition and characterization of signs/symptoms. This has in turn significantly limited development of animal models of this disorder, and has slowed development of treatment paradigms. Just consider that there is still considerable controversy over the diagnosis of nocturnal frontal lobe seizures versus parasomnias. In spite of this, important data have been gleaned from nocturnal animal studies which has translated to human studies.

Chapter 17
Human Nocturnal Epilepsy

Nocturnal frontal lobe epilepsy is a seizure disorder sometimes difficult to diagnose because the seizures occur predominantly or exclusively during night time sleep. They are often misdiagnosed as parasomnia. Parasomnias are a group of abnormal, undesirable motor, verbal, and/or experiential phenomenon. These include night terrors, sleep walking, teeth grinding, rhythmic movements, etc. Parasomnias will be examined later in this chapter because it is essential to distinguish between these sleep (psychiatric) disorders and highly treatable nocturnal frontal lobe epilepsy.

While there are certain aspects of nocturnal frontal lobe seizures which are quite similar, both clinical and electroencephalographic features can be highly variable. While nocturnal seizures are usually less than 5 min in duration, longer episodes and auras are not common. The seizures are frequently tonic–clonic, but may be tonic or clonic. Tonic contraversive movements of the arm or head are indicative of the dorsal lateral frontal cortex anterior to the motor cortex.

In addition, ictal head version may occur, resulting in complex movements due to the dual origin of the sternocleidomastoideus muscle (Jayakar et al. 1992). Young children are likely to have secondary generalization. Frequently, patients with nocturnal seizures exhibit repetitive stereotyped motor patterns such as bicycling or running movements. Electrical activity of anterior/orbital frontal cortex is linked to automatisms (Riggio and Harner 1995). Changes in trunk posturing are also features seen in nocturnal frontal lobe epilepsy. Because of the very nature of nocturnal seizures, visualization of the phenotypic features is less common.

A relatively "new" version of nocturnal frontal lobe epilepsy was recently defined by Scheffer et al. (1994, 1995). This form of nocturnal epilepsy was originally misdiagnosed as a sleep disorder, and was subsequently defined in an Australian kindred of English and French-Canadian descent. It is defined as an independent distinctive clinical disorder. Other examples have been described (Magnusson et al. 1996; Oldani et al. 1998). This form is rare, but the numbers will surely increase as it is looked for.

Several families have been described and documented with this disorder, and having a CHRNA4 gene mutation, making it a well-defined clinical entity. These cases are frequently overlooked due to the easy dismissal of symptoms as being the

D.W. McCandless, *Epilepsy: Animal and Human Correlations*,
DOI 10.1007/978-1-4614-0361-6_17,

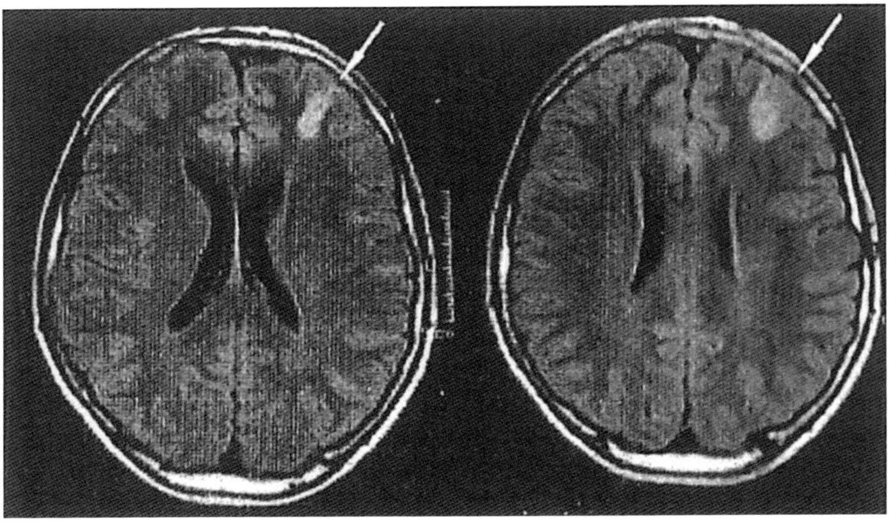

Fig. 17.1 MRI FLAIR: left anterior hypersignal (cortical dysplasia). Nocturnal frontal lobe epilepsy. Published in Crespel, A., et al. Atlas of Electroencephalography Volume 2: The Epilepsies, EEG and Epileptic Syndromes. With permission of John Libbey Eurotext

usual childhood nightmares, or restless sleep. Onset of nocturnal epilepsy is in childhood, or in adolescence, and features include clusters of brief nocturnal motor seizures with dystonic or hyperkinetic characteristics. These persist into adult life, unless treated with AEDs. In some cases, the gene defect is linked to chromosome 20g, but not in all cases.

The frequency of nocturnal frontal lobe epilepsy is not exactly known, but about 30% of all patients evaluated for surgery each year with intractable epilepsy are nocturnal frontal lobe epileptics. The seizure type is usually tonic–clonic, but can also be partial complex. Neither sex nor age seems to have significance, although children seem to be initially affected. Most cases of nocturnal frontal lobe epilepsy may be symptomatic, but many patients have normal MRIs, while others do not (see Fig. 17.1).

In the 1995 full length Scheffer paper (Scheffer et al. 1995), the autosomal dominant nocturnal frontal lobe epileptic families were described in some detail. This consisted of five families from three countries: Australia, England, and Canada. There were 47 patients in these five families who were affected. The largest family group contained 25 patients who were spread over six generations. These patients experienced short clusters of nocturnal motor seizures. They had significant manifestations of hyperkinetic or tonic features. There may or may not have been an aura, and consciousness was maintained in most cases.

The clusters occurred an average of eight episodes per night, usually upon the patient's awakening. The age of onset in the majority of patients was during childhood, and lasted into adult life. As is frequently the case with nocturnal seizures, the disorder was misdiagnosed as parasomnias, and psychiatric/medical syndromes. Interictal EEG studies were unremarkable, as were neuroimaging studies; however, video EEG suggested that the seizures were of a partial nature, having frontal lobe semiology. AED treatment with carbamazepine reduced seizure frequency.

The authors state that these family patients represent an autosomal dominant inheritance pattern. This is nocturnal frontal lobe epilepsy, but a clearly distinct variation. The correct diagnosis is critical because of treatment considerations and because the patients need genetic counseling. The diagnosis is important for the patients who heretofore were considered to have psychiatric problems.

Many of the authors of the above paper participated in a later study examining the genetic heterogeneity, and demonstrating a second locus at 15q24 in cases of nocturnal epilepsy (Phillips et al. 1998). The gene for the Australian family referred to above was mapped to 20q13.2 (Phillips et al. 1995). The Norwegian family had a different mutation locus. The current study shows that the above two mutations are relatively uncommon in patients with nocturnal frontal lobe epilepsy. Seven additional families were studied.

Results showed that none of the seven families had either of the two known mutations in the previous families. These seven families were excluded from the CHRNA4 region, but one family showed localization to chromosome 15. However, some of the seven families do not have a linkage to either CHRNA4, or to the proximity of CHRNA3–CHRNA5 of chromosome 15q24. This region is homologous to the mouse region containing the El-1 mouse locus. The El-1 mouse has spontaneous complex partial seizures.

The authors state that there are at least three loci for autosomal dominant nocturnal frontal lobe epilepsy. The hypothesis that the mode of action is to impair presynaptic nicotinic cholinergic transmission is plausible. There appears to be significant molecular heterogeneity; however, the clinical phenotype is quite homogeneous.

Another paper examining the electro-clinical aspects of autosomal dominant nocturnal frontal lobe epilepsy in a Norwegian family has been published (Nakken et al. 1999). The family had a mutation in the acetylcholine receptor, and ten family members were included in the study. The results were compared to an Australian family with a different mutation at the same locus. In the Norwegian family, six out of ten were interviewed, and interictal and ictal EEGs were performed in eight and three patients respectively. CT/MRI was performed in five patients, and blood taken from seven for genetic analysis.

The results showed the mean age of onset was eight, range 1–11 years. Most had good seizure control with AEDs. One patient has weekly nocturnal seizures in spite of taking three AEDs. While most seizures were at night, some occurred during naps. The seizures lasted about 15–45 s, and some patients had clusters. One patient had 30–40 seizures/night. Some said stress or sleep deprivation or increased frequency of the seizures. A common pattern was for patients to have seven seizures per night for a few days followed by some weeks of seizure freedom, then a repeat of the cycle.

The phenotype consisted of motor and verbal features. Some patients had arm tonic stiffening, others had clenched fists or running/cycling leg movements. Most had

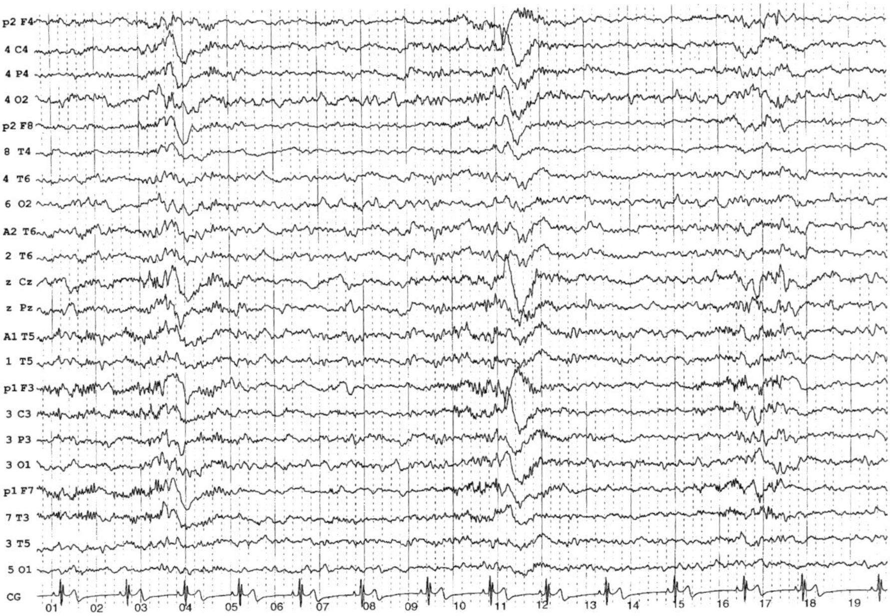

Fig. 17.2 nREM sleep – fast activity typical of dysplasias (cortical). Frontal lobe epilepsy. Published in Crespel, A., et al. Atlas of Electroencephalography Volume 2: The Epilepsies, EEG and Epileptic Syndromes, 2006. With permission of John Libbey Eurotext

verbal outcries consisting of shouting, moaning, etc. Interictal EEGs were normal in seven-eighths of the patients. In video EEGs, the results were complicated by the motor artifacts. The EEGs showed shallow arousal in one patient, and low-voltage epileptiform discharges in the left frontal cortex in two patients. The results were consistent with the seizure semiology of frontal lobe epilepsy (see Fig. 17.2).

The authors comment that clinically the patients had a frontal appearance, but without intracranial EEG data, the seizure focus could not be localized. This results in the common misdiagnosis of parasomnias and/or psychotic disturbances. In fact, one member of the ten patients in this group had been treated for 2 years for a psychiatric disorder.

Comparisons of the current ten patients with the previously described Australian family (776ins3 mutation vs. Ser248Phe mutation) showed many similarities and some differences in phenotypes. The current patients had a clear impairment in consciousness, whereas the Australian group had preserved consciousness. Most Australian patients had auras, the Norwegian group did not. The duration of seizures varied in that four Norwegian patients' seizures were in childhood only. In terms of mechanisms of "toxicity" of the mutation, evidence supports an effect on the nicotinic acetylcholine receptor. Why this affects the frontal lobe preferentially is unclear. These receptors are found throughout the brain. The chromosomal location of the mutation is also not clear, but there may be several locations, each capable of producing nocturnal frontal lobe epilepsy (Sander et al. 1998). These data may shed light on possible new therapeutic AED approaches.

Other families worldwide were demonstrated to have autosomal dominant nocturnal frontal lobe epilepsy (Kobayashi et al. 2000). These cases consisted of members of a Japanese family who were worked up using electroclinical methodology to define the disorder. Mutations were found in the neuronyl nicotinic acetylcholine receptor alpha-4 subunit. Results also showed all affected members had clusters of short tonic seizures occurring during sleep and/or upon awakening. None had auras or were aware they had a seizure. While all had a similar gene mutation, there were variable phenotypes with regard to onset, frequency, and AED response. This represented the first description of autosomal dominant nocturnal frontal lobe epilepsy in a Japanese family.

Yet another example was reported in a Korean family (Cho et al. 2003). This family group involved nine affected patients from three generations. The patients were fully investigated genetically, clinically using imaging methods, and also using EEG.

The results showed a childhood onset of nocturnal episodes which included motor movements, bizarre speech, etc. Neurological examination and MRI were unremarkable. The seizure symptoms were similar in the patients who had video-EEG monitoring. The seizures were brief (20–40 s) and consisted of arm flexion, tonic head extension, and mouth movements. Some had secondary generalized tonic seizures.

Six patients had interictal FDG-PET studies. This consisted of fluorodeoxyglucose-F18-PET imaging. All appeared normal except for one patient with hypometabolism in the left temporal area, which correlated with a previously noted cyst. Using statistical parametric mapping, however, five of six patients showed a decrease in glucose uptake in the superior and middle frontal gyri, left central region (four patients), anterior parietal area (four patients), and right anterior superior frontal gyri (four patients). Genetic results indicated a mutation in CHRNA4, exon 5 at amino acid 252, as shown previously (Hirose et al. 1999; Phillips et al. 2000).

The authors comment that this is the first family to be reported with this syndrome in Korea, and all have the same genetic mutation. All of the Korean patients had mild to moderate mental retardation, similarly noted in some patients in the Japanese study. In most previous studies of autosomal dominant nocturnal frontal lobe epilepsy, carbamazepine was successful. In the Korean study, five of the seven affected members did not respond to 600–900 mg of carbamazepine.

The results of FDG-PET were consistent with the EEG and clinical results. A previous study (Hayman et al. 1997) presented a case with left frontal hypometabolism. Many patients were taking, or had just taken, carbamazepine usually resulting in global, not regional, hypometabolism. The authors conclude saying that the pattern of phenotypic characteristics in autosomal dominant nocturnal frontal lobe epilepsy seem related to the actual mutation site as opposed to the family genetic background (see Figs. 17.3 and 17.4).

The difficulty in the differentiation between nocturnal frontal lobe seizures and pavoe nocturnes (night terrors) has been alluded to before (see above). In a case study and review, this problem has been closely examined (Lombroso 2000). The subject of the differences between the two, and even the existence of nocturnal seizures, has been the subject of much long debate.

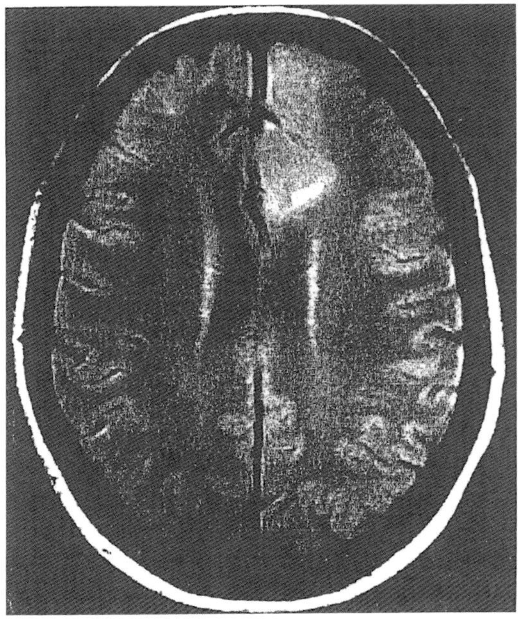

Fig. 17.3 Left mesial frontal lobe hypointense with mild edema. Anoplastic oligodendroglioma. With permission of Elsevier-Kuzniecky, R., and Jackson, G. Magnetic Resonance in Epilepsy 2nd Ed., 2006

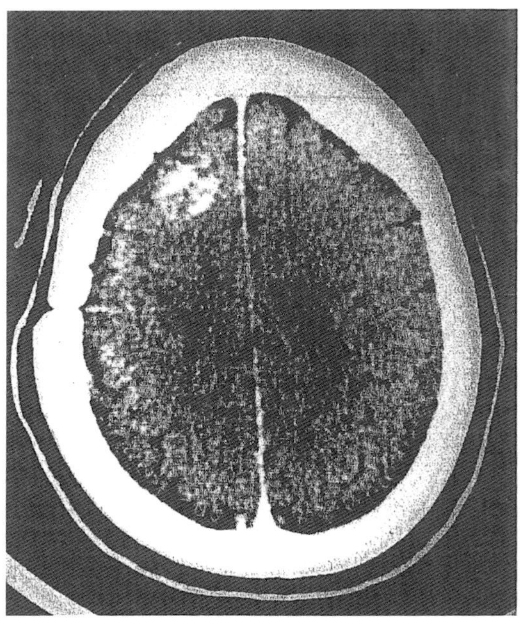

Fig. 17.4 Axial CT frontal lobe scan in a patient with cavernous angioma and intractable epilepsy. Published in Crespel, A., et al. Atlas of electroencephalography Volume 2: The Epilepsies, EEG and Epileptic syndromes. With permission of John Libbey Eurotext

Table 17.1 Characteristics of specific non-REM sleep disorders and seizures

	Seizure	Sleep drunkenness	Sleep terrors	Somnambulism	Somniloquy
Incontinence	+	–	–	–	–
Tongue biting	+	–	–	–	–
Confusion	+	+	+	+	+
Tonic-clonic movements	+	–	–	–	–
Drooling	+	–	–	–	–
Amnesia	+	+	–	+	+
Occur while awake	+	–	–	–	–

Adapted from Brazil, C. 2004 Seminals in Neurol. Vol 24: p. 293

The case was of a 6-year-old boy referred due to nightmares, which caused sleep deprivation, and were stressful for the family. Medical and neurological exams were negative. The seizure onset was at age 4, and family history revealed a brother, aunt, and cousin as having had a few simple febrile seizures. Blood, metabolites, and EEGs were normal. A scalp long-term monitoring was able to record no less than 15 stereotypical episodes during sleep. During the attacks, he sat up, coughed but did not talk. His right arm seemed dystonic. The episodes lasted 1–2 min, after which he went back to sleep. Ictal EEG showed the event started as a rapid arousal pattern followed by diffuse low-voltage 4–5 Hz activity with bilateral frontal lobe activity. An ictal SPECT displayed hyperperfusion in the left frontal parietal lobes.

Invasive long-term monitoring displayed low-voltage spikes originating from the rolandic cortex. Within 5–10 s the electrical phenomenon had spread to the parietal cortex and first temporal lobe gyrus; the frontal/subfrontal cortices were not affected. Surgery served to remove the affected areas, with a 90% control of seizures for as long as 5 years postoperatively. There were no surgical sequelae.

The author comments that this was an unusual case in that the primary focus in the motor cortex did not appear to induce primary motor characteristics, but other emotive-like symptoms. Most sleep terrors are parasomnias of a benign nature. Over the years, some patients are noted to have features distinct, implying a separate syndrome. Ultimately, the existence of a separate seizure disorder, nocturnal frontal lobe epilepsy was defined. The exact number of patients with this form of epilepsy, misdiagnosed as parasomnias, is impossible to say, but must be significant. Table 17.1 lists the main differences.

Complex partial seizures can arise from the frontal lobe, but have features which distinguish them from nocturnal frontal seizures such as longer duration, loss of awareness, automatisms, meaningless vocalization, incontinence, and tendency to occur in daytime. They frequently have an aura, and can generalize into tonic–clonic seizures. Although with some variability, the description of autosomal dominant nocturnal frontal lobe epilepsy differs from nocturnal seizures in that it manifests as vocalizations, prominent bilateral motor activation, and usually with an aura. The EEGs from these patients show diffuse discharges.

The author observes that the current patient awoke from non-REM sleep, as has been described before (Montagna 1992). In these cases, sleep walking (wonderings) also occurred. This group of patients was thought to have atypical complex partial seizures instead of parasomnias. The author says his study does not elucidate the precise pathophysiology of the 6-year-old's seizures. The peculiar cough could have been an autonomic nervous system response, as in incontinence. The arm dystonic movement is unclear as to mechanism, but might have involved basal ganglia nuclei. It is also unknown if the arousal precipitates the seizure, or vice versa. These questions and others await further investigation.

A recent commentary has emphasized a major problem occurring in nocturnal frontal lobe epilepsy, and that is the fact that they occur at night, making observation of the seizures much less likely (Brazil 2004). This especially means the onset of the seizures, no matter how dramatic, is largely missed. Patients rarely have meaningful recollection of events.

The differential diagnosis of sleep disorders and nocturnal seizures depends significantly on a careful history. Seizures and parasomnias have very similar clinical features, in fact, this is a common misdiagnosis. Specific sleep disorders which most closely resemble nocturnal epilepsy include cataplexy, sleep attacks, and REM behavior disorder.

Nocturnal seizures disrupt sleep, and there is almost always at least a brief period of awakening during the onset of the seizure. Normal sleep is not a feature of postictal depression, so treatment of the seizures results in increases in REM sleep (Besset 1982). Even with daytime seizures, REM sleep is disrupted on the subsequent night. Thus, seizures can affect sleep even after the postictal phase has passed. The electrical activity in brain between the sleep states and awakefulness is different, which influences the onset of various seizure status. For example, one study showed a 5× increase in night time onset of nocturnal frontal lobe epilepsy when compared to temporal lobe onset seizures (Crespel et al. 1998). One feature of nocturnal seizures is the paucity of seizure onsets during REM sleep. The mechanism whereby REM sleep protects against seizure onset is unclear. Possibly the hypersynchrony in non-REM sleep facilitates the onset or spread of some partial seizures. This needs further research as this might suggest new treatments.

Nocturnal frontal lobe epilepsy is a sleep-related event, with an absence of clear EEG abnormalities as shown by scalp electrical recordings, both interictal and ictal (Ferri et al. 2004). Because of this, video polysomnography has assumed a key role in the diagnosis of nocturnal frontal lobe epilepsy. Use of this diagnostic technique permits description of three stages of nocturnal epilepsy: paroxysmal awakening, nocturnal paroxysmal dystonis, and episodic nocturonal wanderings (Provini et al. 1999).

The present paper by Ferri presents a case of nocturnal epilepsy in a 30-year-old man. Analysis of history showed no relatives with seizure disorders. The patient had been diagnosed as a child with a sleep disorder. At the time of study, the patient was having 10–15 seizures per night, which caused considerable insomnia, and daytime sleepiness. The patient was otherwise medically and neurologically normal. Video polysomnography was used to evaluate this patient, and results showed at least 15 separate ictal episodes. All occurred during non-REM sleep, and lasted around 20 s.

The stereotypic attacks were characterized by a sudden awakening, lifting of the head, and then sitting upright, followed by a dystonic right leg elevation. The leg was extended, and exhibited a slight tremor. At the end, the patient fell back, emitted a moan, and experienced a period of wakefulness. The patient was started on carbamazepine, which greatly reduced the frequency of the seizures.

The authors speculate that a self inhibiting complex mechanism could be involved with cessation of seizures. This mechanism might be cortical, based in part on the efficacy of carbamazepine. The episodes, which could have been classified as paroxysmal awakenings, were actually nocturnal seizures as shown by neurophysiological increases in electrical activity at about 8 Hz, which was different to the normal alpha rhythm which was present during wakefulness.

A useful technique for diagnosis is phosphorus magnetic resonance spectroscopic imaging. Previous studies with this imaging method have demonstrated that patients with temporal lobe seizures have increased inorganic phosphate, and decreased phosphomonoester levels in the epileptogenic region. The present study extends these findings to patients with frontal lobe epilepsy. Results from eight patients showed interictal alkalosis in the focus as compared to the contralateral area. Seven of eight patients had a decrease in phosphomonoester levels in the epileptic focus as indicated by phosphorus magnetic resonance spectroscopy. The levels of inorganic phosphate, however, were unchanged in the frontal lobe epilepsy patients. These data suggest that magnetic resonance spectroscopy could be a valuable tool to evaluate frontal lobe epileptics with regard to possible surgery.

Another study looked at non-lesioned frontal lobe epilepsy in 21 children, 8 of whom were having nocturnal seizures. The purpose was to examine outcomes in nonlesioned cases of frontal lobe epilepsy.

The results from this retrospective study showed a mean age of onset of 6.6 years. They were followed for a mean time of 9.4 years. Histories were unremarkable, except that 40% had some developmental delay such as verbal and/or motor delays. Seizures were variable, with six children generalizing to tonic–clonic seizures. Many had partial seizures, and eight of these were only nocturnal. Well over half (57%) had daily seizures, five had ten or more seizures per day. AEDs included carbamazepine, valproate, phenytoin, phenobarbital, and clobazam. Complete seizure control was achieved in ten children in a mean period of 14.6 months. The EEG returned to normal in 48% in a mean time of 20 months. Sixteen of 21 patients developed a learning disorder. Six of those with a learning disorder were seizure free.

The major stated result of this retrospective study in non-lesioned frontal lobe epileptic patients was that seizure control does not assure absence of severe comorbidity. The comorbidity may precede the onset of seizures. Since the normal development of the frontal lobes probably continues into adolescence (Hernandez et al. 2003), the impact of seizures may be quite significant for unfavorable outcomes. Thus, the existence of nonlesional frontal lobe epilepsy, while not necessarily associated with a poor seizure outcome, may be associated with a poor behavioral outcome plus learning disabilities.

In another study of nocturnal frontal lobe epilepsy in two cases, intracerebral stereo-EEG ictal recordings were obtained. Six seizures were studied, and sustained activity was seen in the frontal sleep spindle frequency (12 Hz). The duration was compared to preceding "non-seizure" sleep spindles (Picard et al. 2007).

All the nocturnal frontal lobe seizures occurred during non-REM sleep, consistent with other studies cited above. These were immediately preceded by a 12 Hz frontal predominant sleep spindle. The sleep spindle was of long duration, and actually continued after the seizure started. It was much longer than the sleep spindles in the previous "non-seizure" state.

The authors interpret these results as a possible indication of thalamic involvement. The cessation of sleep spindles when arousal starts could be related to a brainstem-thalamic cholinergic effect which disrupts spindle activity. This has been independently demonstrated in vivo (Lee and McCormick 1997). Thus, this paper suggests that nocturnal frontal lobe seizures might have a thalamic defect component involved in the seizure generation. Further studies with much higher numbers of patients seem warranted.

The actual correct diagnosis of nocturnal frontal lobe epilepsy is of great importance for reasons stated above, but still an area of debate as how to render the proper diagnosis. Some important methodologies such as video polysomnography is expensive and not available everywhere (Tinuper et al. 2006). These authors were interested in the possibility that a thorough well-designed history might prove valuable in discerning nocturnal frontal lobe epilepsy from parasomnias.

In this study, three categories of patients were identified: patients with nocturnal epilepsy, patients with parasomnias, and those with other sleep disorders which could be confused with nocturnal epilepsy. The study was a retrospective one in that data from clinical, neuroradiological, and interictal/ictal video polysomnological recordings were reviewed. Epileptologists classified patients as either having nocturnal epilepsy, or not. On the basis of criteria for nocturnal seizures, a questionnaire was developed and used to evaluate patients by a physician unaware of the previously established diagnosis, or the purpose of the study.

Results showed that of 44 nocturnal epilepsy patients vs. 59 non-nocturnal seizure patients, sensitivity was moderate (range 0.2–0.6), and specificity was high (0.9–1.0). The highest diagnostic accuracy was obtained by combining the first criterion (semiology of the seizure event) with one of the following seizure characteristics: vocalization, seizure under 2 min, presence of an aura, personal history, stereotypy of motor symptoms, and nightly recurrence.

They conclude saying that the criteria listed above provide moderate sensitivity and high specificity for being able to accurately diagnose nocturnal frontal lobe epilepsy from a refined history. Further studies on larger numbers of patients are needed. It has also been pointed out, and noted above, that not all nocturnal frontal lobe epilepsies have a seizure onset in the frontal lobes (Nobili et al. 2004). It is difficult to localize the exact focus in patients with frontal lobe seizures due to the frontal lobe's large surface area, and rapid propagation of seizure activity.

Other workers have developed criteria and predictive scales in order to better aid in the diagnosis of nocturnal epilepsy (Derry et al. 2006), and state that a careful, well-designed history is invaluable in the diagnosis of this disorder.

In cases of nocturnal epilepsy, at least 30% are refractory to AEDs. From a surgical group of 522 patients, 21 were selected as having nocturnal frontal lobe seizures (Nobili et al. 2007). The mean age of onset was 6.2 years, and mean age at surgery was 24.7 years. Seizure frequency ranged from 20/month to over 300/month. Almost half reported daytime sleepiness. All patients had auras, and clinical signs included hyperkinetic automatisms, tonic posturing, asymmetric posturing, and mimetric automatisms.

Interictal and ictal EEG was useful in lateralizing or localizing a focus. All patients received a microsurgical resection of one frontal lobe based on the presurgical work up. Histological examination of resected material showed a Talor type cortical dysplasia in 16 patients, and 4 had an architectural focal cortical dysplasia.

At a 12-month follow-up, all 16 with Taylor type cortical dysplasia were in Engel's class 1a, and the rest were in Engel's class II or III. By 6 months, the six with excessive daytime sleepiness no longer had this complaint. The authors conclude that surgery for this type of epilepsy was successful in providing positive results in both nocturnal frontal lobe epilepsy, and in elimination of excessive day sleepiness.

As an overall disorder, nocturnal frontal lobe epilepsy is considered somewhat benign. As many as 20% of patients decline further treatment (after AEDs) because their seizures were tolerable. Successful surgical treatment should also act to reduce the frequency of sudden unexplained death in epilepsy. Since both memory and quality of life depend in some measure on improvement of sleep, successful surgical outcomes should also improve life quality.

The objective of another paper was to describe results of investigations on the reliability of video recordings in the diagnosis of nocturnal frontal lobe epilepsy (Vignatelli et al. 2007). In this study, 66 patients with suspected nocturnal epilepsy underwent nocturnal video polysomongraphic recording. Special attention was paid to the inter-observer reliability. Four major criteria have been defined regarding nocturnal epilepsy: paroxysmal arousals (brief sudden recurrent behavior), hyperkinetic seizures, asymmetric tonic seizures, and epileptic nocturnal wonderings.

Results showed that, based on the four criteria, the level of agreement between pairs of observers ranged from 63 to 79%. When just agreeing on seizures or not, the level of agreement ranged from 77 to 93%. These results suggest that the overall inter-observer reliability in video analysis is not always satisfactory. The judgements on paroxysmal arousals were only fairly to moderately reliable. Evaluation of hyperkinetic features was high at substantial to almost perfect, with regard to inter-observer reliability.

The authors comment that features such as multiple paroxysmal motor attacks should be recorded to establish if they are stereotyped. Even experienced clinicians do not always diagnose brief arousals consistently. The problem would appear in difficulty in the classification of such fleeting motor activity, and distinguishing it from normal bodily sleep movements. The authors conclude saying that video recording is probably adequate for a reliable diagnosis of nocturnal frontal lobe epilepsy. Inter-observer reliability is inadequate for diagnosis in which minor attacks such as paroxysmal arousals are seen. Diagnosis really needs recording of multiple stereotyped movements for a positive diagnosis.

A recent new scale called the frontal lobe epilepsy and parasomnias scale (FLEP) was developed (Derry et al. 2006) and evaluated for its diagnosing abilities

(Manni et al. 2008). The FLEP scale is purported to distinguish nocturnal frontal lobe epilepsy from parasomnias. Criteria for inclusion were a final diagnosis of arousal parasomnia, nocturnal frontal lobe epilepsy, or REM sleep behavior disorder (RBD). The diagnoses were secure, and based on in-lab all-night video EEG polysomnography recording-full scalp EEG. The international 10–20 system was used.

The FLEP scale was translated into Italian by bilingual experts. The scale form was filled in by a medical doctor based on patient and relatives' reports. The person filling in the scale was trained in sleep/epilepsy diagnosis, and was blind as to the patient's identity, and their final diagnosis. FLEP scores of "0" or less indicated episodes likely to be parasomnias, scores of "0" to +3 indicated possible epilepsy, and scores over +3 are highly likely to be nocturnal frontal lobe epilepsy.

The results showed the median FLEP score was +2 in the nocturnal epileptic patients, –4 in the arousal parasomnias group, and −1 in the RBD group. Of the 14 patients in the nocturnal frontal lobe epilepsy group, 4 had scores strongly indicative of nocturnal epilepsy, and 6 were suggestive of nocturnal epilepsy. In total, the FLEP scale gave an incorrect diagnosis in 4/71 cases (5.6%), and uncertain results in 22/71 (30.9) of the patients.

The authors note that the FLEP scale is reliable even when used by clinicians with limited experience. The use of the FLEP scale could reduce the need for expensive and complicated video-polysomnography for the diagnosis of nocturnal paroxysmal episodes. The authors further comment that some items in the scale such as nocturnal wandering episodes do not adequately address nocturnal frontal lobe epilepsy. There is a need for extensive statistical analysis of individual FLEP scale items in order to test reliability. The FLEP scale holds promise as a useful tool for triage in determining what additional courses of action might be needed to accurately diagnose nocturnal frontal lobe epilepsy.

The video EEG monitoring method is the usual definitive technique for the diagnosis of many seizure types, including nocturnal frontal lobe epilepsy. The method is, however, cumbersome in that scalp electrodes are used. A new method has been devised (Cuppens et al. 2009) in which accelerometers are attached to the wrists and ankles, and recordings are made. Nocturnal seizures often include running and bicycling movements, and these are easily detected.

In this study, data from three patients with nocturnal frontal lobe epilepsy were examined. Movement epochs were detected in a pre-data collection step, and the device was "zeroed". Next, thresholds are set and data from the arms and legs are collected. The data, after statistical manipulation, had a sensitivity of 92% and a specificity of 84%. The authors comment that the accelerometer technique could reduce the necessity for video EEG recordings to diagnose nocturnal frontal lobe epilepsy.

Although (autosomal dominant) nocturnal frontal lobe epilepsy is considered a somewhat benign form of epilepsy, severe cases have been reported which are associated with both psychiatric symptoms and intellectual disabilities (Derry et al. 2008). This paper describes two families with autosomal dominant nocturnal frontal lobe epilepsy. The records from 17 family members were available for study. Living affected family members were given detailed assessment such as neurologic examination, EEG, and video EEG. Some members of one of the families were the subjects of original reports (Scheffer et al. 1995).

The results showed that the age of onset was 7 years. Frequency when not well controlled was 30 seizures per night, and status epilepticus occurred in four patients. Seizure symptoms were typical of frontal lobe epilepsy, and included prominent vocalization, brief duration, hypermotor automatisms, and dystonic posturing. In this group, well over 50% were refractory to AEDs. Nearly all seizures were nocturnal, but a few were of an atonic nature, and occurred during daytime. Otherwise all seizures were complex partial.

Nine of 17 patients had psychiatric and/or behavioral problems. In adults, depression was a frequent disturbance, whereas in children the disorder was one of aggression and destructive behavior. Children were easily distracted and overactive. In addition to depression, adults also displayed personality disorder and paranoia. Intellectual disability was also present in four patients. In some cases, developmental regression was prominent. There was no evidence or history of psychiatric disorders in non-epileptic family members.

The authors note that patients from these two families have autosomal dominant nocturnal frontal lobe epilepsy which has a more severe phenotype than heretofore described. The severity is emphasized by significant numbers of patients developing status epilepticus, seizures refractory to AEDs, and early onset. In addition, patients had decreased intellectual capacity and psychiatric disorders. The psychiatric problems ranged from depression to outright paranoid schizophrenia.

The authors speculate that the genetic mutation(s) underlying these patients' nocturnal epilepsy could confer a predisposition to cognitive and psychiatric comorbidity. The association of cognitive and neurological sequelae has been reported in patients with complex partial seizures, which could suggest the frontal lobe seizures per se were involved in the adverse outcomes (Thomas et al. 1999). The authors suggest that more than one factor may result in severe psychiatric, behavioral, and cognitive disorders in these cases of autosomal dominant frontal lobe epilepsy.

Part IV
Miscellaneous Epilepsies

Chapter 18
Febrile Seizures

The International League Against Epilepsy (ILAE) has defined febrile seizures by various criteria (Commission on Epidemiology and Diagnosis 1993). The criteria include age, seizure history, etc. (see Table 18.1). Febrile seizures can be divided into two categories, simple or complex. Simple febrile seizures are those in which there is a generalized seizure, with loss of consciousness, without focal localization. These include tonic clonic, tonic, atonic, with a duration of less than 10 min. Tonic clonic seizures are focal, and have a 15 min or longer episode. Repeat seizures may occur. These occur within 24 h of the first attack.

At least one-fifth of children with febrile seizures experience the first seizure within one hour of the fever onset (Berg et al. 1992), and over half in the first day. Most febrile seizure patients have the first seizure between 1.5 and 5 years of age. Febrile seizures over 7 years of age are rare. Febrile seizures may become febrile status epilepticus, and account for about one-fourth of all status patients (Maytal and Shinnar 1990).

The treatment of febrile seizures depends in large measure on the characteristics of the episode. The usual approach is, if a single seizure is all that is recorded, not to attempt any AED treatment. The risks of a long-term treatment regime outweigh risk of repeat episodes, which are fortunately rare (Steering Committee on Quality Improvement and Management 2008). Risk factors for recurrence include a family history of febrile seizures, suggestive of a genetic factor, and age of the first seizure (under 18 months) (Annegers et al. 1990). Children with complex febrile seizures are much more likely to have recurrence, which can be 50% (1987).

Animal studies related to long-term effects of febrile seizures show an age-dependent effect (Baram et al. 1997). In vitro experiments show that temperature elevation can cause increased spike wave discharges in the hippocampus. This in turn may cause changes in hippocampal circuits (Toth et al. 1998), but there may be permanent changes (Dube et al. 2000). Even so, more than 20 min was required to produce long-term changes. As with humans, rats and mice had increased seizure susceptibility with pre-existing neurological conditions.

In contrast, children with simple febrile seizures have little or no sequelae, and continue to develop at a normal rate. Their IQ, socialization skills, etc. are normal.

D.W. McCandless, *Epilepsy: Animal and Human Correlations*,
DOI 10.1007/978-1-4614-0361-6_18, © Springer Science+Business Media, LLC 2012

Table 18.1 Criteria for febrile seizures

Criteria for febrile seizures
Age older than 1 month
Without prior a febrile seizures
Temperature greater than 38.4°C
Simple febrile seizure = isolated, brief
Complex febrile seizure = focal, multiple, and prolonged
Prior neurological condition not essential

Adapted from I.L.A.E. Commission on Epidemiology and Diagnosis 1993

There is no evidence, unlike complex febrile seizure patients, of any structural cerebral damage. This form is, therefore, a true metabolic encephalopathy. There does not appear to be an increased risk of sudden death in children with simple febrile epilepsy. As mentioned above, it is important to rule out potentially serious acute neurological disorders such as meningitis.

The idea has recently emerged that febrile seizures may not have to result in structural cerebral damage in order to become temporal lobe epilepsy. It has for example, been shown that there may be a role of cytokines and interleukin-1 beta (IL-1B) in animal models of febrile seizures (Heida and Pittman 2005). In fact, IL-1B is necessary in order for febrile seizures to occur in the rodent hyperthermia model (Sanchez et al. 2001; Brewster et al. 2005). The action of IL-1B would appear to be directed toward hippocampal neurons in the developmental febrile seizure experimental model. With regard to mechanisms, IL-1B release acts to increase neuronyl hyperexcitability (Dube et al. 2005). These changes in expression act to leave the hippocampus in a state of hyperexcitability. It is speculated that this may be a feature of all seizures occurring in early development. Other early developmental models of epilepsy, such as the pilocarpine model, show similar mechanisms (Raol et al. 2003).

Retrospective studies suggest that some patients with intractable temporal lobe epilepsy have an early history of febrile seizures (Falconer 1971). This has not been borne out looking at the long-term sequelae of children who have had febrile seizures (Verity et al. 1993). In patients who do show an association between mesial temporal sclerosis and early childhood febrile seizures, the length of time of the episodes was very long, and represents a small percentage of the patients. Such long lasting episodes of febrile seizures are rare and usually focal. Depending on the methods of clinical investigation, mesial temporal sclerosis may or may not be connected to febrile seizures, but there is probably a connection to hippocampal pathology (Lewis et al. 2005).

In recent years, a gene mutation has been linked to more than one type of epilepsy (Mulley et al. 2005). The gene is called the neuronyl sodium channel alpha subunit gene (SCN1A). The mutation exists in the first transmembrane segment of domain in one of the gene. This results in a blockage of normal function of SCN1A.

A recent paper reports results of an electro/clinical study of a family with a history of febrile seizures, temporal lobe epilepsy, and a SCN1A mutation

(Colosimo et al. 2007). Most of the mutations are associated with generalized epilepsy/febrile seizures, and myoclonic seizures. Some phenotypes are associated with milder forms of seizures. The family studied in this investigation had data from 35 family members covering four generations. Results showed 14 members were affected with the mutation, and 13 were still alive. No spouse had the genetic defect. This was an autosomal dominant genetic alteration. The 13 affected family members had thorough clinical evaluations, plus EEG studies, CT scans, and MRIs.

Further results show all 13 patients had febrile seizures in childhood, the majority age at onset was under 1 year. Nine of the 13 had simple febrile seizures. Neurological and psychiatric examinations were normal in all patients. Results from 10 of the 13 in which MRIs were performed showed that seven were normal, whereas two had hippocampal sclerosis, and one had post-traumatic epilepsy. Most EEGs were normal or had slowed sharp waves.

The authors note that their data show that the mutation associated with SCN1A results in a phenotype of simple febrile seizures. The patients later exhibited temporal lobe epilepsy. These late affected three patients had unprovoked seizures. Their EEG results were consistent with complex partial epilepsy. Extending this, the data show that the SCN1A mutation can lead to hippocampal sclerosis and temporal lobe epilepsy. The authors state that in spite of considerable similarities between those patients who went on to develop pathological damages and those who did not, no differences could be delineated. Speculation is that other as yet unidentified (genetic?) factors must impact the severity or mildness of phenotypic expression. Therefore, familial febrile seizures are a risk factor for structural pathological alterations. Simple febrile seizures are not without risk, and the genetic mutation is no doubt strongly pathologic.

Another paper examines febrile seizures/SCN1A alterations and epileptic patients (Marini et al. 2007). While febrile seizures are associated with the gene mutation, other seizure types such as severe myoclonic epilepsy also have the SCN1A mutation. As many as 100% of myoclonic epileptic infant patients have the SCN1A gene alteration (Mulley et al. 2006). The purpose of the present study was to better define phenotypes associated with the SCN1A mutation. Three groups were studied: (1) those with severe myotonic epilepsy of infancy, (2) generalized epilepsy with febrile seizures, and (3) a group with fever precipitated febrile seizures, but for whom the febrile seizures were not characteristic.

Results showed that 132 patients had a mean age at study of 9 years. Of the 132 patients, HPLC showed 40 genetic mutations. Genetic mutations were present in 78% of patients with severe myoclonic epilepsy of infants. This is totally consistent with earlier estimates of the frequency (Wallace et al. 2003). The severe myoclonic epilepsy in infants with genetic mutations all manifested early as repeated febrile seizures, which became status.

Multiplex ligation-dependent probe amplification (MPLA) performed in 18 severe myoclonic epilepsy patients showed genomic deletions in two. This powerful tool permitted the recognition of a subset of severe myoclonic epileptics without the SCN1A alteration. The authors state that MPLA is a quick cost-effective method for finding mutations. In this study, of the 28 mutations, 86% were de novo, whereas only four patients were inherited. The possibility exists for a partial inherited mode, but less likely, as most mutations were de novo. Further studies with increased numbers would be helpful. Generalized epilepsy patients with febrile seizures had an SCN1A mutation rate of 11.5%.

The authors conclude saying that the high correlation between severe myoclonic epilepsy and SCN1A mutations indicates a phenotypic specificity of SCN1A. It remains to be seen what is the effect of SCN1A mutation in other seizure types. And, there appears to be a genetic factor, but its exact significance awaits elucidation.

Another recent study examines these phenomena in Chinese families with generalized epilepsy with febrile seizures plus (Sun et al. 2008). This study was designed to evaluate the roles of SCN1A, SCN1B, and ligand gated gamma aminobutyric acid receptor genes (GABRZ). A total of 23 separate Chinese families were evaluated over a 2.5-year period. From the 23 families, 127 affected members were identified. Of these patients, 49% had typical febrile seizures only. Febrile seizures plus equaled 32% of patients. Other subtypes constituted the remainder of the patients.

Results showed that a new SCN1A mutation p.N935H was described in this Chinese population. This is not too surprising, since over 170 SCN1A mutations have been identified in severe myoclonic epilepsy in infants, and at least 16 missense mutations are seen in generalized epilepsy with generalized seizures (Harkin et al. 2007). In the present study, most mutations were de novo, while almost all SCN1A mutations in febrile seizures plus were inherited missense mutations. GABRGZ showed six new mutations not previously described. This raises the known mutations of GABRGZ mutation to six. The new mutation involves p.W390X. The first GABRGZ mutation was described in a French family (Baulac et al. 2001).

The authors conclude saying they have described two new mutations in Chinese families with generalized epilepsy with febrile seizures plus. This indicates that this disorder is a number of channelopathies. The number of mutations in these patients was rare; mutations were not found in most patients studied.

Another idea emerged that perhaps a mutation in the gene SCN1A might have an influence on the age at onset of generalized epilepsy with febrile seizures plus. The current study (Sijben et al. 2009) looks at this possibility. Twelve families were identified and incorporated into the study. Eleven of the 12 families had been studied before: SCN1a mutations (Scheffer and Berkovic 1997), SCN1B mutation (Scheffer et al. 2007), and GABRGZ (Marini et al. 2003). A total of 105 patients who had one of the above mutations were studied.

Results showed the median age of onset of febrile seizures and febrile seizures plus was 12 months. Patients with SCN1A were lowest with regard to age of onset. This was in contrast to the median age of 18 months in patients without these three mutations.

The authors note that these data represent the first published observation that the gene associated with phenotypes such as febrile seizures/febrile seizures plus may affect age of onset of the seizures. This may be related to age-related developmental differences such that affected patients are susceptible to seizures at less than 1 year of age. Patients with the GABRGZ mutation also had an earlier onset, but not those with the SCN1B mutation. While the exact causes of these differences are unclear, the authors suggest studies of the development of expression of ion channels (sodium) might shed light on this problem.

The Dravet syndrome is a very severe form of myoclonic epilepsy, with intractable seizures (Dravet et al. 1992). A recent paper looks at the role of SCN9A in both febrile seizures, and in the Dravet syndrome (Singh et al. 2009). This was a study on a family in which at least 21 members were affected with febrile seizures. Studies showed that there was a mutation in the gene SCN9A, This gene is located adjacent to SCN1A, and is a mutation previously shown to be associated with febrile seizures.

In addition, one of the febrile epileptic children developed Dravet syndrome, and the authors sequenced the SCN1A gene. Results showed a heterozygous mutation. Analysis of the Dravet syndrome patients showed a missense variation in several patients. The authors comment that the study is significant because of the association of mutations in both SCN1A and SCN9A in patients with Dravet syndrome. In addition, the study shows a mutation in SCN9A, not previously described. The finding was confirmed by creating a knockout mouse model, and looking at seizure susceptibility. This showed the SCN1A knockout mice had a lowered threshold for seizures. Seizures in these mice had a faster rate with subthreshold stimulation. This was a combined patient/animal model study, the results from each serving to complement each other. Translation was in each direction.

Most mouse models of febrile seizures utilize the warm air (hair dryer) method to induce febrile seizures. This model (Holtzman et al. 1981) has shown that the hippocampus and amygdala are altered by hyperthermia. These effects are not limited to immature mice. The present study (van Gassen et al. 2008) looks at febrile seizure susceptibility and long-term effects in seven strains of mice. Rats were also utilized.

Results showed that the warm air method successfully induced febrile seizures in all seven strains of mice when applied between the 10th and 14th postnatal days. Rats had similar results with regard to the hypothermia method of febrile seizure induction. Days 10–14 were chosen since this time roughly corresponds to the period when children are most susceptible to hyperthermia. Duration of seizures in mice was an average of 22.5 min, with a low mortality rate.

The authors comment that febrile seizures are a common feature of patients later showing temporal lobe epilepsy. While a variety of gene mutations occur in familial febrile seizures, there are scant data on sporadic febrile seizures. The behavioral characteristics of one mouse strain, C57Bl/6j mice and the tested Sprague–Dawley rats were similar. When mice were exposed to hyperthermia and then cooled, the seizures stopped.

The current study showed that febrile seizures in the neonatal period served to lower the long-term pentylenetetrazole threshold for seizures. These data are in agreement with previous long-term rat data (Dube et al. 2000).The mouse strain A/J was least susceptible to pentylenetetrazole-induced seizures later in life, but they were susceptible. The authors speculate that this difference may result from slight differences in development between strains. These results warrant further investigation as to mechanisms of hyperthermia-induced seizures in different strains of mice and rats. Another recent study (Lemmens et al. 2008) using Sprague–Dawley rats examined the effects of early postnatal hyperthermia exposure on the fate of cells in the dentate gyrus. These studies are important in that newborn febrile seizures are relatively common in children, and their association with later life complex partial seizures remains unclear. The occurrence of febrile seizures during a critical period in brain development is ominous. One hypothesis of post-hyperthermia mechanisms is that there occurs a proliferation of dentate cells (demonstrated by bromodeoxyuridine labeling), leading to an increase in network connection, which affects hippocampal physiology.

Results from this study showed that newborn (postnatal day 10) rats were rendered hyperthermic using the hair dryer method, and 65% displayed seizures consisting of clonic limb contraction. The authors found that hyperthermia-induced seizures (but not hyperthermia alone) led to an increased number of bromodeozyuridine labeled cells. The numbers of cells which survived hyperthermia-induced febrile seizures in these rats was decreased. Finally, the number of GAD67 positive GABAergic cells was not affected.

Since glycolysis is often associated with hippocampal sclerosis, this feature was assessed in hyperthermic induced febrile seizures. There was no increase in glial cells, the increased bromodeoxyuridine being all due to increased numbers of mature neurons. Double labeling studies resulted in a reduction in the numbers of bromodeoxyuridine/excitatory amino acid transporter 3 labeled cells. Excitatory amino acid transporter 3 is thought to be involved in the synaptic elimination of glutamate. Lack of removal would be conducive to a heightened excitatory state. Future studies are indicated to confirm these results.

In another study (Qu and Leung 2008), effects of hyperthermia in slice preparations were studied regarding effects on GABAergic synaptic transmission in immature rat hippocampus. For electrophysiological recordings, whole cell voltage clamp recordings were performed on CA1 hippocampal neurons.

Results showed that hyperthermia has both pre- and post-synaptic effects on GABAergic CA1 hippocampus. The present study also showed that hyperthermia decreased spontaneous inhibitory post-synaptic currents and miniature inhibitory post-synaptic currents (sIPSCs and mIPSCs). This is an indication of suppression of GABA release from presynaptic nerve terminals caused by the hyperthermia. Further results showed a hyperthermia-related decrease in the decay time constant of the mIPSCs, which in turn led to a significant decrease in charge transfer of mIPSCs during hyperthermia.

The present study shows an inhibition of GABA release in the slices, caused by hyperthermia. This may be the result of a suppression of a signaling pathway regulated by adenylyl cyclase protein kinase A. This is a selective effect on presynaptic neurotransmission (Bouron 2001). The authors also comment that their study shows that presynaptic 4-aminopyridine sensitive potassium channel depression causes IPSC reduction.

The significance of this study is that it shows for the first time that hyperthermia has an effect on both pre- and post-synaptic GABAergic synapses. The further suggestion from these animal experiments is that AEDs might be efficacious if designed to increase cyclic AMP dependent protein kinase A activity and/or increasing post-synaptic GABA A receptor function.

One "pause" in interpretation of data from the commonly used hair dryer method for inducing hyperthermia-related febrile seizures is that it may cause hyperventilation (Schuchmann et al. 2008). In this study, rectal temperatures in Wistar rat newborns were measured, and seizures started at a temperature of 41.2°C. During the 60 s preceding the onset of seizures, respiration rates rose from 163 to 246 breaths/min. The respiration rate was measured using a piezo crystal sensor placed on the rat's abdomen. The increased respiration rate is in contrast to the results by Dube et al. (2007).

The authors note that other studies use a warm/cool temporal method to try to prevent overheating which may confound results. The idea is advanced that a continuous rise in temperature in an animal model as opposed to an off/on again elevation should more closely approximate the onset of febrile seizures in children.

A report looks at the temporal features of the onset of severe myoclonic epilepsy in infancy (Davet syndrome) (Oakley et al. 2009). In this study, three postnatal day periods were examined: 17–18 postnatal days, 20–22 postnatal days, and 30–46 postnatal days. Results were monitored by video EEG. Elevated body temperature produced no effect at postnatal days 17–18, but almost all mice with mutations in the alpha subunit of the type 1 voltage-gated sodium channel examined had seizures in the other two time groups. The seizures were those of severe myoclonic epilepsy in infancy. Spontaneous seizures were not seen until after postnatal day 32, indicating the elevated temperature played a critical role.

Further, most postnatal day 20–22 mice showed interictal spike activity with an increase in core temperature. The authors state that there is a critical period in development during which an increase in bodily core temperature is sufficient to induce febrile seizures. There is a significant correlation between the temperature-induced onset of severe myoclonic seizures (Dravet syndrome) in human newborn infants, and in mice with the described mutation. The correlation of age and temperature between the two examples is striking, state the authors.

Yet another excellent study employed chromosome substitution strains (CSS) to try to identify new febrile seizure quantitative trait loci (QTLs) (Hessel et al. 2009). In this study, C57Bl/6J mice served as a host strain, A/J mice were the donor strain, and CSS were examined to determine febrile seizure latency. Video EEG monitoring served to define phenotype.

Results showed six strains – CSS-1, CSS-2, CSS-6, CSS-10, CSS-13, and CSS-x – were more susceptible to febrile seizures than were C57Bl/6J mice. Video EEG demonstrated that tonic clonic latency correlated with spike wave discharges, and it is an effective method to determine febrile susceptibility.

The authors state that several febrile seizure genes are identifiable in human familial febrile seizure syndromes. These include both ion and non-ion channels. Some of these genes mapped to homologous mouse regions in the screen. Mouse chromosome 2 was shown to have a strong febrile seizure quantitative trait loci. The mouse chromosome carries febrile seizure genes SCN1A, and SCN2A among others. Chromosome X in mice carries a quantitative trait loci, and is homologous to Xg22 in humans. Xg22 has been shown to be linked to epilepsy and mental retardation (Scheffer et al. 2008).

Two other studies involving C57Bl/6J-Chr#A/NaJ panels have been conducted. One, looking at nocturnal epilepsy identified CSS4 and CSS7 as involved in the seizures (Strohl et al. 2007). Another study using the same panel examined the chemical seizure producing agent pilocarpine. This study showed chromosomes 10 and 18 as involved in quantitative trait loci.

In conclusion, the authors say that their study describes a method for determining febrile seizure susceptibility genes using a forward genetics approach. They have identified six quantitative trait loci from A/J mice, which altered febrile seizures in C57Bl/6J mice. Three were protective – CSS1, CSS10, and CSS13. All the rest increased seizure susceptibility. Future studies looking, for example, at F1 generations in mice will provide new data which may increase knowledge such that new effective AEDs might be developed.

Reviewing salient clinical features of febrile seizures in children, it is important to remember that febrile seizures, while relatively common, have a very low mortality rate. Further, the risk for additional seizures is sufficiently low that prophylactic AEDs are usually not recommended. Risk factors predicting recurrence of seizures after an initial febrile seizure include age under 15 months, complex febrile seizures, and a family history. Even considering the above risks, 97% of children having febrile seizures will not develop epilepsy (Nelson and Ellenberg 1976).

While there is a tendency to prescribe antipyretic drugs to treat almost all fevers in children with previous febrile seizures, little data support the concept that this will reduce the incidence of repeat febrile seizures (Rutter and Metcalfe 1978). One study examined the effects on fever of ibuprofen syrup vs. placebo, and results showed no statistical difference between the two treatments regarding febrile seizure recurrence (Van Stuijvenberg et al. 1998).

For most children who have had a febrile seizure, the most effective treatment is probably parental education and/or counseling. It has been reported that in parents with some knowledge about febrile seizures, anxiety levels are quite high (Flury et al. 2001). It is important to provide complete and accurate information to parents. The amount of information should be balanced depending on parental anxiety levels and education/ability to understand the concepts surrounding febrile seizures.

In terms of medical evaluation of febrile seizures, serious medical illness which may have initiated the seizures must be ruled out. Otherwise, simple febrile seizures may be over diagnosed, over investigated, etc. (Dunlop and Taitz 2005). It is felt, however, that children under 18 months should be admitted. If the diagnosis is simple febrile seizures, the emphasis, as stated above, should be reassurance and education of parents.

The following is a case report of a more serious outcome of a patient with complex febrile seizures (Merkenschlager et al. 2009). In this case, a healthy boy had a complex febrile seizure at age 3. Following the seizure, he had transient right side flaccid hemiparesis. An MRI showed normal results with no discernable structural alterations in the hippocampi, and no evidence of microdysgenesis, Results from the EEG showed unspecific left side slow waves without spikes. Otherwise all examinations were normal.

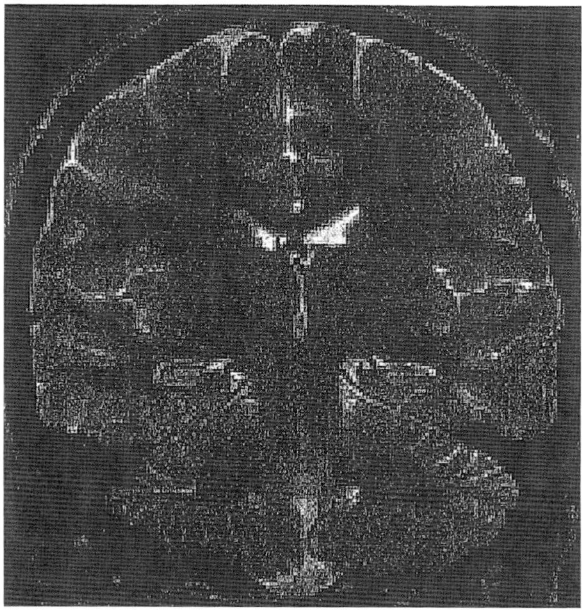

Fig. 18.1 Typical hippocampal sclerosis on the left side. Published in Crespel, A., et al. Atlas of electroencephalography Volume 2: The Epilepsies, EEG and Epileptic Syndromes. With permission of John Libbey Eurotext 2005

A year later at age 4, the patient had another febrile seizure. This episode lasted about 25 min. This time, there was a postictal flaccid hemiparesis involving the left side. Within 24 h, an MRI showed hyperintense signal (T2 weighted) in the right hippocampus. There was, in addition, acute edema in the right hippocampus, clearly shown in coronal FLAIR sequences. MRI was repeated 16 months later with the result that the right hippocampus was smaller in size than before, and the signal was compatible with right hippocampal sclerosis. The patient was subsequently treated with valproate therapy as prophylaxis, and 6 years later was developing normally without further seizures.

In this case, there was little doubt as to the complex febrile epilepsy diagnosis. Blood/CSF tests were negative for a variety of infections including measles, rubella, HSV, borreliosis, and CMV. There were no vascular anomalies. The EEG slowing following the first event was normal by 4 weeks. Thus, the hippocampal sclerosis seems to have developed with the second febrile seizure. This case is of special interest since the first febrile seizure documented a normal appearing hippocampus.

About 50% of pediatric patients with complex partial seizures (temporal lobe epilepsy) have hippocampal sclerosis and have had febrile seizures. One previous study (Maher and McLachlan 1995) found a strong association between febrile seizures and temporal lobe epilepsy, and the length of the febrile seizures was positively linked to the risk for hippocampal sclerosis. The current case showed no familial connection with febrile seizures, however, nearly 60% of patients with familial febrile seizures have hippocampal sclerosis (Kobayashi et al. 2001) (see Fig. 18.1).

The association of hippocampal sclerosis and febrile seizures is not 100%. The possibility exists for a preexisting hippocampal anomaly causing the seizure (Scott et al. 2006). It is possible that there is more than one type of hippocampal sclerosis. In an MRI study of children about to have a temporal lobe resection, half had a prolonged febrile convulsion and half did not. Those who had a prolonged febrile seizure showed more side to side asymmetry of the T2 relaxation time, and smaller hippocampal volume as compared to those who had not had a prolonged febrile seizure (Scott et al. 2001). Another study (Murakami et al. 1996) published similar data, with similar conclusions. The authors thus state their belief that their case documents the development of hippocampal sclerosis in conjugation with a prolonged second febrile seizure episode. This occurred in a patient with MRI documentation of a lesion-free hippocampus after a single initial febrile seizure.

One controversial issue regarding the role of lumbar puncture for febrile seizure patients under 1 year old has recently been addressed (Tinsa et al. 2010). This study was undertaken based in part on the American Academy of Pediatrics recommendation that lumbar puncture is recommended in children under 1 year of age with febrile seizures. The study was a retrospective one, and included data over 8 years.

Results showed that 106 cases were examined and divided into two groups: the first group presented with bacterial meningitis, and the second group presented with febrile seizures. Predictors of meningitis were aged less than 7 months, seizures greater than 5 min, and seizure recurrence the same day. In patients over 7 months, indicators were short stature, and with a CRP less than 20 mg/l.

The conclusions state an extremely low incidence of meningitis if the patient is over 7 months old, has had a single short (less than 5 min) febrile seizure, and has a normal neurological exam. In these cases, a lumbar puncture is probably not necessary, thereby eliminating some patient risk. The patient should, however, be hospitalized for at least 24 h for observation and monitoring.

Chapter 19
Infantile Spasms

Infantile spasms were first described in the early 1840s by West (1841), whose son had the disorder. West's syndrome, or infantile spasms, is a catastrophic type of epilepsy, with an early onset, often occurring in the first postnatal month. A rough estimate of prevalence is 1 in 2,000 to 1 in 6,000 live births (Saliba et al. 1996). The earlier the onset, generally, the worse the disorder. If infantile spasms appear in the first several weeks, it can be difficult to diagnose and treat. Even the terminology surrounding infantile spasms is complex. There are dozens of names besides "infantile spasms" which have been used or suggested over many years.

Clinical signs and symptoms are different than those seen in infants and children. In infantile spasms, seizures can be disorganized in that they have unusual patterns of spread, and can occur in clusters. These differences may likely develop because of the immaturity of the newborn. The seizures carry a poor prognosis, and so treatment is usually aggressive and immediate. The idea is to try to change the course of the disease as early as possible. The course of infantile spasms usually involves severe developmental delay, and 70–90% have severe mental retardation. As many as 30% also develop cerebral palsy. Frequently infantile spasms will have less organized, and exaggerated, motor behavior, which has been termed motor automatisms (Mizrahi and Kellaway 1987). The types of spasms are motor, and are flexor, extensor, and mixed, based on postural manifestations. These movements involve neck and trunk muscles. Other seizure types are present in one-third to half of all infantile spasm patients.

Infantile spasms are very hard to treat at least in part due to the multiple seizure types. It is an evolving problem, with as many as 20–40% developing the Lennox–Gastaut syndrome. Adrenocorticotropic hormone (ACTH) and/or oral steroids may result in decreased seizure reduction. There are some methodological problems with the ACTH studies. Nevertheless, most believe that ACTH is effective in treating infantile spasm cases. Steroids may also be effective, but not as much as ACTH. The drug vigabatrin has also been used for infantile spasm treatment.

There are many causes of infantile spasms. These include tuberous sclerosis, hypoxia, ischemia, inborn errors of metabolism, anatomic malformations of the brain, genetic disorders, etc. Evidence supports the brainstem as being the area of

D.W. McCandless, *Epilepsy: Animal and Human Correlations*,
DOI 10.1007/978-1-4614-0361-6_19,

origin of aberrant hypsarrhythmic EEG recordings. The implication is that descending brainstem pathways are altered such that spinal reflex activity is disrupted. Ascending brainstem pathways are also affected in this hypothetical model. The finding that ACTH is effective in suppressing seizures in infantile spasms is consistent with the idea that ACTH suppresses central serotonergic activity (Pranzatelli 1989). Other mechanisms have been proposed, which remain to be tested. It is certain that over a dozen genetically based syndromes have infantile spasms as a feature, so the association of certain chromosomes (X) is a possible link to the pathology of infantile spasms.

From the standpoint of EEGs, hypsarrhythmia is the most commonly seen interictal pattern. Hypsarrhythmia is usually seen early and in patients less than 1 year of age. Hypsarrhythmia is a characteristic background pattern consisting of high-amplitude slow waves and spikes. As would be expected, there are several variations seen in a 24 h video EEG. It seems to be more pronounced in non-REM sleep (Hrachovy et al. 1984). Ictal patterns include generalized slow wave transients, sharp and slow wave transients, and episodes of attenuation. Various combinations can occur, but most common is the generalized slow wave transient, followed by attenuation of background activity. Ictal EEG events may last from a second to more than one minute. Asynchrony, nonrhythmic, and variability characterize the EEG pattern.

There can be a spontaneous remission of infantile spasms (Jeavons et al. 1973). There were 28% seizure-free patients with infantile spasms at the end of 1 year after start of treatment. Another study has confirmed the rate of spontaneous remission and disappearance of hypsarrhythmia (Hrachovy et al. 1991). The mechanism of spontaneous remission is unknown; however, investigating this interesting phenomenon could have significant influence over future infantile spasm treatment direction.

There has been a suggestion that infantile spasms may be related to routine immunizations, however, careful statistical analysis shows that the apparent relation between DPT immunization and infantile spasms is merely coincidental with no causal relation (Melchior 1977).

In refractory cases, there may be some benefit to surgery. This requires a successful localization study, which is usually unsuccessful. If lesions (tumors, etc.) are localized, then surgery may result in seizure freedom (64%) although the number of patients who are good candidates is small.

Since infantile spasms have multiple etiological factors, studies have been done examining the relationship of etiology to the seizures (Matsumoto et al. 1981). In this study, 200 patients with infantile spasms were studied with regard to etiology, seizures, and prognosis. Forty-eight patients (24%) died, the remaining patients were followed up at 6 years or more. Patients were divided into three groups: prenatal, perinatal, and postnatal with regard to etiology.

Results showed that cases could be categorized according to severity of mental/motor activity from 1 (normal) to 5 (mental and physical severe handicap). The prenatal group had signs of infantile spasms at birth. The perinatal group consisted

Table 19.1 Etiology of prenatal infantile spasms (West syndrome)

Etiology	Number of cases	Percentage
Brain malformations	42	58
Tuberous sclerosis	11	15
"Small for date"	10	14
Intrauterine infection	3	4

Adapted from Matsumoto, A., et al. Eur J Pediatr. 38: p. 456, 1981

of cases such as perinatal asphyxia, and they were normal at birth. Actual anatomic brain malformations were diagnosed in forty-two patients with infantile spasms. Normal mental and physical development was found in only 10% of the prenatal and perinatal groups. There was no correlation between etiology and persistence of seizures. Delayed development, convulsions, and neurological symptoms were correlated only with the cryptogenic etiology group. Cluster formations of spasms did not correlate with any etiologic group. In terms of treatment, the positive response to ACTH was over 80%, and was similar in each etiologic group.

The authors comment that symptomatic cases represented 67% of the entire group of patients, and 54% of patients had evidence of prenatal damage of some type. There was a 40% incidence of family history in the cryptogenic group, and only 9% in the perinatal group, indicating a possible genetic factor. There were a high number of patients with laughing attacks in the postnatal group. More studies seem warranted on the factors influencing course and outcomes of infantile spasms (see Table 19.1).

A brief paper on treatment of infantile spasms with vigabatrin originated from Thailand where ACTH is not available (Visudtibhan et al. 2004). This retrospective study covered 6 years in which 46 patients were treated with vigabatrin out of an infantile spasm group of 57 infants and children. Seizure freedom was achieved in 76% of those given vigabatrin. The mean duration of responders was 38 days from treatment onset to seizure cessation. Examination showed normal eye features (no field loss) in 20 patients, and cortical blindness in 6 patients.

Mean follow up was 4.5 years, and 19 patients were still seizure free on follow up. Vigabatrin has been used in Europe and elsewhere since 1990, while valproate, topiramate, etc., are effective. Vigabatrin is an excellent option when available. ACTH continues to be effective, but it is not available everywhere.

Another brief paper reports a study looking at length of time for infantile spasm treatment with vigabatrin (Kroll-Seger et al. 2007). The problem is that there is significant retinal toxicity associated with vigabatrin (Vanhatalo et al. 2002). The present paper is a report of four patients with infantile spasms.

The patients had cortical dysplasia, and one had tuberous sclerosis, confirmed by MRI. All were treated successfully with vigabatrin. The seizures were controlled, and development was normal. Vigabatrin was discontinued between 1 and 5 years. Seizure relapse occurred in all cases. The subsequent spasms were refractory to vigabatrin, and the result was severe mental retardation in two cases.

This report demonstrates clearly the difficulty of the balance between preventing infantile spasms and saving eyesight. Seizure onset is a serious risk factor for mental retardation when it starts under the age of two (Vasconcellos et al. 2001). When infantile spasms reoccur, the relapse is extremely difficult to control.

There may be a correlation between the tendency for relapse following vigabatrin cessation and visual field defects. From this study, it would seem that focal cortical dysplasia is associated with the risk for relapse when vigabatrin is stopped. With the risks of vigabatrin, consideration of surgery might be appropriate at an early stage in patients with infantile spasms.

Hypothalamic hamartomas are rare lesions (tumors) of the inferior hypothalamus. Two types occur, and the one attached to the hypothalamus is associated with infantile spasms; the other type, the parahypothalamus form, is not associated with seizures (Kerrigan et al. 2005). In the first instance, the hypothalamic hamartoma can be surgically removed, with corresponding seizure remission.

In this study, small number of (six) patients with hypothalamic hamartomas and infantile spasms were studied. The age of onset of infantile spasms was 6.2 months. Four of the six had infantile spasms as the first seizure type, and two had gelastic seizures. Five of 6 had MRI identifiable hypothalamic hamartomas. Five patients received ACTH, and in four it produced complete seizure – gelastic seizures continued in two patients. The mean age for surgery was 13.8 years, and three patients had mental/developmental retardation. Surgical results showed four patients had a 90% or better seizure reduction. The authors conclude that hypothalamic hamartoma should be looked for in infantile spasm patients, and early surgery should be considered.

A retrospective study examined the efficacy of the ketogenic diet and ACTH for infantile spasms (Kossoff et al. 2008). The speculation was that the ketogenic diet would have good efficacy and better tolerability than ACTH. In this study, 13 patients with infantile spasms were administered the ketogenic diet and 20 patients received a high dose of ACTH.

Results of this study showed the mean age of spasm onset was 5 months. Twenty-two of the 33 patients had identifiable etiologies. Treatment results showed that after 1 month of treatment with the ketogenic diet, 8 of 13 patients were spasm free, compared to 18 of 20 seizure free in the ACTH group ($p=0.06$). Of five not responding to the diet, four were switched to ACTH, and three of them became seizure free. One patient was treated with topiramate and became seizure free in 7 days. Patients receiving the ketogenic diet had a lower recurrence rate. Only one ketogenic diet patient had a relapse, at 3 months.

Developmental outcomes were similar between the two treatment groups, with about one-third in each group having a poor outcome. Adverse effects included irritability, weight gain, and insomnia in the ACTH group; the adverse effects in the ketogenic diet group included weight loss, gastroesophageal reflux, and low tolerability.

The authors conclude that the ketogenic diet was safe, effective, and reasonably well tolerated. The ketogenic diet required a longer time to normalize the EEG.

Fig. 19.1 Schematic representation of EEG of an animal model of infantile spasms using tetrodotoxin to induce spontaneous seizures in newborn rats

There was a similar result with ACTH. The authors advocate the ketogenic diet as a first line therapy for infantile spasms, but with a 2 week time limit for success before trying something else such as ACTH.

The search for animal models for infantile spasms has been somewhat slow, but a recent report describes a new animal model for infantile spasms (Lee et al. 2008). The etiology of infantile spasms is difficult to identify, and over 200 associated conditions (!) may accompany the spasms.

In this study, the sodium channel blocker tetrodotoxin was chronically infused into the developing cortex and thalamus via an indwelling subcutaneous pump. EEG electrodes were implanted, and the pump was removed after 4 weeks. Video EEGs were obtained and analyzed. Recordings were performed for 28 days after EEG electrode implantation (see Fig. 19.1).

Results showed that about one-third of tetrodotoxin (TTX) infused rats showed spasms. The number of animals exhibiting spasms was similar between cortical and hippocampus infused rats. Spasms were observed as early as day 11 during the infusions.

Spasms were brief (1–2 s), and consisted of extension or flexion of body muscles, and were either symmetrical or asymmetrical. Ictal EEG patterns were generalized high amplitude, slow wave transient, followed by generalized voltage attenuation. Most episodes occurred during the awake period, while some were associated with sleep. Clusters occurred, with as many as 33 ictal episodes occurring within a 6 min time frame. Interictal EEGs were abnormal in all 35 TTX infused rats. They were characterized by multifocal spike and sharp wave discharges. The results show a high correlation with features of human spasms, slight differences may be attributable to differences in electrode placement (rats, intracortical; humans, scalp).

The authors comment that the new rat model is essentially identical to the symptoms and EEG features seen in human infantile spasms. As in humans, in the rats, there is a latency period between the brain perturbations and spasm initiation. There are still to be determined similarities such as response to ACTH and cognitive impairment. The exact pathophysiological basis of infantile spasms is speculative. This newly developed rat model of infantile spasms is speculative. This newly developed rat model of infantile spasms should speed up an understanding of still speculative features of human infantile spasms. This model is highly exciting and important in terms of translation.

Vigabatrin is an effective treatment modality for infantile spasms which has been in use in Europe for many years (approved in Aug 2009, in the US for infantile spasm treatment). The problem lies in an adverse effect which is characterized by a potentially permanent visual field loss. It involves bilateral concentric constriction of the visual field (Wild et al. 1999). The frequency of the visual field loss is 40% (Kalviainen and Nousiainen 2001). The present study was done in order to examine the effects of vigabatrin in children who were treated with the drug in infancy.

The patients studied were all treated with vigabatrin for infantile spasms, mostly within the first 8 months of age. The duration of vigabatrin treatment was a mean of 24.4 months. The first Goldmann perimetry test was administered at a mean age of 5.6 years after vigabatrin treatment, and nine children were normal, six were uncertain, abnormal, or borderline. The results were essentially normal later when the test was readministered at a mean age of about nine. One child had a mild vigabatrin attributed visual field loss.

The authors comment that theirs' is the first study examining visual field results of infants undergoing vigabatrin therapy for spasms in infancy. Results showed only one patient of 16 had a mild field of view loss associated with vigabatrin. This incidence is lower than others' results (Vanhatalo et al. 2002). The data do not measure the effect on visual fields of vigabatrin while the patient is actually undergoing treatment. This would be interesting to do using an electroretinogram.

A limitation of the study, the authors comment, is that the 16 patients were not representative of all spasm patients in that they did not have symptomatic etiology or mental retardation. The current study suggests the risk for permanent eye damage is low provided there is a good seizure outcome, and the patient is in good health. The authors further state that there is no controlled data regarding the best duration of treatment for vigabatrin. It is possible that only a few months are sufficient for efficacy. Relapse is a serious consequence (see above), and the risk for eye damage, risk of relapse, and risk of severe mental deficiency must be carefully evaluated and balanced.

Another brief paper examining the visual field deficits associated with vigabatrin use has been published (Wohlrab et al. 2009). The authors note that vigabatrin, an inhibitor of GABA transaminase, is effective in treating infantile spasms. It has, however, had some equivocal reports regarding visual field constrictions. Some reports speak of an incidence of visual field defects of up to 20% (You et al. 2006), whereas other reports show very little visual problems after vigabatrin treatment (Gaily-Seicebore op cit). The present study was intended to look at the presence or absence of visual field deficits following vigabatrin treatment of infantile spasms.

This was a retrospective study in which 15 children were studied (4 girls, 11 boys). Vigrabatrin was initiated for spasms between 2.5 and 12 months. The mean treatment duration was 1 year, 8 months. All patients were over 8 years old at the time of ophthalmologic exam except one who was age 6.5 years old. None of the 15 had evidence of visual field constrictions. No doubt patient cooperation is a key element in obtaining good results from the Goldmann Kinetic Perimeter examination, these results are in complete agreement with the Gaily et al. (2009).

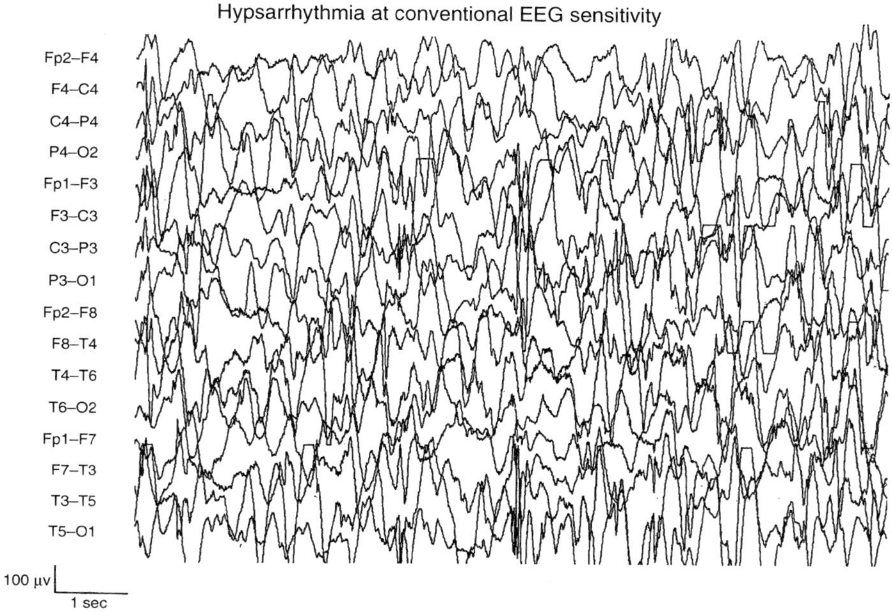

Fig. 19.2 Typical hypsarrhythmia seen at standard sensitivity. With kind permission from Springer Science+Business Media: Epileptic Syndromes and their Treatment, 2010 p. 281 Panaylotopoulos, C., Fig. 10.2a

Adolescents and adults have a higher incidence of visual field constriction than is seen in children.

It is not known why children are less susceptible to the visual field deficit than older patients. Neurological plasticity at a younger age may be a factor. There was no correlation between duration of treatment, dose, or cumulative dose and outcomes. The authors conclude saying the low risk for visual field defects in infants justifies vigabatrin use to treat infantile spasms.

In another retrospective study (Rener-Primec et al. 2006), the correlation of hypsarrhythmia duration and mutual outcome in infantile spasm patients was examined. The medical records of hundreds were screened, and 48 cases of infantile spasms were identified. The patients had 12 channel EEG recordings taken during the first 2 months or until remission of spasms. Detailed written clinical observations were obtained from nurses and/or parents. The duration of hypsarrhythmia was estimated from the infants' age, and EEG results (see Figs. 19.2 and 19.3).

Data relating to age of onset of spasms, history, development up to onset of spasms were obtained from medical records and interviews. CT and MRI results were reviewed. During follow up, a psychomotor test was administered, as well as IQ tests, and developmental and maturity tests. All except three were tested who were severely retarded.

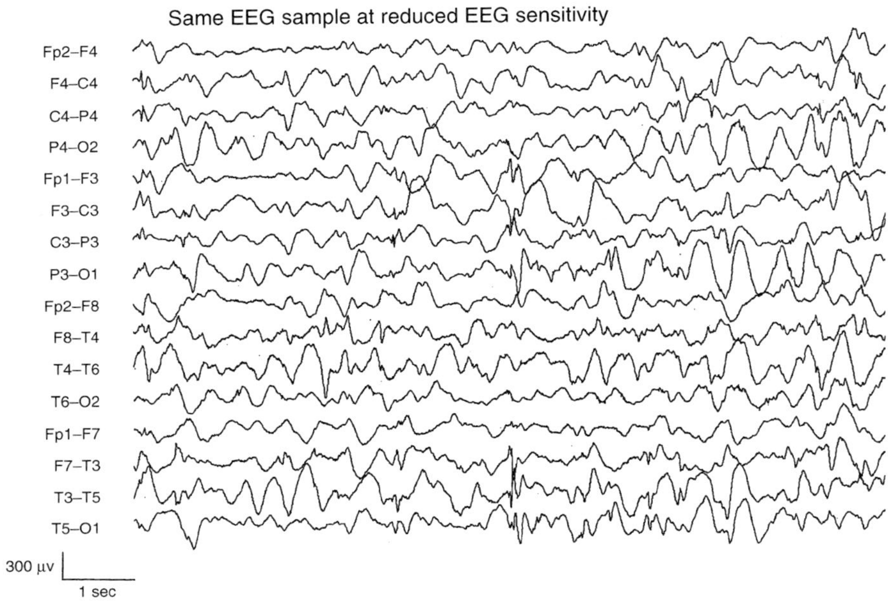

Fig. 19.3 Typical hypsarrhythmia seen at reduced sensitivity. With kind permission from Springer Science+Business Media: Epileptic Syndromes and their treatment, 2010 p. 281 Panaylotopoulos, C., Fig. 10.2

Results showed that the mean age of spasm onset was 6.1 months. Symptomatic etiologies included Down's syndrome, tuberous sclerosis, etc., and EEG data showed all symptomatic patients had atypical hypsarrhythmia. All 18 cryptogenic infantile spasm cases had typical hypsarrhythmia. Treatment lags ranged from 1 to 4 weeks, with a mean of 2 weeks. Most infants had the initiation of treatment within 1 month. Treatment consisted of AEDs in 21 infants, ACTH in 20 patients, and vigabatrin in 7 patients. At follow up, 28 children were seizure free, and of these, 18 were medication free. Sixteen had refractory epilepsy, and four died.

With regard to mental outcomes, a normal outcome was found in 15 children. Patients with typical hypsarrhythmias had a better outcome whereas only four patients with atypical hypsarrhythmia had normal mental outcomes. Mental retardation was present in 33 of the infantile spasm patients. Fourteen cases had IQs under 35. Of those four who died, three were severely retarded.

The authors state that the aim of the study was to correlate, if possible, hypsarrhythmia and mental retardation outcomes in infantile spasm patients. Using logistic regression, the authors note that after 3 weeks of hypsarrhythmia, the risk for mental retardation increases. These results are in keeping with previous studies of infantile spasms in cases of tuberous sclerosis, and Down's syndrome (Goh et al. 2005; Eisermann et al. 2003). The authors state that early treatment is essential, and has been stressed in early studies (Lombroso 1983). The exact mechanisms are unclear as to whether brain damage occurs due to hypsarrhythmia or if hypsarrhythmia merely reflects other underlying brain pathology.

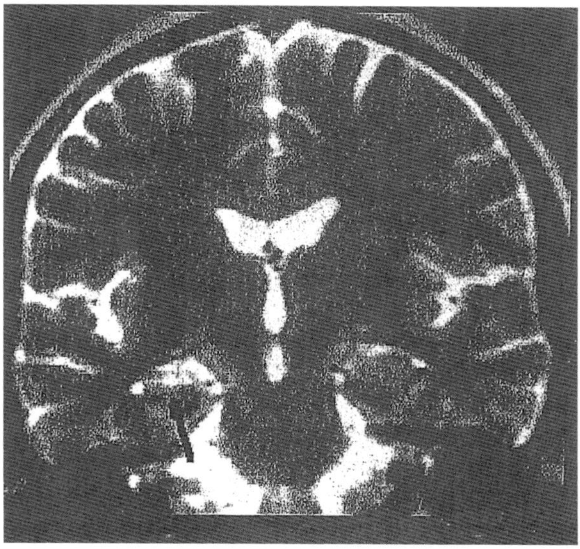

Fig. 19.4 Typical hippocampal sclerosis MRI showing increased signal in heavily T@ weighted image. With permission of Elsevier-Kuzniecky, R., and Jackson, G. Magnetic Resonance in Epilepsy, 2nd ed., 2006

Tuberous sclerosis, mentioned above, is a genetic disorder which manifests in multiple organs and sites, including the brain. It is associated with epilepsy and mental retardation: the TSC1 gene on chromosome 9q34, and the TSC2 gene on chromosome 16p13.3. Infantile spasms occur in 30–60% of tuberous sclerosis patients, depending on the study (Au et al. 2007). While many patients have both tuberous sclerosis and infantile spasms, leading to a poor prognosis, some of these patients have favorable outcomes. It was these patients who were retrospectively studied in this report (see Fig. 19.4).

Two hundred fifty-four patients seen in a tuberous sclerosis clinic over 10 years were initially reviewed. After questionnaires were analyzed and a total of 45 patients were included in the study. EEG data were obtained from the first recording made after diagnosis. Another written questionnaire was used to collect data such as spasm history, use of ACTH, use of vigabatrin, seizure frequency, etc. For mental assessment, standardized IQ tests, infant development tests, etc. were utilized.

The results showed that the average age of infantile spasm onset was 7.1 months, and the average age of cessation of spasms was 15 months. Forty-one patients had gene mutational data, and 35 had TSC2 mutations. Eighteen had ACTH treatment, with an average treatment onset of 2.3 months after diagnosis. Total seizure freedom occurred in six, partial efficacy in nine, and no efficacy in three patients. Vigabatrin was given to 33 patients, mean treatment onset was 4 months after spasm onset. Of these, total efficacy was seen in 18 patients, partial seizure control in 12, and only 2 patients had no help from vigabatrin.

In follow up for epilepsy, results showed the mean age of follow up was 9.6 years. Results showed that 13 patients were seizure free for at least 1 year, 7 had less than one seizure per week, 6 had at least one seizure per week, and 14 had at least 1 seizure per day.

The cognitive outcome showed that of 33 patients available for study, the average IQ was 53, range 7–103. Lower IQ was associated with a higher hypsarrhythmia severity score and EEG disorganization.

The authors note that, as in other studies, patients with tuberous sclerosis and infantile spasms did not achieve a favorable clinical outcome. On the other hand, the authors state about one-third of patients had at least 1 year of seizure freedom, and about one-fourth had normal or borderline mental cognition.

One hypothesis as to the mechanism of cognitive improvement states that with successful vigabatin treatment, the elimination of seizure activity is conducive to a cognitive benefit (Jambaque et al. 2000). Regarding the EEG data, one-third did not specify amplitude, and 11% were performed during sleep. Since hypsarrhythmia disappears during sleep, both these estimates may be low in this study.

The authors state that the findings stress the need for early recognition of infantile spasms, and aggressive treatment. Future studies may elucidate reasons for treatment success in some cases, but not others.

Chapter 20
Gelastic Epilepsy

Gelastic seizures are an uncommon seizure type associated most often with hypothalamic hamartomas. They have a prevalence rate of about 0.5 per 100,000. Lesions can be present in the frontal lobes, temporal lobes, limbic lobes, and most commonly, in the hypothalamus. The laughter is pathological in nature and can be spontaneous without obvious cause. The term gelastic comes from the Greek word gelos meaning laughter. During laughter, SPECT shows a bilateral area of hypoperfusion in the frontoparietal regions and in the cerebellum.

Gelastic seizures associated with hypothalamic hamartomas are easily the most common form, and can begin at birth. The hamartoma is a benign tumor most frequently located in the interpeduncular cistern, or between the tuber cinereum and the mammillary bodies. Gelastic seizures affect nearly all patients with hypothalamic hamartomas, and consist of a sudden onset of laughing, giggling, or chuckling. The seizures last only a few seconds, and crying seizures are also seen. More than one half of the patients evolve and develop focal, tonic/atonic, and tonic–clonic seizures (Tassinari et al. 1997). Some patients develop Lennox–Gastaut syndrome (see Fig. 20.1).

Cognitive and behavioral abnormalities are associated with gelastic seizures and hypothalamic hamartomas. This is especially important in patients who develop symptoms early, and whose seizures are refractory. This sequence leads to severe intellectual disability (Delalande and Fohlen 2003). Central precocious puberty is a feature in 30–40% of hypothalamic hamartoma patients. It can be effectively treated with gonadotropin-releasing hormone agonists.

In patients with mild symptoms, interictal EEG may be normal leading to missed diagnosis. Abnormal interictal EEG patterns may not actually appear until the appearance of other seizure types. In patients developing partial seizures, the ictal EEG will show focal rhythms or spike wave activity. Depth electrodes show the EEG abnormality associated with gelastic seizures originate from the hypothalamic harmatoma.

Gelastic seizures are largely refractive to AEDs, and surgery is the mainstay of treatment. The approach is difficult, and several operative methods are used, including open craniotomy or disconnection of the harmatoma. The long-term success

D.W. McCandless, *Epilepsy: Animal and Human Correlations*,
DOI 10.1007/978-1-4614-0361-6_20, © Springer Science+Business Media, LLC 2012

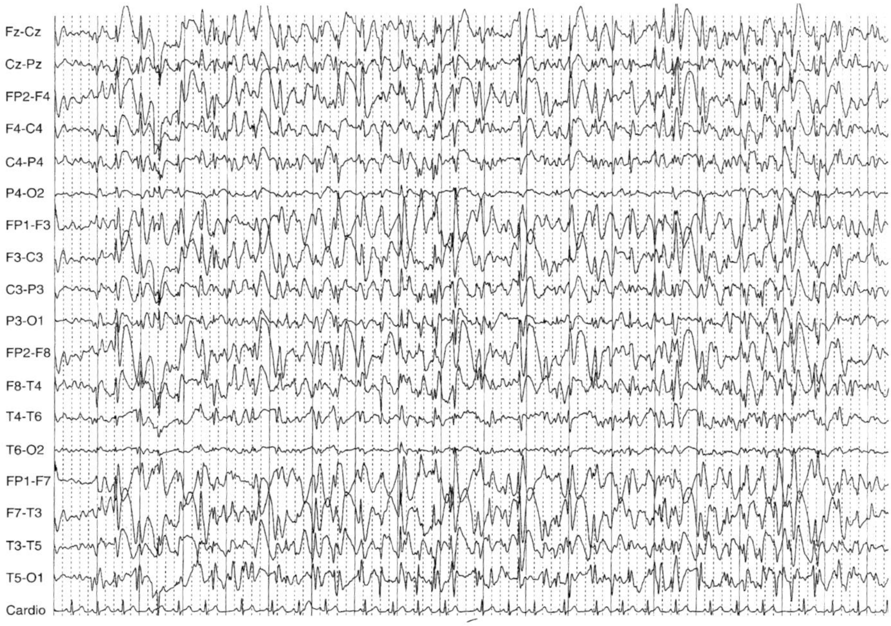

Fig. 20.1 Lennox–Gastaut EEG. Published in Crespel et al. Atlas of electroencephalography Volume 2: The Epilepsies, EEG and Epileptic Syndromes. With permission of John Libbey Eurotext

seems reasonable, with improvement in school, behavior, and development. EEG improvement correlates with the cognitive amelioration.

In a paper on gelastic seizures (Pearce 2004), comments were made regarding the site of origin. The author states only a handful of cases have described with a frontal lobe origin (<20 cases). Seizures began late before age 6 years old in most cases (six) and in adulthood is in four cases. The laughter was described as unnatural, and lasted under 30 s. MRI showed no abnormality in the frontal cortex. SPECT showed a small area of hypoperfusion in the right frontal lobe.

Temporal lobe involvement in gelastic seizures is much more common than is the frontal lobe. The age of onset in these seizures is in childhood or later. The seizure usually lacks any sensation, but some state a feeling of humor (Gascon and Lombroso 1971). Interestingly, in about half of temporal lobe gelastic seizures, an aura is reported, typical of complex partial (temporal lobe) seizures. The clinical descriptions in these seizures of temporal lobe origin are highly variable.

In the clinical spectrum of epilepsy in patients with hypothalamic hamartomas has also been explored. Hamartomas in the hypothalamus result in a seizure state characterized by gelastic seizures, precocious puberty, and multiple seizure types.

The hamartomas are intrinsically seizure processes (Kuzniecky et al. 1997), and a surgical procedure (ablation) is efficacious.

In this study, 19 patients were examined over a 10-year period. All dates were reviewed, including histories, EEGs, videos EEG, etc. The relation of the hamartoma to the hypothalamus was noted. Seven patients were surgically treated, and the harmatomas examined histologically. Patients were divided into two groups based on age.

History results showed that all patients except one previously had severe epilepsy. Sixteen patients had gelastic seizures; other types of seizures include atypical, atonic, complex partial, etc. Gelastic seizures were the first to appear in all patients except two.

The first group of patients consisted of eight children and eight adults. The diagnosis age range was 2–54 years old. Gelastic seizures were observed in 14 patients in the first group. Later seizure types such as complex partial started after a 1.3-year mean time frame. The older groups of patients were between 19 and 54 years at the time of diagnosis. The gelastic seizures completely disappeared in four patients in the adult group. Seven of the adult group were able to take care of themselves, and four were employed.

Another group of three patients had epilepsy starting in adult life. All were employed, and the seizures were not as severe. Some of this group did not start seizures with gelastic seizures, but all had gelastic seizures in the course of their disease process. The gelastic seizures lasted several seconds.

Results of EEG studies showed a background of slowed slow waves in patients younger than 10 years of age. One had infantile spasms. Slowing in background activity was also seen in older patients, but not as often. The epileptiform abnormalities were multifocal in 10 patients. Of the seven patients who underwent surgery to resect the hamartomas, all showed histological features of a neuronal harmatoma: gray matter with large and small neurons (resembling normal hypothalamic tissue), presence of large multipolar neurons, streams of myelinated fibers, and aggregates of cells with hyperchromic nuclei.

The authors note that this study of 19 patients with gelastic seizures and hypothalamic harmatomas showed that the disease process could start in adulthood, and when starting in childhood, moves to a milder form with age. The adult form is milder and does not have the catastrophic behavioral and cognitive problems (Fratelli et al. 2001).

The authors state a pattern of epilepsy types evolves with age, usually starting with gelastic seizures, then progressing to complex partial seizures, tonic seizures, and a typical absence seizures. This last group of epilepsies probably contributes to the cognitive decline of children with harmatomas. Whatever factors contribute to the development of additional seizures from gelastic epilepsy is not known.

The relation of the harmatoma to the rest of the brain is not well defined. There is a birth defect called the Pallister-Hall syndrome, which has a hypothalamic hamartoma, and is a lethal defect (Clarren et al. 1980). The neoplastic nature of some cell types in a hamartoma is controversial. One of the present adult patients had the controversial cells, with no signs of malignancy.

The authors summarize saying that hypothalamic hamartoma-induced epilepsy ranges from severe to mild. Cognitive and behavioral problems decrease with age. Gelastic seizures, prominent out in early-onset seizures from hypothalamic harmatomas, decrease in frequency and duration with age of the patient. In cases of preserved cognition, minimally invasive surgical procedures such as stereotactic thermocoagulation may be superior to resection or detachment.

The precise lobe participation in gelastic seizures is unclear. This chapter examines this issue in adult patient who had seizures consisting of hypermotor and laughing behavior. Stero-electro-encephalographic recordings were used in order to identify the epileptogenic zone. The supplementary motor area (SMA) was the target of this study (Chassagnon et al. 2003).

The case reported was of a 24-year-old right-handed patient who had a refractory partial epilepsy. The patient's first seizure was at age 4, then following treatment, reappeared at age 10. Seizures were then successfully treated with carbamazepine until age 18, when he became intractable to AEDs, and was having seizures at a frequency of 30 per day. The seizures consisted of an initial painful sensation of the right shoulder. Then followed arising of the right arm, swinging movements of the trunk, and uncontrolled laughter. Interictal examinations were negative, and MRI was normal.

Interictal scalp EEG recordings showed paroxysmal slow waves, sharp waves, and spike wave complexes. Actually, results showed low voltage fast activity and spike discharge bilaterally over the precentral regions. The laughter occurred at the time of trunk swinging and was accompanied by a smile. Stero-electro-encephalograhic (SEEG) was used in order to locate the focus, and in preparation for surgery. Eleven electrodes were used to map the left premotor region, and two on the right explored the SMA and anterior cingulated gyrus (see Fig. 20.2).

Results showed interictal discharges in the left hemisphere at the SMAp, pre-SMA, and cSMA. Low-intensity stimulation of SMAp induced seizures similar to the spontaneous seizures reported by the patient. No discharges were seen in the pre-SMA, premotor lateral cortex, or cingulate gyrus. The patient was surgically treated using steroataxic monopolar electrode radiofrequency lesions, and is now seizure-free.

The authors note that the patient's lesion was small and limited to SMAp and dorsal cingulate gyrus. Data indicate the epileptic zone was ahead of the paracentral sulcus. The clinical semiology did not have the typical findings of SMA seizures (Bancaud and Talairach 1992). The authors also comment saying their data speak to the role of the anterior cingulate gyrus in gelastic seizures. The data suggest the laughter in the present patient was contributed to by the supra- and infra-cingulate-sulcus fields.

Hypothalamic hamartomas are rare, and result in gelastic seizures. The hamartomas associated with this seizure form are located on the inferior hypothalamus and tuber cinerum. They are developmental malformation and independently associated with the Pallister-Hall syndrome, which is also characterized by polydactyly and bifid epiglottis. This chapter describes a patient with a serious "status gelasticus" who successfully underwent an emergency transcallosal resection of his hypothalamic hamartoma.

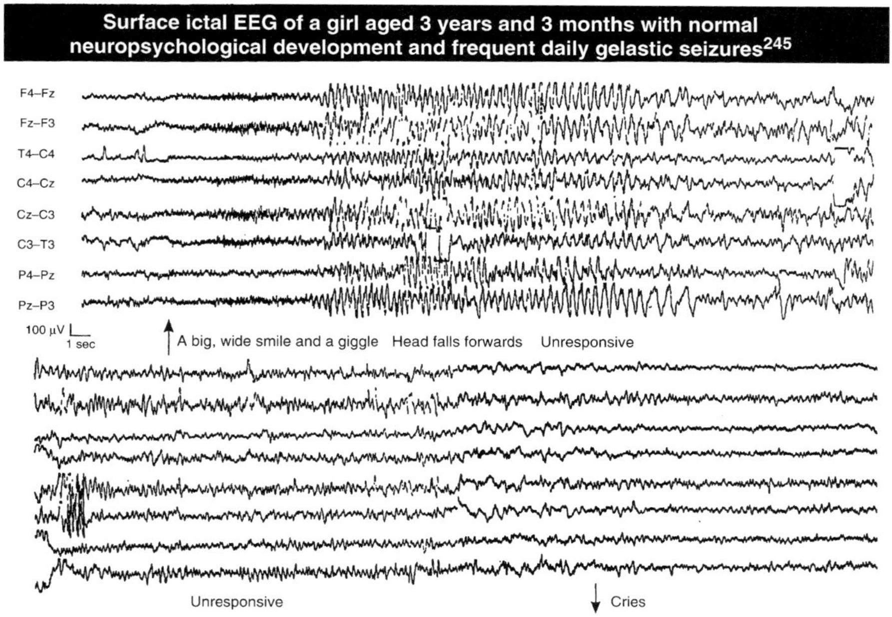

Fig. 20.2 EEG of a 3-year-old girl with normal neurophysiological development and gelastic seizures. The condition was first noted at age 2 when the patient had a "big smile" and goggles. With kind permission from Springer Science+Business Media: Epileptic syndromes and their treatment. 2010, pp. 313, Panaylotopoulos. Fig. 10.14

This was a case of a 30-month-old boy who had gelastic seizures initially at 4 months. MRI showed a hypothalamic hamartoma. The seizure was characterized by sucking motions, then laughing. At times, the patient would become violent, and occasionally had postictal depression. In the 6-week period prior to admission, he had an upturn, experiencing a gelastic seizure about every 5 min, even persisting during sleep.

Of many AEDs, none helped. A 24 h scalp video EEG and MRI showed the hypothalamic hamartoma. The video EEG confirmed ten gelastic seizures per hour. The patient underwent a corpus callosotomy. The hamartoma was localized and resected. Pathology confirmed the hypothalamic hamartoma, which microscopically showed disorganized neurons and glia.

The authors note that this surgery served to essentially eliminate the status gelasticus in this patient. At a 1-year follow-up, he has a brief gelastic seizure about once every 2 weeks.

The authors state this case represents the first to be treated for status gelasticus by an emergency hypothalamic hamartomectomy. While somewhat rare, hypothalamic hamartoma is a potentially devastating seizure disorder, and can be successfully treated surgically.

Another paper has described a patient with a dual pathology of complex partial seizures and hypothalamic hamartoma-related gelastic epilepsy (Palmini et al. 2005).

The case is of a 35-year-old man whose seizures started at age 10. The patient's seizures were initially infrequent and consisted of secondarily generated tonic–clonic episodes. A epigastric/nausea aura was present. The seizures were AED refractory. The patient was prepared for surgery, and an arachnoid cyst was localized, and SPECT located a temporal lobe area of hyperperfusion. He underwent a resection of the left temporal cortex and marsupialization of the cyst.

The immediate postoperative period was complicated by status gelasticus. An MRI showed a small 8 mm hypothalamic hamartoma. FDG-PET showed an intense hypermetabolism in the hypothalamic hamartoma. The patient was treated with carbamazepine with success.

The authors state this is the first case of a patient with complex partial seizures being treated by cortical resection, only to later find a hypothalamic hamartoma. This only showed after temporal lobe resection. Postoperative ictal FDG-PET and SISCOM both showed the hypothalamic hamartoma was the source of the gelastic seizures.

Hypothalamic hamartomas are a well-accepted intrinsic source of seizure activity resulting in gelastic epilepsy (Munari et al. 1995; Berkovic et al. 1997). Hypothalamic hamartoma-induced epilepsy is considered an epileptic encephalopathy, this secondary epileptogenesis may occur independent of the hamartoma. The present study (Ng et al. 2006) is a prospective one in which patients with refractory hamartoma epilepsy were enrolled for evaluation for possible surgery.

A total of 26 patients were identified. Mean age was 10.0 years, range 2.1–24.2 years old. All had AED refractory seizures, and the mean age of onset was 8.6 months. All except one patient had daily seizures. One patient had status gelasticus with over 200 seizures per day. Seventeen (65%) of patients had mental retardation with an I.Q. below 70. Behavioral problems consisted of rage (89%), autism, etc. Central precocious puberty was prevalent in 42% of the patients.

Surgical results showed 54% of patients to be seizure-free for all seizure types. The cessation of seizures was immediate in most of these patients. Twelve patients continued to have seizures. Nine had a seizure reduction of greater that 90%; two experienced no decrease in seizure frequency. Seizure outcome correlated with lesion size, and efficacy of seizure control correlated with extent of surgical resection.

The authors note that there were no mortalities, and the most common adverse effect was a partial loss of short-term memory, seen in 58% of patients. This study states the authors prove that hypothalamic hamartomas can be safely operated on, via transcollosal resection, and reduce or eliminate seizures. Parental perception of an overall cognitive and behavioral improvement was positively viewed. The potential for secondary epileptogenesis predicts that the potential for seizure freedom will decline.

The authors conclude saying that they think earlier surgery is suitable for patients with refractory gelastic epilepsy. This serves to diminish the likelihood of cognitive and behavioral problems, plus limiting the opportunity of secondary seizure development. Future studies are clearly warranted.

In addition to gelastic epilepsy resulting form hypothalamic hamartomas are the cognitive deficits which are produced (Quiske et al. 2006). The authors point out very few studies exist in which any of these problems outside of epilepsy are even mentioned.

In this study, 13 patients with hypothalamic hamartoma-related gelastic epilepsy were studied. The patients received a full presurgical work up including history, neurological examination, high-resolution MRI, long-term video EEG, and full neuropsychological examinations. The neuropsychological tests included: I.Q. testing, tests of attention and executive functions, verbal memory tests, and visuospatial abilities.

The authors state that this study is the first detailed report on cognitive performance in a large group of adolescent and adult gelatic epilepsy patients. These studies showed high prevalence of defects. Most frequent were impairments in working memory and visual/verbal learning. Intellectual performance was also compromised in these patients. Cognition was unimpaired in only two patients.

Severe global memory dysfunction was seen in about half of the patients. A correlation between hypothalamic harmatoma volume and cognitive flexibility. These results partially agree with other studies showing cognitive deficits in gelastic epileptic children (Frattali et al. 2001).

The authors state their study shows a correlation between partial seizure frequency and cognitive flexibility and mental retardation. These both involve frontal lobe functions. Since ictal discharges involve the frontal lobes, a correlation between seizures and frontal lobe functioning is not surprising.

Size and location of hypothalamic hamartomas are important. Large hamartomas tend to have a poorer outcome and have a worse outlook for cognitive deficits. Small hamartomas have the opposite effect. In addition, hamartomas are located near the mammillary bodies. Proximity to the mammillary bodies can induce Korsakoff's psychosis. Because of connections between the hypothalamis and the hippocampus, it is possible some memory problems might relate to hippocampal dysfunction.

Once again, more studies using larger numbers of patients would be beneficial. Longitudinal studies using consistent cognitive assessment would be highly valuable. While relatively rare, a multicenter approach should be advantageous.

A recent paper (Wu 2007) looks at possible abnormalities in the GABA A receptor in patients with gelastic epilepsy and associated hypothalamic hamartomas. The relation between other epilepsy forms and GABA A structure and function is well documented. In the present study, dissociated single neurons from surgically resected hypothalamic hamartomas were obtained. The pharmacological properties and subunit characteristics were determined.

Results showed that GABA induced an inward current (IGABA) which was mimicked by muscimol and was blocked by the GABA receptor bicuculline. The EC_{50} was 6.8 μM and the current voltage curve was linear. IGABA had a high sensitivity to pentobarbital and pregnanolone. When compared to control hypothalamic cells, those of hypothalamic hamartoma were similar as regards GABA A receptor subunits.

The author suggests that the data are suggestive that the GABA A receptor subunits on small hypothalamic hamartoma neurons are normal as regards pharmacosensitivity and subunit composition. The author concludes saying this may have a bearing on mechanism of inhibitory neurotransmission in human hamartoma tissue. Another interesting paper by the same group on GABA A receptors further delineates the above experiments (Kim et al. 2008).

An interesting paper from Argentina describes clinical features, treatment, and prognosis in a group of patients with gelastic seizures in a developing country (Papayannis et al. 2008). The gelastic seizures are usually associated with a non-neoplastic tumor, the hypothalamic hamartoma, which contains glia and neurons in a disordered fashion. The hamartoma is intrinsically epileptogenic, producing gelastic seizures.

In this study, eight patients were seen over a 10-year period with a diagnosis of hypothalamic hamartoma and gelastic epilepsy. The diagnoses were made using MRI, video EEG, neurological examination, and the I.L.A.E. classification. The hypothalamic hamartoma was defined as "small" if it is less than 1 cm in diameter, and also classified as either pedunculated or sessile depending on its mode of attachment to the hypothalamis.

Results showed a mean age of onset as being 4.5 years (range 0.3–13 years). All patients had gelastic seizures, and all patients also had complex partial seizures. Other seizure types included generalized tonic–clonic, atonic, myoclonic, and partial motor seizures.

All patients had normal developmental milestones prior to the onset to epilepsy. Afterwards, half developed mild to severe mental retardation and cognitive defects. Video EEG was done in three patients, and these showed an ictal recording of diffuse attenuation. In all eight patients, the gelastic seizures were brief with no change in EEG.

Results from MRI showed that seven patients had a small hamartoma, while one had a large pedunculated hamartoma. All received AEDs either as monotherapy or polytherapy. Both "old" and "new" AEDs were used. Fine patients underwent surgery in order to correct the seizures. The outcomes were mixed, with some improvement early, but most patients had their seizures return.

The authors state that in hypothalamic hamartoma, the first symptom is usually gelastic seizures, as seen in their study. Multiple seizure types are characteristic in patients with hypothalamic hamartomas, and all seem to arise from the hamartoma since its ablation eliminates seizure activity. Patients with mild to moderate mental retardation had complex partial seizures.

Some authors propose that the hamartoma should be removed early, then treat the remaining neurologic issues (Striano et al. 2005; Berkovic et al. 2003). This aids the further treatment of the epileptic encephalopathy. Only five of the eight patients had surgery due to the difficulty of attempting the surgical techniques.

The authors conclude saying that this disorder has a wide spectrum of semiology. Longer follow-up is essential for estimation of efficacy of treatments.

Treatment options are limited in developing countries, complicating further this difficult-to-treat syndrome.

Striano et al. (2009) have published a paper looking at the potential reversibility of hypothalamic hamartoma and gelastic seizures. The authors note that gelastic seizures rarely occur in frontal or temporal lobe epilepsy, whereas they almost always occur in hypothalamic hamartoma patients. The development of a catastrophic epileptic encephalopathy, with severe mental retardation, and equally severe cognitive deficits was soon widely recognized (Berkovic et al. 1988). The hamartoma itself generates gelastic seizures, which can secondarily generate additional seizure types. The secondary generation may be an anatomical feature of hypothalamic connections (Freeman et al. 2003).

Hypothalamic hamartomas are benign developmental malformation located between the mamillary bodies and the infundibular stalk. The relation between the hamartoma and gelastic seizures is conclusive. The prevalence of hypothalamic hamartomas is about 1–2 per 100,000 people. Hamartomas are closed as sessile or pedunculated. Precocious puberty is a feature of hypothalamic hamartoma.

Gelastic seizures are often the first symptom of hypothalamic hamartomas. The seizures are brief, usually occur in day time, and may occur in clusters. Crying (dacrystic) seizures are occasionally seen in hypothalamic hamartoma patients. In gelastic seizures, there is no loss of consciousness, no sense of mirth, but autonomic signs (facial flushing) are often seen. In the movement to a catastrophic syndrome like the Lennox–Gastaut syndrome, multiple seizure types can secondarily develop. Slowing of background EEG, as well as interictal local and generalized paroxysmal activity, is seen.

After the onset of epilepsy, most patients show cognitive degeneration, as well as behavioral problems. There is a larger spectrum of cognitive and behavioral features associated with large hamartomas. Conversely, small hamartomas are often in patients with no appreciable cognitive/behavioral disorders.

AEDs are not effective in gelastic seizures, and they also cannot alter the course of the cognitive and behavioral problems. Surgical ablation of the hamartoma is effective. Early surgical approaches were not without risk, so nonconventional surgeries such as the gamma knife radiosurgery is a safe effective technique for treatment. Methods which do not employ a stereotoxic frame and a cyberknife may offer surgery for hamartomas devoid of risk.

From a microscopic standpoint, examination of surgical specimens of hypothalamic hamartomas removed from patients show nodules of small neurons, associated with gelastic epilepsy. The following paper describes a case in which the patient had gelastic seizures, and sectioning of this hamartoma showed nodules containing glial cells, but scant few randomly arranged neurons.

The case is of a 5-year-old boy who had laughing seizures since birth. He was placed on small Levetiracetam doses, and had a significant reduction of seizures (from 15–20 per day to 3–5 per day). At 2 years of age, he was found to have a hypothalamic hamartoma. Video monitoring staring spells and bilateral spike wave pattern (1.5–2.5 Hz).

The resection took two attempts to eliminate gelastic seizure and to remove all of the hamartomas. Histopathological examination showed nodules basically with only a few disarrayed neurons. This pathology is associated with precocious puberty. The neurons were not organized and the processes extended into the masses of glia.

The authors comment this descriptive nodular feature is seen in patients with gelastic epilepsy having an abundance of small neurons. These small neurons are thought to be GABAergic interneurons driving larger ganglion output cells (Fenoglio et al. 2007).

The possibility exists that the gelastic epilepsy in this case was generated by these glial cells. Astrocytes may play an important role in seizure activity (Tian et al. 2005). The authors conclude saying this case seems to be the first showing gelastic seizure without neurons of significance in the hamartoma. The authors comment that it is important to closely examine hamartomas microscopically so as not to misdiagnose the seizure condition as pilocytic astrocytoma. A neuronal stain should be utilized.

Chapter 21
Landau–Kleffner Syndrome

The Landau–Kleffner syndrome is a form of epilepsy, which is classified as an epileptic encephalopathy. This indicates an encephalopathy, which is made worse by epilepsy, just as bilirubin encephalopathy is made worse by bilirubin. A closely aligned condition is called epilepsy with continuous spike waves during slow-wave sleep (CSWS). These two syndromes were once thought as one, but are now considered to be two separate but related epilepsy entities. Several other epileptic encephalopathies are described, including infantile spasms and Lennox–Gastaut syndrome.

Both Landau–Kleffner syndrome and CSWS are rare, with around 300 cases of Landau–Kleffner syndrome described in the literature, and less than one fourth the number of CSWS cases. Overall, the syndromes are responsive to AEDs, and seizure control is usually achieved, but the conditions are associated with significant neuropsychiatric comorbidities, such as behavioral problems, hyperactivity, ADD, and mood instability.

Clinical features include presentation between age 2 and 8, with a mean around 4.5 years. The first symptom may be an apparent language deficit which is progressive. It is central in origin and manifests as word deafness or the inability to decipher words (verbal auditory agnosia). This can reach a stage where a car horn honking or a door slamming would be an unknown sound to the patient.

Following the onset of word deafness, are usually seizures, almost all of which are idiopathic. Seizure types include atypical absences, partial clonic, atypical tonic, or generalized tonic–clonic seizures. EEG features during wakefulness show centrotemporal spikes and spike wave discharges. These may be lateralized or bilateral. In non-REM sleep, generalized spike wave discharges occur very frequently and contribute electrical status epilepticus in sleep (ESES).

Children with Landau–Kleffner syndrome or CSWS may not have frequent seizures, but all have severe regression. There is a severe loss of I.Q. In one series of CSWS patients (Tassinari et al. 1992), alterations include severe loss of I.Q., aggression, mental deterioration, hyperactivity, difficulties interacting, and two patients became psychotic. Some patients seem borderline autistic. The language

D.W. McCandless, *Epilepsy: Animal and Human Correlations*,
DOI 10.1007/978-1-4614-0361-6_21,

disturbance in CSWS is not like the Landau–Kleffner syndrome of verbal auditory agnosia, but is an expressive aphasia.

The neuropathology of the Landau–Kleffner syndrome is largely unknown. Pathology has been noted in the left posterior Sylvian fissure. Fluoro-deoxyglucose PET shows some involvement of the temporal lobes. In keeping with the epileptic encephalopathy concept, even Landau and Kleffner (1957) note that the continuous convulsive discharges in brain language areas act to render the tissue "functionally ablated." The neuropathologic changes are still, however, uncertain.

From a treatment standpoint, AEDs can act to suppress overt seizures, but ordinarily do not affect ESES or language. In some cases, ESES are steroid-sensitive, and can be treated with prednisole once per week. It goes without saying that all patients need a thorough medical/neurologic workup. Speech/language evaluation should be very important and therapy should start immediately. There may be improvement in the language deficits when the seizures come under control. Data exist showing that corticosteroids can act to control both seizures and language deficits, probably more often than AEDs alone. Surgical approaches may be warranted in some cases.

Outcomes can be variable. Risk factors cannot always be identified and it may therefore be impossible to predict outcome at the onset of treatment. The epilepsy component outcome is usually favorable; the cognitive outcome is rather the opposite. The age of onset is a partial predictor, in that onset before 4 years of age usually signals a less favorable result (Bishop 1985). In another group of patients, 88% had residual language problems (Shinnar et al. 2001). In some studies, I.Q.s were normal in the Landau–Kleffner syndrome, but nearly always present in CSWS was significant mental retardation (Scholtes et al. 2005).

As stated above, there are many similarities between the Landau–Kleffner syndrome and CSWS as regards clinical course, but the outcomes are different. One study has focused on these different outcomes (Praline et al. 2003). Many workers postulate that these two clinical entities are highly similar and may be slightly different expressions of the same pathologic mechanisms (Rossi et al. 1999). The poor prognosis of these two syndromes is usually based on cognitive outcomes rather than the epilepsy, which usually abates in adulthood. The study goal was to assess adulthood cognitive defects in Landau–Kleffner and CSWS and determine which childhood features of each might influence outcomes.

In this study, seven young adult patients were enrolled, five with CSWS, and two with the Landau–Kleffner syndrome in childhood. Questionnaires and medical records were able to supply early information, and a variety of cognitive tests were administered to assess mental function in adulthood. Patients with the Landau–Kleffner syndrome were classified as regards aphasia on five different levels. Results showed that the two patients in the Landau–Kleffner syndrome group were seizure-free for at least 5 years; the CSWS group had one patient member with active epilepsy. The age range of these seven adult patients was 16–26 years. Three of five patients with CSWS were mentally deficient with I.Q.s in the mid to upper 50s. Two were in the normal I.Q. range. The two patients in the Landau–Kleffner group had normal I.Q.s at a nonverbal intellectual level. As regards language in the adult patients, those in the CSWS group of five were normal in the naming tests. The two in the Landau–Kleffner group were severely pathologic in the naming tests. All patients in the CSWS except one were deficient in reading and writing. Four of five in the CSWS group exhibited slow thinking and speech, and inability to control thoughts.

The authors note that in the CSWS group, two subgroups could be distinguished. The first one consisted of two intellectually normal patients who were integrated into society. The other three were poorly integrated into society due to psychological and neurological sequelae.

Executive functions were assessed using the Stroop test and the trail marking test. The authors found no evidence of a dysexecutive syndrome. The authors state this could indicate that the tests were not sensitive enough. Another factor might have been the lack of any frontal lobe lesion foci usually associated with lack of executive function. Age of CSWS onset did not play a role in long-term prognosis. The two patients with the best long-term prognosis were those without a focus on waking EEG.

Length of epilepsy may have been a predictor of outcome. Two patients with long duration of seizures had the most severe outcomes. One of two patients with long duration of seizures had the most severe outcomes. One of two patients who had learned sign language remained poorly integrated. These patients had poor aftereffects as regards language, but maintained good intellectual capacity. The authors state that while Landau–Kleffner syndrome and CSWÍ are similar clinically, the two have different outcomes. The overall number of patients in each group prevents a fully meaningful comparison.

In an interesting case of Landau–Kleffner syndrome, a 3.5-year-old girl developed stuttering. She had an early history of benign myoclonis epilepsy at 3.5 months, and she was treated with valproate. She was withdrawn from the AED after being seizure-free for 2 years. Because of her previous epilepsy history, she had a sleep EEG, which showed multiple spike and wave discharges on the left tempo-cortical and frontal regions. Background was normal. This case had myoclonic seizures at an age of 3.5 months, where the usual seizures are complex partial, atonic, and generalized tonic–clonic seizures.

This patient was successfully treated for stuttering with ACTH and immunoglobulin therapy after the diagnosis of Landau–Kleffner syndrome was made. There was a rapid (2 months) recovery of speech with ACTH treatment, and the EEG returned to normal. The patient's early history prompted an EEG evaluation, leading to the Landau–Kleffner diagnosis, prompt therapy, and reversal of the speech impediment. This potential diagnosis should be considered in cases of stuttering.

In another case study (Pedro and Leisman 2005), a multimodal approach aimed at facilitating interhemispheric communication was attempted. This case is a 14-year-old right-handed girl, delivered normally from an uncomplicated pregnancy. Development seemed normal, but at 3 years old, she was noticed to have disrupted language comprehension. The patient had an abnormal EEG showing right centrotemporal spike discharges, a normal MRI, and was diagnosed with

Landau–Kleffner syndrome. She was treated with valproate, which improved her language skills.

Prior to the novel treatment paradigm, the patient was fully tested neurologically and physiologically. At that time, she was noted to have left hemisphere and right cerebellar deficits. A 12-week training course lasting 4.5 h per week of optokinetic, visual, vestibular, olfactory, auditory, and somatosensory stimulation was initiated. One goal was to facilitate interhemispheric communication.

Results showed that the 12-week training period significantly improved the patient's auditory processing. The patient also had gained in language, auditory, and motor skills. The patient was better able to make concise statements. She was able to consistently listen to parents and follow directions.

The authors note that their patient had an abrupt loss of language function, abnormal EEG recordings, improvement of ability to read, and motor skills disruption/incoordination. The above-described multimodal approach aimed at increasing interhemispheric communication seemed efficacious as evidenced by more controlled EEG, and improvement in reading, language, and behavior/social measures.

The overall management of the Landau–Kleffner syndrome, including current thinking as regards surgery, is still not clear, probably due to relative lack of the lesser number of patients. A paper reports studies of traditional AED treatments, steroids, I.V. immunoglobulin, and surgery as treatment modalities in the Landau–Kleffner syndrome (Mikati and Shamseddine 2005).

The major clinical features of both Landau–Kleffner syndrome and CSWS are presented above. Clinical seizures do not always appear in children with the Landau–Kleffner syndrome, but EEG abnormalities are always present. Seizures may vary, but the most frequent are complex partial, atonic, and generalized tonic–clonic seizures. The clinical picture sometimes resembles autism, and this should be termed the Landau–Kleffner variant. The EEG spikes of autism are usually of a lower frequency than those of the Landau–Kleffner syndrome.

Treatment with AEDs is generally effective in controlling seizures in Landau–Kleffner syndrome patients, but aphasic improvement is variable. Valproate has been somewhat (40%) effective in improving language (Holmes and Riviello 2001). From a diagnostic standpoint, the rectal diazepam test can differentiate between CSWS and hypsarrhythmia. CSWS patients with hypsarrhythmia show 0% response.

ACTH is supposedly effective in reversing language changes, cognitive, and behavioral alterations in Landau–Kleffner syndrome patients. Oral prednisone is a typical drug of choice. Adverse effects are of concern and include hip necrosis, hypertension, immunosuppression, and serious infection (see Table 21.1).

I.V. immunoglobulin was first administered in 1997 to a Landau–Kleffner patient (Fayad et al. 1997). This represented the first successful I.V. immunoglobulin attempt. The patient had had 3 years of AED attempts as well as prednisone. I.V. immunoglobulin therapy resulted in a complete clinical and EEG response.

A surgical approach for Landau–Kleffner syndrome called multiple subpial transection was developed to section horizontal interneurons, thereby interrupting the epileptiform activity while maintaining physiologic function. This has resulted in a 50% success rate (Morrell et al. 1995), in that there was an age-appropriate

Table 21.1 AED treatments and their effectiveness for Landau–Kleffner syndrome

Usually effective	Maybe effective	Not effective
Valproate	Topiramate	Phenobarbital
Ethosuximide	Vigabatrin	Phenytoin
Sulthiame	Felbamate	Carbamazepine
Benzodiazepines	Ketogenic diet	

Adapted from Lagae, L. Epilepsia 50: p. 59, 2009

speech recovery. There have been other similar or better results from this procedure. Speech therapy (as well as behavioral therapy) obviously plays a major role in the overall treatment of Landau–Kleffner syndrome patients.

The authors conclude saying the initial therapy attempt should be AEDs. Lack of response by 2 months should prompt analysis of other treatments mentioned above. Surgical methods are a final treatment modality.

The evaluation of cognitive performance in Rolandic epilepsy has been examined (Metz-Lutz and Filippini 2006). Rolandic epilepsy is considered to be a very similar seizure syndrome to Landau–Kleffner syndrome. In this study, a cohort of 18 girls and 26 boys were enrolled. The diagnosis age ranged from 4 to 7 years. Based on EEG, they were divided into typical and atypical cases. Typical cases were shown to have a focal spike wave discharge against a normal EEG background, and atypical cases had a slow spike wave discharge, asynchronous foci, or generalized 3 cps discharges.

A variety of neuropsychological tests were conducted two times per year. Time was divided into three phases: onset of seizures, an active period (time of most spike and wave discharges), and recovery, 1 year after withdrawal of treatment. Results of this study showed that the atypical group had statistically lower full-scale I.Q. and verbal I.Q. as compared to those in the typical group. Performance I.Q.s were significantly lower only at onset and active phases. Verbal short-term memory and verbal learning tests were worse in the atypical group. The results indicate a strong group effect, showing a predictive value for typical vs. atypical Rolandic epilepsy. The authors note that clinically, in Rolandic epilepsy patients, it is possible to see early on a more severe decline in cognitive function correlated with EEG features.

CSWS is a rare seizure disorder, related to, as stated above, the Landau–Kleffner syndrome and is an example of an epileptic encephalopathy. It is characterized by spike waves during slow sleep and evidence of neuropsychological and behavioral disorders. The relation between CSWS and the Landau–Kleffner syndrome is close enough to suggest to some that they are a continuum (Tuchman 1994). The language disorder in Landau–Kleffner syndrome is a verbal auditory agnosia, and the language disorder of CSWS is not as well understood. These two clinical entities show common features such as both are age-related, appearing in childhood, and have a mild epilepsy and a severe neuropsychological component.

In this paper (Debiais et al. 2007), ten CSWS patients were examined. The patients were either in an active phase or in remission. Only patients who strictly fulfilled the I.L.A.E. criteria and patients who established electrical status epilepticus with a spike wave index greater than 85% during sleep were included. Symptomatic cases were not included. Major interictal focus was frontal for four patients, temporal for four, and parietal in one, and occipital in another patient. Age range of the patients examined was 9–23 years.

In terms of seizures, four were considered cured by treatment, two never had seizures, and four had a right-sided EEG interictal focus. All had taken AEDs including carbamazepine, valproate, and benzodiazepine. Language and literacy skills were evaluated by several standardized language tests. In some cases, tests were de novo developed for assessing pragmatic performance (McDonald 2000). An example, called the letter test, asked a subject to imagine that they were going to write a letter to a friend, then tell the examiner everything needing to be done to prepare and mail the letter, and achieve the goal.

Results showed that nine of ten patients had very low scores on tasks of lexical and grammatical judgment. All ten subjects had normal results on oral comprehension. Prognostic performance was lower overall. In the letter test, for example, over 50% produced a decreased number of essential steps compared to a group of 8-year-old children. Nonverbal tests were scored lower in CSWS patients as compared to controls.

The authors state no correlation was noted between language performance and age of onset of epilepsy, age of initial EEG spike waves, duration of CSWS, or topography of the focus. Active vs. remission phase patients was also not correlated. The patients in remission still had severe language problems. These patients also had severe language disorders. These data support the concept that in CSWS patients, the greatest deficiency is in lexical and grammatical judgment. This impairment is different than that seen in the Landau–Kleffner syndrome. Language judgment, significantly changed in the present study, needs further examination with greater number of patients. Pragmatic abilities were also altered in patients in this study.

Speech disorders have suggested that CSWS patients have pragmatic deficits. Some hypothesize that the right hemisphere might play a role in pragmatic function since patients with right-sided lesions tend to demonstrate pragmatic disorders. The authors note that the exact pathophysiology of CSWS is unclear. These unknown mechanisms are persistent since they persist even after the disappearance of EEG alterations. The degree of recovery may relate to the maturation of language development and the child's age. The authors conclude saying that a better understanding of the specific nature of cognitive disorders in CSWS should help improve treatment.

A paper from Finland (Paetau 2009) discusses and analyzes the technique of magnetoencephalography (MEG) as a method for localizing spikes. There seems to be a consensus that there is a correlation between epileptoform discharges during non-REM sleep and language regression in the Landau–Kleffner syndrome. The discharges are bilateral and located in the temporal lobes, centrotemporal, and parietooccipital regions.

The concept exists that prolonged epileptiform discharges can serve to produce a contralateral focus. This mirror focus is initially dependent on the original focus, but over time may become independent (Morrell et al. 1989). The primary focus can be identified by ensuring latency between the two sides' discharges. The latency is usually 20–45 ms. Use of both EEG and MEG can facilitate this identification process.

MEG detects very weak magnetic fields outside the head and is sensitive to fissural currents, thereby identifying fissural generators (Hari and Lounasmaa 1989). The use of both EEG and MEG is important in localization studies, for example, prior to surgery.

The authors note that they have studied 28 Landau–Kleffner syndrome patients. Results showed the age range of patients was 3.5–12 years. Twelve had normal auditory evoked responses, four had the response in only one hemisphere, and seven had no auditory responses. Analysis of epileptiform spike sources occurred bilaterally in 21 patients, unilaterally in four patients, and three had no spikes. Eleven patients had bilateral spikes in both Sylvian cortices and six had a single pacemaker. The six with unilateral pacemakers were judged as good surgical candidates.

The authors comment that verbal auditory agnosia is a key feature of the Landau–Kleffner syndrome. Results of EEG and MEG and other techniques indicate that focal epilepsy of auditory cortical sites contributes to the language dysfunction of the Landau–Kleffner syndrome. Most children with Landau–Kleffner syndrome had spikes in the intra-Sylvian and perisylvian cortices. These are adjacent to the auditory association regions. More than one half of the patients had abnormal AERs. The AER technique is a relatively early noninvasive method for analyzing central auditory activity.

After surgery (multiple subpial transection, MST), the MST-treated cortex no longer generated spikes. One key to success with MST is to have good localization, and focused MST on one sole pacemaker. The occurrence of transient recovery after unilateral MST for bilateral pacemakers supports the hypothesis stated above by Morrell. The use of MEG focuses on bilateral latency differences between perisylvian regions. The use of both EEG and MEG has the advantage of recording both sulcal and gyral activity.

Treatment for Landau–Kleffner syndrome and CSWS is still uncertain. This study (Arts et al. 2009) examines results of the administration of immunoglobulin to these epilepsy patients. Corticosteroids are a widely used therapy, with success, but the long duration of treatment and the potential for serious adverse effects limit the appeal. Several other studies have shown efficacy of I.V. immunoglobulin (Fayad et al. 1997; Mikati and Shamseddine 2005). This study examines immunoglobulin treatment in six children with either the Landau–Kleffner syndrome or CSWS, over a 5-year period.

This study included six children, who received a standardized protocol, and was followed for 1 year. Initial workup included history, physical, neurological, and neuropsychological examinations, and 24 h EEG studies. Neuropsychological evaluations were performed at 6 weeks, 12 weeks, 6 months, and 1 year.

Results showed two girls and four boys, with an average age range of 4–9 years. Three were diagnosed with Landau–Kleffner syndrome, and the other three with CSWS. Three patients completed the entire course of immunoglobulin treatment, and two children showed no favorable response to immunoglobulin. One of the three had an obvious improvement, which deteriorated with cessation of therapy. Improvement returned upon reinstitution of therapy.

The three who withdrew were then treated with prednisone. One of these three made a complete clinical recovery, while the other two improved by the end of the first year. The long delay precluded the assignment of cause of the partial recovery. During the 12-month period of study, all continued to have severely abnormal EEG results. In only one case did neuropsychological evaluation show any improvement.

The authors comment that most studies with immunoglobulin have focused on forms of epileptic encephalopathies: the Lennox–Gasteaut syndrome, infantile spasms, Rasmussen encephalopathy, Landau–Kleffner syndrome, and CSWS. The authors correctly note that there is a lack of controlled randomized studies. Only small anecdotal positive studies have been published; negative results frequently do not reach print. Not only this, but also cases are difficult to interpret due to the highly variable course and ever changing definitions and criteria. Descriptions oft begin when the patient is at a low ebb, and when improvement follows, it could be due to variation in course, not treatment-based improvement.

The authors conclude saying that there is currently no conclusive evidence to support any specific treatment modality (or combination thereof). Most currently believe that corticosteroids are the preferential treatment. It will take large multicenter-controlled randomized studies to resolve this problem.

Another paper (Lagae 2009) looks at AED treatment (valproate, ethosuximide, and benzodiazepines) in Landau–Kleffner syndrome patients. Specific to this classification of seizures and its variations (CSWS) are significant nocturnal EEG abnormalities, and usually associated cognitive/behavioral problems (Nickels and Wirrell 2008). Any goal of treatment must take into account the language and behavioral problems associated with these childhood epilepsy encephalopathies. The correlation relations of the spike wave index and clinical states are somewhat mixed.

The findings of correlations are reflected by studies showing that the language/behavioral problems are less intense when the EEG shows little or no epileptic abnormalities (Lagae et al. 1998). So treatment should consider not only effects on seizure phenotypes, but also on the frequency of EEG changes. Also, newer AEDs (see above) have recently been shown to be efficacious. Topiramate and levetiracetam have a substantial positive effect (Rigoulot et al. 2003; Bouwman and van Rign 2004).

The author quickly/accurately points out that no systemic, controlled, randomized studies exist regarding AEDs, and only clinical case studies and class IV data are available. Outcome data are short-term. As stated before, multicenter-controlled studies are essential. The lack of adequate studies is most certainly related to small overall numbers of patients (on several levels).

A couple of early studies on five children with the Landau–Kleffner syndrome stated that valproate, ethosuximide, and clobazam were effective in treating the Landau–Kleffner syndrome. They also noted that phenobarbitol, phenytoin, and carbamazepine might exert a negative effect. The three AEDs advocated in these early papers were then tried in multiple studies. Later results using a sequential additive AED approach achieves good success in the Landau–Kleffner syndrome, but not in CSWS.

In terms of newer AEDs, data are scanty. Levetiracetam, of the newer AEDs, has been shown efficacious. In one study (Capovilla et al. 2004), results showed that two of three cases had a positive response. Another study (Aeby et al. 2005), which looked at seizures, EEG, and neuropsychological data, showed a positive evaluation in 9 of 12 children, but, in the long-term follow-up, four of eight had a relapse, and levetiracetam had to be discontinued.

The ketogenic diet has been shown in animal studies to decrease spikes (Raffo et al. 2008), which should translate to more clinical trials. The ketogenic diet has actually been tried in a few patients, with some positive results for as long as 26 months.

The authors conclude stating that lack of controlled data makes difficult concrete recommendations. Little data are available on the ketogenic diet, yet it appears about as effective as AEDs, and seems a good choice possibly after steroids or immunoglobulin treatment, for an attempt in refractory AED cases. Remember, the modified Adkins diet is equal to the ketogenic diet as regards efficacy (see above). The author suggests a large multicenter study, especially a placebo-controlled study. One wonders if a placebo study is actually realistic in a pediatric setting.

The issue of genetic factors in Landau–Kleffner syndrome is a subject of conjecture (Rudolf et al. 2009). The epileptic encephalopathies from the Landau–Kleffner syndrome, CSWS, and benign rolandic epilepsy are suspected of being a continuum, with a genetic component. Twin studies suggest a genetic factor (Berkovic et al. 2006). Idiopathic focal epilepsies include a range of epileptic syndromes, which have seizures plus a variety of cognitive, behavioral, and social problems. The range varies from very mild to severe. There are characteristic focal sharp centrotemporal spikes, and slow-wave spikes are the hallmark of these seizure forms.

The strongest evidence for a genetic susceptibility seems to be in benign rolandic epilepsy patients. Some evidence exists for chromosome 15q14 being the involved site. The CSWS and the Landau–Kleffner syndromes constitute disorders at the severe end of the mild to severe range of these seizure types. While over 300 cases of the Landau–Kleffner exist, only two twin sets of patients with this seizure type have been described.

The authors conclude that these above-described syndromes are different, but are a part of a single continuum of disorders. The conclusive data proving a genetic component have been elusive. Much further work is needed in order to conclusively link a genetic mutation/alteration to this interesting spectrum of epileptic encephalopathies.

The relation of non-REM and REM in sleep and various aspects of seizures, especially in the Landau–Kleffner syndrome, has been studied and described (Mascetti et al. 2009). The Landau–Kleffner syndrome is associated with acquired aphasia and verbal auditory agnosia. Behavioral abnormalities are frequent and often severe. During nocturnal non-REM sleep, abnormal interictal EEG recordings can increase to become more or less continuous. PET shows changes in glucose metabolism, which were often seen in the perisylvian region.

There is an enhancement of paroxysmal activity during non-REM sleep, which may lead to bilateral spike and wave discharges. Activation of spike and wave discharges during sleep may indicate that whatever maintains non-REM sleep is acting to significantly influence paroxysmal activity and the seizure disorder. Since sleep may be involved in brain plasticity (Maquet et al. 2003), the paroxysms could theoretically alter plasticity during the critical period for development,

Non-REM sleep involves thalamocortical spindles, delta rhythm, and slow oscillations. These waves each originate at a specific location and can be traced over the scalp in a repeatable pattern. These slow waves are associated with activity increases in cortical areas such as the inferior frontal, medial prefrontal, and posterior cingulated cortices (Dang-Vu et al. 2008). Animal studies have shown that slow-wave sleep oscillations can lead to self-sustained paroxysms (Steriade 2006). These discharges are cortical in origin and are associated with thalamocortical loops.

The effects of sleep on neural plasticity are significant and have been studied. One example is that masking one eye in kittens results in a rapid remodeling of the visual cortex. Sleep of the non-REM type facilitates the plastic process (Frank et al. 2001). It is unclear whether these neuronyl processes play any part in the cognitive defects of the Landau–Kleffner syndrome. The spike wave discharges may only coexist and may not be associated with the Landau–Kleffner syndrome symptoms.

The authors suggest that many as yet unanswered questions regarding sleep and the Landau–Kleffner syndrome can be answered by functional imaging studies. Results from noninvasive methods such as high-density EEG and MEG, plus EEG and fMRI recordings, and PET/SPECT elucidate mechanisms on regional effects in the Landau–Kleffner syndrome. In addition, animal model studies can serve to suggest hypotheses applicable to the human disorder (see Fig. 21.1).

The association with the Landau–Kleffner syndrome of an acquired aphasia with verbal auditory agnosia constitutes an extremely difficult clinical problem. It is, of course, a central problem, which is sometimes difficult for parents and others to understand. Initially, no sounds can be recognized, but environmental sounds may recover at a later stage. If nonverbal capabilities are present, then initiating sign language training is appropriate. It is important to have an infrastructure including parental acceptance, availability of a sign language teacher, and a school accustomed to the special needs student.

The patient should be assessed as to disease duration, extent and success of medical therapy, the patient's nonverbal cognitive and motor skill level, and level of residual (if any) oral language. During sign language training, possible speech recovery should be watched for, so as to incorporate whatever skills might return into the child's communicative skills. Results from sign language can vary widely depending on the multiple factors which impact outcomes.

The authors comment that full participation of family members is essential. Parental responses to sign language can vary from full acceptance to full rejection. Successful acquisition of sign language can significantly change behavior, mood, well-being, etc. Sign acquisition does not change any ability to recover oral language.

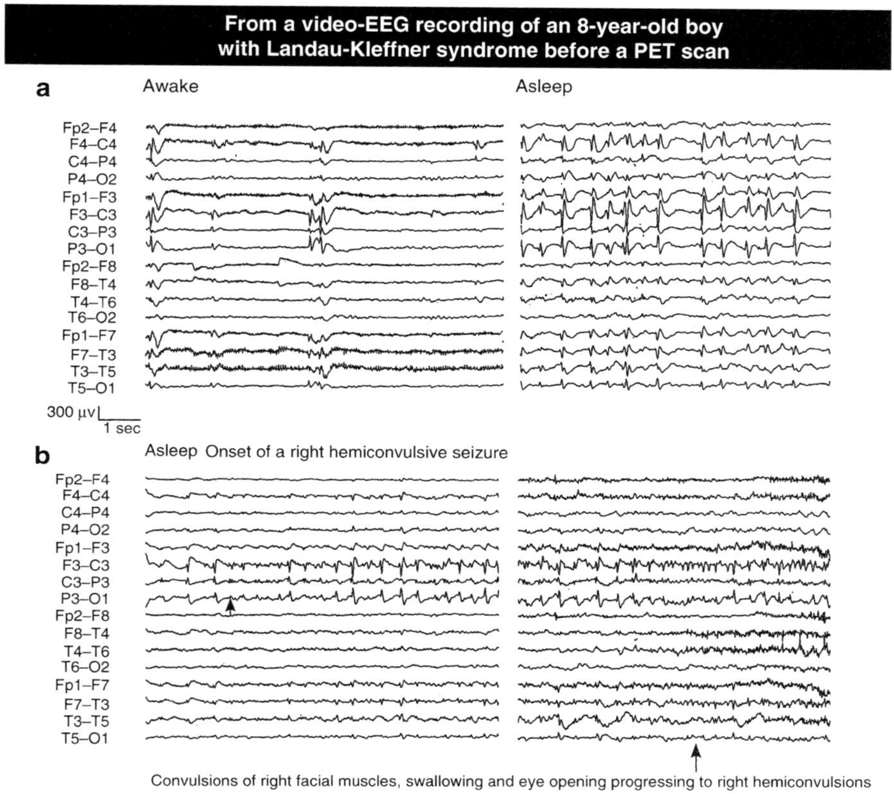

Fig. 21.1 Video EEG of an 8-year-old boy with Landau–Kleffner syndrome. With kind permission from Springer Science+Business Media: Epileptic Syndromes and Their Treatment, 2010, p. 289 Panaylotopoulos, C. Fig. 10.10

One difficult aspect of sign language concerns ambivalence noted on the part of children and families as regards the culture of the deaf. This includes clear prejudices toward the deaf and toward sign language. It should be emphasized that language is a key defining feature of any culture.

The authors conclude that as knowledge of the Landau–Kleffner syndrome and similar syndromes permits earlier diagnosis and treatment, the unfortunate fate of these centrally deaf children will occur less and less. However, refractory cases of the Landau–Kleffner syndrome will always exist, as well as a need for communication, so sign language will still be an important feature of an overall treatment program.

Surgery usually always reflects a last attempt at seizure control. In the case of the Landau–Kleffner syndrome, the seizure is usually controlled by AEDs, but the behavior and language problems are often very difficult to control. In these cases, the surgical procedure of multiple subpial transections is an option. It is important in this procedure to demonstrate lateralization, and the surgical procedure should include Wernicke's area. The current paper outlines the procedure and looks at outcomes (Cross and Neville 2009).

The first suggestion of possible surgical treatment for the Landau–Kleffner syndrome came in 1995 (Morrell et al. 1995). The technique is called multiple subpial transections and involves the sectioning of intracortical transverse fibers and the sparing of vertically directed fibers. The idea was that by cutting transverse fibers, epileptogenic activity spread could be stopped without compromising important cortical function.

In Morrel's first treatment attempt, 14 children received this procedure. All except one had bilateral spike wave discharges in sleep. They were almost continuous and were diagnostic for ESES. Multiple subpial transections were performed (including Wernicke's area) deep into the sylvian fissure using EEG for localization.

Results/follow-up showed six children with normal speech and no seizures. Five had improved speech and no seizures, while three showed no change. There was some indication that the length of time with the Landau–Kleffner syndrome adversely affected long-term surgical outcomes.

The authors note that a clear definition of the clinical syndrome must be certain before surgery. Behavioral/cognitive problems need to be completely assessed before surgery. Drug intractability should include the usual AED trials plus corticosteroids. A trial of prednisolone should include a 6-week trial time frame. The usual drill is to wait 2 years (trying AEDs, etc.) before surgery, but the authors state that after intractability has been proven, time delay serves no purpose.

A key point is the unequivocal indication of a single surgical target. This is a problem solved by neurophysiologic methodology. The methohexitone suppression test is one possible localizing test, but the focus is not always successfully demonstrated. MEG is another technique used in localization. Invasive EEG monitoring can be used, but is not necessarily better than noninvasive procedures.

The experience of the authors of the current paper (Cross and Neville) is largely the same as others in the literature. This study with ten children receiving subpial cortical transection showed that seven had language improvement, but none had a return to complete normal language. Seizure improvement was seen in five patients, and some (but not all) had some level of behavioral improvement. One patient had significant loss of autism symptoms.

The authors conclude that while a promising surgical method has been devised for the treatment of the Landau–Kleffner syndrome, more data are needed. While as many as two third of patients see improvement with this surgical procedure, full recovery is not likely. The risks and side effects seem to be low, so a decision must be made including all pertinent data possible. The usual suggestion is made as regards the Landau–Kleffner syndrome and is that large multicenters must coordinate definitive well-designed controlled studies in order to elucidate reliable information on this interesting epileptic encephalopathy.

Chapter 22
Rasmussen's Syndrome

Rasmussen's syndrome is a childhood developing syndrome, usually characterized by refractory focal motor seizures (epilepsia partialis continua). Most cases originate between the age of 1 and 10 years, mean age about 5 years. Symptoms are attributed to chronic pathogen-free inflammation of both gray and white matter. There is a progressive unihemispheric atrophy (Rasmussen 1978). In 50% of patients, the onset of Rasmussen encephalitis is preceded by an inflammatory episode such as an U.R.I.

Although everything points to a chronic viral infection, no virus has ever been consistently isolated from cerebral tissue of Rasmussen's patients. The finding of inflammatory changes in cerebral tissue from resected brain suggested a viral encephalitis; this progressive disorder was early called Rasmussen's encephalitis with epilepsy. Indeed, status epilepticus is common in about 20% of patients.

Seizures associated with Rasmussen's syndrome are usually refractive to AEDs, and show little to no response. The seizures progressively become more frequent and severe. The seizures are focal motor in 75% of children and half of the patients lapse into epilepsia partialis continua (Oguni et al. 1992). The seizures nearly always involve the same side of the body. As the encephalitis continues, neurological deficits become prominent; these include visual field defects and hemiparesis, taking from 3 months to 10 years to materialize.

Progressive cognitive impairment is a common feature of Rasmussen's syndrome. This can be associated or not associated with the epilepsy progression. The overall progression of the syndrome is relentless, but seems to "burn itself out" at a stage in which neurologic deficit is moderate to severe, and death directly from this disorder is uncommon.

As stated above, the search for some evidence of cerebral viral factor to explain the inflammatory reaction has not been consistently demonstrated. There have been some positive results (Walter and Renella 1989). They have reported the presence of the Epstein–Barr genome in the encephalitic infiltrations. Other studies lend support for a possible role of Epstein–Barr virus in Rasmussen's syndrome (Power et al. 1990). In contrast, others found negative evidence for any presence of viruses in brain specimens in Rasmussen's patients (Rasmussen 1978; Mizuno et al. 1985).

D.W. McCandless, *Epilepsy: Animal and Human Correlations*,

DOI 10.1007/978-1-4614-0361-6_22,

Another study was based on a link between circulating antibodies of a ligand-gated ion channel CNS receptor in rabbits and seizures (Rogers et al. 1994). In light of these results, plasma exchange was tried in one patient with Rasmussen's syndrome, with positive effects on seizures, but that result was temporary (4 weeks).

EEG results show background abnormalities, and the changes are usually asymmetrical. Even when bilateral, predominance is unilateral. Frequently seen were multiple independent foci lateralized over one hemisphere. As a rule, as the disease progresses, bilateral abnormalities become more frequent. EEGs associated with plasma phoresis improved, then again worsened, as was shown clinically (see above).

The neuropathologic findings in Rasmussen's patients can be grouped into four stages. The first is an inflammatory stage, in which perivascular lymphocytes and microglia are seen in brain tissue. In the second stage, the histological findings of stage 1 become worse, affecting all six cortical layers. In stage 3, the neurons are decreased in number, and severe cortical, and degeneration and gliosis predominate. In the fourth stage, the cortical atrophy is severe, gliosis and vacuolation of neuropil exists and cavitation is seen. Often seen are areas of totally normal structure only a few microns away from devastated areas.

The occipital cortex is usually less affected than other regions. Subcortical white matter may display evidence of axon injury, possibly Wallerian degeneration. The unique unilateral appearance of brain from Rasmussen's syndrome defines this encephalitis from all other immune-related CNS disorders. Gene expression profiling of brain from a Rasmussen's patient shows upregulation of inflammation genes, and downregulation of several GluRs.

In terms of prognosis, the seizure phenomenon can be characterized as an initial stage, seizures are simple partial seizures with motor signs, plus complex partial seizures without automatisms. Later, the seizures increase in frequency, and hemiparesis becomes more permanent. In the second phase, all gets worse, including frequency of seizures, more fixed neurological semiology, and increasing disability. As mentioned above, in the third final stage, the progression seems to "burn out," and seizures diminish in both frequency and severity.

The relentless progression of the seizures has prompted various other treatment modalities. These too have not been successful. Apparently, only hemispherectomy has shown any tendency to stop the ravages of this syndrome. This suggests consideration of hemispherectomy earlier rather than later.

Rasmussen's encephalitis is often difficult to diagnose based on neuroimaging due to the gradual, insidious nature of the progression of symptoms. A retrospective study carefully examined neuroimaging results on four patients with a histologic diagnosis of Rasmussen's encephalitis (Tien et al. 1992).

In this study imaging was achieved using CT, MRI, PET, and xenon CT. The age of the four patients was 8.5–21 years. All had a history of intractable focal motor seizures. Average age of seizure onset was 6 years. Of the four patients, two had partial hemispherectomies, and the other two had total hemispherectomies. The two with partial hemispherectomies had been seizure-free, then had seizure recurrences. The other two remained seizure-free.

All four patients had normal CSF findings. EEG results were consistent with a diagnosis of Rasmussen's encephalitis, as were the pathology results. They showed perivascular lymphocytic cell filtration, and clusters of inflammatory cells were noted in the brain parenchyma. Abnormalities of EEG corresponded to imaging study results. Thus, in two patients, xenon CT CBF studies showed decreased blood flow to the affected hemisphere. In one patient, both CT and MRI showed atrophy of the left hemisphere, as well as reduced FDG uptake.

The authors comment radiological studies as performed at the onset of seizures are frequently normal, CT scans from 1 to 4 weeks begin to show patchy low attenuation and swelling in the right temporal cortex. Xenon CT scans may begin to show a decrease of CBF to the affected hemisphere. PET scans showed hypometabolism of the affected hemisphere. SPECT is capable of showing hypoperfusion which correlates to the anatomical location of the epileptogenic focus. Proton MR spectroscopy may reveal a decrease in *N*-acetylaspartate concentration. This decrease corresponds to the loss of neurons, and with brain atrophy.

The authors comment that imaging results can contribute to the sometimes difficult problem of diagnosing Rasmussen's encephalitis. The presence of large areas of functional abnormalities should signal a possible Rasmussen's case, thereby speeding appropriate treatment measures before severe damage occurs.

The possibility of positive outcomes with steroid treatment for Rasmussen's syndrome has been evaluated (Chinchilla et al. 1994). No AED has proven to be efficacious in Rasmussen's encephalitis as regards the control of seizures. The multiple subpial transection method has similarly proven to not have any long-term effects (Morrell et al. 1991). The treatment which seems to have positive effects is hemispherectomy. This surgical approach is difficult and has several associated problems. This prompted the current study looking at high-dose steroid treatment.

This study of eight patients, six girls, and two boys had an onset of seizures at 3.2–13.5 years of age. Initial seizures were brief. Epilepsia partialis continua began between 3 years, 8 months and 13 years, 5 months. When high-dose steroid treatment was instituted, the seizures involved an extremity in three patients, the face and upper limb in one patient, upper and lower extremity in one patient, and was bilateral in the other two patients.

The time lag between onset of epilepsia partialis continua and steroid treatment ranged from 2 weeks to 8 years. It was less than 15 months in five patients. Results showed that epilepsia partialis continua resolved in five patients in that three improved, while two had a complete recovery. In the long term, however, there were relapses of varying degree, prompting the authors to caution that the effects of steroids are hard to assess in the present study, although all patients had a positive initial effect from steroids. Also, steroid-related adverse effects occurred in all eight patients. The adverse effects consisted of Cushing's syndrome in all eight, osteoporosis in three, hypertension in one, and infection in another patient.

The idea that patients with Rasmussen's encephalitis need a quick diagnosis so as to consider treatment has again been examined in terms of neuroimaging (Geller et al. 1998). The authors note that Rasmussen's encephalitis has an abrupt onset with focal persistent motor seizures and epilepsia partialis continua, hemiplegia, and rapid cognitive deterioration. This paper presents two cases and discusses the role of neuroimaging.

The clinical onset of seizure activity began in the two patients at 5 and 6 years of age. They both experienced a rapid deterioration during which AEDs and gamma globulin and prednisolone were ineffective (minor transient improvement in one case). One patient had plasmapheresis. This actually resulted in a significant improvement and cessation of seizure activity for about 3 weeks, then came the inevitable recurrence of seizure and cognitive decline, and hemiparesis.

The authors comment that early in the course of Rasmussen's syndrome, cerebral CT and MRI imaging may be normal as was the case in one of the two patients described in this study. Progressive atrophy is a common feature seen with CT and MRI imaging (Rasmussen and Andermann 1989). Rasmussen and Andermann reported progressive atrophy in 17 of 19 patients undergoing serial imaging examinations.

The authors note the lesions are almost always unilateral, and a key feature of Rasmussen's syndrome. Findings on CT and MRI often parallel each other. Brain PET and SPECT show diminished cerebral perfusion and metabolism which corresponds to and may exceed in size areas shown to be actually atrophic.

The anatomic areas of atrophy seen by SPECT correlate with EEG, MRI, and clinical assessment of the anatomical location of epileptogenic foci (English et al. 1989). SPECT using 99Tc-HMPAD was a critical imaging technique used in the diagnosis of a patient with a progressive deterioration (Burke et al. 1992).

Proton MRI spectroscopy shows decreased *N*-acetylaspartate concentration in Rasmussen's patients. This too correlates with the cerebral atrophy and neuronal loss. *N*-acetylaspartate was decreased in both patients in the present paper. Interestingly, creatine was seen to be decreased in the first patient described in this paper, and creatine plays an important role in seizures, and protection from seizures (see Chap. 29).

The authors conclude that the treatment of Rasmussen's encephalitis needs early diagnosis and monitoring. This means that anatomic and functional neuroimaging are critical for success. Neuroimaging has an important role to fill in that the early diagnosis leads to early treatment, including surgery, which seems to be the currently most efficacious treatment.

In an interesting review (Tran et al. 2000), various salient features of Rasmussen's syndrome are discussed. The etiology of the syndrome was estimated to be due to one of three possible causes: (1) persistent chronic infection, or (2) an earlier acute viral infection with a local immune response, or (3) an independent autoimmune process. Pathological features of Rasmussen's syndrome have features highly suggestive of a viral etiology, yet no virus has ever been isolated, which calls into question this postulate.

Clinically, onset mean age is about 5 years of age, yet scattered cases start in adolescence and adulthood. Simple partial motor seizures affect 75% of cases. Epilepsia partialis continua occurs in 50% of cases and suggests Rasmussen's syndrome. Hemiparesis occurs early and is progressive. Progression leads to global hemispheric dysfunction. There is an accompanying cognitive dysfunction, which includes mental retardation (I.Q. lower than 50 to as high as 103) (Taylor 1991). Low I.Q. (below 50) indicates bilateral hemispheric involvement. While the cognitive decline mirrors motor decline, whether or not the refractory seizures contribute to cognitive dysfunction is not known.

In terms of neuropathological features, the low numbers of patients and variable nature of the course act to prevent conclusive delineation of processes. Perivascular cuffs of lymphocytes extend into the neuropil (Robitaille 1991). Inflammation is a pathologic feature usually limited to the subarachnoid space. Spongiosis is also seen.

Treatment has been discussed elsewhere in this chapter and antiviral therapy has been suggested. Treatment with zidovudine was tried in a Rasmussen's encephalitis patient of 4 years. Results showed a temporary improvement in seizure control and a slowing in neurological deterioration, but the favorable response was short lived. If cytomegalovirus is present in a biopsy, antiviral therapy should be considered, otherwise it is not indicated.

High-dose corticosteroids have had mixed results. The appeal is based on the inflammatory processes seen in Rasmussen brains. The corticosteroids are attractive because they are: antiepileptic, enhance the blood–brain barrier function, and are anti-inflammatory and immunomodulating. Some studies have shown a favorable response to corticosteroids, yet others have not (see above). A trial might be justified before surgery.

Ultimately, surgery may offer the best choice statistically. Functional hemispherectomy is the procedure most commonly used. Some studies show an over 80% success with this method. The exact timing of the surgical process is open to much discussion, One theory is that early surgery is important in order to take advantage of cerebral plasticity before the "critical period" expires.

Another paper has used FDG PET in order to assess cerebral glucose metabolism in early and late stages in Rasmussen's encephalitis (Lee et al. 2001). In this study, 15 children were evaluated with 2-deoxy-2-F-fluoro-D-glucose as a functional imager.

Results showed the mean duration of epilepsy was 3.1 years at the time of PET study. All 15 patients showed an FDG PET area of altered glucose metabolism corresponding to the areas of change as shown by clinical assessment and EEG. Frontal lobe involvement was seen in all patients with abnormal MRI results. Five patients also had temporal lobe involvement. In nine patients with interictal PET, metabolic changes were in one hemisphere only, and in ictal PET studies, the changes were still only in one hemisphere, but the other hemisphere was suggestive.

In six patients whose epilepsy was less than 1 year in duration, frontal and temporal regions showed metabolic change, whereas the occipital area was preserved. The fluorodeoxyglucose PET scans show rather striking hypometabolism in affected frontal and temporal areas. As time between seizure onset and time of FDG PET increases, the areas of hypometabolism increase as well as the severity of the hypometabolism.

The authors note there is a predilection for frontal/temporal lobes to be affected in early Rasmussen's stages, and an early sparing of the occipital lobes. Initially the areas of involvement are scattered, meaning early biopsy results can be compromised by inclusion of adjacent less affected areas. Delineation of the affected areas prior to biopsy could guide the biopsy sampling. There was an excellent correlation between interictal PET scan findings and pathologic results. The functional PET results showed an increase in area in patients with longer duration of seizures.

The authors state that their studies show different patterns in early as opposed to late stages of Rasmussen's encephalitis patients. Some areas adjacent to cortical hypometabolism showed hypermetabolism. This could be explained by the presence of interictal spike wave discharges during FDG uptake, leading to an increase in glucose uptake (Denays et al. 1988). These adjacent areas of hypometabolism and hypermetabolism should be considered in biopsy site.

A paper closely examining the natural history of Rasmussen's syndrome has been published (Bien et al. 2002). This paper focuses on the time course of the clinical history of the brain destruction in Rasmussen's syndrome. In this study, a new method of assessment of the degree of atrophy called the hemispheric ratio is used.

This study focuses on 13 patients who were histopathologically studied over an 11-year period. Data were collected using medical record examination, MRI scans including both quantitative and qualitative analysis, biopsy evaluation, etc.

Results showed that careful examination of all data allowed for a clarification of stages in the progression of Rasmussen's syndrome. The first described stage, the prodromal stage, is characterized by low seizure frequency and rarely a hint of hemiparesis. The duration of the initial stage has a mean of 7.1 months. This phase lasts longer in adolescent and adult patients.

The second phase is the acute disease phase. This is a phase characterized by simple partial motor seizures and epilepsia partialis continua. Hemiparesis develops during this period. MRI scans in this phase show inflammatory lesions in all patients, and had a monofocal onset. Lesions were localized in most patients between the Rolandic and temporo-medial areas. The lesions then spread. The hemispheric ratio (HR) was determined to be 0.72 (HR = ratio of pixels of the affected areas/area of pixels unaffected.

The residual stage follows the acute phase and is characterized by stabilization of hemiparesis. Most patients had a grade 3–4 hemiparesis. Two adult patients had no hemiparesis. The course of the hemispheric volume loss correlated with the evolution of hemiparesis. During the relatively stable stages (prodromal and residual stages), seizure frequency was low. The authors comment that this chronological study shows an initial short (8 months) period in which all patients had a low seizure frequency. The following acute phase consisted of relentless seizure increase, including epilepsia partialis continua. The authors also note two types of patients presented: children with a more rapid and severe disease, and adolescents/adults with a longer chronic and milder course.

Earlier studies did not distinguish between the two time courses (child vs. adolescent/adult) in descriptions of their cases (Oguni et al. 1992). Later a "late onset group" was described (Hart et al. 1997). The two represented different expressions of the same disease and were not based on age.

The authors conclude the study sheds light on the course of Rasmussen's syndrome in children and adults. These data can aid in predicting prognosis; it helps to define the time frame during which the most extensive cerebral damage occurs, and the last "burned out" period starts. It suggests therapies should be applied during the second active stage (or before) rather than waiting for the burned out phase, when it is likely too late. In addition, therapies applied at the start of the "burned out" period may appear to be efficacious, when in fact they have no effect. This is a very interesting paper, well worth acquiring.

The clinicopathologic features and immunohistochemical profiles of seven Rasmussen's patients were examined (Prayson and Frater 2002). All had refractory seizures of 6 months to 7 years duration (mean 3.5 years). These cases represented those seen over a 10-year period. At the time of surgery, age ranged from 4 to 15 years (mean age 12 years). The affected brain area was localized by radiography and EEG studies. All seven patients underwent a functional hemispherectomy of the diseased cerebral hemisphere. Results of the hemispherectomy showed four patients had no seizures postoperatively, one had a decrease in seizure frequency, two had no follow-up.

Histologically, all cases showed focal leptomeningeal chronic inflammation. Parenchymal perivascular chronic inflammation was noted in both gray and white matter. Focal cortical atrophy was present in five cases. Viral inclusions were not seen. In four cases there were areas of cortex which appeared histologically normal.

Immunoperoxidase stains were performed on each case as follows: CD3, CD4, CD5, CD7, CD8, CD10, CD20, CD56 (killer lymphocytes), CD68 (monocytes and macrophages), CD79a (B lymphocytes), and latent membrane protein. Increased CD3 and CD7 immunoreactivity was seen. Most lymphoid cells had a T-cell immunophenotype. The majority of lymphoid cells had a T-cytotoxic/suppressor immunophenotype. CD56 immunoreactivity was not seen. CD68 antibody was observed in all cases.

The authors comment that the diagnosis of Rasmussen's syndrome comes from characteristic clinical and radiologic features. The histologic results resemble those seen in viral encephalitis. The authors state their study details the features of the lymphoid cells. The majority seemed to have a T-cytotoxic/suppressor immunophenotype shown by CD8 immunoreactivity. CD56 was negative and only CD10+ cells were observed.

The histological appearance of Rasmussen's syndrome brain resective tissue resembles those tissue samples from viral encephalitis, prompting an intense search for a viral causative agent.

Vinters et al. (1993) found evidence of cytomegalovirus and Epstein–Barr virus gene expression in excised cerebral tissue from patients with Rasmussen's syndrome. Others were not able to demonstrate Epstein–Barr, or cytomegalovirus, or herpes simplex virus in the tissue of Rasmussen's patients. Other investigators have produced experimental data suggesting an immunological basis for Rasmussen's encephalitis. Rogers, for example (Rogers et al. 1994) showed auto-antibodies of glutamate GluR3 in three of four Rasmussen's patients. One patient received plasma exchange with resultant seizure diminution, neurological improvement, and a GluR3

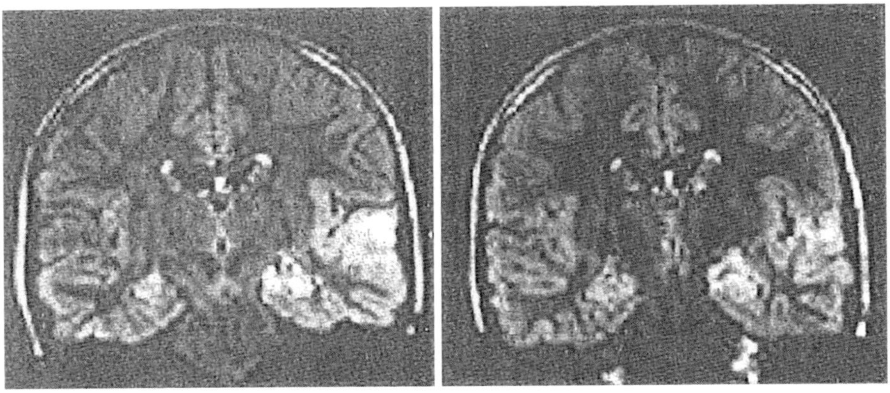

Fig. 22.1 MRI images of Rasmussen's encephalitis. The *left* image is of status epilepticus, the *right* is of more mild seizure activity. With permission of Elsevier-Kuzniecky, R., and Jackson, G. Magnetic Resonance in Epilepsy, 2nd ed., 2006

antibody decrease. The authors correctly state that a viral origin for Rasmussen's syndrome remains unknown.

A study looking at the way the apparent cerebral inflammatory changes progress has been published (Maeda et al. 2003). The idea was that perhaps examining the pathological progression in the resected brain half might provide clues about possible etiology. The inflammation seen in Rasmussen's syndrome has always suggested an infectious etiology.

This was a case study of a 6-year-old girl with normal birth, family history, and development until age 5 years, 3 months when she had an afebrile left seizure while asleep. The patient subsequently had autonomic seizures consisting of facial flushing, a feeling of nausea, and lip smacking. At age 5 years, 10 months, the patient developed epilepsia partialis continua in the left lower extremity. Following admission, she was shown to have normal intelligence, and a minimal paresis in the left foot.

Interictal EEG demonstrated asymmetric background activity differences in regular delta and slow theta waves in the right hemisphere. Video EEG monitoring showed many subclinical EEG seizures from the right temporal, parietal, and occipital cortical regions. MRI showed mild right-sided cerebral atrophy. Fluid attenuated inversion recovery (FLAIR) images showed many small high signal intensity foci scattered in several areas of the right hemisphere. FDG-PET showed an increase in glucose uptake in the right hippocampus and posterior temporal/medial occipital lobe. The patient was discharged following prednosolone treatment and a temporary decrease in seizures (see Fig. 22.1).

The seizures again increased, and an MRI at 6 years, 8 months (5 months later) showed an increase in left frontotemporal lobe atrophy. EEG seizure activity arose from the right central region, and the patient had lost right finger movement, and had difficulty walking. She underwent a left functional hemispherectomy 1 month later. After surgery, she had cessation of seizures, epilepsia partialis continua, and only had a few autonomic auras.

The authors note that neuroimaging technique progress aids in the clinical diagnosis of Rasmussen's syndrome. The introduction of MRI, and T2-weighted and FLAIR imaging allows the description of foci of high-signal lesions. Some have described high-intensity T2 lesions appearing and disappearing in short time periods, then involving adjacent areas (Nakasu et al. 1997). Foci of abnormally increased T2 signal intensity have been reported by others (Fiorella et al. 2001).

The authors state the underlying mechanism of Rasmussen's syndrome is that of progressive encephalitis with neuronal loss and gliosis, and the details of possible inflammatory processes, or how spreading throughout the hemisphere, are not yet known. In the time frame, the current patient developed refractory epilepsy and ipsilateral hemiplegia in 1.5 years after seizure onset. This correlated with the staging time frame described above. The FLAIR method provided good resolution of the images allowing detection of small lesions. The observation of decreased glucose uptake in the entire right hemisphere is attributed to seizure activity.

The authors postulate that the inflammatory process spreads either multifocally or at the same time. Spread was viewed as occurring at the same time in this study. Spread may also be from one area to adjacent tissue. The authors suggest more studies in larger numbers of patients and using FLAIR could further elucidate the problem.

In another study, over 18 years, 45 patients underwent hemispherectomy, and the neuropathology of these resections has been described (Pardo et al. 2004). All patients had a typical semiology and clinical course. Mean age at onset was 6.8 years (range 1.5–16.5 years), mean age of surgery was 9.6 (range 3.8–20.6 years). The surgical procedure was hemidecortication (Carson et al. 1996). A mean of 25.7 cortical areas per patient were processed for histology and immunocytochemical methods. The patients were staged according to the criteria of Robitaille described above (Robitaille 1991).

Results showed that areas of normal cytoarchitecture were immediately adjacent to areas of late stage destruction. Also, there was considerable variation of destruction within a given hemisphere. A normal area might have a moderate area of destruction on one side and a severe area on the other side. There were overall differences between lobes as well, the occipital lobe being consistently less affected at the same time destruction was more advanced in the temporal lobe.

A correlation was found between the age of onset of symptoms and the level of pathology. Thus, an early onset correlated with a higher degree of pathology in the occipital lobe. Similar (not statistically significant) trends were seen in other lobes. There were also areas of white matter which showed significant involvement. White matter, for example, showed a pallor reflecting loss of myelin and axons.

The authors note that this comprehensive study of 45 hemispherectomy patients showed that the neuropathology of Rasmussen's syndrome is multifocal and progressive. The authors state the progression seems like an immune-mediated disease. The progress in magnitude of the syndrome was striking. The "extreme" heterogeneity within a single hemisphere suggests the disease process is ongoing, affecting different hemisphere regions at different times over weeks to months. The author's observations support the concept that the syndrome is initiated by T lymphocytes which initiate the inflammatory process which progresses to produce neuronal injury. The overall appearance of neurons in early stages appear healthy.

The authors note that at any given time point, in a single histological sample, all levels of destruction can be seen. This indicates a multifocal intrahemispheric process. It seems an inciting process causes the immune mechanism to "attack" brain regions over a variable time frame. This implies an autoimmune process, affecting various areas over variable times, as opposed to a single viral encephalitis.

The authors conclude saying the variability of pathology can lead to erroneous biopsy results. Also, the increase in pathology in younger children may reflect the maturity of young brains as compared to that of older patients, or greater inflammatory responses. In addition, the duration of Rasmussen's syndrome affects surgical outcomes, and this argues in favor of early diagnosis and treatment. The presence of T lymphocytes and neurological reaction may also suggest improved treatment paradigms. This study is excellent.

In a study from the same group as the previous paper, the cognitive outcomes of hemispherectomy were evaluated in 71 children who underwent hemispherectomy for intractable seizures (Pulsifer et al. 2004). The patients studied underwent hemispherectomies due to Rasmussen's syndrome, cortical dysplasias, or vascular abnormalities/stroke.

Methods involved included cognitive testing. These tests included I.Q. tests, visual motor tests, receptive and expressive tests, developmental tests, and psychosocial/behavioral tests. The testing was performed before surgery and upon follow-up. Not all patients received all tests, in part because some patients were outside the age limits of some of the tests.

Results showed the sample was 65% female, and equal between right and left handedness. Rasmussen's patients composed one half of the cases. Age at surgery ranged from 0.25 to 20.6 years of age. Age at follow-up was 2.4–37.5 years.

Almost all patients had a statistically significant decrease in seizure frequency. Only three patients who had hemispherectomy had no improvement in seizure control. These three were from the cortical dysplasia group. About one half of all patients were on AEDs. Only six patients were not satisfied with the surgical outcomes. Half of the children were in school, this pathway was lower in patients who had cortical dysplasia. Overall cognitive functioning was high for respondents than for nonrespondents (I.Q. 71.4 vs. 43.7).

Cognitive results showed significant differences between etiology groups. Both Rasmussen's patients and those in the vascular groups scored higher than those in the cortical dysplasia group. Several patients in the cortical dysplasia group could not complete all of the language/visual motor tests. Again, those in the dysplasia group scored lowest.

In the developmental functioning tests, patients in the Rasmussen's group and those in the vascular group scored higher overall than those in the dysplasia group. Overall, scores in behavioral tests were not significantly different between the three groups. There were deficiencies in social activity scores, but all three groups were about equally affected.

Cognitive defects were also correlated with the side of hemispherectomy, and the result was that significant differences were noted between right and left in Rasmussen's syndrome patients. In receptive/expressive language tests, the right hemispherectomy Rasmussen's patients scored higher than the left-sided counterpart patients. The right-sided patients also scored better in I.Q. tests ($P=0.05$). Other right/left comparisons were comparable.

Longitudinal data showed no significant difference between sides in terms of onset age, duration of seizures, age of surgery, etc. Differences were shown in cognition between sides in that the left hemispherectomy group scored lower than the right-sided group. Also, in the adaptive/developmental functioning tests, the left-sided hemispherectomy group scored lower than the right in Rasmussen's patients. There were no differences in I.Q., visual motor skills, or behavior.

In patients from the cortical dysplasia group, no right/left differences were seen in age of onset, age of surgery, duration of seizures, etc. Most other test results showed no difference as regards side of surgery. There was an almost significant difference in the overall adaptive/development test scores. Self-help and academic skills were lower in the right-sided hemispherectomy cortical dysplasia patients.

The vascular etiology group did show significant hemispheric side differences as regards age of seizure onset, age at surgery, and age at follow-up. The small number of patients in the left-side group (2) renders the distinction of differences in this group between right and left hemispherectomies difficult.

The authors comment that the consideration of a hemispherectomy for epilepsy patients certainly suggests the occurrence of a severely devastating disease process. Physical disability and cognitive/behavior changes from normal are prominent as well as intractable epilepsy. These combined deficiencies render independent living and a good quality of life problematic.

In data analysis from this extensive and excellent study, it seemed apparent that the actual etiology was more devastating on follow-up than having a hemispherectomy. Patients with Rasmussen's syndrome faired better in most ways than those in the cortical dysphasia group. Most had either no seizures or only mild seizures, I.Q.s in the 70s, and were overall in good health. Patients with cerebrovascular disorders were about as well off as the Rasmussen's patients.

Those with cortical dysphasia had the worst outcomes. These children almost all had seizures in the first year, and for many, the disease process started prenatally. Children with cortical dysphasia have continuing seizures, lower I.Q.s, lower language skills, etc., as compared to those in the other two groups. The postsurgical problem of cortical dysplasia children has been recognized by others (Devlin et al. 2003), and might indicate an organic problem in the remaining hemisphere. Rasmussen's patients showed a statistically significant decrease in I.Q. and language tests when the degeneration was left-sided. The impairment of language in left-sided Rasmussen's patients is consistent with language lateralization models.

The authors conclude stating that one problem in this study related to assessing extremely low-functioning patients. In some cases, the patients were unable to take certain assessment tests. The authors note that more data are needed regarding language outcomes. The results from this excellent study should be beneficial in improving outcomes and in suggesting postsurgical treatments.

Rasmussen's syndrome was originally termed chronic encephalitis or Rasmussen's encephalitis. In a recent review, the history and possible future of Rasmussen's syndrome are analyzed (Freeman 2005). The author notes that the disorder was originally thought to be a chronic viral encephalitis in large part due to inflammatory characteristics seen in brain slides. It is now believed to be an autoimmune disorder. The syndrome's main features include intractable seizures, hemiparesis, progressive mental retardation, and severe cortical atrophy. The disorder strikes young children predominantly, usually around age 5.

The seizures which characterize Rasmussen's syndrome are focal throughout the course of the disease. Epilepsia partialis continua is also a feature of the seizure component of Rasmussen's syndrome. The seizures are highly refractory to AEDs. The same result awaits immunotherapy – the possibility of a temporary arrest of symptoms, then the relentless decline begins again. These occasional cessations in progression may be due to the natural course of the disease, not treatment.

Surgery has evolved as the most likely treatment to afford any hope of amelioration. Initially, surgical attempts included focal cortical resection rarely had a positive effect, and hemispherectomy became the surgical treatment of choice. The neurosurgery group at Johns Hopkins were early advocates of performing the hemispherectomy early in the course of the syndrome in order to minimize increased deleterious effects from continuous seizures (see above). The surgery is done, however, only when the patient/family realize it is currently the only hope.

The author notes that in terms of outcomes, at the time of this paper, 111 hemispherectomies had been performed and 46 were on Rasmussen's patients. These enjoyed a 65% freedom from seizures and most were not taking AEDs. Four had continued seizure activity due to incomplete hemispherectomy. Following surgery, most patients can walk, but have a paretic arm. The patients may have a loss of sensation, thereby somewhat limiting use of the hand.

Complications of hemispherectomy include intraoperative bleeding, hydrocephalus, and superficial cortical hemosiderosis. The bleeding complication most frequently occurs in cases of cortical dysplasia, not in Rasmussen's patients. Hydrocephalus occurs in about 25% of hemispherectomy patients. Shunts correct the problem, which usually abates spontaneously. Superficial cortical hemosiderosis may be a result of bleeding due to shifting and/or trauma to the empty side of the cranium. CT and MRI have resulted in the early recognition of these problems and their effective treatment.

Early examination of cerebral tissue from Rasmussen's patients showed perivascular lymphocytic cuffing and gliosis which led to the conclusion of a viral etiology. No clear-cut association with a virus was ever shown, and analysis of 45 hemispherectomy tissue samples (see above) led to the conclusion that the syndrome was one of an autoimmune process of cellular origin. This led to the conclusion that early diagnosis and treatment can decrease overall pathology, and the age of onset can also negatively increase the amount of pathological damage. Therefore, the recommendation, says the author, is early and aggressive diagnosis/therapy to guarantee the most optimal outcome for Rasmussen's patients. The author also notes that immunoablative therapy may prove effective soon as an alternative to surgery.

Another review comment from a member of the Hopkins group has been published (Vining 2006). The author comments that the problems associated with diagnosing and determining the etiology have provided plenty of frustration. The obvious histologic picture resembles that of cerebral inflammatory lesions, leading many to propose a viral origin to the syndrome. This concept was maintained for many years, while investigators searched in vain for a consistent finding of a viral cause of increased GluR3 antibodies in some Rasmussen's patients (see above). One interesting finding relating to this theory was that plasmaphoresis did serve to provide a rapid improvement in the clinical picture. It certainly suggested a transient reduction in some blood borne "agent" which was eliminated by plasmaphoresis. Theories to explain this stated that a local (outside the brain) infection or trauma led to a blood–brain barrier breakdown which allowed GluR3 antibodies to attack neurons.

While certain interest still surrounds the above potential mechanism, most recent interest centers on potential T-cell-mediated toxicity. One feature arguing in favor of this concept is that inflammatory cells seen in the brains of Rasmussen's patients are T cells. The T cells are actually cytotoxic CD8+ lymphocytes. These are known to be attackers of neurons.

Another striking feature of the neuropathology of Rasmussen's syndrome is that microscopically, areas of vast destruction can be found adjacent to areas which are completely normal. The transition from devastation to normalcy may occur over a space of just a few microns. This patchy lesion feature also produces a clinical response in which seizure activity skips around, first producing facial symptoms, then leg symptoms, then hand symptoms, etc. This reflects the scattered development of cerebral lesions. This type lesion placement is reminiscent of those seen in other metabolic encephalopathies such as thiamine deficiency and bilirubin encephalopathy, among others.

The seizures of Rasmussen's syndrome are refractory to AEDs. Immunotherapy has also largely been unsuccessful. Plasmaphoresis shows short-term efficacy, but not in the long term. Trials of immunomodulators may have a positive effect, for example tacrolimus may serve to stabilize neurons. Early diagnosis is important in order to try to stop the progressive nature of Rasmussen's syndrome.

The author comments that the patient with Rasmussen's syndrome currently is offered little hope outside of hemispherectomy. This is compounded because children are normal for several years before the onset of the disease. Parents want a cure which will return their children to the earlier period of good health. This review by Vining is a thoughtful and cogent statement.

A recent paper (Takahashi et al. 2009) has examined various immunologic molecules in the CSF of patients with Rasmussen's syndrome, which is considered to be an autoimmune disease by many workers. The present study attempts to characterize the immunologic molecules IgG, CD4+ T cells, TNF-alpha, and Granzyme B, in 27 Rasmussen's patients.

The study took place over 5 years during which samples were collected from all except three patients prior to surgery. Clinical features of the patients included intractable partial seizures, and EEG results showing progressive unilateral hemispheric involvement. Histologic studies confirmed the diagnosis following surgery in nine patients. The usual histologic criteria of microglial nodules and perivascular cuffing were present.

Results showed the average age of onset was 7.5 years, and CSF sampling time ranged from 1 to 288 months after onset. Sixteen patients had epilepsia partialis continua. Control subjects were epileptics without infectious etiologies or progressive symptoms. At the first CSF examination, features such as cell counts, protein, glucose, albumin, and chloride were all similar between Rasmussen's syndrome patients and controls. Conversely, IfG values were higher in Rasmussen's patients than in controls ($P=0.01$).

CD4+ T cells were also elevated significantly in Rasmussen's patients ($P=0.02$), while CD8+ T cells were not elevated in CSF. Additionally, CD3+ cells, TNF-alpha levels, and Granzyme B levels were all higher in Rasmussen's patients than the values in control subjects. IgG levels were elevated in 2/3 of patients in all stages of the syndrome.

During the evolution of the disease process, CD4+ T cells were elevated during all stages, as were CD3+ t, TNF-alpha levels, and Granzyme B levels. Anti-GluR2 (NR2B) auto-antibodies were found to be present in 50% of patients and IgM auto-antibodies were present in 12.5%. As the clinical course evolved, those positive for auto-antibodies against NR2B disappeared.

The authors note that in the first CSF examination, IgG, CD4+ T cells, TNF-alpha, and Granzyme B were elevated and might contribute to a diagnosis of immune-mediated Rasmussen's syndrome-associated epilepsy. The authors state that measurement of these immune factors early can lead to a diagnosis of Rasmussen's syndrome, and further, following these factors serves to confirm the diagnosis.

The cross-sectional data from their study imply that IFN-gamma and IL-12 are produced early after onset, and that CD4+ T cells and the TNF-alpha levels are elevated into the programmed stages. IFN-gamma activates macrophages to secrete IL-12 and TNF-alpha. IL-12 promotes the proliferation of Th1 cells, which indicates the ratio of CD4+ T cells in CSF. TNF-alpha is involved directly in epileptogenesis by reducing GABA receptor (Stillwagon et al. 2005).

The authors note that their data show that auto-antibodies against GluR2 were detected in half of the patients and may therefore be involved in the pathogenic mechanism from early stages of Rasmussen's syndrome through to several years after onset. In addition, auto-antibodies against GluR2 may contribute to cognitive and behavioral changes in Rasmussen's syndrome due in turn to hippocampal apoptosis (Kowal et al. 2006).

The authors say that in later stages of Rasmussen's syndrome, CSF levels of albumin are associated with astrocytic uptake of the protein, and may induce NMDA-receptor-mediated neuronal hyperexcitability. This in turn would increase potential for epileptiform activity and refractory epilepsy. The authors believe their data could contribute to new therapeutic targets for intractable seizures in Rasmussen's syndrome patients.

Chapter 23
Status Epilepticus

Status epilepticus is one of the true medical (neurological) emergencies. Convulsive as well as nonconvulsive status epilepticus can affect epileptic patients with essentially all forms of seizures from absence seizures to tonic–clonic seizures, and progressive myoclonic epilepsy in infants. It is a situation requiring fast and correct treatment, as the mortality rate is high. Various studies set the mortality rate as being from 10 to 30%, depending on treatment (Hauser 1990; Aicardi and Chevrie 1987).

The actual numbers of status epilepticus patients in the USA are only an estimate. It appears that around 1% of the population has epilepsy – three million patients. Roughly 0.5% of these have a status epilepticus episode, so that represents about 150,000 episodes per year. This number is higher than earlier estimated (Hauser et al. 1990). Estimates are that as many as 70% of children having seizures before age 1 will experience status epilepticus at some point. Of patients first presenting with status epilepticus, further seizures occur in 41% of patients (Hesdorffer et al. 1998). There appears to be a bimodal age frequency with those under 1 year old, and over 60 years having high incidences (DeLorenzo et al. 1996). With an aging population in this country, these numbers will increase.

From a world health concern standpoint, as many as one million cases of status epilepticus probably occur each year. There are probably many more considering reporting methods in third world countries. With a mortality rate of at least 30% in these countries, the numbers of deaths are considerable (see Chap. 1).

Even in the USA, there are controversies over issues as how to define status epilepticus. For example, it is frequently defined as a seizure or multiple seizures without recovery lasting 30 min or more. The 30-min figure is based on time after which cerebral damage occurs, since the brain cannot sustain the energy utilization rate. Yet, one study (DeLorenzo et al. 1999) showed a ten times greater mortality rate in patients having status epilepticus for 30 min or more as compared to status epilepticus patients with a seizure duration of 10–29 min. Another study showed that in patients seizing 7 min or more, the majority were still seizing after 30-min duration (Shinnar et al. 2001a, b). These data favor a shorter time to intervention than 30 min.

The neuropathology and associated mechanisms are not fully understood. It is clear that continued seizures lead to brain injury. The excitatory amino acid glutamate

D.W. McCandless, *Epilepsy: Animal and Human Correlations*,
DOI 10.1007/978-1-4614-0361-6_23, © Springer Science+Business Media, LLC 2012

binds to *N*-methyl-D-aspartate (NMDA), a neuronal receptor. This results in calcium influx, depolarization, and further seizure activity. Neuronal damage probably occurs because of this excessive electrical neuronal activity. Indeed, cell damage/death occur in areas with high concentrations of NMDA/glutamate receptors such as the hippocampus. This serves to explain neuropathological features seen microscopically such as hippocampal sclerosis. Clearly, overlying medical problems in status epilepticus patients like cardiac problems, hypoxia, core temperature, oxygen/glucose use, etc., all will contribute to the unfavorable prognosis of status epilepticus.

In terms of prognosis, if about 150,000 patients have a status epilepticus episode per year, and the mortality rate is 30%, then 45,000 patients die per year. This number depends on the severity of each case, as well as the efficacy of emergency treatment. As education of health care workers of the importance of rapid effective treatment increases, the mortality rates should drop. Morbidity from status epilepticus is also significant, at about 30%, and includes residual ataxia, incoordination, possible language difficulties, cognitive disorders, etc. Synergistic effects of, for example, stroke and status epilepticus have been recorded (Waterhouse et al. 1998).

MRI T2-weighted studies show focal areas of damage, which may persist, and be recognizable by EEG abnormalities. Mesial temporal sclerosis is a feature seen in cases of febrile/status epilepticus.

In order to prevent brain damage and morbidity/mortality, the primary goal of therapy is to stop the status epilepticus as soon as possible. AED therapy should be aimed at using drugs which are fast acting, have a broad spectrum of efficacy, and are easy to administer. Rectal diazepam is frequently the initial AED tried. It is important to administer an appropriate dose on the first attempt. Lorazepam is another AED with an increasing initial use, in some cases replacing diazepam as the first AED used. Further attempts to stop status epilepticus include IV administration of AEDs. Maintenance of respiration and attention to other medical problems is essential.

The longer the duration of status epilepticus, the worse is the prognosis. Any delay increases the likely hood of not stopping the seizure easily. If two (or three) AEDs are not effective in stopping the status, then anesthesia is a last chance treatment option. Complications such as renal failure, cerebral edema, hypothermia, etc., must always be looked for.

Tonic–clonic status epilepticus is the most frequent type of status seen in emergency rooms. Nonconvulsive status epilepticus accounts for about one fourth of all status cases. Nonconvulsive status consists of complex partial status epilepticus and absence status epilepticus. The most common single cause of status epilepticus is noncompliance as regards AEDs.

Generalized tonic–clonic status epilepticus is thought of as being perhaps the most life threatening form of status, requiring the most rapid accurate diagnosis, and aggressive treatment. A paper describing the retrospective review of 66 patients treated over a 10-year period for generalized tonic–clonic status epilepticus was published (Sagduyu et al. 1998). Ultimately, the outcome depends on the time between onset of seizures and effective treatment, seizure type, age of patient, and underlying condition of the patient (DeLorenzo et al. 1996; Towne et al. 1994) (see Fig. 23.1).

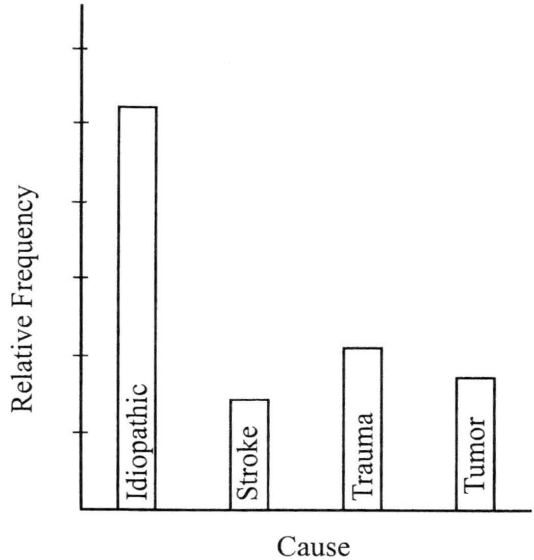

Fig. 23.1 Relative frequencies of status epilepticus by causes. Adapted from Sagduyu A., et al. J Neurol. 245:640, 1998

In this retrospective study, 66 patients who had been diagnosed with status epilepticus of 30-min duration or longer were identified. Only those patients with a clear seizure type of either partial or generalized status epilepticus were included in the study. Records were carefully evaluated in terms of possible confounding coexisting conditions such as alcohol/drug toxicity, AED withdrawal, metabolic disorders, duration of status before treatment, complications, outcomes, etc. Treatments were standardized and consisted of initial diazepam IV. If status persisted 30 min after the first injection, phenytoin was infused. If seizures still persisted, phenobarbital was given.

Results showed a 66-patient cohort with 28 females and 38 men. The mean age was 33.9, range 6–77 years. Patients having generalized tonic–clonic status epilepticus showed primary (71%) or secondary (29%) status epilepticus. There were only three patients with partial status, and that they were excluded from the study due to the small numbers. There were no significant differences between primary and secondary generalized status as regards causation.

Seventy three percent of the patients had a history of epilepsy, and the number one cause of status epilepticus was attributable to AED withdrawal. Other causes included infections, trauma, isoniazid, etc. In patients without previous history of seizures, the causes were similar except of course, for AED withdrawal. Regarding treatment, all received diazepam as first therapy, and almost all had status control, except three who died. Twenty patients entered the second treatment regime and 80% (16 patients) had seizure control. Four patients in this group died because of cardiac problems. Of the remaining 11 refractory patients, IV pentobarbital treatment ended the status epilepticus in 8, rectal chloral hydrate was successful in 1 patient, and 2 died.

Complications consisted of uremia, hypotension, tachycardia, GI bleeding, aspiration pneumonia, apnea, leucocytosis, etc. Fourteen patients died during status epilepticus treatment, and status control could not be maintained in eight of these patients. Features of fatal outcomes included age – significantly higher in the fatal group versus survivors (46.0 vs. 30.6 years). There was no difference between preexisting seizure patients and those without preexisting seizures. There was a statistically significant association between duration of status and fatality ($p=0.0003$). Only CNS infection correlated with fatalities as regards preexisting etiologies.

The authors state that 73% of their patients had a preexisting seizure disorder, and that the main cause of status epilepticus in those with previous seizure disorders was lack of compliance with AED therapy. AED discontinuation brings more than one half of status patients to the emergency room. It is imperative for physicians to stress the importance of absolute compliance to AED therapy. This carries over to even therapies such as the ketogenic diet.

The authors note that about half of E.R. patients responded well to IV diazepam therapy. Three patients in this group died, but the relation of the deaths to status was not clear. Pentobarbital is a drug of choice for refractory status epilepticus (Osorio and Reed 1989) and was successful in this study.

The authors note that a fatal outcome of status epilepticus was statistically linked to age of the patient, duration of status, and CNS infection. The mortality rate for status epilepticus is inordinately and unacceptably high. In this study, pentobarbital was very effective in suppressing refractory status.

A study focused on refractory status epilepticus in terms of frequency, risk factors, and outcome has been published (Mayer et al. 2002). The study involved 74 patients, mean age 63 years, with status epilepticus. The authors state at the outset that inconsistencies in criteria for refractory status and lack of data on morbidity and mortality were stimulants for this study.

The retrospective study involved patients admitted to a neurological ICU over a 4-year period. Status epilepticus was diagnosed when criteria were met, such as tonic–clonic seizure activity for 10 min or intermittent seizures for over 30 min without recovery of consciousness. These criteria resulted in 83 separate episodes of status epilepticus in 74 patients.

Results showed that the most frequent cause of status was low AED levels or a change in medications. Toxic metabolic encephalopathy, stroke, hypoxia, etc., were also contributors. Treatment drugs included diazepam, lorazepam, midazolam, phenytoin, and phenobarbitol. In 31% of status epilepticus cases, seizures continued after the second AED failed to stop seizures, and these patients all had seizure duration greater than 60 min.

Statistical analysis showed patients with refractory status epilepticus were less likely to have a generalized convulsion status than nonconvulsion status or focal motor seizures. EEG findings showed that only periodic lateralized discharges were associated with refractory status epilepticus. The initial EEG in all cases was performed in less than 72 h posthospital admission.

The mortality rate appeared to be linked to other overwhelming medical complications. Of six deaths, only two were having uncontrolled seizures at the time of death.

The authors note that alcohol withdrawal was uncommon in this group of patients. The length of seizure duration of 1.3 h before admission was high, and this was reflected in a low rate of only 31% responding to initial treatment.

This retrospective study showed a higher frequency of refractory status in nonconvulsive status epilepticus than in convulsive cases. Previous studies showed a high mortality rate is associated with complex partial epilepsy (DeLorenzo et al. 1998; Krumholz et al. 1995). Independent risk factors included nonconvulsive status and focal motor seizures. The authors note that failure to generalize quickly might be an indication of more severe underlying brain pathology. This could block propagation of the discharge. Periodic lateralized epileptiform discharges on EEG are associated with a poor outcome in status (Nei et al. 1999).

The authors note that serious medical complications such as fever, pneumonia, hypotension, and anemia contributed to refractory status epilepticus. Morbidity after status results from neurological deficits and physical deficits following the episodes. The authors state that better treatment strategies are needed in order to reduce the morbidity. Prospective randomized controlled trials are much needed in order to define optimal treatment of status epilepticus.

In light of the morbidity and mortality associated with status epilepticus, it is not surprising that many papers have been published examining facets of this problem. One such paper looks at controversial aspects of status, and appraises recommended management approaches (Lawn and Wijdicks 2002). The author of this review starts with the usual problems of defining status epilepticus. They allow as how the 30-min period of time seizing before status is declared, may stem from baboon studies showing neuron survival starts decreasing at 25 min of seizures (Meldrum and Brierly 1978). The authors correctly point out that a time frame of less than 10–15 min probably means treatment may be started in an "out of hospital" setting, with concomitant management issues. There is no evidence that overtreatment or early treatment results in increased morbidity. In addition, is the problem of correct diagnosis in status and its etiology? This may be best done in a hospital setting.

Nearly all status epilepticus patients will initially be evaluated and therapy started by ER physicians. The key is to have a team approach with previously agreed upon protocols in place. There is no data available on outcome in relation to first diagnosis/treatment, but neurologists should be consulted at an early time, and the patient transferred to a neurology ICU. Ongoing EEG monitoring is an important feature of treatment.

EEG analysis is difficult, and should be in the purview of an experienced EEG evaluator. Difficult patterns include periodic epileptiform discharges. There is no clear definitions of these, but they are usually under 3 Hz and do not show a spatiotemporal evolution. The presence of burst suppression in conjunction with postanoxic myoclonus is an ominous predictor of outcome.

Both morbidity and mortality are linked to the etiology of the status epilepticus. Duration of status is a good predictor of outcome, and as earlier stated, a duration greater than 30 min is worse than under 30 min (DeLorenzo et al. 1999). The goal of treatment is to stop the seizures, and AEDs are the early choices. These can be given rectally or by IV. As a last resort, anesthesia can be used. Both morbidity and mortality are frequently linked to confounding other neurological and medical complications.

A recent study (Rossetti and Bromfield 2005) was a retrospective one in which status epilepticus was identified in patients over a 7-year period. The outcomes were evaluated in status patients treated initially with levetiracetam or in those treated without levetiracetam. The criteria for status epilepticus was uncontrolled seizures lasting 30 min or more. Patients under 16 years of age, or with equivocal diagnosis, etc., were excluded. The final cohort was a group with 127 episodes in 107 adult patients. From these, 12 patients given levetiracetam and 25 patients not given levetiracetam were identified for study. Status epilepticus was considered resolved if there was no recurrence within 3 days of seizure cessation.

Results showed that levetiracetam was administered to 12 patients (13 episodes). Twenty six control (nonlevetiracetam) episodes were matched. The groups showed similar characteristics such as seizure type, duration, age, etc. Overall, the patients receiving levetiracetam did the same as regards efficacy and outcomes as did those not receiving levetiracetam. The authors note that their study had relatively low numbers, so analysis by seizure type could not be performed. Further studies analyzing results with many more patients are warranted.

The term "dialeptic" is used to refer to seizures with alteration of consciousness as a main feature (Luders et al. 1998). Dialeptic status epilepticus therefore represents a group of patients whose status is characterized by a nonconvulsive clinical phenotype. By nonconvulsive it means an absence of major positive motor signs. The duration is frequently over 30 min, and in some cases, dialeptic status can last for several days before coming to medical attention. There exists some confusion as to exactly what falls into this category besides the "generic" nonconvulsive forms of status epilepticus.

Clinically, cognitive/behavioral alterations are frequently seen in children with nonconvulsive status epilepticus. These changes may fluctuate, sometimes resembling dementia. Decreased levels of consciousness may be present, as well as defects in speech. Clinical features often include ataxia, myoclonic jerks, hypotonia, automatisms such as fumbling with clothes, mouth movements, etc. Autonomic signs may be present.

EEG monitoring frequently serves to confirm the diagnosis. The results of complex partial status epilepticus may show spike waves, spikes, and fast spiking. Generalized EEG forms show 1–2 s spike wave or sharp slow complexes. The features of dialeptic status are usually better defined and more sustained than others. The dialeptic status EEG traces do not correlate with level of consciousness or behavioral alterations (see Fig. 23.2).

The pathophysiological mechanisms of dialeptic status epilepticus are not known, but probably involve glutamate/GABA imbalances. The clinical ramifications of dialeptic status are usually mild compared to the obvious EEG changes. This dichotomy between clinical features and EEG results often aids in the correct diagnosis. Overresponse to treatment is variable making management of such patients difficult.

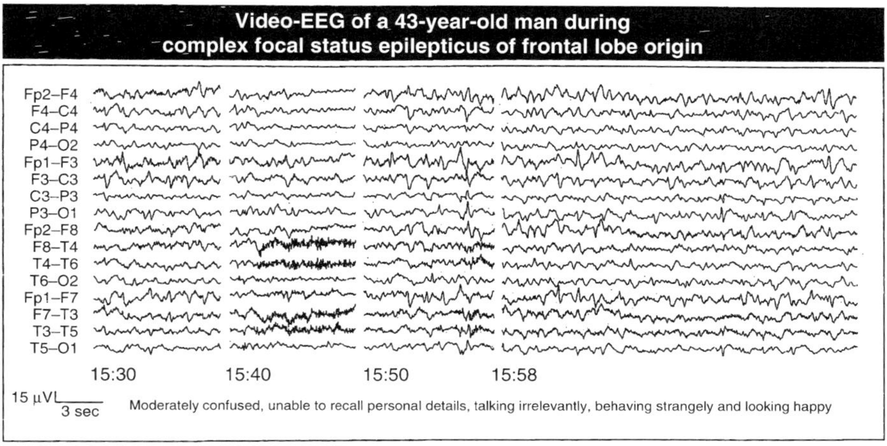

Fig. 23.2 Video EEG of a patient during complex focal status epilepticus from the frontal lobe. With kind permission from Springer Science+Business Media: Epileptic syndromes and their treatment. 2010, p. 80 Panaylotopoulos, C. Fig. 3.2

A new onset refractory status epilepticus (NORSE) syndrome, not previously described has been defined (Wilder-Smith et al. 2005). This new syndrome was characterized because of three similar cases which had a NORSE. All had severe status epilepticus, and all died in 3 months of presentation. A retrospective study resulted in the discovery of four more cases. The "30 min" criteria were used to define status. Refractory was defined as seizures not responding to treatment following at least two treatment regimes.

Results showed that of 355 records screened, seven cases were identified, all females. Mean age was 33 years, six were Chinese and one was Malay. Six patients had generalized tonic–clonic seizures, one had nonconvulsive seizures. EEG monitoring showed ictal discharges in all cases. Six of seven had EEG results showing frontotemporal discharges, which generalized. Five of seven had fever, and the length of time in the ICU averaged 32 days (range 7–92). The outcomes included 5 deaths and 2 with a vegetative state with frequent seizures.

Treatment consisted of IV benzodiazepines initially, followed by IV midazolam, then phenytoin and valproate. Other drugs tried included levetiracetam, topiramate, propofol, and immunoglobulin therapy. Those who died had multiorgan failure. Neurohistological results in two patients who succumbed showed reactive gliosis plus neuronal cell loss. FLAIR hyperintensities were noted in all patients in the temporal lobes. WBCs were noted in the CSF.

The authors comment that the febrile onset of the NORSE syndrome suggested an infection or inflammatory etiology. This seemed not to be the case, however, due to the absence of other inflammatory markers. The neuropathologic mechanisms underlying this disorder are only speculative. The authors note that increased recognition and characterization of the NORSE syndrome is key to being able to develop better treatment paradigms. Its disastrous outcome encourages more studies.

The low numbers of patients will surely increase as recognition of NORSE as a clinical entity increases.

A retrospective study (Thomas et al. 2006) has examined the precipitation of absence and myoclonic status epilepticus by antiepileptic drugs. This generally occurs in idiopathic generalized epilepsy, by the administration of the wrong AED, and is rarely reported. A few previous papers have been published on this interesting phenomenon (Lerman 1986; Bauer 1996).

From a total of 3,874 patient records over an 8-year period, 22 patients were identified as having had an episode of status epilepticus and diagnosed with idiopathic generalized epilepsy. Of these, 14 had results showing a worsening of seizures from AEDs. Status was confirmed by video EEG monitoring. The diagnosis of idiopathic generalized epilepsy was made based on EEG, history, etc.

Results showed that all patients had experienced an AED-induced aggravation of seizure activity many months before referral. Absence/myoclonic seizures were increased in five cases, and new seizure types were induced in six other cases. In five cases, a significant deterioration was seen after initiation of treatment with the drug later identified as causing increased seizures. In some cases (7), the increasing seizures prompted the sequential addition of even more AEDs.

As for status epilepticus, five patients had typical absence status, while five other patients had atypical absence status. Four patients developed myoclonic status epilepticus. Three of the four with myoclonic status showed unusual features. IV benzodiazepines stopped status in eight patients, and three cases ended spontaneously. Three required combination AEDs.

The authors note that seizure worsening due to AEDs is not easy to diagnosis. The aggravation of seizures is not the same as "normal" fluctuations in seizure presentation. Myoclonic status is a relatively rare occurrence in patients with juvenile myoclonic epilepsy, yet occurred in three of four juvenile myoclonic patients. Absence status epilepticus is similarly rare in absence epileptic patients, yet a high percentage occurred in this study.

The underlying neuropathological mechanisms are not clear, but altered GABA metabolism may be implicated. The authors state that their study emphasizes that severe pharmacological AED-induced aggravation of seizures can be seen as either absence or myoclonic status epilepticus, often with infrequently seen variation in presentation. Carbamazepine is the AED identified most frequently as causing the increase of seizures and status epilepticus, and should be contraindicated in absence epilepsy and in juvenile myoclonic epilepsy.

Another paper (van Rooij et al. 2007) examined neurodevelopmental outcomes in term infants with status epilepticus. A total of 311 neonates monitored using amplitude integrated EEG (aEEG) were identified. Eighteen percent (56) of these patients developed status in the neonatal period. Forty two (75%) of these status patients had a poor outcome. Thirty five died, and of the seven survivors, all had severe disability. Twenty-five percent were normal, or only mildly abnormal at follow-up. Ninety-five percent of those with a poor outcome were diagnosed with hypoxic-ischemia encephalopathy. In the group of 14 with a good outcome, 12 were normal, 1 had slight hemiplegia, and 1 had a speech delay.

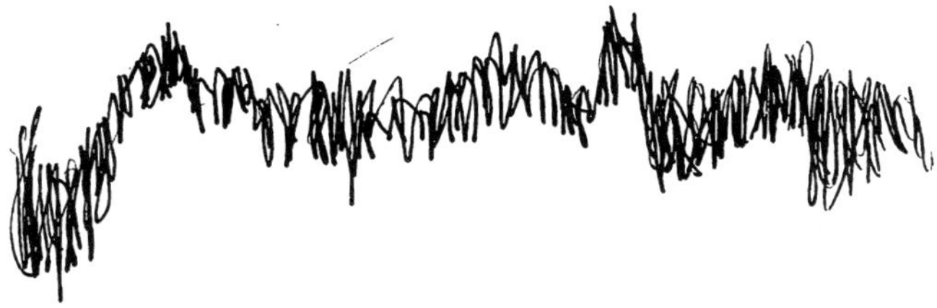

Fig. 23.3 Schematic representation of status epilepticus on an EEG (amplitude integrated EEG). Adapted from van Rooij, L, et al. Pediatrics 120:354, 2007

The authors note that the best predictor of outcome seemed to be the aEEG background pattern. Frequently seen was a discontinuous normal voltage and a continuous deterioration of the background pattern. In the patient group with a good outcome, there was no deterioration of the background pattern. The actual duration of the status did not correlate with outcome. Thirteen patients with hypoxic-ischemic encephalopathy had status at the initiation of EEG studies. In patients with hypoxic-ischemic encephalopathy, status can and does start within 24 h of birth. If cerebral injury is antenatal, seizures may start in just a couple hours after birth (Younkin et al. 1986) (see Fig. 23.3).

Postmortem examination of brain showed extensive damage in deep gray nuclei. The author states that recent studies show that neonatal seizures are not good, especially if associated with hypoxic-ischemic (Miller et al. 2002). The authors note that some studies suggest cerebral energy metabolism deficits may adversely affect outcomes. Presence of intracerebral hemorrhage can complicate treatment.

Studies examining the use of levetiracetam in 34 patients have been published (Gamez-Levya et al. 2009). In this study, 19 men and 15 women (age range from 11 to 90) diagnosed with status epilepticus were given IV levetiracetam. Status in this study was defined as continuous seizures for at least 30 min or repetitive seizures for over 30 min without return of consciousness. All patients had received phenytoin and/or valproate.

Results showed a stoppage of seizures with IV levetiracetam in 24 patients (71%). Some patients (63) received levetiracetam after IV benzodiazepines, and 37 after phenytoin. Twenty-nine percent of patients receiving levetiracetam had no efficacious outcome, and three died. AEDs previously being administered were not changed – levetiracetam was added.

The authors comment that a large controlled randomized study is needed, but in the meantime, their study is suggestive of good results using levetiracetam, especially in older adult patients with concomitant medical issues.

A recent paper (Baracskay et al. 2009) reports studies looking at cells of the pontine reticular formation in rats which had status epilepticus produced by a combination of amino pyridine, pilocarpine, or kainic acid. The idea was to further investigate previously noted dark neuron populations in the pontine reticular formation (Baracskay et al. 29008).

Results showed that cortical 4-amino pyramidine crystal application to rats produced dark neurons in the pontine reticular formation. Four rats were IV injected with 4-amino pyramidine, and the rats that went into status had the dark neurons in the reticular formation. Systemic injection of pilocarpine and of kainic acid also resulted in dark pontine neurons in rats developing status epilepticus.

The authors point out that the pontine reticular formation was structurally affected by a variety of status epilepticus producing agents in the rats. Pontine reticular formation neurons are implicated in rat generalized tonic–clonic seizures (Raisinghani and Faingold 2005), and in humans (Joo et al. 2006). The actual fate of these dark neurons is unclear, but they may actually recover (Atillo et al. 1983).

Early studies (Moruzzi and Magoun 1949) showed that high frequency stimulation of the reticular formation results in arousal. Lesions of the reticular formation, including biochemical lesions, produce stupor and coma. This resembles, state the authors, what they see as slow oscillations in connection with status epilepticus.

The authors further speculate that the pontine reticular formation giant neurons receive intense activation during repeated seizures, and this promotes generalization. These giant neurons are glutamatergic and send long projections to the thalamus, which in turn projects to the cerebral cortex. Ultimately, the giant pontine reticular formation neurons become dysfunctional and compacted, forming the visualized dark neurons. This depression of the ascending reticular activating system yields slow synchronized oscillation of cortical EEG activity.

The authors note that slow neocortical activity is associated with EEG changes in coma (Parvizi and Damasio 2003). This could be related to postictal coma seen in some experimental animals or postictal depression in humans. Interestingly, reticular formation lesions have been seen in patients with postpolio fatigue syndrome (Bruno et al. 1995). The authors end stating that the described pontine reticular formation lesions could well be key features of the progression of slow oscillations in status epilepticus.

A brief review (Scott 2010) examines convulsive status epilepticus and its relation to damage to the hippocampus. It is certain that convulsive status epilepticus is a common and serious neurological emergency, and is associated with morbidity and mortality. A major predictor of outcome is the etiology of the convulsive status. Supposed predictors such as seizure duration and seizure type were not predictors in this study.

One hypothesis regarding neuropathology is that convulsive status epilepticus causes damage to the hippocampi, and mesial temporal lobe sclerosis, commonly seen, is a result. Febrile seizures are associated with mesial temporal sclerosis, and focal temporal lobe lesions are the objective of neurosurgical attention. It is well known that febrile seizures can produce MRI changes in the hippocampi even within a few days of elevated temperature (Scott et al. 2002). While prolonged febrile seizures can cause hippocampal damage, the exact percentage who then develop mesial temporal sclerosis is unclear. Further studies should clarify this question.

Another paper on status epilepticus looks at the balance of GABA and glutamate (Naylor 2010). A loss of 50% of postsynaptic GABAa receptors would have a negative effect on synaptic inhibition. Immunocytochemistry shows a movement of

GABAa subunits away from synapses into the cell body (Naylor et al. 2005). Animal models of status epilepticus have features in common which involve either decreased inhibition or increased excitation. This revolves around GABA and glutamate. Some status producing agents, such as bicuculline, pentylenetetrazole, etc., act on GABA, while others (kainic acid) act on excitatory processes. More than one half of animal models of status epilepticus have sequelae of recurring seizures (Suchomelova et al. 2006).

In terms of glutamate receptors, both NMDA and non-NMDA receptors show an increase at excitability synapses, thereby accentuating GABA loss. In fact, NMDA-mEPSC increase by 23% in status. This results in a normal change in both inhibitory and excitatory synapses during status. Early loss of synaptic inhibition changes the balance between excitation, encouraging self-sustaining status. This helps explain the loss of efficacy of benzodiazepines since sites for agonist binding are reduced. This leads to a cycle of self-sustaining status epilepticus.

The author notes the acute changes may have long-term effects due to cell death and the development of epilepsy. The data suggests therapy should be directed at both problems: decreased GABAa-related inhibition and a decrease in increased excitatory phenomena.

This question raised in the above paper regards predicting which particular patient might develop spontaneous recurrent seizures following convulsive status epilepticus. Use of long-term video EEG is not always practical, and biomarkers which could predict which patients might be expected to develop chronic epilepsy are needed. The present study (White et al. 2010) looks at rat models of chronic epilepsy using continuous radio telemetric EEG and video monitoring, and automated computer detection of EEG spikes to see if EEG spikes precede spontaneous recurrent seizures.

In this study, Sprague-Dawley adult rats were implanted with depth recording electrodes, and EEG recorded by radio telemetry. Baseline EEG was recorded 1 week before kainic acid treatment. Kainic acid was administered IP to rats, controls received saline. IP kainic acid was administered until the rats had convulsive status epilepticus. Spikes were recognized based on a definition of human EEG spikes which states spikes are clearly identified from background, and last between 20 and 70 ms. Amplitude/duration criteria served to identify spikes.

Results showed that the frequency and temporal clustering of EEG spikes which were present prior to the first spontaneous seizure correlated with subsequent epilepsy. This finding was considered to be predictive only in proportion to the duration of monitoring. A 24-h spike count correlated with the prognosis of future epilepsy, but short interval results had considerable variation. The use of digital EEG and spike detection algorithms act to increase the utility and sensitivity of EEG, thereby giving a more accurate analysis of spikes and future epilepsy. These EEG recordings are readily available, and amenable to spike detection, and are therefore more available than other potentially correlated markers such as cell death.

Chapter 24
Epilepsy and Autism

Autism was first described in 1943 by Kanner (1943). This landmark paper detailed 11 children whose "condition differs so markedly and uniquely from anything reported so far, that each case merits… a detailed consideration of its fascinating peculiarities." Kanner saw symptoms which included an inability to relate to others, unusual responses to the environment, stereotyped motor mannerisms, odd communication peculiarities, and a tendency to echo language. Kanner saw only one case of epilepsy in his cohort of 11 patients. The case reports by Kanner are notable for their thoroughness of clinical description.

Interestingly, Jackson in 1870 in describing epilepsy, allowed as how during a seizure, there was a disorderly discharge of cerebral nervousness on muscles. Jackson says there is an instantaneous loss of consciousness, altered perception, impaired psychic function, convulsant motor behavior, sensation disturbances, etc. These represent fast symptoms similar to chronic seizures seen in autistic patients.

Autism ordinarily evolves in two different ways. Patients who may show some signs of autism at or shortly after birth are diagnosed as having an autism spectrum disorder. This form, starting at birth (or maybe before?) is termed nonregressive autism. On the other hand, regressive autism starts from about 18 months to 24 months of age. In these cases, development is rather ordinary until regression begins. These children typically start by losing speech and then other developmental milestones begin to deteriorate. This condition is known as "regressive autism," and may be rapid in onset. It has been noted that both epilepsy and abnormal EEGs are more common in regressive autism than in nonregressive autisms.

It is fair to say that autism is a pervasive developmental disorder. Seriously impaired social interaction and abnormal communicative skills are noted in all cases by age three. Dysfunctional behavior such as lack of eye contact, rocking, hyperactivity, sleep problems, etc., are common in autism (Filipek et al. 1999). Severe mental retardation is seen in over one third of patients.

While the precise etiology of autism is unclear in 90% of cases, some comorbidities are present in some cases, such as PKU, tuberous sclerosis, Rett syndrome, etc. Some autistic patients on MRI investigation show a hypoplasia of the cerebellar vermis, but the exact relation to autism is not clear (Courchesne et al. 1988).

D.W. McCandless, *Epilepsy: Animal and Human Correlations*,

DOI 10.1007/978-1-4614-0361-6_24,

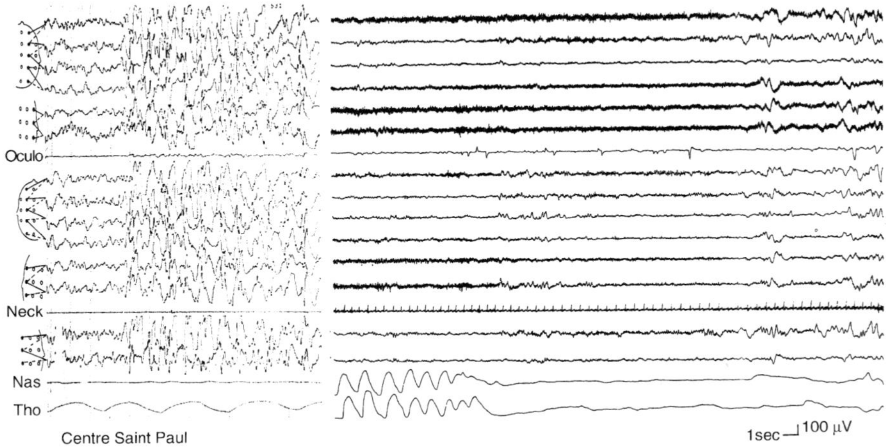

Fig. 24.1 Hippocampal sclerosis EEG. Example of autism with epilepsy plus Rett syndrome. Published in Crespei, A., et al. Atlas of electroencephalography Volume 2: The Epilepsies, EEG and Epileptic Syndromes. With permission of John Libbey Eurotext

Abnormalities also have been noted in the amygdala and hippocampus, and a monkey study shows damage in these two regions can result in autistic behavior (Bachevalier and Merjanian 1994) (see Fig. 24.1).

EEG changes have been seen in from 27 to 77% of autistic patients, depending on the study. The incidence of clinical epilepsy might be as high as 40% in autistic children, but this result was from a cohort of more impaired cases, and the actual incidence of epilepsy in autism overall is surely lower (Gabis et al. 2005).

Treatment of autism has focused largely on behavioral problems, language training, and education. Many AEDs such as valproate have been used to treat autism. Many workers think that AEDs may be of benefit for both behavior problems (autism) and seizures (Trevethan 2004). AEDs have been used in some autistic patients who do not have seizures, with some success. There may be a subgroup of autistic patients in whom both autism and seizures respond positively to AEDs (Tuchman 2004).

The incidence of epilepsy in autistic children is different between studies and depends on several variables including the age of the patients. In addition to age, the cognitive level and degree of language disorder are factors in the incidence of epilepsy (Tuchman and Rapin 2002). These studies show that the incidence of epilepsy in autistic patients is highest in adolescence and adulthood. These patients with severe mental retardation and language deficits are at greatest risk.

The concept that epilepsy might have a negative effect on cognition, language, and/or behavior is equivocal. The opposite – that regressive autism might somehow influence epilepsy – is also without substantive proof. They correctly suggest that this interesting concept needs validation with double blind studies.

A review of the relationship between tuberous sclerosis and autism has been published (Wiznitzer 2004). The author notes that the coexistence of tuberous sclerosis and autism has been recognized. The question is can this be an unrelated occurrence, or is there a cause–effect relationship between the two disorders. The existence of tuberous sclerosis in the autism spectrum disorder is about 1–4%, whereas the existence of autism in patients with tuberous sclerosis is somewhere between 25 and 50% (Fombonne et al. 1997; Hunt and Dennis 1987). Problems with some of these studies include questions about criteria for inclusion in the autism spectrum group and lack of thoroughness in confirming tuberous sclerosis.

The question of association between tuberous sclerosis and autism spectrum disorder may ultimately hinge on the cerebral location of the tubers. Studies have shown that tubers in the temporal lobe have a high association with seizures (Jambaque et al. 1991).

The author states that because of the clear association of tuberous sclerosis and autism spectrum disorder, when one is diagnosed, clinicians should be on the lookout for the other. There is a need for more controlled double blind studies. Seizures well defined are essential in order to permit interpretation of data.

Another paper examines the long-term outcomes of a large group of patients with or without seizures (Danielsson et al. 2005). In this study, epilepsy was defined accordingly to the standard ILAE definitions and autism by the DSM-IV criteria (Am. Psychiatr. Assoc., 2004).

In this study, nearly 80,000 children were screened for autism using various psychological tests. One hundred twenty people were identified as meeting the autism criteria. These patients were followed up, and 108 were actually participants. These were again evaluated by psychological and I.Q. tests. All except one were still diagnosed as having autism or autism-like disorders.

Epilepsy was suspected in 48 of 108 patients. Six were excluded for various reasons, leaving 42 patients for study with epilepsy and autism. Detailed descriptions were sought of the characteristics of the seizures. EEG, CT, MRI, etc., were all obtained when possible.

Seizures were classed into four groups, including (1) simple or complex partial seizures with or without secondarily generalization; (2) generalized seizures such as tonic–clonic, myoclonic, atonic, tonic, etc.; (3) mixed seizures when both partial and generalized seizures were occurring; and (4) unclassified seizures. Appropriate statistics were used to evaluate data.

Results showed that 38% of the 108 autistic or autistic-like patients had epilepsy. A similar number had mild to severe mental retardation. In severely mentally retarded patients, age of onset of seizures was 3.5 years, and in mild retardation, age of onset was 7.2 years. Almost all (96%) of the severely retarded group had autism.

In the 48 suspected seizure patients, 23 had partial seizures, 19 had generalized seizures, 6 had mixed, and 3 were unclassified. Three patients had infantile spasms. Epilepsy age of onset was before 18 years in all cases except for 4 patients. At follow-up, active epilepsy was seen in 34 of 42 cases. The number of patients taking one AED was 15. Ten patients were on two AEDs, while the remaining patients were

Table 24.1 Types of seizures seen in autistic children

Types of seizures seen in autistic children
Complex partial seizures
Infantile spasms
Generalized
Tonic
Atonic

Adapted from Canitano, R., et al. J. Child Neurol. 20:27, 2005

taking three or more AEDs. Two of the 34 had surgery, one successful and the other not successful. Ten patients had less than one seizure per year (see Table 24.1).

The authors note that the frequency of epilepsy in autistic patients appears to be about one third (Olsson et al. 1988) and that correlates with their study (38%). Patients with both autism and epilepsy were of a more severe cognitive impairment than those with autism only. Most autism only patients continued to have poor social skills and were unemployed.

The authors state their data suggest the risk of epilepsy in autistic patients is highest in early aged patients, then decreases, although 10% of cases begin after 18 years of age. One study (Camfield et al. 1993a, b) showed that over half of childhood epilepsy remits. Remission predictors include only one type of seizure, lack of a lesion, and normal development. The authors note that the remission rate was 16%.

The authors comment on the difficulty in diagnosing seizure types in patients with autism and mental retardation. Patients with autism and severe mental retardation, for example, rarely can comment on aura symptoms. It is sometimes difficult to properly examine autistic patients without overly stressing them. However, the authors correctly point out that this must not be an excuse for not correctly diagnosing and treating epilepsy in these patients.

Follow-up showed that patients with the combination of severe mental retardation, autism, and epilepsy are severely disabled. The authors note that frequently even physicians do not have a complete understanding of the situation. Collaborative care involving several specialties is needed throughout life for these patients.

The symptoms of autism include impairment of social skills, disorder in communicative and language skills, and patterns of repetitive behavior including stereotypies and restricted activity. A study (Canitano et al. 2005) was published looking at epilepsy in autistic children, with special emphasis on regressive autism. This group represents approximately 30% of autistic children who are essentially normal until age 2, when they reverse development, losing developmental milestones.

This study involved 46 children referred to a center specializing in autism and pervasive developmental disorders. The patients mean age was 7.8 years. All received a neuropsychiatric evaluation of language, social interactions, play skills, etc. All were diagnosed with autism based on history and clinical evaluation. Regression was diagnosed when a patient had acquired several words, then ceased to pronounce them for 3 or more months. The language regression was associated with loss of emotional and social behavior, including attention, contact, facial expression, etc.

EEG recordings were taken in both awake and asleep states. Paroxysmal abnormalities were noted. Epileptic disorders were described using the revised classification of epilepsy (Commission on Classification and Terminology of the I.L.A.E. 1989). Cognition was assessed using standard I.Q. tests.

Results showed, using laboratory tests, that none of the children had metabolic disease or genetic involvement. The only significant neurologic abnormality was epilepsy. Results from EEG recordings revealed that 65% of the group had normal EEGs. Paroxysmal abnormalities contributed 35% of patients. Six patients had epilepsy, and ten had EEG changes without seizures. Of the six, three had complex partial seizures, secondarily generalized, two had complex partial seizures generalized seizures, and one had complex partial seizures presenting as atypical absences and generalized seizures. All were treated with AEDs, and two became seizure free.

Regressive autism was found in 52% of patients, and nine patients had paroxysmal discharges shown on EEG. No significant differences were shown between children with and without regressive autism as compared to those with and without overt seizures. Mental retardation was present in all patients, with moderate to severe retardation present in 78% of patients.

The authors note that the major factors associated with epilepsy in autistic patients include neurological impairment, extent of language disorder, and patient age (Tuchman and Rapin 2002). The relatively low rate of epilepsy in this study could have been related to the age (mean 7.8 years) of the patients. Children with neurological and intellectual issues were more likely to develop epilepsy. The authors state that patients with autism, epilepsy, and other neurological problems might represent a distinct subgroup, but this is not clear.

The authors point out that the significance of paroxysmal discharges without seizures is controversial. While obviously seizures should be treated with appropriate AEDs, the decision to treat or not treat paroxysmal discharges is far less clear, and usually not treated (Tuchman 2000).

Some studies suggest that epilepsy and/or paroxysmal abnormalities are associated with an increase in the autism regression rate, but that was not substantiated in this study. The authors note that more studies are warranted regarding this issue. The authors further note that in the evaluation of autistic children with EEG alterations, the Landau–Kleffner syndrome should be considered. Early onset of the Landau–Kleffner syndrome can be confused with autistic regression. Differences in EEGs as regards cerebral locations of activities in Landau–Kleffner syndrome versus autism can be diagnostic (Shinnar et al. 2001a, b, c).

The authors conclude by saying that epilepsy in autistic children is a confounding clinical problem. Patients with both disorders have highly significant interferences with development, and further studies are warranted to determine the effects of the disorders on each other.

While the association between autistic spectrum disorder and an increased prevalence of epilepsy seems clear, the prevalence of epilepsy in children who have been diagnosed with autism is less well described. The present study (Clarke et al. 2005) sought to delineate the comorbidity between autism spectrum disorders in epileptic children. In this study, children with epilepsy were evaluated for autism using in part, autism screening questionnaires. Answers to questions regarding sleep disorders, behaviors, seizure features, and AED treatment were evaluated.

Results showed that 97 patients' questionnaires were properly completed, allowing evaluation. Thirty-two percent of the children fit the criteria for autistic spectrum disorder. Bad behavior and daytime sleepiness were present in children with high risk of autistic spectrum disorder. Epileptic seizures occurred earlier (at 2 years of age) in those at risk for autism spectrum disorder.

The authors conclude that children with epilepsy are at greater risk for autistic spectrum disorder than children without epilepsy. The authors state that autism should be looked for in epileptic children rather than supposing that autism symptoms are the result of AEDs and seizures. As usual, the idea that early identification and treatment may lead to better outcomes is valid (Aldred et al. 2004). Further study is warranted to see if autism can be reduced in children with epilepsy, when diagnosed early.

A study of tuberous sclerosis complex and epilepsy and autism was undertaken (Wong et al. 2006). In this study, the MRI findings in children with tuberous sclerosis complex were correlated with a variety of potential related features including age, sex, mental retardation, autism, seizures, etc. Twenty-two patients were identified who had MRI data, and categorized by features listed earlier.

Results of this retrospective study showed that of the 22 cases, 10 were males and 12 were females. Mean age in the autism group was 15 years old, and the non-autism was similar. The mean number of tubers was 12 (range 0–50), with no significant difference between autistic and nonautistic patients. Cerebral areas with the most tubers were the subependymal nodes, frontal lobes, parietal lobes, temporal lobes, and occipital lobes. There was no statistical difference between autism and nonautism as regards tuber location. Mental retardation, epilepsy, onset of seizures, all had no correlation with tuber number or site of occurrence.

The authors comment that only tubers in the parietal or occipital regions had a relation to the history of infantile spasms. They also note that epilepsy was present in 92% of cases with MRI detectable tubers, and recurrent seizures in 33% of autistic patients. This correlates well with results from other studies (Curatolo and Cusmai 1987). The authors state that tuberous sclerosis complex is a broad spectrum disorder with many phenotypes besides autism and epilepsy, although these two are most serious.

Another group (Stobbe et al. 2006) has looked at the AED levetiracetam as a treatment for both EEG epileptiform abnormalities and the autism spectrum disorder. In this study, 11 patients diagnosed with autism spectrum disorder and with epileptiform discharges on EEG were investigated. Results showed the maximum benefit of levetiracetam was at the lowest dose (5–10 mg/kg). Higher doses had similar results plus behavioral adverse effects such as aggression. Some behavioral features such as play/imagination and awareness improved with levetiracetam.

Children with left focal epileptiform discharges had positive effects in expressive language results. The authors conclude that levetiracetam is safe in children

Table 24.2 Age, diagnosis, and percent epilepsy in autism

Age	Diagnosis	Percentage with epilepsy
5.2	AD	14
7.0	AD	46
7.8	AD	13
8.1	AD	33
9.0	AD	7.4
9.1	AD	22
9.5	ASD	7.6
10	ASD	20
12	ASD	28
14.1	ASD	21
17.2	AD	38

Adapted from Canitano, R., Eur. J. Adol. Psychiatr. 16:61, 2006
Ad autistic disorder, *ASD* autistic spectrum disorder

with autism spectrum disorder and epileptiform discharges. Also shown was that some behavioral phenomena improve with the AED at low doses. Higher doses do not correlate with any further improvement and are associated with withdrawal from the trial.

Conclusions from a brief review (Canitano 2007) state that there is still some controversy regarding epilepsy in autism. It is certain that epilepsy is a feature of the autism spectrum disorder and must be considered and treated. The exact relationship between epilepsy and autism is not certain, but the incidence of epilepsy in autistic children is at least ten times that of the general population. The risk of epilepsy in autism is quite high in some subtypes of autism spectrum disorders (childhood disintegrative disorder and Rett syndrome) (see Table 24.2).

The possibility of a genetic similarity is boosted by the finding of alterations in chromosome 15 in some epileptic patients, and in chromosome 15 in autistic children. Only about 10% of autism cases are linked to genetic causes (Fombonne 2003). Several types of seizures are seen in autistic children including partial, generalized tonic–clonic, myoclonic, absence seizures, etc. Autistic regression occurs in children about 18–24 months old who reverse development, losing all developmental milestones. Epileptiform abnormalities may be present with resultant seizures. Language regression is seen as an ominous sign. Landau–Kleffner syndrome is the classic example of the language regression syndrome. Treatment of epilepsy in autism, as stated earlier, must be undertaken and is similar to that of other seizure types. This review emphasizes the need for awareness of the comorbidity of epilepsy and autism, and the aggressive treatment of each. Many factors of the two disorders are unclear.

Another paper (Saemundsen et al. 2007) describes a study aimed at examining autism spectrum disorder in children with unprovoked seizures in the first year of life. One hundred two children were identified with early seizures, and 84 wound up participating in the study. First, the parents filled out a questionnaire and were interviewed as regards their children's autism. In order to be diagnosed with autism, developmental abnormalities must be present by 36 months of age. These include impaired social interaction, impaired communication, and stereotyped/repetitive actions.

The patient's parents (31) expressed concern as regards development and therefore their children were investigated for possible autism. Of these 31 patients, all but two had more than one seizure. Seventy-one percent were classified as cryptogenic. Based on the questionnaires, 13 children were actually examined further for autism spectrum disorder, and six were diagnosed as having the disorder.

The authors comment that the prevalence of autism spectrum disorder in this cohort was from 6.0 to 7.1% depending on the possible inclusion of a couple of patients. This exceeds the general population rate of 0.6–1.0%, and approximately in keeping with other results, yet somewhat lower. This is probably due to different criteria for inclusion, including that of mental retardation. The authors note that a better experimental design including some degree of random sampling might have been better. Also, the numbers of subjects are small, so data are only suggestive.

The authors conclude that the estimation of autism spectrum disorder with a history of seizure activity in the first year of life exceeds that of the general population. All children with autism spectrum disorder had mental retardation, some severe, and epilepsy. The presence of seizures in the first year of life should be an alert to the possibility of autism. Further investigation in this area is highly warranted.

An interesting paper has presented data examining comorbidities in several cases of folate deficiency (Moretti et al. 2008). In this study, seven children with folate deficiency affecting the CNS were all noted to have a variety of ancillary abnormalities. Previous studies (Ramaekers et al. 2002) had shown CNS deficiency of folate to be associated with psychomotor retardation, cerebellar ataxia, dyskinesia, and seizures. Other reports showed patients with heterogeneity of symptoms, but most had both autistic features and seizures.

In the present study, seven patients were studied. CNS folate was measured in CSF using HPLC. Results were compared to over 450 normal samples including all age groups. DNA sequencing, chromosome analysis, and genomic hybridization were performed. Five patients received psychological testing, two were too severely affected to be tested. Cognitive/motor skills were evaluated.

All children in this study showed developmental delay and regression. All were low in CSF methyltetrahydrofolate, the biologically active form of folate. Other tests were within normal limits. Six of seven patients had epileptic seizures. All patients exhibited some form of regression. This was characterized by more or less normal development until 2–3 years of age, then psychosocial regression occurred, rapidly in some cases. This included language skills, locomotion, increased seizures, etc. All were diagnosed as autistic. The heterogeneity of CNS folate deficiency was indicated by various "final" diagnoses of patients, including Rett syndrome, Landau–Kleffner syndrome, autism spectrum disorder, possible Angelman syndrome, and severe mental retardation with autism. Folate treatment provided some improvement in symptoms in some patients.

The authors comment that the study shows autism spectrum syndrome present in five patients with cerebral folate deficiency. Two children had severe neurological abnormalities which precluded formal psychobehavioral testing. Overlap of symptoms between patients was more evident in those with higher cognitive and language development. Patients with folate deficiency were clearly different in semiology than patients without folate deficiency in terms of seizures, hypotonia, movement disorders, and ataxia. Regression occurred at different time points, whereas in autism alone, the timing of regression is consistent.

The authors note that different timing of clinical features argues in favor of different pathways of folate alterations. Possibly some degree of compensation would permit variability. There may also be CNS folate abnormalities in idiopathic autism which escape detection. In these cases, folate determination in CSF is warranted. Further investigations are needed.

Another case study examines the correlation between tuberous sclerosis and autism (Kothur et al. 2008). In this study, two children are presented who have tuberous sclerosis complex with autism. Autism is present in patients with tuberous sclerosis complex in 17–58% of patients depending on the study (Curatolo et al. 2004). The two patients in this study also had epilepsy.

Both patients presented with symptoms of autism. In one case, regression of milestones started at 9 months, the other patient showed developmental delay especially as regards language development. Seizures in the first case above consisted of infantile spasms followed later by generalized and focal seizures. This same patient had severe mental retardation (I.Q. 27), speech delay, no social interaction, and stereotypic movements such as lip tapping. The other patient had seizures characterized by shouting and generalized shaking of the body. This occurred four to five times per day without loss of consciousness. The patient also showed a decrease in social interaction plus stereotypies. Both patients had neurocutaneous lesions.

The authors note that results from MRI studies showed that both patients had tuberous sclerosis located in the temporal lobes. Previous studies showed a correlation between temporal lobe tubers and autism (Bolton and Griffiths 1997). The authors state that early temporal lobe discharges may disrupt brain development leading to autism (Zuddas et al. 2002). Both of the present cases had temporal lobe EEG abnormalities. Tuberous sclerosis complex disease is a useful model system for further investigations.

The relation between autism spectrum disorder and infantile spasms was investigated in terms of whether unprovoked seizures as compared to infantile spasms may increase the risk for autism (Saemundsen et al. 2008).

In the study, children with seizures in the first year of life were divided into two groups: one with unprovoked seizures and the other with infantile spasms. There were 25 children in the infantile spasm group and 96 children in the unprovoked (other than infantile spasm) group. Not all children participated due to moving abroad, and deaths.

Results showed that of 95 participants, 17 children had infantile spasms and 78 had other unprovoked seizure types, all occurring initially during the first year. Of this group, one fourth had intellectual retardation. Six children with autism had infantile spasms, and seven in the nonspasm group had autism. The nonspasm seizure group had one more case of autism than the infantile spasm group, yet had 4.5 times as many patients in the group.

The authors conclude that following statistical analysis, infantile spasms were not an independent risk factor for autistic spectrum disorder. The clinical importance of this study is that if seizures, especially of symptomatic origin, are present in the first year, then autistic spectrum disorder should be considered.

Another paper recently published compares developmental and psychological features in two groups of autism spectrum disorder – one group with epilepsy and another group without epilepsy (Turk et al. 2009). In this study, 60 children with autism and epilepsy were compared to 60 others with autism, but not with epilepsy.

After consent, families were questioned about their children's problems. Questions asked included age at diagnosis, age of onset, types of seizures, features of autism, etc. Clinical interviews of parents were also conducted gathering information on social and communication disorders. Children were seen by school psychologists, and developmental quotients were developed.

Results showed the I.Q. was lower in the autism plus epilepsy group when compared to the autism only group. This could have been in part due to participant selection. Regarding comorbidities, the most frequent was dyspraxia in the autism only group. Tuberous sclerosis occurred in four patients in the autism plus epilepsy group. Age of diagnosis in each group was under 8 years of age. Results between groups in terms of activity, social interaction, passive behavior, etc., were not significantly different. Children with autism plus epilepsy were significantly more likely to have gross and fine motor problems. They also lagged in obtaining daily living skills such as bathing.

The authors conclude that their investigations support the idea that autism spectrum disorder is not just one syndrome, but a range of disturbances. This indicates that in autism, there are a range of clinical symptoms/syndromes manifesting in common pathways. Differences in autism only and autism with epilepsy were not overwhelming as might be expected. Obvious differences such as impaired motor skills, delayed daily living skills, etc., were present. Further studies examining these differences are warranted.

A paper looking at GABAa receptor downregulation in autism has recently been published (Fatemi et al. 2009). Previous studies have suggested abnormalities in both glutamate and GABAergic pathways in the brains of autistic patients (Blatt et al. 2001). In the present study, the GABAergic system in autism is studied in three brain areas associated with autism, including the cerebellum, superior frontal cortex, and the parietal cortex (Brodmann's area 40).

Results showed the expression of four subunits of GABAa, GABRA1, GABRA2, and GABRB3 were all reduced in Brodmann's area 40 of the parietal cortex. GABRA1 and GABRB3 were changed in the cerebellum, and GABRA1 was altered in the superior frontal cortex in patients with autism as compared with controls. The presence of seizures in autism in this study did not affect the significant reduction in GABAa receptor subunits.

The authors comment that there were significant reductions in GABAa subunits in brain regions of autistic patients. While there was no difference between autistic and autistic plus epilepsy patients, all had mental retardation which, as the author's state, might have influenced subunit values. The authors further note their data are the first on GABAa subunit expression in the superior frontal cortex, parietal cortex, and cerebellum in autistic patients. The authors suggest future studies should assess other areas such as the hippocampus. Increased numbers and lack of possible confounding mental retardation should be taken into account. Younger patients including children should be evaluated when possible.

Chapter 25
Reactive Epilepsy: Phenylketonuria

The first case of phenylketonuria (PKU) was described in 1934 by Folling. Only 5 years later it was shown that PKU was an inherited condition, transmitted through an autosomal recessive gene homozygotes. Very high levels of phenylalanine accumulated in the patients. The enzyme deficiency was identified as that of phenylalanine hydroxylase (Jervis 1953).

Both parents must be carriers of the genetic defect; their enzymatic defect carries no sequelae. One of four offspring from two carriers will be homozygous for the defect, and if untreated, suffer a severe outcome.

PKU symptoms and signs appear very shortly after birth, first by the elevation of serum phenylalanine levels. The levels can reach 30× normal, and phenylpyruvic acid is excreted in the urine. By 6 months of age, mental retardation, seizures, and other neurological evidence are present. There are pigmentation changes in skin and hair. In adolescents and adults, the progression of the disorder diminishes and stops.

The mental deficiency can be severe. The incidence of PKU is thought to be about 1 in 20,000. Patient's urine has a dark green color, which is diagnostic, and offers a ready diagnosis. Most cases are in infants and older patients are much rarer, probably due to an increased death rate. There is no difference in incidence between males/females. More males than females have elevated phenylalanine, but not all of these have PKU.

The extent of mental retardation from untreated PKU is severe. As an example, an early study showed untreated PKU patients with an I.Q. of around 15 contributed 25% of all PKU patients. As the I.Q. rose, the percentage of cases dropped. Coupled with significant behavioral problems, nearly all PKU patients were institutionalized.

Clinically, few signs/symptoms are noticeable in the first month. By the 2nd to 3rd months, developmental mile stones are delayed. The developmental delays are progressive, producing further delay. About 65% of untreated PKU patients could not talk at the normal time for vocalization. The change from a normal baby into a highly deficient infant in the 1st year of life is a dramatic change.

The only other abnormalities besides those of the CNS are related to skin, teeth, and bones. The skin shows eczema, with one third classed as severe. The eczema is

D.W. McCandless, *Epilepsy: Animal and Human Correlations*,
DOI 10.1007/978-1-4614-0361-6_25, © Springer Science+Business Media, LLC 2012

sometimes the initial complaint, and may be related to a light and sensitive skin. Decreased pigmentation with resultant blue eyes and blond hair are characteristic. In one study, 60% of PKU infants had blue eyes and blond hair.

As regards brain in PKU patients, about half were microcephalic with a small head circumference. In some cases, bone spicules are seen extending from the bone into adjacent cartilage. Of interest are the anatomic changes (skin, teeth, and brain), all of which have developed from ectoderm. Both brain and teeth continue to develop simultaneously through the period in untreated PKU during which the most severe deficits develop.

It is reasonable to note that the most severe alterations due to PKU relate to the CNS. Epilepsy appears in about 25% of untreated cases of PKU. Absence attacks are also seen, bringing the total number of patients with epilepsy to about 50%. The onset of seizures is between 1.0 and 1.5 years of age, and generally stops by the mid-late teens. There is an even higher incidence of abnormal EEGs in PKU patients than there are overt seizures. EEG characteristics consist of spike wave complexes, dysrhythmias, and absence seizure EEG profiles. The EEG is not diagnostic of PKU, but reflects pending deep midline brain damage.

As indicated earlier, there is a significant altered behavior problem. This can consist of restlessness, fearful behavior, destructive and psychotic behavior, and the presence of night terrors. Severe uncontrolled temper tantrums are often seen. Even after treatment, PKU patients are seen as clumsy, awkward, and hypersensitive. Peculiar stance and sitting postures are present.

Hypertonicity is a common finding, and is shown as an unrelaxed attitude, which in turn contributes to a rigid gait. Two thirds of PKU patients show hyperactive tendon reflexes. Many patients show a variety of different abnormal body movements. Sometimes a fine upper extremity tremor exists, usually involving the hands and fingers. The tremor may become more severe over time. Other neurological deficits can occur, including cerebral palsy.

From the neuropathology standpoint, deficient myelination may be seen. Myelination is active during the exact time that PKU is progressing. The myelination period is critical, and once the time frame is passed, "catch-up" is rare. Defects in myelination are present in dozens of other abnormal conditions from malnutrition to severe inborn errors of metabolism. All are associated with mental retardation. Chemical analysis of myelin substantiates alteration in its production (Crome et al. 1962). Intense gliosis and myelin breakdown products are seen in PKU brains. The neuropathologic changes correlate closely with functional deficits in untreated PKU patients.

Since early treatment with proper dietary restrictions is essential to restrict neurological damage, screening protocols were developed and implemented. In one screening modality, whole families are closely watched after an initial child is diagnosed with PKU. Examination of subsequent children early could lead to early onset of treatment. With an incidence of 1:20,000, many must be screened in order to identify early treatable cases. Large geographical areas were screened with the common FeCl3 urine test.

One finding (Cahalane 1968) was that about 1:1,000 newborns had increased phenylalanine levels in the 1st year of life (over 4 mg/100 ml), which returned to normal within a couple of weeks. Elevation of phenylalanine levels for a couple of months or more is indicative of PKU. Problem patients are those with elevated levels of phenylalanine but lower than most PKU patients.

In cases of suspected PKU, and with levels of phenylalanine over 20 mg/100 ml, the question is how and when to treat the patient. Historically, early treatment consisted of useless attempts using multivitamins, minerals, hormone supplements, etc. Ultimately, a low phenylalanine regime was adopted.

Initial dietary attempts were made (Armstrong and Tyler 1955) in which a mixture of pure amino acids was purified and administered to patients. Another attempt used charcoal treatment of casein hydrolysate to remove phenylalanine. A patient was maintained on the diet for 6 weeks, during which the reversal of major biochemical alterations occurred (Bickel et al. 1954). Longer treatment time frame, plus greater numbers of patients confirmed the above findings. A simple lowering of phenylalanine intake is insufficient for treatment efficacy.

The complete elimination of phenylalanine may produce dietary insufficiency. Frank phenylalanine deficiency and starvation was inadvertently produced in some early treatment patients. Adequate control hinges on a balance between the phenylalanine requirement while meeting caloric and other nutritional needs. Overtreatment (mild phenylalanine deficiency) may occur in young, early treated patients with higher phenylalanine requirements.

All of the major biochemical changes associated with untreated PKU are eliminated with proper dietary treatment. Treatment also eliminated the peculiar and obnoxious odor of PKU urine, and blood serotonin levels rose. Abnormal components disappeared. Skin eczema cleared, and pigment was restored.

Of significant importance was that the correct treatment reversed the neurological semiology except, of course, for those caused by anatomic structural deficits. Hyperactive reflexes, hyperkinesis, and tremors decreased. Treatment served to normalize EEG findings and decreases or eliminates seizures. The exceptions, again, are in cases in which irreversible structural cerebral damage has occurred. This correlates perfectly with the concept of a pure metabolic encephalopathy, in which early aggressive treatment can more or less completely reverse the disorder, and in these cases, normal development resumes.

In terms of intelligence, the reversal of lower I.Q. in older patients does not occur with correction of phenylalanine imbalance. However, the low phenylalanine diet prevents the impairment of intelligence during the early months in untreated PKU. The diet can reverse the most recently acquired intellectual impairment (Woolf et al. 1958).

Subsequent studies in large numbers of patients have unequivocally shown the efficacy of the diet on intelligence (Berry et al. 1967) and the correlation between age of onset and final I.Q.

There was no correlation between time of treatment and I.Q., further suggesting that most cerebral damage occurs during the first year. Treatment effects seem to correlate with the concept that damage is postnatal and the newborn is normal at birth. Avoiding the deleterious accumulation of phenylalanine is essential at the earliest sign of PKU.

A paper describing mouse models of PKU and their effect on seizure susceptibility has been published (Schlesinger et al. 1969). Mice were treated by one of three methods: parachlorophenylalanine injection, phenylalanine, and diets varying in tyrosine. The purpose was to examine seizure susceptibility to metrazol-induced seizures in mouse models of PKU.

Results showed that acute administration of parachlorophenylalanine on the cerebral concentrations of 5-HT and NE lowered levels (50%) of 5-HT, but had no effect on NE. This effect was similar between 21-day-old DBA/2J and C57BL/6J mice. The results were also similar at 28 days of age. Furthermore, injection of phenylalanine lowered brain levels of 5-HT by 20%, with no effect on NE.

Susceptibility to audiogenetic seizures was increased by parachlorophenylalanine, methyl tyrosine, and a combination of the two in DBA/2J mice. In C57BL/6J mice, only the combination of the two drugs affected susceptibility by raising it slightly. Chronic injection of DBA/2J mice increased seizure susceptibility to four audiogenic-induced seizures. Manipulating dietary levels of phenylalanine and tyrosine served to increase susceptibility to all phases of seizures in DBA/2J mice. In C57BL/6J mice, only one instance of seizures occurred, otherwise there were no responses in C57BL/6J mice.

The authors note that the phenylketonuric symptoms of seizures can be minimized in mice by increasing phenylalanine or decreasing phenylalanine hydroxylase. The result was more pronounced in some mouse genotypes as compared to others. These models show that symptoms other than mental retardation can be studied with these mouse models.

The authors suggest that, using previously published data, certain mouse strains are generally more likely to display seizures with stimulation. The authors note that 5-HT is lowered by parachlorophenylalanine. The authors suggest that seizure susceptibility and nervous system excitability may be associated with a cerebral decrease in 5-HT.

The possible effect on seizure production of aspartame or phenylalanine was examined in the baboon *Papio papio* (Meldrum et al. 1989). Aspartame with a phenylalanine component is a widely used sweetener. Previous studies have suggested that aspartame in high doses may be involved in epileptogenesis (Wurtman 1985).

In Meldrum et al. 1989, photosensitive baboons were given phenylalanine at doses of 50, 159, or 450 mg by gavage. The myoclonic seizure response to photostimulation was graded before and after drug administration. Blood samples were taken for amino acid measurement. In the assessment of aspartame as a possible seizure inducer, doses of 300 mg/kg or 1,000 mg/kg were administered. Blood samples were also taken. Results showed that both phenylalanine and aspartame increased blood levels of plasma phenylalanine concentrations. In contrast, neither phenylalanine nor aspartame resulted in any different response in seizure susceptibility as compared to the results in control baboons. These results are in contrast with previous studies showing a positive increase in seizure activity produced by manipulation of the monoaminergic system.

There are reports of the consumption of large amounts of aspartame producing seizure activity in humans (Wurtman 1985). This effect might have been due to rehydration and hyponatremia, known to induce seizures. The authors conclude there is no evidence of a convulsant action by phenylalanine or aspartame in this primate model.

As the effective dietary treatment for PKU continues, and efficacy well documented, nevertheless some patients escape treatment, are diagnosed late, or have spotty treatment, or just stop treatment. All of these may lead to a more progressive course of the disorder. The paper presented was a review of these concepts and evaluation of adult PKU (Brenton and Pietz 2000).

Some authors speculate that progressive deterioration is a constant feature of PKU. In a reassessment of untreated PKU patients (Pitt and Danks 1991), 46 patients were followed up after 20 years. Of these patients, 5 had died, and 21 had I.Q.s of less than 35. This paper lends credence to the concept of neurological deterioration. In another follow-up study of treated patients (Pedersen and Birket-Smith 1974), the authors concluded that deterioration (severe upper motor neurons) occurs in the second to third decades. One third of patients in this study had epilepsy.

In a series of nine patients with late onset of neurological symptoms, eight had spastic paraparesis, and four had a late onset of epilepsy. Only three had an I.Q. over 80. Evidence shows a correlation between the progress and evidence of dysmyelination as shown by MRI in periventricular white matter. There was seen to be improvement in the MRI images in terms of dysmyelination, when the diet was reinstituted. This again emphasizes the concept of an at least partially reversible metabolic encephalopathy.

In patients treated very early (7 days) and without interruption, there were still some neurological signs such as tremor, ataxia, hyperreflexia, fine motor affects, etc. These represent minor signs of some minor CNS damage.

The conclusions are that the diet will result in progressive deterioration, and even in a well-controlled continuing diet, neurological signs may be present. Unanswered questions relate to the dysmyelination, including mechanisms, sites, and relation to MRI changes/symptoms. The question of the degree of in utero damage from hyperphenylalanine is another area needing study.

In another study (Martynyuk et al. 2007), audiogenic seizures were studied in homozygous PKU (Pah BTBR) mice. PKU is the most common amino acid metabolism disorder, with a frequency of 1:20,000 to 1:10,000. The disorder has two forms, the classic type and another less serious form, hyperphenylalaninemia. The PKU mouse has a missense mutation in the Pah gene, created by germline ethylnitrosourea mutagenesis of BTBR mice.

The age of the mice in the audiogenic testing was 36–336 days. Testing took place in a plexiglass cage, permitting video recording. The diet consisted of a phenylalanine free diet supplemented with phenylalanine. Serum phenylalanine was measured using fluorometric assay.

Results showed the phenylalanine levels to be ten times that of heterozygotes and wild-type control mice. Young 5–7-week-old homozygous mice did not seize in response to a 118-dB sound stimulus, whereas adult mice (18–20 weeks) with lower phenylalanine levels than the young mice had a full range of audiogenic seizures. The response included wild running, tonic–clonic seizures, respiratory arrest, and death. The mice exhibited a clonic phase with violent limb kicking, followed by a severe tonic phase. When the seizure started, the stimulus tone was stopped. Following the seizure there was a 1–2 min period during which the mice were unconscious.

The authors comment that the PKU mice are susceptible to audiogenic seizures. Within the hyperphenylalanine range, seizures were negatively correlated with seizures in the lower serum ranges, the seizures were worse than in young mice with higher phenylalanine serum levels. The authors comment that since high levels of phenylalanine depress glutamatergic synaptic treatment (excitatory). This could serve to decrease the audiogenic responses.

The study also showed a resumption of susceptibility in the PKU mice taken off the diet. This finding correlates well with humans who are not compliant. As many as 90% of adult patients are not compliant at some point in their therapy. The authors state that further studies are needed, and that the PKU mouse is an excellent model.

In a paper examining various treatment modalities for PKU, it was noted that compliance with the original diet was sometimes difficult to maintain. Various newly developed or developing strategies are emerging (dos Santos et al. 2006). PKU is a relatively simple inborn error of metabolism for which a straightforward efficacious treatment is available. The problem is that the diet is difficult to adhere to and different phenotypic levels exist rendering treatment not so simple.

In terms of the conventional dietary therapy, tolerance depends on genetics of the patient, plus overall health and metabolic status. This means the diet for PKU must be customized for each patient in order to prevent phenylalanine to reach toxic levels. The diet is a low protein diet supplemented with an amino acid mixture without phenylalanine. Because of its restrictive nature, compliance is infrequent.

Gene therapy would theoretically be ideal in that a viral vector could insert a corrected gene into liver cells, which would then make a corrected tyrosine hydroxylase enzyme. Unfortunately, the in vivo transduction efficiency is low. Placing a recombinant adenovirus expressing PAH cDNA into the portal vein restored 10–80% of PAH activity. This normalizes plasma phenylalanine levels. Recombinant adenovirus-associated vectors worked well, but were less effective in females (mice).

The authors comment that PKU can be a dangerous disorder, and needs correction in order to prevent devastating effects, and refractory epilepsy. The dietary correction may mean a loss of quality of life. Estimates are that by adolescence, 60–80% of patients have partly or completely abandoned the diet (Merrick et al. 2003). The diet suffers from bad taste and smell, plus strong restrictions in protein intake.

The authors note that gene therapy may be some years away, and the dietary method in which phenylalanine is internally degraded should be a satisfactory solution. Genetically modified organisms might be more efficient in the intestinal breakdown of phenylalanine. An exciting time in PKU therapy is the development of new strategies, each with advantages and disadvantages, that are intended to eliminate the associated mental retardation and epilepsy.

A recent paper examines the prevalence of seizures in PKU patients (Karimzadeh et al. 2010). The aim of this study was to look at the frequency of seizures in a cohort of PKU patients, as well as to relate these findings to the EEG findings and behavioral disorders. The authors note that three levels of PKU severity can be discerned, all related to blood phenylalanine levels: classic PKU-levels over 1,200 μM. Mild PKU levels were between 600 and 1,200 μM, and mild HPA levels above normal, but below 600 μM. Results were classed according to the above scheme.

Results were based on a total of 94 patients seen over a 2-year period. Mean age was 8.5 years of age. Age of onset ranged from 1 month to 23 years of age. All patients had blood taken for analysis, and EEGs were recorded after 8–12 h of sleep deprivation. Results showed 43% of patients had seizures, the rest were seizure free. The male/female seizure ratio was about equal.

EEG results were abnormal in 81% of those showing seizures. The relation between EEG abnormalities and seizure presence was statistically significant (0.001). There was no statistical significance between phenylalanine serum levels and clinical seizures. The difference in seizures between the three levels was also not significant. Overt seizures occurred less frequently than EEG abnormalities. The authors conclude saying there was a significant correlation between EEG abnormalities and behavioral disorders.

Untreated PKU is characterized by an abnormal phenotype including growth failure, seizures, developmental delays, and mental retardation. The present paper examines such a case of maternal PKU (Bouchlariotou et al. 2009). During pregnancy, the diet must be reinstated in PKU women because phenylalanine crosses the placenta and can cause craniofacial malformations, microcephaly, heart disease, and mental retardation (Koch et al. 2003).

The case presentation in this paper was a 33-year-old woman who was diagnosed as having PKU based on prenatal screening. Maternal PKU was suggested when the fetus showed a ventricular septal defect. The first phenylalanine measurement was 900 μM. Examination of the maternal PKU woman was largely negative. The patient had no history of poor school performance or behavioral problems. She was put on a phenylalanine free diet, conceived again, and delivered a normal nondysmorphic child.

The authors comment that not all cases of PKU result in severe neurological outcomes and behavioral problems. I.Q.s are not always drastically reduced. This means that health care workers, especially in obstetrics, should be aware of the existence of nocturnal PKU. This knowledge, as in the above case can prevent additional embryopathy.

In order for pathology to occur, there needs to be a genetic absence of phenylalanine dehydrogenase and an exposure (dietary) to phenylalanine. As phenylalanine levels rise, it crosses the blood–brain barrier and causes neurological damage (Huijbregts et al. 2002). Treatment currently involves a strict adherence to a relatively unpleasant phenylalanine free diet. Compliance, as detailed earlier, is a major problem. The usual progression of untreated patients, or spotty treatment, is pronounced mental retardation, behavioral problems, and epilepsy (25–50%).

Of interest is the oft observed phenomenon of at least partial reversal of some neurological problems when ceased dietary treatment is reinstituted. These reversals consist of a decrease in behavioral issues, reduction in frequency and severity of seizures, improvement in white matter (myelin) problems, etc. Improved dietary features are being sought, as well as other possible treatments. Gene therapy should be an efficacious treatment in the future (see Chap. 31).

The offspring of maternal PKU are carriers of the PKU genetic deficiency and are exposed to increased levels of phenylalanine from very early in the pregnancy. Phenylalanine is toxic and teratogenic due to its inhibition of the transport of free L-amino acids. The frequency and severity of teratology depends on the concentration of phenylalanine (Levy and Ghavami 1996).

The main concern with cases of maternal PKU is to prevent fetal pathology. This is accomplished by keeping phenylalanine levels in the normal range (120–360 μM). This prevents teratology such as mental retardation, etc., and ensures normal growth/development. Alert observation can act to prevent pregnancy complications by placing the mother on a phenylalanine free diet as soon as possible. The authors state that it is hoped that increases in scientific knowledge, and awareness, will lead to better treatment paradigms, and future development of new effective treatments for PKU and associated epilepsy.

A recent paper describes seizure aggravation by the AED levetiracetam in an adult PKU patient (Dericioglu and Saygi 2010). This is a case report of a 20-year-old female who was compliant as regards the PKU diet. The patient was diagnosed with PKU at age 26 months, and therapy started at 3 years 9 months.

The patient had two siblings, one normal, another was a heterogeneous carrier. The patient had the first seizure at the age of 18, and it was a generalized tonic–clonic seizure. She was placed on levetiracetam after the second seizure, but they continued. The drug dose was increased, but the seizures actually became worse.

At hospital admission, the patient was noted to be moderately retarded, but otherwise normal. An overnight video EEG revealed generalized 2.5–3.0 Hz high amplitude delta paroxysms lasting 3–5 s. The patient was switched to topiramate, and the levetiracetam was slowly decreased. This resulted in a complete cessation of seizure activity within a few months.

Generalized seizures in PKU patients are usually stated to occur in about 25% of patients, but some studies show a frequency of 50%. Certainly the incidence is related to presence or absence in delays in initiating a phenylalanine treatment regime. At a high level of phenylalanine in blood, there is a decrease in excitatory glutamatergic synaptic transmission. This is reversible when there is a decrease in phenylalanine concentration, then an increased susceptibility with hyperphenylalaninemic state. Discontinuation of the diet is associated with epilepsy in PKU patients (Villasana et al. 1989).

The authors note that most reports of seizures in patients with PKU were studied earlier, and therefore their treatment was with "older" AEDs. In this case report, one of the new AEDs, levetiracetam, was used, and there is little data regarding efficacy/risk in this type of patient (PKU). It is well known that some newer AEDs are contraindicated in some metabolic situations (Lin and Thajeb 2007). It has been recently noted that levetiracetam may lead to worsening of seizures in mentally retarded epileptic patients (Szucs et al. 2008). The mechanism for this is unknown. The authors conclude this AED may not be appropriate for epilepsy in PKU patients.

Part V
Medical Treatment

Chapter 26
Diagnostic Technology

The first-line treatment for patients suffering from epilepsy is pharmacological therapy with antiepileptic drugs (AEDs). However, 20–30% of patients with epilepsy are resistant to medical treatment, despite efforts to find effective combination therapy (Lowenstein 2008). For patients who suffer from pharmacoresistant, localization-related epilepsy, resective surgery offers the best chance for a reduction in seizure activity or a lifetime free of seizures. The focus of this chapter will highlight the presurgical evaluation, imaging technologies, and intraoperative monitoring techniques used to aid surgical teams carry out successful procedures designed to help both adult and pediatric patients afflicted with intractable epilepsy.

The presurgical evaluation of patients includes a detailed history and physical exam and the classification of prior seizures and epileptic syndromes, if present. An important chapter in Nelson's Essentials of Pediatrics (Kliegman et al. 2006) lists the following as different classifications of seizures: generalized, tonic–clonic, absence, tonic, atonic, partial, simple partial, complex partial, and status epilepticus. This chapter also highlights the common epileptic syndromes, which include benign focal epilepsy (benign rolandic, benign centrotemporal), juvenile myoclonic, infantile spasms (West Syndrome), Lennox–Gastaut syndrome, acquired epileptic aphasia (Landau–Kleffner syndrome), and benign neonatal convulsions. Documenting the age of onset of seizures and evolution over time, the presence/absence of auras, the patient's past medical history, past surgical history, allergies, medications (focusing on the AEDs used), family history (exploring a genetic component), social history, and a complete review of systems (including a detailed neurological exam) will help to paint a clear clinical picture of the patient's illness.

According to the work of Kliegman et al. (2006), common etiologies of epilepsy include perinatal hypoxia, infectious causes, metabolic conditions, poisonings, neurocutaneous syndromes, systemic disorders, trauma, tumors, intracranial hemorrhage, CNS structural anomalies or malformations, degenerative disease, febrile or idiopathic causes. If the etiology can be determined this information along with that of the history and physical exam can help to determine whether or not the seizures are indeed intractable and whether or not a surgical intervention is warranted.

D.W. McCandless, *Epilepsy: Animal and Human Correlations*,

DOI 10.1007/978-1-4614-0361-6_26,

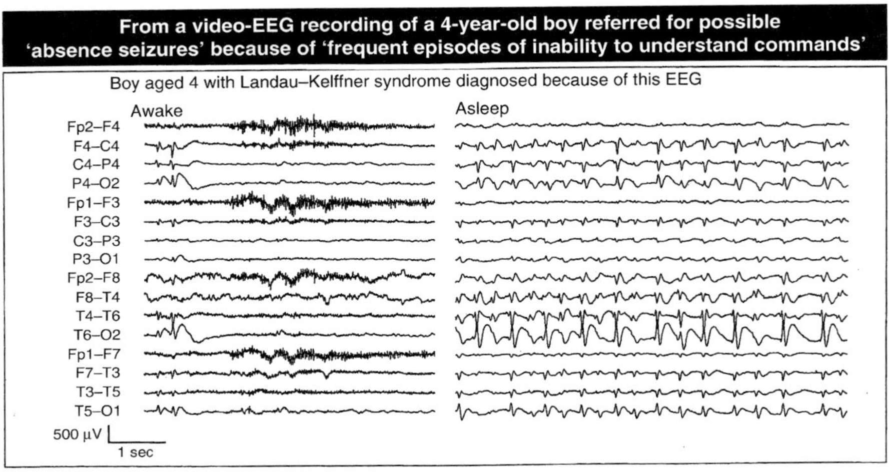

Fig. 26.1 Video EEG of a 4-year-old boy with Landau–Kleffner syndrome. With kind permission from Springer Science+Business Media: Epileptic Syndromes and their Treatment, 2010, p. 307, Panaylotopoulos, C., Fig. 10.11

Included in the initial workup of epilepsy patients are routine laboratory studies. Tests such as CBC and CMP should be ordered during the initial workup of epilepsy in order to identify common metabolic causes of seizures, such as abnormalities in electrolytes, calcium, magnesium, or glucose (Lowenstein 2008). These studies can also point to hepatic or renal disease and provide as a baseline of liver function prior to AED administration. These tests can also be of value in the evaluation of the presurgical patient. Additional studies such as blood and urine toxicology studies or a lumbar puncture may be necessary if no identifiable cause is determined or if patients fall into particular risk groups for infectious etiologies.

An important paper regarding the preoperative evaluation of intractable epilepsy in children (Go and Snead 2008) suggests that presurgical studies go from least invasive (patient hx, physical, and video-EEG monitoring) to the most (Wada testing), and that the data is discussed amongst a multidisciplinary epilepsy surgery team to decide on how to proceed on a case-by-case basis. The authors discuss the importance of video-EEG monitoring in the documentation of ictal (and intericatal) events, the lateralization and localization of electrographic onset of seizures, and the interpretation of the features of the seizure activity. The paper highlights the importance of prolonged evaluation upon admission to an epilepsy monitoring unit, in order to properly evaluate ictal activity during wakefulness and sleep (see Fig. 26.1).

According to the work of Ferrie et al. (1998), video-EEG allows for the precise timing and replay of clinical and EEG events. The authors state that the most useful phenomena detected with video-EEG are the sudden jolts of limbs from myoclonic jerks, rhythmic myoclonic movements of the face, the timing of events, automatisms and eye opening, and the impairment of consciousness. This information can

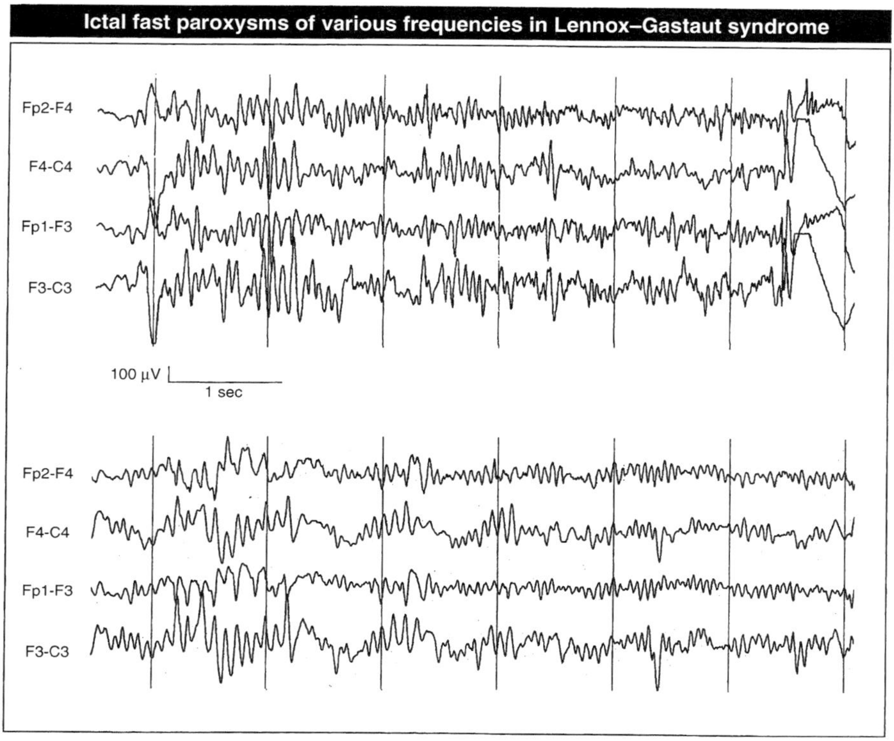

Fig. 26.2 EEG from Lennox–Gastaut Syndrome. This EEG is easy to identify, facilitating diagnosis. With kind permission of Springer Science+Business Media: Epileptic Syndromes and their Treatment, 2010, p. 296, Panaylotopoulos, C. Fig. 10.9

assist in establishing a definitive diagnosis of the epilepsy and can also assist in localization of the epileptogenic focus, which will be targeted in the case of a surgical resection (see Fig. 26.2).

Another technology used to map epileptogenic foci is magnetoencephalography (MEG). According to the work of Nordi and Pedley (2008), MEG detects and measures the magnetic activity of the cortex using superconductive quantum interference devices (SQUIDs). The results of several researchers have compared this technology to be favorable with EEG localization. MEG detects the magnetic fields created by the electrical activity of the brain. This information can be measured to provide a localization of a neuronal source, such as epileptiform activity. MEG is an innovative and noninvasive neuroimaging technology, which has the ability to accurately localize dipoles oriented parallel to the surface and the recordings are minimally affected by overlying bone and tissue. In addition to MEG serving as a supportive adjunct to EEG monitoring, the work of Go and Snead (2008) states that the combination of MEG and MRI data creates magnetic source imaging data that has been proven helpful in the localization of extratemporal foci and in the case of >1

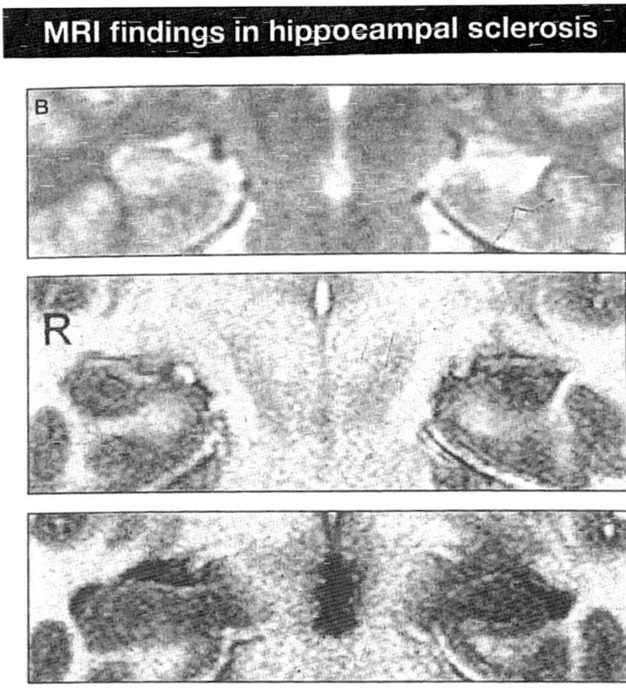

Fig. 26.3 MRI findings in hippocampal sclerosis. Coronal T1-weighted MRI scan showing right-sided hippocampal sclerosis. With kind permission of Springer Science+Business Media: Epileptic Syndromes and their Treatment, 2010, p. 447, Panaylotopoulos, C., Fig. 15.5

epileptogenic focus. They also state that the data collected from MEG can assist in ensuring the accurate placement of subdural electrodes on the cortex during invasive EEG monitoring.

Neuroimaging studies assist in the identification of structural abnormalities within the brain and their relation to the epileptogenic focus and the eloquent cortex of the brain, which is responsible for motor, language, and sensory function. Magnetic resonance imaging (MRI) has the ability to reveal structural features, structural lesions, the presence or absence of hippocampal atrophy, and hippocampal volume. MRI is the primary imaging modality for patients with epilepsy according to the International League Against Epilepsy guidelines (Go and Snead 2008), most likely due to the fact that it has the ability to display brain pathology more accurately than CT scans. As stated by Johnston et al. (2008), routine imaging should include high-resolution T2-weighted, fluid attenuation inversion recovery (FLAIR), T2-weighted gradient echo, 3D high-resolution T1W gradient echo, and diffusion-weighted imaging. In the case of a structural anomaly, the authors also suggest that studies with gadolinium contrast be carried out in order to differentiate neoplasia from dysplasia. As stated earlier, MR imaging is sensitive and specific for detecting abnormalities such as hippocampal sclerosis (see Fig. 26.3). A recent study by

Jutila et al. (2001) states, "Signs of unilateral hippocampal damage or unilateral foreign tissue lesions on (MR) imaging probably best predicts successful postoperative outcome in patient undergoing surgery for drug refractory temporal lobe epilepsy."

In addition to structural neuroimaging studies, complementary functional neuroimaging studies such as positron emission tomography (PET) can be performed. This nuclear medicine study allows for the visualization of brain glucose metabolism with the use of fluorine-18-labeled fluorodeoxyglucose positron-emitting ligand (Johnston et al. 2008). This procedure is carried out during the interictal phase and compares the relative glucose utilization in suspected epileptogenic areas to those within the unaffected contralateral structures of the brain (Go and Snead 2008). According to the article by Passaro and Jobe (2010), hemispheric abnormalities amenable to focal cortical resection include those that display ictal EEG onset that localizes to one temporal lobe, in combination with the (FDG-PET) showing unilateral temporal hypometabolism or the MRI showing unilateral hippocampal atrophy.

Single-Photon Emission Computed Tomography (SPECT) is a noninvasive modality, which can be used to measure cerebral blood flow changes associated with epileptic discharge (Johnston et al. 2008). This nuclear medicine technology uses agents such as technetium-99m ethyl cysteinate dimmers or iodinated radiotracers, which have the ability to cross the blood–brain-barrier. According to the aforementioned study, SPECT can be used to show ictal hyperperfusion within the epileptogenic zone if the injection occurs during seizure activity; hypoperfusion during the postictal state; or baseline perfusion levels during the interictal phase. This technology can be useful in MRI negative partial epilepsies and epilepsy with focal cortical dysplasia as the etiology. It should be noted that the sensitivity of SPECT for temporal lobe epilepsy is higher than that of extratemporal epilepsy, 97% and 50%, respectively (Johnston et al. 2008). A newer methodologys called the subtraction ictal–interictal SPECT co-registered to MRI (SISCOM) has been show to be more accurate than either ictal or interictal SPECT scanning at localizing the cortical area of ictal onset. The authors note that scans must be obtained after 48 h of one another in order to allow for radionucleotide washout, during an interictal period and within seconds of seizure onset (see Fig. 26.4).

Another imaging study used in the assessment of cortical activation, which drives cerebral metabolism and cerebral blood flow, is the functional MRI (fMRI). This technique allows for the visualization of active brain regions based on changes in cerebral blood flow and oxygen metabolism. The chapter by Johnston et al. (2008) explains that fMRI relays important data dealing with the lateralization of memory and language activation, the localization of seizure activity and areas of eloquent cortex. The authors also point out that this technology can be used to assist in the proper placement of subdural grid electrodes, influence the perimeters of cortical/subcortical resection, and in some cases play an instrumental role in positive surgical outcomes (see Fig. 26.5).

Another test that can determine language dominance is the Wada test. This invasive test involves injecting sodium amobarbital into each internal carotid artery, one

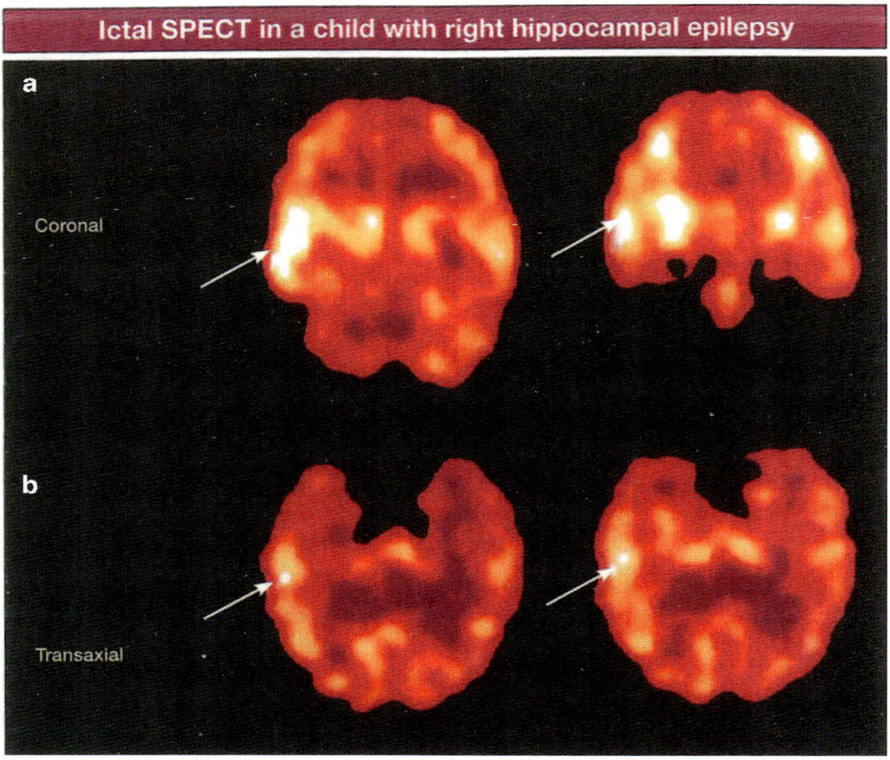

Fig. 26.4 Coronal (**a**) and axial (**b**) 99mTc HMPAO SPECT images with increased right anterior temporal lobe increased perfusion. With kind permission of Springer Science+Business Media: Epileptic Syndromes and Their Treatment 2010, p. 167, Pannaylotopoulos, C., Fig. 6.11

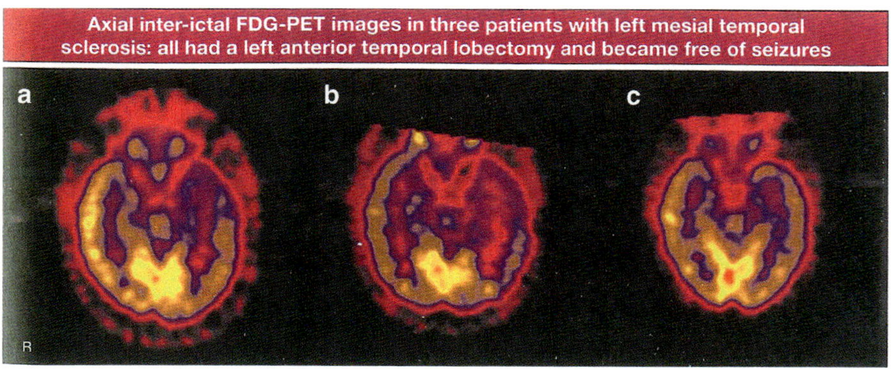

Fig. 26.5 (**a**) Unilateral glucose hypometabolism shown by FDG uptake in mesial temporal lobe and temporal neocortex in the left hemisphere, (**b**) bilateral glucose hypometabolism in both temporal areas, and (**c**) equivocal glucose metabolism without laterality. With kind permission of Springer Science+Business Media: Epileptic Syndromes and their Treatment, 2010, p. 167, Panaylotopoulos, C., Fig. 6.12

at a time. This substance causes the patient's contralateral limb to become anesthetized (impedes hemispheric function), and once this occurs the clinician can test for affected speech areas. According to the work of Go and Snead (2008), this test can also be used to assess the risk of postoperative amnestic syndromes in temporal lobe epilepsy. Because of the invasive nature of this testing, noninvasive tests such as MEG, MRI, and neuropsychological testing should be attempted first.

Neuropsychological testing assists in the preoperative evaluation of cognitive functional and the potential impact of surgery. It is comprised of a comprehensive series of tests, which measure concentration, language, executive abilities, personality and emotional functioning, verbal memory, associations, delayed story recall and visual memory including the delay recall of figures. This testing also assists in the lateralizing of cerebral hemisphere dominance for verbal and nonverbal function in older children (Go and Snead 2008).

Once the presurgical evaluation is complete, a multidisciplinary team including the neurosurgeon, neurologist, neuroradiologist, and neuropsychologist work to develop a surgical plan. If the noninvasive testing has failed to establish a conclusive localization of the epileptogenic focus then preoperative, intracranial EEG monitoring, also referred to as chronic electrocorticography (ECoG), will be performed. Indications for this type of monitoring include nonlesional epilepsy, lesional epilepsy with discordant data for the localization of the epileptic focus, or if any portion of the epileptogenic zone resides within the cortex responsible for motor, language, or sensory function (Go and Snead 2008).

A recent study by describes the equipment used in this test including implantable depth, strip, and grid subdural electrodes, which are placed to record interictal and ictal activity and provide stimulation in order to map/define the boundaries of the functional cortex. According to the author, strip electrodes are usually implanted through a burr hole in the skull after the patient is placed under general anesthesia. Fluoroscopy is then used to confirm proper placement. Patients are then monitored in an inpatient unit for 2–7 days with the electrodes in place in order to record their seizure activity. Once the relevant data is obtained, the electrodes are removed through the skin with gentle traction on the leads, under conscious sedation or general anesthesia.

The subdural strip is comprised of inert substances such as platinum and silastic and is limited in its sensitivity because it only provides up to 1×8 contact surfaces. In more complex cases, subdural grid electrodes with arrays of 5–8 rows (20–64 contacts) are required in order to localize the epileptogenic focus. In this case, a craniotomy is required for placement of the grid. The author of the previously mentioned study suggests the use of prophylactic antibiotics and dexamethasone in these procedures. The bone flap is either replaced or frozen under sterile conditions. The patient is then transferred to a video-EEG monitoring unit where they received constant nursing supervision. The grid is then removed when the appropriate data has been collected. It is important to note that if a resective surgery is scheduled, the relationship of the grid to the underlying cortex must remain undisturbed until the day of surgery. The grid must then remain in its exact location as the craniotomy is re-opened and then the relationship of the grid's electrode contacts to the underlying cortical locations must be documented. If recordings from the hippocampus or amygdala are necessary, then depth electrodes can be placed in conjunction with subdural strip or grid electrodes.

Another form of extraoperative monitoring includes cortical mapping (language or motor). This is similar to the intraoperative direct cortical electrical stimulation (DCES), which will be discussed in greater detail. Presurgical motor mapping is accomplished by providing stimulation with bipolar current to the cortex with an Ojemann stimulator probe or two adjacent contacts (~1-cm apart) on the previously implanted cortical grid. This monitoring is carried out by a neurologist, neurophysiologist, or physician and serves to map the eloquent regions of the cortex. The disadvantage of performing this extraoperatively is that it can be associated with pain or discomfort; however, the advantages include being able to assess and then discuss the probable benefits vs. risks of the surgery with the patient and their family. This information can also help the surgeon, anesthesiologist, and OR staff to decide whether or not an awake-approach or a procedure under general anesthesia would be more appropriate because many of the anesthetic agents used in the operating room can interfere with the intraoperative recordings of ECoG, Language, Motor, and Sensory mapping.

In addition to the long-term presurgical intracranial ECoG, intraoperative ECoG is also available to surgeons who feel that epileptiform discharges recorded during surgery indicate which part of the brain is epileptogenic and they use this data to guide the cortical resections. Other intracranial studies include somatosensory phase reversal studies which help to define the central sulcus, the pre and postcentral gyrus, and DCES, otherwise known as motor mapping.

Phase reversal testing includes the stimulation of the posterior tibial nerve (PTN) or more commonly the median nerve (MN), contralateral to the side of the grid placement. The contacts of the grid are placed across the presumed central sulcus and are referenced to an electrode on the scalp such as Fpz (10–20 system). Somatosensory evoked potentials (SSEPs) are then recorded directly from the cortex. The hand region, over the sensory cortex, generally yields the largest amplitude response to MN stimulation. This response typically occurs at a latency of 20 m (N20 response) and is displayed as an upward or downward deflection according to how the contacts and the grid and the reference electrode are plugged into the differential amplifier of the monitoring system. In contrast, a response from the motor cortex in the precentral gyrus commonly occurs at a latency of 22 ms (P22) and appears as a waveform with a deflection that is inverted compared to the sensory response. After the motor regions are confirmed using this feedback, some surgeons opt to take the monitoring one step further, with the use of DCES of the motor cortex, to map functional motor cortex.

An excellent paper by Schuh and Drury (1997) summarizes the goal of DCES very concisely, to provide information on the location of eloquent areas of the cortex that should not be removed during surgical procedures for epilepsy, thereby decreasing the risk of additional neurological deficits. A common setup for such monitoring includes placing recording electrodes into various muscle groups: face, upper arm, lower arm, thigh, leg, and foot, on the contralateral side of the body from the craniotomy. If for example, the epileptogenic focus is closer to the region of the hand/face, more specific electrode montages can be devoted to these areas of the body. Stimulation is provided with a sterile Ojemann probe or with two electrode contacts of the subdural strip/grid. Constant current can be applied in the classic bipolar, 50-Hz fashion (Penfield technique) or with a pulse-train stimulation (Tanaguchi method). Communication between the surgeon, neurophysiologist, anesthesiologist, and the remainder of the OR staff is crucial to the success of these procedures. If DCES yields motor responses from any of the areas of cortical stimulation, these areas will be marked in the surgical field and resection will be carried out sparring these areas of functional cortex.

A similar technique is employed to map out the functional areas of the language cortex including Broca's area and Wernicke's area. An awake approach is necessary, at least during the time of monitoring, because patient feedback is required during this type of monitoring. According to the work of Shuh and Drury (1997), the following stimulation parameters are commonly used to stimulate the cortex: biphasic rectangular pulses with an intensity between 0.5 and 15 mA, a pulse duration of 0.3 ms, at 50 Hz, with a train duration of 2–5 s. The stimulation intensity is gradually increased until language interference: speech arrest (Broca's), or anomia (Wernicke's), or after-discharges are produced. It is important to note that after an initial episode of language disruption occurs that it must be reproduced in order to confirm the findings. The ideal threshold to stimulate at is just below the threshold of after-discharges. It is suggested that ECoG be recorded during the time of cortical mapping in order to detect after-discharge activity. If after-discharges do occur, it is suggested that ice-cold saline or lactated Ringer's solution be applied to the area immediately, this will diminish the activity and will allow for mapping to proceed.

In conclusion, many patients suffering from intractable epilepsy can benefit from epilepsy surgery in terms of a reduction or elimination of their seizure activity. Careful consideration, including a thorough presurgical workup, performed by a multidisciplinary team must be carried out in order to identify candidates who could benefit from surgical resection. Of utmost importance, is the correct presurgical diagnosis of the patient's condition following an extensive workup. This allows clinicians to ascertain whether or not surgery is truly indicated for each individual patient. Modern technological advances in the way of diagnostic laboratory, imaging, and neurophysiologic testing have enhanced the ability of clinicians to appropriately diagnose and surgically treat patients. These contemporary methods contribute to reduce risks of negative neurological sequela during the procedures and improved outcomes in patients with intractable epilepsy.

Chapter 27
Antiepileptic Drugs

The issue of choosing an effective antiepileptic drug (AED) is in many cases a trial and error process. There are significant clues based on the seizure type, etiology, age of onset, etc., but there is no clear cut equation to be applied for success. Multiple considerations including drug cost, age of the patient, physician experience and bias, patient sensitivity to adverse effects of AEDs, relative ability to treat a particular classification of seizures, etc. are all in consideration to some degree for a first AED choice.

It is useful to examine Table 27.1 for common AEDs in current use.

Narrow spectrum drugs are thought of as working well on specific seizure classifications, whereas broad spectrum drugs have effects on several types of seizures.

It is important to note that some AEDs are available in generic form, which are usually less expensive. These generic drugs usually work well, but site of manufacture, methods of manufacture, constitution, etc. of the generic product are different. They may not be metabolized the same as the original brand, and therefore may produce different side effects, or breakthrough seizures. Most AEDs should be initiated at a low dose in order to assess any adverse effects. This process may take days to weeks to reach optimal dose. Not all AEDs are approved for monotherapy, and are listed as adjunctive therapy to other AEDs.

Side effects should be monitored closely. AEDs can produce allergic reactions and side effects such as skin rashes, fever, mental changes, etc. All seizure medications carry a warning about depression/suicide. AEDs may have adverse effects on liver function, and therefore should be monitored periodically.

Evidence based medicine is a descriptor to a basis for AED selection based on three groups of knowledge. The first is physician based experience, training, knowledge, etc. The second are similar features for the patient/family, which include age, comorbidities, insurance, etc. Finally are all the scientific knowledge regarding AEDs – efficacy for a seizure type, tolerability, etc. All these features together suggest an AED. Cultural beliefs and personal values/biases of either the physician and/or patient are sometimes suggested, but should be kept to a minimum and good science, whenever possible must be utilized.

D.W. McCandless, *Epilepsy: Animal and Human Correlations*,
DOI 10.1007/978-1-4614-0361-6_27,

Table 27.1 Commonly used AEDs showing generic and brand names

Narrow spectrum		Wide spectrum	
Generic name	Brand name	Generic name	Brand name
Carbamazepine	Tegretol	Clonazepam	Klonopin
Gabapentin	Neurontin	Lamotrigine	Lamictal
Lacosamide	Vimpat	Levetrigine	Keppra
Oxcarbazepine	Trileptal	Rufinamide	Banzel
Phenobarbital		Topiramate	Topamax
Phenytoin	Dilantin	Valproate	Depakote
Pregabalin	Lyrica	Zonisamide	Zonegran
Vigabatrin	Sabril		

Early epilepsy drug treatments over 150 years ago were based on erroneous theories. Later (1910), phenobarbital was found to have antiepilepsy properties. Animal models were developed in order to test various compounds for antiepileptic activity, and by 1940, phenytoin was shown to be a major drug for epileptics (Putnam and Merritt 1937). Over the years, three approaches have been developed for AED identification. First is a random search and screening for efficacy of large numbers of potential AEDs. Second is rational predetermined drug design in which some level of chemical modification and synthesis occurs, and lastly are choices for trial based on mechanisms of action.

The idea that seizures represent an imbalance between excitation and inhibition has survived the test of time, but many intracerebral mechanisms have influence over this simple concept. From a mechanistic standpoint, many opportunities exist for design of AEDs which could return the balance, thereby modulating seizure activity. With this in mind, AEDs can be conveniently grouped according to their mode of action. Some have more than one mechanism of action, others have an unknown mechanism of action. Groups include those that involve mechanisms of sodium channel blockers, GABA enhancers, glutamate blockers, carbonic anhydrase inhibition, etc.

Firing of action potentials occurs through sodium channels, which in an active state, allows increased influx of sodium. After the active state, an inactive state occurs (refractory period). During this stage, the axon is not able to propagate an action potential. AEDs targeting this process stabilize the inactive stage, thereby reducing or eliminating axonal repetitive firing.

GABA binds to GABA-A receptors, allowing chloride to enter the neuron thereby increasing the negative charge, and decreasing the action potential. Some AEDs modulate glutamic acid decarboxylase, enhancing GABA (an inhibitory neurotransmitter) production. AEDs of this type act to either enhance chloride influx or increase GABA production or block GABA uptake/metabolism thereby increasing GABA levels, conducive to inhibition.

Glutamate receptors bind glutamate, which is an excitatory neurotransmitter. Its presence (increased amounts) is conducive to an excitatory neuronal state. This occurs when both sodium and calcium flow into the cell, and potassium ions exit the cell. AEDs which modify this process are antagonistic to glutamate, thereby decreasing the excitatory state. The glutamate receptor has five binding sites, including the: AMPA site, NMDA site, glycine and kainate sites, and the GluR site.

The actual presence or absence of anticonvulsant activity resulting from an AED in humans needs to be tested in clinical trials. Animal studies are useful predictors of anticonvulsant activity, and are extensively used, especially in maximal electroshock, in electrokindling, and pentylenetetrazole models (White et al. 1998). In the development of new AEDs, potential efficacy should be examined in as many animal models as possible.

Maximal electroshock and kindling models represent the two models usually used to check a drug's ability to modulate either tonic clonic seizures or complex partial seizures.

Use of the pentylenetetrazole seizure model was undertaken in order to test for efficacy against generalized absence seizures. In some ways this initially in some measure was a reverse process in that drugs effective against absence seizures blocked the clonic phase of pentylenetetrazole induced seizures. Other models with genetic components (the Genetic Absence Epileptic rat of Strasbourg-GAERS, and the lethargic lh/lh mouse) are also being used to assess AED effects against absence seizures (Marescaux and Vergnes 1995; Hosford et al. 1992).

It should be noted that the ability of an AED to modify physiological phenomena such as sodium and calcium channels, increase GABA inhibition, and decrease glutamate excitatory neurotransmission may also alter normal neuronal function. This means the AED may have unwanted adverse side effects. Therefore, some degree of cognitive impairment may occur with channel blockers such as phenytoin.

It is certain that a better understanding of the mechanisms of actions of existing AEDs has led to the design and synthesis of new AEDs which have better antiepileptic efficacy and/or less adverse effects. The reverse has also occurred in that new drugs with less efficacy and greater side effects can be avoided. Further development in both genetics and molecular biology should lead to further and better development of new AEDs.

The relatively new AED levetiracetam is a well used drug in epilepsy patients due to its relatively broad spectrum of efficacy. It has a low level of adverse effects in terms of cognition or drug interaction. Several recent studies have examined various aspects of the use of levetiracetam.

In one study (Stefan et al. 2006), the technique of video EEG was used in order to evaluate the onset of action of levetiracetam. Results showed that antiepileptic action was noted only two days after initiation of the drug treatment. Use of video EEG clearly confirmed the previous results in a larger number of self-reporting patients.

Another similar study (French et al. 2005) had previously shown essentially the same set of results in that levetiracetam showed a significant level of efficacy in the first week of treatment, which lasted unaltered over the three-month trial period.

In a separate study (Tsai et al. 2006), the efficacy and safety of levetiracetam were evaluated in a cohort of Taiwanese patients with refractory partial seizures. This was a multi-center, randomized, double blind, placebo controlled study. The authors note that levetiracetam has had extensive good testing results, and is registered in over 50 countries. The authors wanted to confirm these results in a non-Caucasian population.

In this study, 94 Taiwanese patients with refractory epilepsy were enrolled to evaluate levetiracetam as compared to placebo. The testing period lasted 14 weeks, and almost one-half had a reduction in seizure frequency when levetiracetam was used as an add on drug. Only mild to moderate side effects (fatigue, dizziness, headache) were reported. The authors state their study shows levetiracetam is safe and effective as an adjunctive drug in Taiwanese patients with treatment resistant complex partial epilepsy. The data confirm previous results in Caucasian patients.

Another levetiracetam study examined its effect on nocturnal sleep and daytime vigilance (Cicolin et al. 2006). Since many patients report sleeplessness at night and drowsiness during daytime, the present study looked at the possible effect that levetiracetam might have had as a drug side effect. This present study was performed on a group of adults who were healthy and non-epileptic. The study was double blinded, some volunteers receiving a full dose of levetiracetam, the others received placebo.

Results showed that the subjects, studied with polysomnography, sleep logs, and sleep tests such as the Epworth Sleepness Scales, were evaluated. After treatment, significant increase in sleep time, sleep efficiency, and time in non-REM sleep were noted. No changes were observed between levetiracetam and placebo treated subjects. The authors conclude their study shows that in healthy volunteers, levetiracetam actually consolidates sleep, and has no adverse effect on daytime vigilance.

As stated above, there is a need to test AEDs in patients with various types of classifications of epilepsy in order to determine human efficacy. That however does not rule out animal studies, which have proven to be invaluable. An animal study examining non-convulsive seizures in a rat model of focal brain ischemia has been published (Williams et al. 2004). Non-convulsive seizures occurring after traumatic and ischemic brain damage may be refractory to AEDs. The rat model using middle cerebral artery occlusion (MCAo) to induce non-convulsive seizures was proven valuable for evaluating AED efficacy. In this study, a number of AEDs from different classes were analyzed.

It had been previously shown that if AED therapy prevented non-convulsant seizures, then the mortality rate decreases from 27% to 7%. The incidence of convulsive vs. non-convulsant in brain trauma patients is estimated at 27–34% respectively (Jordan 1993, 1999). The occurrence and duration of non-convulsive seizures appear key in predicting outcomes (Young et al. 1996). Delay and duration of seizures also predict a poor outcome. Without continuous EEG, NCS may be overlooked.

In this study, male Sprague Dawley rats (300 g) were utilized. Indwelling catheters were placed in the right jugular vein, and EEG electrodes were implanted into the skull. After 2–3 days of recovery, the middle cerebral artery was occluded (Williams et al. 1999). The rats were transferred to chambers which permitted continuous EEG recordings. AEDs tested in this experimental model included: dextromethorphan, ethosuximide, gabapentin, midazolam, phenobarbital, phenytoin, and valproate.

Results consisted of continuous EEG recording and NCS quantification after a 24-h period following the occlusion of the middle artery. NCS EEG events started as rhythmic spike/sharp wave discharges. These developed into large amplitude rhythmic spike, spike wave, or polyspike discharges. No overt motor seizures were observed clinically.

The incidence of non-convulsive seizures was reduced significantly by treatment with high doses of ethosuximide, gabapentin, phenytoin, and valproate. There was also a statistically significant difference between the number of NCS per rat and duration of NCS between groups. The number of NCS per rat and frequency were lowered by ethosuximide, gabapentin, and phenytoin. Dextromethorphan and midazolam had no effect on incidence or duration of seizures. In fact, the above two AEDs plus phenobarbital may actually worsen the EEG abnormalities. The AEDs gabapentin, phenobarbital, ethosuximide, and dextromethorphan all reduced the infarction produced by MCAo. Morbidity rates were also reduced by the AEDs.

The authors note that this animal study sheds light on the question of efficacy of several new AEDs on non-convulsive seizures resulting from brain injury. After final analysis, gabapentin and ethosuximide proved most effective in protecting the animal from non-convulsive seizures, as well as reducing the degree of brain injury at both high and low doses. Gabapentin protected against both non-convulsive seizures, as well as reducing the degree of brain injury. Gabapentin is a cyclic analog of GABA, and is used to treat neuropathic pain.

Ethosuximide modulates T-type calcium channel function, and blocks generalized spike wave discharges seen in absence seizures. Treatment with ethosuximide also had a neuroprotective effect, and lowered the mortality rate. Valproate was another effective AED at the higher dose. Valproate is highly prescribed, inhibits sodium channel activity, and increases GABA levels.

The authors conclude saying that use of continuous EEG monitoring and neuronal AEDs have clarified efficacy of treatment of non-convulsive seizures in brain damage. Attenuation of NCS was obtained with high dose (2× ED50) of ethosuximide, gabapentin, phenytoin, and valproate (in decreasing order). The author hopes these data will translate to humans, thereby improving brain trauma outcomes.

Another study examined the concept that positive results occur with prophylactic antiepileptic drug treatment in patients with acute traumatic brain injury. In this retrospective study, data from six trials involving 1,218 randomized patients were used.

Results of mortality data showed no prophylactic use of AEDs after head injury had any effect on reducing mortality or morbidity. There was evidence that prophylactic use of AEDs reduced early seizures, but not later seizures. The authors state that insufficient evidence exists to evaluate the benefits of treatment any time after head trauma (post-traumatic epilepsy).

The rare epilepsy disorder Lennox–Gastaut syndrome is a catastrophic type of epilepsy of children. The present study (Benedict et al. 2010) examined the efficacy and financial costs of a new AED, rufinamide, in the treatment of the Lennox–Gastaut syndrome. This disorder is characterized by retardation, a high frequency of seizures, and a difficult to treat seizure disorder which is often refractory to drug therapy.

The study aimed at a comparison of the efficacy and costs of three AEDs: topiramate, lamotrigine, and rufinamide in children with Lennox–Gastaut syndrome. The Lennox–Gastaut syndrome is a particularly severe epilepsy form, affecting children at a young age (3–5 years old), and is generally refractory to AEDs. Seizures are variable, and interfere with both intellectual and cognitive development. Seizure types may be tonic, clonic, atonic, myoclonic, etc. The risk of trauma from falling is high. Usually treatment is initially limited to an older AED such as valproate. Next, newer drugs such as topiramate and lamotrigine are tried.

Rufinamide has been shown to reduce the frequency of falls, and also decreases seizure frequency. Rufinamide is approved for adjunctive therapy; adverse effects include headache, dizziness, fatigue, nausea, and vomiting. The introduction of new AEDs serves to increase costs significantly, and the cost/efficacy must be balanced. The cost effectiveness analysis was the stated purpose of the study,

Data from the pivotal phase III trial served as the material to establish a cost effectiveness model for the adjunctive AED treatment for Lennox–Gastaut syndrome patients. Rufinamide as a first choice adjunctive therapy followed by standard therapy was examined. These results were compared to topiramate plus standard therapy, and standard therapy alone. A variety of features were evaluated including outcomes, clinical data, costs, etc.

Results showed that adjunctive treatment with rufinamide costs more than topiramate or lamotrigine. However, the treatment efficacy was superior: over a 3-year time, drop attacks were successfully treated in 11.3% of rufinamide patients, compared to a success in 7.2% and 5.2% of topiramate and lamotrigine patients respectively. As regards seizure reduction, the percentage of patients with seizure reduction was 7.7% for rufinamide adjunctive treatment, and 5.6% and 6.9% reduction for topiramate and lamotrigine respectively.

The authors note that although rufinamide is more costly than the other two AEDs, the number of patients with a 50% or greater reduction in seizure frequency is greater. The assessment of drop attacks is important because of the potential for injury from these attacks. In a survey of patients with intractable epilepsy, 90% agreed that any small improvement in seizure control is meaningful (Wheless 2006). The reduction of seizures is the goal of AED therapy, and a significant increase in quality of life accompanies this goal.

The authors conclude saying that while rufinamide is more costly, the reduction in seizures, reduction in drop attacks, and improvement in life quality seems worthwhile. The extra cost is a small fraction of total costs of Lennox–Gastaut syndrome treatments. An increase of about $400 per patient per year, leading to a 50% reduction in drop attacks seems cost effective.

Another recent paper presents results in which pregnant women taking a single AED for epilepsy were evaluated as regards the cognitive function of their offspring (Meador et al. 2009). It had been known that from animal studies in utero exposure to subteratogenic doses of AEDs can produce cognitive changes in offspring (Fisher and Vorhees 1992).

In this study over a 5-year period, pregnant women taking monotherapy (carbamazepine, lamotrigine, phenytoin, or valproate) for epilepsy were enrolled in a prospective multicenter study to compare neurodevelopmental outcomes after exposure in utero. There are always risks from AEDs, but in pregnancy, this must be balanced against the risks to mother and newborn of seizures (Adab et al. 2004).

In this study, multiple variables were obtained: age at delivery, maternal I.Q., AED, dose, frequency of seizures, use of alcohol, etc. A variety of tests were used to evaluate cognitive level of the offspring. The tests were performed up to three years of age: another assessment is planned at the 6-year point in time.

Results from 258 children at age 2–3 showed 73 exposed to carbamazepine, 84 to lamotrigine, 48 to phenytoin, and 53 to valproate. The different numbers reflect different usage AED rates between the centers involved. I.Q. scores were lowest in the valproate group (92), which was statistically lower than any of the other three groups. Predictors of this decreased I.Q. included: AED, maternal I.Q., maternal age, etc. The authors also note that the newborn I.Q. is usually linked to the mother's, except for mothers taking valproate for seizure control. The present study was consistent with previous studies showing a similar link in lower I.Q.s with valproate (Adab et al. 2001). There has also been an association between valproate exposure in utero and actual congenital malformation.

The authors clearly state that valproate should not be used in pregnant women, or those who might become pregnant. If a pregnant woman is already taking valproate, it should not be stopped as this could result in more seizures and adverse consequences.

Epileptic falls represent a risk factor for broken bones, and significant morbidity, especially in older patients. Many such patients are taking AEDs in order to control seizures, and the present study was undertaken to determine if newer second generation drugs influence patient balance (Sirven et al. 2007).

Data from 16 randomized controlled studies were used which had compared adjunctive therapy with a second generation AED (gabapentin, lamotrigine, levetiracetam, oxcarbazepine, pregabalin, tiagabine, topiramate, or zonisamide).

In discussing the results, the authors note that most tested secondary AEDs were associated with a dose dependent risk of both imbalance and ataxia. Only gabapentin and levetiracetam did not affect balance or produce ataxia at any dose. The mechanisms of imbalance associated with the other AEDs were not clear. A wide variety of potential mechanisms such as somatosensory dysfunction, metabolic effects, cerebellar toxicity, vestibular function, etc. may be involved, and await further study. These results could serve as a guide to AED choice especially when there is a specific concern regarding falls, or concern in general about reducing falls, and associated morbidity.

In 2008, the Food and Drug Administration published an alert that indicated AED treatment is associated with an increased suicide risk. A recent paper examines this conception in older V.A. patients taking AEDs (Van Cott et al. 2010). Cases were selected from a V.A. database, and were focused on older patients (over 65 years of age) with new AED use.

The patients were selected based on first time AED monotherapy for seizures, and age over 65. From this group, patients with suicide behavior were identified. Twelve controls were matched to each patient with seizures and suicide thoughts or actions.

Several characteristics were compared in looking for risk factors for suicide tendencies. These included age, gender, race, depression, anxiety, bipolar disorder, substance abuse, etc. Physical conditions were also categorized such as chronic pain, dementia, stroke, diabetes, etc. Gabapentin was chosen as the comparator because 75% of the seizure patients were taking gabapentin.

Results showed that 7,445 patients over a 5-year period had new onset epilepsy, and were receiving monotherapy. Of these, 64 individuals had suicide related behavior. A total of 832 patients (64 epilepsy/monotherapy patients plus 768 controls) were examined. The sample was 97.5% men, and 86.4% were white. Half the sample was over 75 years old. Neither their chronic pain nor chronic disease burden was significantly associated with suicidal behavior. The diagnosis of dementia was however associated with suicidal related behavior.

The authors comment that their analysis of data found the risk for suicide was about tenfold lower than previously described in an FDA study (Avorn 2008). While gabapentin was the most frequently administered AED in this study, levetiracetam and lamotrigine were also administered, and showed the strongest link to suicide behavior in epileptic patients. Both these two AEDs have been previously shown to influence suicidal behavior (Kalinin and Polyanskiy 2005).

The authors note their study identifies affective disorders as the most reliable predictor of suicidal tendencies in older V.A. patients using monotherapy for epilepsy. The authors state that these findings might not apply to younger patients. Additional research with larger numbers is needed to confirm these findings, but this possible relationship is noteworthy.

The availability of new, second generation AEDs starting around 1993 has been a welcomed addition to the number of possible treatments available. Many of the first generation drugs have been available for many decades. The main new AEDs developed and marketed since 1993 are described below.

1. Felbamate (felbatol) is an AED released in 1993, and was the first in the second generation AEDs. It is approved for adjunctive and monotherapy for partial seizures in patients over two years old. It has a broad efficacy, being effective in partial seizures, generalized seizures, and may be of some help in juvenile myoclonic epilepsy, and in Lennox–Gastaut syndrome. It carries a risk of aplastic anemia, thereby needing frequent blood cell counts and liver function tests.
2. Gabapentin (Neurontin) is a GABA analog which works as an adjunctive AED in partial and some generalized seizures. The exact mechanism of action is unknown. It may work as a sodium/calcium channel blocker. Gabapentin is contraindicated in primary generalized seizures. Otherwise there is a wide safety margin and good tolerability. It is a good AED if quick titration is needed.
3. Lamotrigine (lamictal) is a broad acting AED with mechanisms of action probably related to sodium and calcium channels. It is effective against refractory complex partial seizures, and secondarily generalized seizures. In some patients, the AED seems to alert patients, instead of producing drowsiness. About 10% of patients taking lamotrigine get skin rashes.
4. Levetiracetam (keppra) is approved for adjunctive use in partial onset seizures. It is a fast absorbing AED, and a steady state can be achieved in two days. Levetiracetam is an analog of piracetam, a nootropic agent. It also has some efficacy in the Lennox–Gastaut syndrome. It has been used as monotherapy in new epilepsy patients.

5. Oxcarbazepin (trileptal) is structurally similar to carbamazepine (analog), and can be used as monotherapy or adjunctive therapy for complex partial seizures. It can be used safely in children. Its efficacy is similar to carbamazepine, but is more tolerable. Its mode of action is unclear but is thought to a sodium channel blocker. Common side effects include nystagmus, dizziness, and ataxia.
6. Tiagabine (gabitril) acts to inhibit the reuptake of GABA which would raise GABA levels, and it augments the inhibition of seizures. Monotherapy using tiagabine may be effective for partial seizures. It can be used as adjunctive therapy in AED refractory cases. Tiagabine is efficacious and well tolerated.
7. Topiramate (topamax) is used for refractory complex partial seizures. It has a broad spectrum clinical effect, and is efficacious for use as adjunctive therapy. It can be used in children, and as monotherapy. It is effective in generalized tonic clonic epilepsy. Topiramate has broad mechanisms of action including on sodium and calcium channels, blockage of glutamate receptors, and facilitation of GABA modes of action. Cognitive side effects may occur.
8. Zonisamide (zonegran) is another broad spectrum AED, being effective for both partial and generalized seizures. It also has efficacy in the progressive myoclonic epilepsy syndrome. Its mode of action seems to be blockage of both sodium and calcium channels, and inhibition of carbonic anhydrase.

The FDA advisory committee has very recently (August 2010) determined that a new AED, Ezogabine, shows substantial evidence for efficacy as adjunctive treatment in adults with partial onset seizures. Several recent reviews have examined these and other new and old AEDs (French et al. 2004; Tatum et al. 2000).

Adverse effects of AEDs can be minimal and easy to deal with, or may be intolerable and predict associated morbidity. A variety of classification systems exist in order to categorize adverse effects. A classification system based on mechanisms is useful, although not all mechanisms are understood or even known. They do however include teratogenic, metabolic, toxic, endocrine, apoptotic, etc. The understanding of mechanistic adverse effects allows physicians to try to avoid potential adverse effects. Certain AEDs, for example, are teratogenic, and can be avoided in most cases.

A variety of factors may act to influence the occurrence of various adverse effects of AEDs. These include age, sex, inherent susceptibility, rapid dose changes (compliance issues), pre-existing conditions, etc. The greatest risk for adverse effects actually occurs at the time of the initiation of the prescribed AED (see Table 27.2).

Idiosyncratic reactions such as hepatic adverse effects and hypersensitivity usually occur within the first 2–8 weeks (Schlienger and Shear 1998). Because of the association of adverse effects with the initiation of therapy, all concerned should be vigilant.

Interestingly, some AEDs actually worsen seizure frequency and severity. For example, carbamazepine can increase absence seizures, and gabapentin may worsen myoclonic seizures. The seizure phenotype seems to change as a new AED acts to bring the seizure under control. Toxic levels of some AEDs might increase seizures previously under control.

Table 27.2 Adverse effects of some commonly used AEDs

AED	Adverse effect
Phenytoin	Peripheral neuropathy, cosmetic changes
Valproate	Hepatic toxicity
Felbamate	Anorexia
Gabapentin	Hyperactivity, aggressive behavior
Lamotrigine	Skin rash
Topiramate	Cognitive deficits
Tiagabine	Tremor
Vigabatrin	Behavior problems
Zonisamide	Kidney stones

Adapted from Greenwood, R. Epilepsia 41: p. 42, 2000

Adverse effects such as growth problems are of concern in children, but not adults. Other age related adverse effects have to do with lower metabolism and liver function in the elderly. This would act to alter (lower) metabolism of some AEDs. The overall rate of blood dyscrasias is significantly higher in patients taking AEDs who are over 60 years old. Children under two years of age are more likely to experience liver toxicity from valproate than those over two years of age.

The term inherent susceptibility refers to genetic differences between patients which may lead to different handlings of AEDs, and possible side effects. AEDs which are metabolized in the liver may in some patients induce porphyria. This could lead to an increase in seizure activity. Other examples include those which have an effect on P450 activity.

Pre-existing conditions (comorbidities) can also affect the efficacy of AEDs, and produce adverse effects. The liver is an example of this. Patients with liver disease and decreased protein synthesis are at risk from drugs which have high protein binding such as phenytoin. AEDs which are metabolized by liver may result in increased levels, which could reach toxicity. Patients with a history of drug related skin rashes are more likely to develop rashes from AEDs.

Rapid dose changes and/or high AED doses are also related to an increased risk of adverse effects. An example of high doses and rapid changes is during status epilepticus, when the primary goal is to stop the seizures. Blood levels of AEDs are not always indicative of tissue levels which produce the adverse effects. Seemingly remote adverse effects such as thrombocytopenia seen in some patients on valproate therapy (May and Sunder 1993) can be AED related.

Patients on polytherapy may have adverse effects related to increased drug interactions. Some AEDs inhibit liver enzymes, while others cause enzyme induction (Tanaka 1999). Drug interactions (AEDs or other drugs) can interact producing multiple adverse effects. Health care workers must not be complacent about these possibilities, as they can happen over time even in the absence of treatment changes.

Chapter 28
Monotherapy and Polytherapy

Over the last 50–60 years, the concepts regarding AED monotherapy or polytherapy have fluctuated back and forth. On one hand, it was suggested that monotherapy served to minimize the chances of an increase in adverse effects possibly associated with multiple AEDs. On the other hand, polytherapy was considered positive if monotherapy did not control seizures because of synergistic effects of more than one AED for seizure control. Using two drugs with different mechanisms of action should increase the chance for seizure control. And there is an emotional factor. If monotherapy does not work, how could anyone just give up? Polytherapy became a foregone conclusion. Sequential monotherapy is an option, but more difficult to achieve.

When going from monotherapy to polytherapy due to failure to control seizures, it is important to carefully consider mechanisms of action of AEDs versus clinical/EEG type of epilepsy. The move should be based on data, not merely trial and error. Both the advantages and disadvantages of polytherapy must be considered. Data from many studies show: (1) as many as 30% of epilepsy patients remain refractory to AEDs and (2) after 1–2 AED trials, only about 5% ever gain seizure control from AEDs.

Problems with the addition or subtraction of AEDs include the concept that this process may alter the pharmacokinetics of AEDs in use at the time. AEDs which inhibit reactions or pathways usually create a change in only a few hours, as opposed to those which induce enzymes, a process usually taking much longer.

Cumulative toxicity from multiple AEDs shows that the effect sometimes can be additive (Levy et al. 2007). The actual effect of additive AEDs may be to render the patients' AEDs in a toxic range. One study (Bourgeous 1988) showed 14% of patients with monotherapy were in the toxic range, 50% of patients were in the toxic range when taking two AEDs, and 100% were toxic with three or more AEDs. One key question is how often is this phenomenon being monitored? In polytherapy, it is unlikely the therapeutic range of any one AED will remain the same as when administered as monotherapy.

Some toxic reactions are idiosyncratic in that they occur when two AEDs are taken together. This may occur, for example, when valproate is taken with certain

D.W. McCandless, *Epilepsy: Animal and Human Correlations*,
DOI 10.1007/978-1-4614-0361-6_28, © Springer Science+Business Media, LLC 2012

other AEDs, producing a stuporous encephalopathy (Marescaux et al. 1982). The neurotoxicity of AED drug combinations has been demonstrated in murine models of seizures (Bourgeois and VanLente 1994). There are scant human clinical studies showing effects on AED combinations based on systematic comparisons.

As stated above, the choice of additive AEDs should be a deliberate thought-out decision based on mechanisms of action, characteristics of the epilepsy, and potential side effects. Again, as stated above, the mechanism of action of the AEDs should be complimentary, not identical, thus increasing the spectrum of antiepileptic targeting. Using a low dose approach to polytherapy might serve to reduce side effects. The disadvantages of polytherapy seem to be well known, so the burden of proof remains with the health care provider to document the therapy individually for every patient.

One paper (Mattson and Cramer 1988) showed that patients with generalized tonic clonic seizures are able to achieve the same or better seizure control and fewer adverse side effects with valproate as a single AED than with polytherapy. The authors state that going from polytherapy to monotherapy (valproate) must be done slowly with monitoring. The authors note that the quality of life which results justifies the effort.

Another paper reports results from a study examining adverse effects of monotherapy and polytherapy (Lammers et al. 1995). The authors note that the concept that monotherapy has fewer adverse effects than polytherapy was assessed using standardized quantitative methodology. Both the daily doses of AEDs and the severity of adverse effects were quantitated.

Results showed that the severity of adverse effects increased with dose, and peaked at a prescribed daily dose/defined daily dose ratio of -3.5. The authors state their data show no difference in adverse effects for equipotent doses, underscoring the importance of quantifiable data. Further, frequency/intensity of adverse effects seemed to be not sensitive to dose change.

Another study using a randomized comparative monotherapy trial for four different AEDs has been published (Heller et al. 1995). The four AEDs tested were phenobarbital, phenytoin, carbamazine, and valproate. The authors state that previously there has been too much use of polytherapy, and after the advent of blood monitoring, it was important to re-evaluate the efficacy of monotherapy.

The patients in this study were 16 years old or older, with newly diagnosed epilepsy. They had agreed to be included in a study using monotherapy in a randomized study. An initial assessment of each patient included a general and neurological examination, EEG, and CT scanning. Plasma drug monitoring was performed at each clinical visit.

Results showed similar criteria of all 243 randomized patients as regards sex, seizure types, mental retardation, total number of seizures before randomization, etc. Ten percent of patients had intolerable side effects requiring withdrawal and substitution of another AED. The overall analysis of outcomes showed that at 3 years follow up, 27 patients were seizure free, 75 patients achieved a 1-year remission, and 42 patients a 2-year remission. Differences between the four monotherapy groups were small, ranging from 21 to 33% seizure free at 3 years follow up. Comparisons of the four drugs with seizure types did not show a correlation.

The authors state that they saw no significant difference in efficacy between the four AED monotherapies in this study in either the time to achieve a 1-year remission or time to the first seizure recurrence. And, the results were similar for each seizure type – tonic clonic, or partial seizures.

The authors further note their study had results very similar to results from other studies (Mattson et al. 1985; Turnbull et al. 1985a, b). Of the group of patients who withdrew from this study (25), half were taking phenobarbital. The authors note that their study showed no differences in efficacy between the four drugs tested. This is a comparative statement, not an efficacy statement from a placebo study.

The authors comment that efficacy studies in placebo-controlled trials are not done for ethical reasons. Results from chronic efficacy studies implies efficacy in newly diagnosed patients. The precise contributions of treatment in newly diagnosed epilepsy patient are uncertain. The patients in this study were randomized on clinical evidence, so conclusions cannot be drawn regarding the EEG findings.

The authors conclude saying that they have not found significant differences in efficacy between the four AEDs (phenobarbital, phenytoin, carbamazepine, and valproate).

Animal studies comparing monotherapy and polytherapy have been performed (Roks et al. 1999). This study uses a combination of valproate and ethosuxamide AEDs to measure various performance tasks in rats to assess neurotoxic effects. The combination drug choice was based on previous experiments showing an additional effect as regards both efficacy and toxicity of valproate and ethosuxamide (Bourgeois 1988).

In this study, male Wistar rats weighing 225–320 g were utilized. Rats were divided into four groups: valproate, ethosuxamide, combination of the two AEDs, and saline controls. Tests following AED injection consisted of a forepaw grip test, a rotarod test (length of time the rat could remain on a rotating rod), and a behavioral test.

Results showed that in terms of the grip strength test, both valproate and ethosuxamide, plus the combination treatment had a negative effect. The decrease in strength was dose dependent. The ability of the rats to stay on a rotating rod was negatively affected by both drugs alone, and the combination polytherapy had a similar effect. The dose in the polytherapy group for 50% toxicity was higher than the calculated theoretical dose, suggesting infra additivity. Both valproate and ethosuximide resulted in less activity (sedation) than controls, and results from the polytherapy combination of the two drugs, indicated that significant infra additivity was present.

The authors comment that behavior was more active in polytherapy rats than in either of the monotherapy groups. The behavioral studies showed infra additivity in toxicity. Since valproate and ethosuxamide have supposedly different mechanisms of action, this may explain human studies in which a synergistic effect was seen (Rowan et al. 1983). In the present study, the infra additivity for sedation suggests that there is an advantage in combining valproate and ethosuxamide in low dosages. The authors suggest future studies of efficacy and toxicity of these two AEDs in combination would be useful.

Gabapentin is a second-generation AED, and the efficacy and safety of gabapentin as monotherapy has been carefully studied (Beydoun 1999). The study reports three separate double-blind, dose-controlled groups of patients with partial onset seizures. In the first study, 275 patients on one or two AEDs were switched to gabapentin monotherapy. In the second study, 82 hospitalized patients with refractory epilepsy were ratcheted down from prescribed AEDs, and gabapentin was substituted as monotherapy. In the third group, 292 patients with newly diagnosed partial seizures were randomized to gabapentin, or to carbamazepine. This group of patients were in the study for 6 months.

Results showed that there was no statistically significant difference between the three groups as regards exit from the studies due to adverse effects. Time to exit was longer for medium and high doses as compared to a low dose. Patients were required to exit if there was a worsening in seizure frequency. The author comments that the results of these trials support evidence of both the efficacy and safety of gabapentin monotherapy for partial onset seizures.

In a paper by Deckers et al. (2001), the comment is made that previous studies had shown that the monotherapy concept was that a single AED could reduce adverse effects, and in some cases, increase efficacy (Reynolds et al. 1976; Shorvon and Reynolds 1979). Multiple subsequent clinical studies support this concept. Other studies, however, have cast some doubt on the concept of advantages of monotherapy stated above (Devinsky 1995). The present study aims to look more critically at the differences (if any) between monotherapy and polytherapy.

In this study, 130 adult patients with untreated generalized tonic clonic and/or partial seizures were randomized to monotherapy (carbamazepine) or polytherapy (carbamazepine plus valproate). Five patients were excluded very early for various reasons. Others were excluded during the trial for reasons such as non-compliance, protocol violations, etc.

Results showed a total of 84 participants – 37 in the monotherapy groups, and 47 in the polytherapy group, actually completed the study. Neurotoxic scores from several tests did not show any significant differences between the two groups. Similarly, there was no difference in systemic toxicities. In terms of efficacy, seizure frequency was reduced in both groups.

The authors comment that the study was focused on tolerability more than anything else. The neurotoxicity scores showed no differences between the monotherapy/polytherapy groups. Monotherapy patients did complain more about sedation at the 12-month period, so the polytherapy group developed better tolerability. Other minor differences centered on weight gain and headache. The authors conclude that this study does not support the concept of higher neurotoxicity rates between polytherapy versus monotherapy.

Another paper looks at monotherapy and polytherapy, and finds similar outcomes (Beghi et al. 2003). In this study, a group of patients with cryptogenic or symptomatic partial epilepsy not controlled by single or sequential AEDs were randomized to monotherapy or to adjunctive therapy with a second AED (polytherapy).

Results showed 76 were randomized to alternative monotherapy, and 81 to adjunctive therapy. In patients randomized to monotherapy, 65% remained on that therapy, and 55% of patients randomized to monotherapy stayed on monotherapy at the 12 month post-initiation of therapy period. The probability of being seizure free was 14% and 16%, respectively. Adverse effects were not different between the two groups.

The authors conclude that alternative monotherapy and adjunctive polytherapy were associated with similar outcomes.

The AED topiramate was evaluated as a monotherapy treatment for localized epilepsy (Gilliam et al. 2003). In this study, adults and children over 3 years of age were studied. Initially, 252 patients were identified, and were either untreated or were receiving one AED for less than 1 month. Topiramate was chosen in part because it has multiple modes of action including increased GABA activity, and blockage of Na and Ca channels. Placebos were not used for ethical reasons, rather different doses were used for comparison.

Results showed that 99 patients completed the study in the low-dose topiramate group, and 93 patients completed the study in the high-dose group. Efficacy was evaluated by time to exit, seizure frequency, and time to first seizure. Tolerability was evaluated by careful monitoring, and neurologic/physical examinations. The time to exit comparison between low- and high-dose groups was not statistically different between these two groups. Time to exit with time to first seizure as a covariate in the high-dose group was significant. The high-dose group had a higher time to first seizure rate than the low-dose group.

The authors note that the severity of adverse effects was lower than previous studies have shown (Faught et al. 1996). The predominant adverse effects were neurologic and paresthesias were most frequent. These adverse effects were associated with increasing doses and generally disappeared when target levels of topiramate were achieved. The authors further state that their study shows topiramate is effective as initial/early monotherapy in patients with localization-related seizure disorders. After titration, two-thirds of patients remained seizure-free 6 months after start of the therapy.

In a recent paper the use of the new antiepileptic drugs to treat seizures as monotherapeutic drugs is examined (Kanner and Balabanov 2005). In a general way, monotherapy has inherent advantages such as easier compliance, it has a lower risk of adverse effects-only one drug, and is usually less expensive. In addition, if monotherapy is ineffective, the chances that polytherapy will be successful drop quickly to 5–10%.

The new AEDs (second generation) are relatively unique in that they are not related to each other, and have varying mechanisms of action. Several studies suggest that the efficacy of the new AEDs are similar; therefore choice should be based on tolerability and lack of significant adverse effects. Results from various studies show a lesser effect of these new AEDs on cognitive function than the older AEDs.

Patients suffering from mood disorders who are epileptic should benefit from lamotrigine as it is a mood-stabilizing drug. Lamotrigine also is not suited for pregnant seizure patients. Gabapentine is a useful AED in elderly patients since it has a low interaction with other medications. Topiramate may cause weight loss and is a useful AED for patients with weight gain problems.

The authors conclude saying data for evaluation of new AEDs is not complete. They state newly diagnosed patients should be tried on at least three different monotherapy drug trials before polytherapy. Failure of three new AEDs would indicate a refractory epilepsy and might indicate a presurgical evaluation. Well-controlled studies have shown the new AEDs are at least as effective as the old AEDs and have superior tolerability.

A ten-member subcommission from the ILAE were assembled in order to evaluate AED efficacy and effectiveness as initial monotherapy choices for epileptic seizures (Glauser et al. 2006a, b). Four classes of studies were defined. Class I studies were defined as those with a 48 or more week duration without forced exit criteria. The requirements included data on 24-week seizure freedom (efficacy), or on 48-h retention data (effectiveness). There needed to be power to detect differences between the two and correct statistical analysis. The "lowest class" was Class IV which consisted of expert opinion and case studies, etc.

Extensive data was collected from 50 randomized controlled trials (RCT), and seven meta-analysis studies. Of these, only four RCTs were Class I. Two more were Class II and the rest were Class III. Three seizure types had level A or B AED efficacy: adult patients with partial onset seizures, children with partial onset seizures, and elderly adults with partial onset seizures. Some seizure types had no level A or level B AED efficacy. These were adult-generalized tonic clonic epilepsy, pediatric-generalized tonic clonic, and absence seizures; and juvenile myoclonic epilepsy, and benign epilepsy with centrotemporal spikes.

The authors conclude saying there is a marked absence of rigorous comprehensive adverse effects data. They further state there is an alarming lack of well-designed properly conducted double-blind RCT-designed studies. A variety of problems with previous studies are detailed including: improper design regarding noninferiority, time lines are too short, titration schedules are fixed, use of multiple patient age groups preclude proper conclusions, some studies are run by pharmaceutical companies, not independent investigators, etc. Other concerns include: the importance of clinical endpoints – often the endpoint of time to first seizure is used and this is not clinically relevant.

Another previous study design problem is the definition of "adequately powered." The authors state that properly designed and conducted multicenter, multinational studies and RCTs are needed which are properly designed, conducted, and statistically analyzed. The reader is especially encouraged to obtain this important, detailed manuscript laying out clearly what is known, and obscure in current (2006) clinical data. And the paper stresses the importance of well done studies, including how to do them, so that the effort can be used productively, and compared to others' investigations.

The second-generation AED topiramate has been evaluated for monotherapy in newly diagnosed epilepsy in children and adolescents (Glauser et al. 2007). This was a double-blind dose-controlled study in 470 patients in newly diagnosed seizures of less than 3 months duration. The study had 151 patients of ages 6–15 years of age. Pediatric patients were randomized into a low-dose (74 patients), or high-dose (77 patients) groups.

The endpoint of time to first seizure was more efficacious in the high-dose group. The 6-month seizure rate in patients who stayed in the study was 68% in the low-dose group, and 90% in the high-dose group. At 1 year, the probabilities of seizure freedom were 4% in the low-dose group and 14% in the high-dose group. These data clearly indicate topiramate as an effective monotherapy in children and adolescent patients with either partial onset, or generalized tonic clonic epilepsy.

Another study has examined the safety and efficacy of zonisamide as a monotherapy AED in epilepsy (Park et al. 2007). Zonisamide is a broad spectrum second-generation AED, effective for treating simple/complex partial seizures, generalized tonic clonic seizures, myoclonic seizures, infantile spasms, and the Lennox-Gastaut syndrome (Leppik et al. 1993). This paper reports results of efficacy and safety in a retrospective study of 60 patients using zonisamide monotherapy for long-term (6–37 months) epilepsy treatment.

Results showed the patients in the study had a mean age of 29.8 (range 16–65). Partial seizure and generalized seizure patients were similar in numbers. Six patients had juvenile myoclonic epilepsy. Twenty-seven patients (45%) became seizure free, and 20 additional patients had a seizure reduction of 50% or more. The positive response rate was 78%. Eight patients had no change in seizure frequency.

In terms of adverse effects, 80% of patients reported adverse effects of some type. In long-term treatment, the most frequent adverse effect was memory loss and attention deficit. The same adverse effects were present in short-term treatment, except that somnolence was most frequent. CNS symptoms such as somnolence, headache, dizziness, and fatigue decreased in frequency after 6 months.

The authors note that this retrospective study shows that after 20 months of zonisamide treatment, 70% of epilepsy patients experienced seizure control. This includes patients with simple partial seizures and generalized seizures. Even all of the patients with juvenile myoclonic epilepsy expressed seizure freedom. This is in keeping with other studies (Kothare et al. 2006; Newmark and Dubinsky 2004). No seizure freedom occurred in patients with complex partial seizures.

The authors conclude saying that long-term zonisamide therapy treats a broad spectrum of epilepsies, but there are adverse effects involving cognition. Cognition defects from AEDs potentially outweigh the positive control shown by zonisamide. The quality of life can be adversely affected by changes in cognition. This aspect of zonisamide treatment needs further clarification. The ultimate goal of AED treatment is to pick an AED with the most effective seizure control and least adverse effects. Zonisamide was an effective monotherapy choice for multiple seizure types; the adverse effects cognition problems need further objective studies.

Another new second-generation AED has recently been reviewed as regards its potential for monotherapy (Gambardella et al. 2008). Levetiracetam has a low potential for interaction with other drugs, and is a tolerable AED. Levetiracetam had already been shown effective as adjunctive therapy (Glauser et al. 2006a, b) and as monotherapy (Glauser et al. 2007, see above).

The mechanism of action of levetiracetam is unclear, and the AED does not have an effect on mouse or rat models of seizures using either maximal electroshock or pentylenetetrazole. It also seems not to have any interaction with either GABA or glutamate (Loscher and Honack 1993). Levetiracetam does have anticonvulsant activity against maximal electroshock and pentylenetetrazole kindled seizure activity, and audiogenic seizures. It is also protective in the GAERS model. Its toxicity levels are low and it is not teratogenic. Levitiracetam does not interact with other AEDs.

Adverse effects of levetiracetam include somnolence, asthenia, headache, dizziness, and ataxia (Ben-Menachem and Falter 2000). There have been reports of neuropsychiatric symptoms, including agitation, hostility, emotional problems, etc. The psychiatric side effects could start at any time, but were usually in the first month. These data suggest a close monitoring of patients on levetiracetam is warranted (Mula et al. 2003).

As monotherapy, levetiracetam has been effective. In an open study (Alsaadi et al. 2005), 46 patients were followed for 1 year while on levetiracetam, and over half remained seizure free. Another study, using randomized, double-blind trials also showed efficacy as compared to carbamazepine (Brodie et al. 2007). Levetiracetam is also well balanced as an AED. The overall conclusions are that levetiracetam is an excellent AED for adjunctive therapy, or monotherapy in adult partial onset epilepsy patients. Efficacy and good tolerability are features of levetiracetam.

A recent study (Canevini et al. 2010) aimed to evaluate the adverse effects of multiple AEDs in adult epilepsy patients and the relation between adverse effects and the number of co-prescribed AEDs. Patients from 11 referral centers were studied through interview and the adverse event profile questionnaire.

Results showed 809 patients had an average age of 40.7 and had epilepsy for over a mean of 23.9 years. Localization diagnosis was established in 87.4% and generalized epilepsy diagnoses consisted of 12.6% of patients. Only about 25% of patients were being treated with monotherapy, the most common AED being carbamazepine. The most frequently used AED in polytherapy was levetiracetum. Patient characteristics (age, gender, etc.) were similar between the monotherapy/polytherapy groups.

Adverse effects included: unsteadiness, tiredness, feelings of aggression, headache, skin rashes, double vision, depression, memory problems, etc. There was no statistical correlation between adverse event profile and monotherapy/polytherapy treatment regimens. Multivariate linear regression analysis showed that the adverse event profile scores were associated with female gender and depressed mood.

The authors conclude their data show that adverse effects are common in refractory epilepsy patients. In addition, the adverse effects are likely related to individual responsiveness, types of seizures, and types of AEDs chosen, than to the numbers of AEDs prescribed, or the AED load. One reason for the lack of difference between monotherapy and polytherapy is that there was effective physician intervention in order to "customize" the treatment regimens to each individual patient. This serves to help reduce adverse effects, especially in the polytherapy group.

Another recent paper looked at monotherapy as the first approach toward drug therapy in new epilepsy patients (St. Louis et al. 2009). The paper is a review looking at important features of monotherapy. The observation is made that the second generation of AEDs are efficacious and seem to have fewer adverse effects. Other features of monotherapy such as increased compliance, etc., make them attractive to physicians and patients alike.

The majority of patients respond well to monotherapy; as many as 47% became seizure-free with the first monotherapy trial. Care must be taken to monitor adverse effects. Well-controlled studies are scant (see above). The short-term outcomes deemed desirable are seizure freedom, and seizure control if seizure freedom is not possible, and as high as possible quality of life. This requires close supervision and monitoring of the patient by the physician.

Monotherapy is usually most effective when chosen for a specific patient type. For example, certain drugs should be avoided in patients with liver disease, or in pregnant women. Over treatment (too high dosage) is another area of concern. If a moderate dose is ineffectual, a high dose might be entering a toxic range. Usually about two-thirds of patients can achieve satisfactory results in terms of seizure freedom or control, whereas some will require polytherapy.

The conclusion is that monotherapy is preferred, especially when efficacy and tolerability are equal or better than in patients on polytherapy. Monotherapy may fail in patients not receiving thorough evaluation and counseling by their physician. Clinicians must consider seizure type, patient type, likelihood for compliance, cost, and the possibilities of therapy change to meet evolving symptoms. Careful monitoring, seizure frequency/severity, and attention to adverse effects can spell success for monotherapy.

Chapter 29
Creatine Treatment

Creatine is a naturally occurring amino acid consumed in meat and fish, and also synthesized in liver, kidneys, and pancreas. Creatine is used and stored in many body organs, and is found in the greatest concentration in muscle (95% of body stores). It also plays a major role in brain. Creatine's active form is as phosphocreatine (PCr). This serves as an emergency donor of high energy phosphate molecules to adenosine diphosphate to form adenosine triphosphate (ATP). ATP is in turn a phosphate donor, and supplies energy for almost all cerebral energy requiring processes. These include synthesis, and neurotransmission.

The synthesis of ATP is the following: PCr + ADP $\rightarrow$ ATP + creatine. Phosphocreatine represents an energy reserve, and is present in brain at about a 1.5–1.0 concentration gradient. When ATP demand is increased, phosphocreatine is converted by the above reaction. PCr may also act as a shuttle system, sending ATP to the cytosol from the mitochondria, site of synthesis.

One key function of phosphocreatine, when in excess, is to increase ATP levels in tissue enough to stabilize neuronal membranes. Mitochondrial creatine kinase, when activated by creatine and PCr, serve to stabilization effect. When mitochondrial creatine kinase is inactivated (by free radicals), the neuronal membranes are less stable, and a conducive to seizure activity occurs. This represents the hypothesis regarding the antiepileptic mechanism of creatine.

This hypothesis states that creatine's mechanism of action is that it stabilizes neuronal membranes, and allows phosphocreatine levels to exceed normal levels. This in turn leads to a small increase (10%) in ATP, thereby making the neuron less susceptible to seizure activity. For the seizure, this represents an inhibitory effect. Phosphocreatine acts to increase removal of excitatory glutamate from the synapse.

Creatine kinase is the enzyme catalyzing the reaction of the phosphorylation of creatine to PCr. Creatine is synthesized from glycine and arginine in specific areas, then transported against a concentration gradient by a sodium/chloride transporter. Creatine and phosphocreatine are involved in the shuttle of ATP from the mitochondrion to the site of use (Bessman and Carpenter 1985). ATP is the high energy source for cellular work, whereas PCr serves as a backup source whenever needed.

D.W. McCandless, *Epilepsy: Animal and Human Correlations*,
DOI 10.1007/978-1-4614-0361-6_29,

Another mechanism of neuronal membrane stabilization is thought to involve the binding of PCr to the phospholipids head groups of membranes. This in turn acts to decrease membrane fluidity and decreases loss of cytosol, and cytoplasmic contents such as creatine kinase. This has been shown earlier in rabbit heart (Sharov et al. 1987). An "over supply" of PCr leads to an over supply of ATP, rendering the neuron more able to resist many toxic or threatening conditions such as ischemia or seizure activity.

There were attempts early to try to correlate creatine metabolism and seizure activity (Glötzner et al. 1979). In this study, the levels of creatine kinase, the enzyme catalyzing the formation of phosphocreatine from creatine and ATP, were measured in serum from patients having tonic clonic seizures.

Results showed that serum levels of creatine kinase were postictally elevated in 14 of 17 patients after tonic clonic seizures. No correlation was found between creatine kinase levels and cerebral ictal injuries. In a series of cats treated with muscle relaxants, the creatine kinase was also elevated to maximal levels 2–4 days postictally. Combined, the animal and human studies suggest that the high levels of the enzyme creatine kinase cannot be completely derived from skeletal muscle.

In a later study (Holtzman et al. 1998), hypoxia produced a high susceptibility to seizures at about day 10–12 in newborn rats. This corresponds roughly to the perinatal period in humans. This is also during the developmental period for energy metabolism in newborn rats (5–25 days).

In this study, rat pups were injected with creatine for 3 days preceding exposure to hypoxia on days 10 and 20. Before and during hypoxia, electrocortical activity and nuclear resonance spectra were measured. Results showed that at 10 days, creatine injections increased phosphocreatine/nucleotide triphosphate ratios. This effect was not seen in 20-day-old rats.

This increase in PCr/ATP ratios served to decrease seizures and mortality associated with hypoxia induced seizures. Further, the creatine induced increase in ratios acted to enhance recovery after the hypoxic insult to the high energy phosphates. The authors state that creatine protected the relatively immature rat brain from hypoxia induced seizures. They further state that this might well protect neurons from cellular damage, and that these results could be applicable to newborn human brain.

A similar study was undertaken by the same group in newborn rabbit pups (Holtzman et al. 1999). One reason rabbits were used was that they have a substantially larger brain, permitting intracerebral localization studies. The human, rat, and rabbit all have brains which have significant postnatal developmental periods, helping to translate results from one study to another.

Previous studies (Holtzman et al. 1998) had shown that creatine prevented hypoxic seizures in that rat pup. Other experiments showed ATP turnover during seizures is high, and that PCr is high in gray matter, and is stable in white matter in pentylenetetrazole seizures. The present study compared the relation of hypoxia induced seizures and regional brain PCr during seizures after creatine supplementation.

Newborn rabbits were studied at 5-day intervals from postnatal day 5–30. EEG recordings were obtained from epidurally implanted electrodes. NMR spectra were

Table 29.1 Hypoxia induced seizures in creatine treated and control rabbits

Age	Control seizures	Creatine treated seizures
5	0/7	0/7
10	2/6	1/5
15	6/8	0/11
20	5/8	2/8
30	0/7	0/8

Adapted from Holtzman, D., et al. J Neurochem 73:2477, 1999, p. 2479

obtained using localized surface coils. They used a 4.7 T, 30 cm diameter horizontal bore super conductor magnet. Newborn rabbits were studied at 5-day intervals from 5 to 30 days, and either EEG or NMR data were acquired. Animals were injected with either creatine or saline. Animals received creatine or saline 2 days prior to the hypoxia period. Hypoxia was induced by exposure to 4% O_2 for 8 min. After the 8-min period, the rabbits were again exposed to room air.

Results showed that creatine injection had no obvious effects on the rabbit's appearance or behavior. Hypoxia produced the highest rates of EEG seizures in 15- and 20-day-old rabbits. At 15 days, 75% showed seizure activity, and half had clonic forelimb movements. At 20 days, 63% had EEG evidence of seizures, but none had overt seizures.

The effect of creatine injections on phosphocreatine showed large increases as represented by 31P NMR at 5- and 15-day-old rabbits. This resulted in a 2× increase in the PCr/NTP ratio. This increased ratio had been previously seen in rats (Holtzman et al. 1991). The PCr/NTP ratio was statistically significantly increased in both gray and white matters (see Table 29.1).

Result from this study shows that the rabbit (like the human) is the second animal model to show an increase in EEG activity and seizures at a specific developmental stage. The sensitivity period is about midway in the time period of increasing PCr/NTP ratio development. The rabbit brain size permits regional 31P NMR measurements. These showed that the PCr signal is first seen in white matter versus gray matter. In gray matter, the ratio appears later than white matter, but rises faster. The increase in rat brain PCr/NTP parallels the increase in oxidative ATP metabolism.

Results from these studies show that creatine treatment completely blocked hypoxic seizures. The mechanisms are not fully understood, but creatine acted to increase the PCr/NTP ratios in both gray/white matters. Susceptibility to seizures in newborn rabbits drops when phosphocreatine levels reach adult concentrations. At this time, white matter PCr levels are still increasing. Systemic administration of creatine raises PCr in white matter to adult levels, and the seizure is prevented. The blockage of the seizure may be related to the PCr/creatine kinase/creatine system, or by increased and enhanced ATP metabolism.

The authors comment that the similar results in both newborn rats and rabbits are relevant to newborn human seizure states. At term, the human brain is about half way through the time course of PCr/NTP development (range 26 weeks prenatal to 3 months postnatal) (Hanaoka et al. 1998). In humans, the newborn period is one of potentially high seizure frequency. One issue to study is to see if the creatine induced elimination of hypoxic seizures is due to seizure prevention, or an effect on hypoxia per se. Regardless, state the authors, the PCr/creatine kinase/creatine system is an important feature in adaptation of energy metabolism to metabolic perturbations, and for a stabilization of neuronal activity.

Another paper describes experiments on creatine kinase B driven brain energy metabolism, and its effect on learning behavior, mossy fibers, and seizure susceptibility (Jost et al. 2002). The maintenance of brain homeostasis requires a tight coupling of energy production and utilization at all times, and especially during activity.

Results showed that mice deficient in brain creatine kinase could be produced using gene targeting. Brain creatine kinase deficient mice showed normal development compared to wild type mice as regards weight gain, life expectancy, and fertility.

The distribution of creatine kinase (Brain creatine kinase, mitochondrial ubiquitous creatine kinase-UbCKmit, and M-creatine kinase) was determined in brain using immunocytochemistry. Brain creatine kinase was present throughout the wild type brain. Attempts to localize M-creatine phosphate were equivocal. UbCKmit was detected primarily in the cell bodies in a few neuronal nuclei. These locations included layer 5 in the cortex, lateral cerebellar nucleus, granular cell layer of the cerebellum, hippocampus, and brainstem nuclei.

Light microscopy and electron microscopy were largely unremarkable. There were subtle and specific changes in hippocampal mossy fiber projection fields. 31P MRI showed no significant differences between creatine kinase deficient and wild types as regards ATP, PCr, inorganic phosphate, and pH. In terms of behavioral testing, brain creatine kinase deficient mice showed diminished open field habituation. In a water maze test, brain creatine kinase mice were slow to acquire the task as compared to wild types, but were ultimately successful.

Of interest, results were from pentylenetetrazole intoxication. Pentylenetetrazole induced seizures increase local glucose utilization, and induce a high ATP turnover. The rapid increase in energy availability is used to fuel the seizure. Involvement of GABA results by a reduction of coupling of GABA and benzodiazepine recognition sites (Walsh et al. 1999). This effect increases and prolonged the seizure state.

The authors note their study provides evidence that mice lacking brain creatine kinase have reduced flux capacity for energy metabolism, have decreased open field habituation, have slower learning acquisition, and have a delay in pentylenetetrazole induced seizure development (because of reduced energy capacity for seizure expression). The authors also note their study does not rule out the possibility of increased flux through glycolysis.

As regards the seizure aspects of this study, the brain creatine kinase deficit mice subjected to pentylenetetrazole induced seizures showed a lengthening of the seizure process. The brain creatine kinase deficient mice had a larger number of preseizure jerks, and a prolonged time line interval, and a prolonged sequence of first and second seizures. This suggests that the energy availability to sustain seizures was lacking.

Neuroprotective effects of PCr/creatine loading through creatine administration in several pathological situations have been demonstrated in animal models and humans. These include Huntington's disease, Parkinson's disease, and seizures. The authors comment that their findings demonstrate the potential of brain creatine kinase as an entry site for neuroprotective regimons and treatment strategies.

Inborn errors of creatine metabolism with associated epilepsy have been described (Leuzzi et al. 2000). As stated earlier, creatine is synthesized from arginine and glycine, and the enzyme for the reaction is arginine and glycine (EC 2.1.41), and a second reaction is catalyzed by guanidinoacetate methyl transferase. The synthesis occurs in the liver and pancreas, and creatine is transported to sites of use by a creatine transporter system. Creatine is used by muscle and brain, and is metabolized by creatine kinase. A product of creatine is phosphocreatine (PCr) which serves as a high energy phosphate donor to ADP forming ATP, the ultimate cellular energy source.

The first primary metabolic disorder of creatine was only described in 1994 (Stockler et al. 1994). The defect is a deficiency of guanidinoacetate *N*-methyl transferase. A few cases of this deficiency have subsequently been described. The present paper (Leuzzi) reviews features and treatment of this enzymatic deficiency and associated epilepsy.

Only six patients had been described at the time of this current paper. Patients were normal at birth, but soon developed psychomotor delay and developmental arrest. Dystonic-dyskinetic syndrome was present in three patients, and seizures in five of six patients. Subsequent symptoms included severe mental retardation, dystonia, autism, hypotonia, etc. Seizures consisted of febrile convulsions, drop attacks, tonic clonic seizures, partial seizures with secondary generalization, etc. EEG findings included bilateral synchronous diffuse slow spike waves, multifocal spikes, etc.

The biochemical changes in this disorder originate because of the deficient enzyme guanidinoacetate *N*-methyl transferase. The result is a reduction of creatine formation combined with increases in arginine consuming guanidinoacetic acid formation. The decreased creatine is mirrored by an increase in guanidinoacetic acid.

The main goal of therapy is to return the creatine deficiency to normal, and to eliminate the excess guanidinoacetic acid. Previous studies had shown that creatine had a positive effect on seizure control, and was tried in these cases of enzyme deficiency. Results showed that in five of six patients, there was a dramatic neurological improvement. The patient who did not respond was the most compromised patient in terms of symptoms. Three patients with severe refractory epilepsy achieved complete seizure control. Neurological improvement was achieved in terms of hypotonia, movement disorders, and even improvement in austic-like symptoms.

A reduction of elevated guanidinoacetic acid by restricting arginine intake resulted in seizure reduction in the one patient who did not respond to creatine. The levels of guanidinoacetic acid were reduced in association with the clinical improvement, including seizure reduction. Mental retardation and cognitive defects were still present.

This paper and review is further evidence for the efficacy of creatine monohydrate in correcting a deficiency, and in other papers, for having a positive effect in seizure control. The reader is encouraged to obtain a copy of this interesting manuscript (Leuzzi 2002).

Several animal models of epilepsy have shown efficacy of creatine, but perhaps not all seizure types benefit. A recent paper (Vielhaber et al. 2003) examined the effect of creatine administration on a pilocarpine induced seizure Wistar rat model. This was a chronic model evaluated by an in vivo 1H-NMR spectroscopy technique. The question is whether the effects of creatine represent neuronal cell loss or metabolic dysfunction.

In this study, 120–170 g Wistar male rats were injected with pilocarpine, and seizures started within 20–40 min. After status epilepticus began, diazepam was injected in order to end the seizure activity. Experimental rats were fed a diet containing 2% creatine. Enzymatic analysis was performed on slices of hippocampus. 1H-NMR spectroscopy was performed on control and experimental creatine rats 26–30 days after pilocarpine.

Results showed that in order to evaluate the pilocarpine model of temporal lobe epilepsy, creatine was given for 30–35 days after the onset of status epilepticus. This represents a latent period between pilocarpine status and the start of spontaneous seizures. The spontaneous seizure onset correlates with hippocampal neuronal cell death. In the hippocampus of pilocarpine animals, there was a 15% decrease of *N*-acetylaspartate (marker of neuronal survival) in the hippocampus. Continuous feeding of the 2% creatine diet showed a 25% increase in hippocampal creatine content.

In evaluating nissl stained hippocampal sections, the authors found a dramatic 50% loss of pyramidal neurons in CA1 and CA3 regions. There was an indication of depleted numbers of mitochondria. There was no significant effect of creatine feeding on seizure frequency.

The authors comment that creatine supplementation shows evidence of neuroprotection in several animal models. This implies a broad efficacy for seizure types. This model shows a latent period between pilocarpine induced status epilepticus and the spontaneous occurrence of seizure activity. This would therefore indicate a time frame during which creatine induced neuroprotection should occur. The authors state that they did see a 25% increase in creatine, and decreased *N*-acetyl-aspartate levels in the hippocampus. These results are similar to those seen in other animal seizure models.

The authors state that in spite of these similarities, the in vitro studies of hippocampal sections revealed adverse effects on the density of hippocampal pyramidal neurons. These results included a decrease in cell density and decreased mitochondrial enzyme citrate synthetase, and respiratory chain enzymes. In addition, a protective effect on seizure frequency was not observed. The authors comment that their study was not in keeping with the concept that creatine has positive effects on neuronal energy metabolism. Another temporal lobe model (Kainate) has had similar negative creatine protection results (Mikati et al. 2004). One possibility is creatine does not protect against temporal lobe (complex partial) epilepsy for an unknown reason.

There is a group of inborn errors of metabolism called methylmelonic acidurias. These deficits are caused by a deficiency of methylmalonyl CoA mutase (EC 5.4.99.2) and also by defective synthesis of 5-deoxyadenosylcobalamin, which is the cofactor for the above enzyme. This inborn error of metabolism has associated clinical features including mental retardation, delayed development, cerebral edema, changes in myelination, and seizures. Patients often show elevated cerebral lactate, suggesting an effect on energy metabolism. Studies have shown depleted ATP and PCr levels. The present study (Royes et al. 2006) was designed to ascertain if creatine was capable of protecting the brain from the effects of methylmalonic acid in this model.

In this study, cannulas were stereotaxically placed in the striatum, and creatine (1.2, 3.6, and 12 mg/kg IP) was administered 3 days later. Thirty minutes after creatine treatment, methylmalonic acid was delivered intrastriatally to the rats. Performance in an open field test, behavior, and seizure characteristics were evaluated. Both creatine and phosphocreatine were measured by HPLC.

Results showed that systemic administration of creatine decreased the number and frequency of seizures. Electrographic recordings of ipsilateral striatum and cortex showed that methylmalonic acid resulted in an ipsilateral seizure focus in the striatum, which spread to the ipsilateral cortex and contralateral striatum. Creatine pretreatment prevented this seizure spread, as evidenced by electrographic recording in the high dose of creatine (12 mg/kg), caused a general increase in striatal creatine and phosphocreatine. The creatine administration only increased creatine and phosphocreatine in animals injected with methylmalonic acid, not saline injected controls.

The authors comment that this study shows that creatine protected a methylmalonic rat seizure model from seizures in that the non-creatine seizures were reduced in frequency and duration by creatine pretreatment. Further, the administration of creatine increased striatal creatine and phosphocreatine tissue levels. This suggests that a similar protection from lactate tissue increases could be attributed to an enhancement of cerebral energy metabolism (Royes et al. 2003). Other roles of creatine such as a direct antioxidant effect could also be involved in the protection from seizures by creatine (Lawler et al. 2002). Other data support the antioxidant effect on mitochondria. The authors note their data supports a possible therapeutic role for creatine in methylmalonic acid induced epilepsy.

More evidence of a role for creatine comes from clinical studies. This report (Mancardi et al. 2007) presents a case study of a 5-year-old boy with severe/refractory epilepsy, and a creatine transporter deficiency (CRTR-D). The usual perturbation in cases of creatine transporter deficiencies is mild, having infrequent seizures and a relatively easy and positive response to AEDs. The present case, in contrast, is of a patient with a severe epilepsy phenotype.

This case is of a 5-year-boy with no previous family history of epilepsy. Birth was normal, with no immediate developmental delays. By two years however, there were delays in cognition and language. At age 4.9, the patient had febrile seizures, which later developed into generalized tonic clonic seizures, and intermittent fever episodes were occurring.

The patient was hospital admitted, with increased hyperactivity and impulsive behavior. He was moderately retarded, with severe language delay. The EEG showed slow, high amplitude activity. Seizures were occurring daily, and were generalized tonic clonic types. Extensive laboratory data (autoimmune, infectious, etc.) were all negative. AED treatment (phenobarbital, valproate, clonazepam, and topiramate) was ineffective. MRI was normal, but 1H-MRS showed a decrease in the creatine peak. DNA analysis showed a missense mutation in the patient, his mother and sister. This missense mutation had been previously reported (Mancini et al. 2005).

Subsequent treatment with prednisone and levetiracetam and valproate proved efficacious in seizure control. The patient still has 2–3 tonic clonic seizures per month, with pronounced version. The EEG showed improvement. The authors comment that this case represents a severe phenotypic presentation of a creatine transporter missense mutation, the first described with CRTR-D.

Another paper looks at the anticonvulsive effects of creatine supplementation and physical exercise on pentylenetetrazole induced seizures in adult male Wistar rats (Rambo et al. 2009). This paper describes studies in which both creatine (300 mg/kg/day) and exercise (6 weeks of swimming training) were given to Wistar rats. Results showed that either creatine or the physical training decreased the pentylenetetrazole induced seizures. In addition, the combination of creatine plus physical training had an additive effect in that they increased onset latency, and were more effective in decreasing duration of pentylenetetrazole induced seizures than either alone.

Additional neurochemical studies included prevention of pentylenetetrazole induced Na, K, ATPase decrease. Parameters of oxidative stress were also prevented such as increase in thiobarbituric acid reactive substances, and decreases in non-protein thiols and catalase activity.

The authors conclude that the combination of physical training/creatine supplementation is efficacious in ameliorating pentylenetetrazole seizure activity. The effect is additive-greater than the individual effects alone. The authors further conclude that their data provide additional support suggesting a role of creatine for epilepsy treatment. In summary, substantial evidence suggests creatine in a worthwhile supplementary treatment form for epilepsy control. Creatine has also been shown to be effective in other neurological disorders such as Alzheimer's disease, Parkinson's disease, amyotrophic lateral sclerosis, ischemia, stroke, etc. The protective effect apparently involves a stimulation of energy in that additional phosphocreatine can be generated through creatine supplementation. This provides additional energy reserve for ATP production very quickly when needed (Andres et al. 2008). The ketogenic diet (and modified Atkins diet) also serves to stimulate energy production. It would be worthwhile to try a combination of these two dietary methods to see if there would be a more efficacious outcome from the combination.

Chapter 30
The Ketogenic Diet

The use of starvation to treat various ailments has a long history, actually going back as far as Hippocrates, where starvation was used to treat epilepsy. There is a reference in the bible to Jesus' curing an epileptic. The description of the epileptic child is reasonably accurate, with a description of symptoms including falling, gnashing of teeth, and wallowing. In the healing, a "spirit" is cast out of the epileptic child, resulting in a cure (St. Mark 9:17–27).

The first modern use of starvation as a potential cure for epilepsy was performed by Guelpa and Marie (Guelpa and Marie 1911), two French physicians. In their study, both children and adults with seizures were treated with starvation, which had some level of efficacy (due to ketone body generation). In the U.S., a physical fitness advocate (Macfadden) published widely on the benefits of healthy eating, and claimed that various fasting regimes could cure almost any disease, including epilepsy. Macfadden published a magazine which expounded his beliefs.

As the name implies, the ketogenic diet is a diet which acts to increase ketone bodies. The diet has a long history, and has been used for seizure treatment over 80 years. The ketone bodies acetone, acetoacetic acid, and beta hydroxybutyric acid are found in the urine of starving people, and in diabetic patients. Ketone bodies are largely produced by hepatic mitochondria from acetyl CoA. Free acetoacetic acid cannot be efficiently reconverted into acetyl CoA in the liver, so it enters the circulation and goes to areas of need such as muscle and brain. At three sites it is reconstituted to acetyl CoA, and is available to enter the citric acid cycle to produce energy in the form of ATP. Beta hydroxybutyric acid undergoes similar metabolism. Acetone is present in very small amounts. In normal individuals, ketone bodies are present in small amounts in urine, and the production/metabolism is in balance.

When liver glycogen is lowered beyond a certain point, ketone body production is increased. Hence, the ketogenic diet is one of low carbohydrate and high fat. When oxidation of liver carbohydrates are reduced, oxidation of fat is increased. Both beta hydroxybutyric acid and acetoacetic acid are strong organic acids, so they are capable of producing metabolic acidosis. The symptoms of severe diabetes include adverse effects on CNS function, which may be due to acidosis, electrolyte deficiency, and/or altered energy metabolism. When glucose is low, brain derives

D.W. McCandless, *Epilepsy: Animal and Human Correlations*,
DOI 10.1007/978-1-4614-0361-6_30, © Springer Science+Business Media, LLC 2012

Table 30.1 Evidence suggesting energy metabolism is involved in the efficacy of the ketogenic diet

Evidence in favor of energy metabolism
1. Increase in mitochondrial density in hippocampus in KD animals
2. Increase in energy metabolites in hippocampus in KD animals
3. Up regulation of transcripts encoding energy metabolism enzymes
4. Hippocampal synaptic transmission maintained 50% longer in KD animals
5. Brain rendered more resistant to metabolic stress

Adapted from Bough, K. Epilepsy 49:91, 2008

energy from ketone bodies only, whereas other tissues have other sources besides ketone bodies. The brain can derive as much as 70% of its energy needs from ketone bodies. The initial suggestion that ketone bodies were fuels for mitochondrial respiration was first made by Krebs (1961). He also noted that acetoacetate was more unstable than is beta hydroxybutyric, therefore the latter is a more suitable and likely to be usable nutrient, carried by blood to various tissues of the body.

In the fasting state, a decrease in glucose in blood acts to increase non-esterified fatty acid flux, which in turn can be metabolized to ketone bodies. Ultimately, acetoacetic acid and beta hydroxybutyric acid are made. The two ketone bodies are in equilibrium, but beta hydroxybutyric is the major ketone body used by brain. Ketone bodies enter the brain cell by a mono carboxylic transport system. They then approach and enter the mitochondria, where they are available for metabolism (Pan et al. 2002).

Animal seizure models have been treated using ketogenic diets, and show that they are able to increase the threshold for seizures in rats (Appleton and DeVivo 1974). Results showed an increase in the average voltage required to produce a minimal convulsion from 69.75 V for controls to 81.25 V in rats on a ketogenic diet for 20 days. After a 48 h reversal by replacing the ketogenic diet with a high carbohydrate diet, the seizure threshold returned to normal.

The exact mechanism of action of ketone bodies in terms of modulating seizures is not clear. Early studies in ketotic fat-fed rats showed no changes in electrolytes or pH, but intracerebral levels of many metabolites such as lactate, pyruvate, glycogen, and ATP, were elevated. Cyclic nucleotides and ADP were lower (DeVivo et al. 1978). These data, confirmed by other workers, indicate that glucose flux is decreased when ketone bodies are being utilized for energy. These data further suggest that there is an increase in energy metabolites such as ATP, when ketone bodies are available, and carbohydrate sources are not (see Table 30.1).

It was also shown from their data that the calculated energy reserves were increased by about 10% in fat-fed rats as compared to controls. Ketone bodies are a more efficient source of energy than glucose because they enter the Krebs cycle directly instead of through glycolysis. It has been shown that ketone bodies (all three) do have some actual antiepileptic properties separate from their effects on cerebral energy metabolism. All these results suffer from not having been performed on highly focal regional brain areas. This will be discussed later.

It has also been shown in mice that there is an up regulation of 34 differentially regulated transcripts which encode enzymes of energy metabolism when the animals are on a ketogenic diet (Bough et al. 2006). This resulted in increased numbers of mitochondria, and a concomitant increase in energy reserves. The implication of this study is that the ketogenic diet produces basic changes in cerebral energy metabolism. The increased energy stores seem to enable brain to better resist demands based on greatly increased electrical activity. This protective mechanism seems mostly independent of the seizure type (Rho and Rogawski 2007).

In a general sense, the ketogenic diet is thought of as being a first choice for therapy in situations such as disorders like pyruvate dehydrogenase deficiency and glucose transporter protein deficiency. Treatment of such disorders as these acts to reduce or eliminate accompanying seizures, if present, and also helps other problems of the disorders. The ketogenic diet is usually thought of as a secondary treatment choice, or as an adjunct choice with various AEDs. It seems to be effective with several classes of seizures, but its effectiveness with complex partial seizures is unclear (Keith 1963).

The initiation of the ketone diet in any patient usually requires a nutrition team, or at least a dietitian. This person(s) act to assess the current status of the patient, and to set nutritional goals. Obviously, potential risks are identified in advance if possible. A usual starting point might be with a 3:1 diet ratio, which means for every 4 g of food, 3 are fat, 1 is non-fat. The protein requirement is dependent on the patient's age. After hospital discharge, patient/parent contact is essential in order to assure compliance.

Adverse effects of the ketogenic diet occur in two groups: the first consists of patients with usually mild symptoms such as loss of appetite, nausea, etc. These usually dissipate in the first few weeks of treatment. The second group of adverse effects is those of renal calculi, metabolic acidosis, changes in blood lipids, osteopenia, vitamin deficiency, etc. All of these more consequential adverse effects should be watched for and treated as soon as possible. Severe adverse effects may occur consisting of pancreatitis, coma, and death. These effects are quite rare.

As indicated above, the ketogenic diet has been one of the longlasting potential treatments for epilepsy. Phenobarbital was discovered in 1910, and much earlier, bromides were used for epilepsy treatment (mid-1800s). At the present time, the effectiveness of the ketogenic diet has been confirmed by multiple studies (Kinsman et al. 1992). Some years after its introduction around 1921 as an epilepsy treatment (Wilder 1921; Lennox and Cobb 1928), the ketogenic diet was considered less, probably because of AED development.

The ketogenic diet was later shown to be highly effective in experimental animals of seizures (Uhlemann and Neims 1972; Appleton and DeVivo 1974). More recently, for human consumption, dietary personnel have developed ketogenic diets which are more palatable, and offer substitutable food groups offering customizable alternatives and flexibility. Lack of compliance is one major problem with the administration of the diet (Keith 1963).

Several early studies looked at the issue of which type epileptic patient might benefit from ketogenic diet treatment. One important study (Livingston 1972) showed that in over 1,000 patients, there was seizure control in over 50% of patients, and seizure improvement in 27%. This paper also spoke to the type seizures benefiting from the ketogenic diet. It was noted that patients with myoclonic seizures benefited greatly. While not as effective, patients with absence seizures also improved. The brain is best equipped to metabolize ketone bodies in young as compared to older patients. Children respond more favorable to the ketogenic diet than do adults.

The clinical aspects of the ketogenic diet and its administration have been reviewed in light of participation of dietary experts (Hartman and Vining 2007). As stated above, the ketogenic diet is high in fat content, and low in carbohydrate and protein. The usual ratio of fat to carbohydrate is about 3:1 or 4:1. Obvious criteria to observe are seizure control and adverse side effects. In children, careful monitoring of growth and behavior are also essential. As is the case in many aspects of seizure treatment, participation of family and patient in decision making is viewed as highly important.

The initiation and implementation of a ketogenic diet may be an important event lasting several days. A complete history is essential, and the team then decides on the fat/carbohydrate ratio. Initially, the carbohydrate intake is decreased, then fasting starts, and blood glucose is monitored. The ketogenic diet is then given in increasing amounts of calories over a 2–3-day period. The usual additives such as vitamins, etc., are given. Out patient visits can be scheduled at 3–6-month intervals, except whenever a problem arises.

The ketogenic diet has been administered to patients with a variety of seizure types. All show benefit, in the range of 35% with a small fraction seizure free, the remainder with as much as 90% reduction in seizure frequency. Another 20% may have a 50–90% reduction in seizures. Seizure types include most except for complex partial seizures. This class seems not to have an early positive change. However, many complex partial epileptic patients do have positive results later, and some become seizure free (Maydell et al. 2002; Than et al. 2005). The ketogenic diet is usually administered after multiple AED failure. It has been shown, however, that treatment results are about the same as a first treatment modality (Rubenstein et al. 2005).

As stated above, the ketogenic diet has been administered in a couple of fat/carbohydrate ratios – 3:1 and 4:1. One study has examined the efficacy and tolerability of the two ratios of the diet (Seo et al. 2007). The authors note that despite world wide use of the ketogenic diet, no international standardized treatment regime has been established. Obviously, the ratio of fat to carbohydrate is one important factor in the ketogenic diet. Results from animal studies (Nylen et al. 2005) suggest that the higher ratio has a better efficacy, but may harbor more complications and decreased tolerability.

The current study had 76 patients with refractory childhood epilepsy. These patients were placed randomly into one of two groups – 3:1 or 4:1 fat/carbohydrate ratio ketogenic diet. Patients with positive results (seizure free) on the 4:1 diet were moved to the 3:1 ratio group; patients without seizure-free outcomes were moved to the 4:1 group.

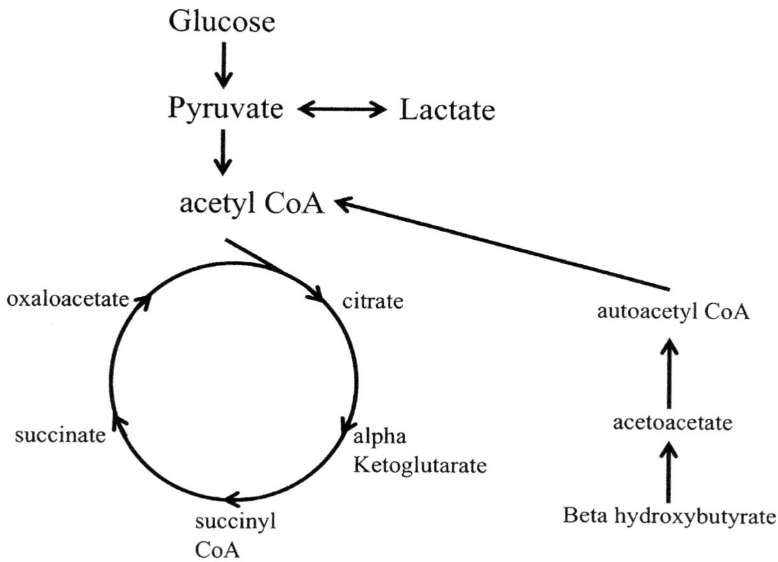

Fig. 30.1 Pathways influenced by the ketogenic diet. Adapted from Nordli, D., and DiVivo, D. (1997) Epilepsia 38:743

Results showed that the 4:1 ketogenic diet was statistically more effective than the 3:1 ratio diet. The seizure-free rate at 3 months after initiation of the diet in the 4:1 group was 55% as compared to 30.5% for the 3:1 group. Patients moved from 3:1 to 4:1 ratio then experienced a 75% seizure reduction. The mechanisms behind these results in patients are not clear, but in addition to effects on energy metabolism, there could be neurotransmitter alterations. The production of both GABA and glutamate are closely related to acetoacetate and beta hydroxybutyric. In animal studies (Bough et al. 2000), a ratio of 6.3:1 has been shown to be even more effective in seizure control than the 4:1 ratio (see Fig. 30.1).

The problem of tolerability is not insignificant. Symptoms such as nausea, vomiting, and diarrhea occurred in about 14% of patients in the 3:1 ratio group, and in 36% in the 4:1 ratio group. Except for the differences in seizure outcome, and tolerability, all other factors were similar between the two groups.

The authors comment that the improvement of efficacy of seizure reduction and tolerability is essential for seizure treatment. While there are diverse reports regarding seizure outcomes with the ketogenic diet, the ratio of fat to carbohydrates is important, and a potential variable between studies. Even though the 4:1 ratio treatment was more disturbing as regards GI symptoms, few patients quit because of that problem. The authors recommend an initial trial of the 4:1 level, to be reduced if adverse effects are intolerable. Perhaps an intermediate ratio (3.5:1) could ameliorate the adverse effects to an acceptable level, maintaining efficacy.

The nature of the anticonvulsant mechanism of the ketogenic diet receives increasing attention (Bough and Rho 2007). The most common ketogenic diet uses a 3:1 or 4:1 fat to carbohydrate ratio, with long-chain triglycerides being key. Over 90% of calories are derived from fat. The authors state that overall data suggest that generally the effects of the diet are related to a change in pathways and "programs" which serve to stabilize cell metabolism, thereby modulating seizures.

The clinical observation that almost any situation (including starvation) will have an anticonvulsant effect if ketonemia is a feature. When a ketogenic diet is stopped, there may be seizures in only a few hours. Such a cessation needs to be avoided, and from a mechanistic standpoint suggests an energy metabolism basis because of the rapidity of the response. This clinical finding should be followed up in animal experiments in which regional cerebral energy metabolites are measured.

Ketone bodies such as acetone, when injected into experimental animals, do produce reduction of seizures. Acetoacetate has similar properties. While these ketone bodies are anticonvulsants, no data exist to indicate that they can directly alter synaptic transmission. Cellular electrophysiological studies have not shown any effect of ketone bodies on ion channels which regulate neuron excitability or inhibition (Thio et al. 2000). All these studies suggest no direct anticonvulsant effects for either acetoacetate or beta hydroxybuterate.

Another proposed mechanism relates to the role of ATP and ATP/ADP ratios in regulating Katp channels. When ATP levels rise, Katp channels close, which in turn is seen as a protective mechanism in preventing or modulating seizure activity (Seino and Miki 2003). A possible anatomical base for this relates to the substanta nigra. This brain area is particularly rich in Katp channels, and acts centrally in propagation of seizures (Iadarola and Gale 1982). Such a pivotal location would be seen as conducive to several seizure types and the ketogenic diet ameliorates several seizure types. This mechanism is not conclusive since low glucose and high-fat ketogenic diets actually promote Katp channel activation.

In terms of energy metabolism, fatty acids act to regulate numerous genes which are involved in energy metabolism (Sampath and Ntambi 2005). The idea is that the ketogenic diet may "reprogram" energy metabolism since the diet induces an up regulation of dozens of genes associated with oxidative phosphorylation (Noh et al. 2004). The ketogenic diet induces an increase in mitochondrial numbers, and cerebral levels of energy metabolites are also increased. In addition, experimental data in ketogenic diet animals show that there is increased metabolic "efficiency," and increased cerebral respiratory rate (Sullivan et al. 2004). Taken together, the data certainly indicate an effect of the diet on energy metabolism conducive to increased seizure protection.

Previous studies show an adverse effect of seizures on energy metabolism (Kunz 2000; Kann et al. 2005). Therefore, the interpretation is made that the above effects of the ketogenic diet are sufficient to act to decrease the seizure state.

How the energy change acts to stabilize synaptic function is speculative at best. One suggestion is that since ATP acts to maintain ionic gradients, the increased ATP levels could prolong the activation of Na/K ATPase (Veech et al. 2001). This in turn could serve to reduce neuronyl firing. There are no studies validating this hypothesis; some evidence also indicates that the ketogenic diet acts to stabilize both excitatory and inhibitory synaptic transmission.

Various other hypothetical mechanisms have been proposed for the efficacy of the ketogenic diet. Some involve an effect on neurotransmission. One such mechanism involves effects on the inhibitory neurotransmitter GABA. The hypothesis is that the ketogenic diet is quite effective against GABAergic antagonists. For example, pentylenetetrazole, bicuculline, etc., are agents which produce seizures, blocked by the diet. Conversely, the ketogenic diet has little effect on experimentally induced seizures which involve glutamate activation (Bough et al. 2006). The actual mechanism for this could involve a shift in amino acid metabolism favoring an increase in GABA production (Yudkoff et al. 2005).

Further, data support an interaction between energy metabolites and GABA metabolism in GABAergic neurons. This comes from the finding that creatine kinase is localized in GABAergic inter neurons (Boero et al. 2003). The hypothesis, not yet tested, is that a ketogenic diet could increase energy reserves in the inter neurons, thereby acting to enhance GABAergic function. This would act to increase seizure control due to increased inhibition. These several hypotheses need confirmation, which should be done using methods permitting analysis at a focal level, not in whole brain.

An interesting side effect of the ketogenic diet is its effect in children on the quality of sleep (Hallbook et al. 2007). In the study, 18 children with drug-resistant epilepsy were started on a ketogenic diet. They were admitted, had a 12 h fast, then gradually a ketogenic diet was implemented. Fifteen patients received a 4:1 ratio diet, three received a 3.5:1 ratio diet. Two patients were changed from a 4:1 ratio to a 3.5:1 ratio. Ambulatory polysomnographic recordings with the patients at home were performed after 3 and 12 months of ketogenic diet.

Results showed that the diet resulted in a statistically significant change in several sleep parameters. There was a decrease in total sleep, and total night sleep. REM sleep increased, while sleep stage 2 decreased. Of 11 children still on the ketogenic diet at 12 months, there was a continued decrease in day time sleep and a further increase in REM sleep. Beta hydroxylate rose fourfold after 3 months on the diet, and was increased another 20% after 12 months.

Monitoring for seizure activity showed significant reduction. Eight children (44%) showed a 90% or better reduction in frequency, and four more showed a 50–90% reduction. Eleven patients continued the diet, with a total of 54% being either seizure free or having a 90% or better reduction in frequency. The seizure severity was improved in those with seizures, and the quality of life was judged to be improved.

The authors state that the observed changes cannot be attributed to age-related alterations in sleep parameters. For example, total sleep time does not change from early childhood to adolescence, and the times (3–12 months) were relatively short. Sleep can affect seizures in two ways: precipitating and protecting. As regards precipitating events, non-REM sleep is associated with tonic-clonic seizures, myoclonic convulsions, and epileptiform discharges. Conversely, REM sleep is considered to be protective from seizures due to the desynchronized low amplitude nature of REM sleep. The changes due to sleep patterns may also have a positive effect on nocturnal epileptic patients.

In this study the efficacy of the ketogenic diet was excellent. This is presumed to be due to the effects of the diet on energy metabolism, decreased glutamate metabolism, and increased GABA synthesis (Schwartzkroin 1999). The authors note that the study is positive in that sleep criteria were improved as well as seizure characteristics by the ketogenic diet in children with drug-resistant seizures.

Seizures are often associated with energy metabolism in the form of respiratory chain complex defects (Kunz 2002). A recent paper examines the use of the ketogenic diet in such cases (Kang et al. 2007). This was a retrospective study in which 14 children with intractable seizures and also with respiratory chain complex genetic defects were examined. They had a variety of different mitochondrial defects, as well as multiple seizure disorders, including complex partial epilepsy, Lennox-Gastaut syndrome, Landau-Kleffner syndrome, infantile spasms, etc.

Results showed that 10 of 14 patients enjoyed an improvement in seizure frequency, and eight were able to stop or reduce AEDs. These results are similar to the results of other studies as to the efficacy of the diet (Freeman et al. 2006). There were no significant complications of the ketogenic diet. There were also no effects of the diet on the various respiratory chain defects, or on any associated enzyme levels.

The authors note that the ketogenic diet has been used more and more frequently over the past 15 years, and may serve to provide ancillary energy sources for cerebral use thereby having a positive effect on seizure reduction (McFarland et al. 2002). The authors speculate that the ketogenic bodies produced by the diet would act to sustain the Kreb's cycle, and minimize lactic acidosis. There were some adverse effects in this group of patients including hypoglycemia, metabolic acidosis, and pneumonia. Two pneumonia patients expired, but these cases were not thought to be directly related to the ketogenic diet.

The use of the ketogenic diet in children less than 2 years of age has always been considered as "dangerous" because of so many developmental processes occurring during the period (Livingston 1972). Much later, a paper appeared reporting the safe effective treatment of seizures with the diet in children under 2 (Nordli et al. 2001). A short review paper (Rubenstein 2008) examines some of the issues.

The review notes that in the Nordli paper, results show that newborn infants less than 12 months old taking the ketogenic diet for intractable seizures enjoyed a 55% elimination, or over a 50% reduction in frequency of seizures. About 96% of the 28 babies on the diet for 3 months or more maintained growth. Adverse effects included gastritis, renal calculi, etc., but were considered to be overall well tolerated. Patients under 12 months should be carefully followed, including metabolic and lipid panels, EEGs, urinalysis, etc., as well as diet adjustments as required.

The author suggests that the ketogenic diet should be initiated in severe epilepsy cases such as Lennox-Gastaut syndrome, seizures associated with CNS malformations, early onset myoclonic epilepsy, Dravet syndrome, etc. In one study in which the ketogenic diet was used as an early therapy (zero or only one AED), 60% of patients on the diet after 6 months showed a 90% reduction of seizure frequency. About 100% of patients still on diet after 1 year had a 90% or better seizure frequency reduction. The author concludes that the ketogenic diet is safe and affective even when administered to newborn infants with catastrophic epilepsies.

While the ketogenic diet has been "revitalized" recently, until just a couple years ago, there were no controlled, randomized studies looking at safety and efficiency of the ketogenic diet (Cross and Neal 2008; Neal et al. 2009). In this study, 145 children with intractable epilepsy (two failed AED trials) were randomized to receive either the ketogenic diet or the medium chain triglyceride (MCT) diet. Seizure frequency was assessed at 3, 6, and 12 months. Results showed no significant differences between the ketogenic diet and the MCT diet, so results were combined for analysis. Generally, at 3, 6, and 12 months slightly less than 25% of patients had a 50% or more reduction of seizures. Side effects were complications consisting of vomiting, diarrhea, abdominal pain, etc., as reported before. The authors state that the study shows no difference between the ketogenic diet and MCT for treating intractable seizures in children.

The variability of reports of efficacy of the ketogenic diet has been eluded to above. The main suspect reason for variability is the inconsistent in the make up of the diet. There had been no consensus on the make up and implementation of the ketogenic diet, especially world wide. This problem, now defined, was dealt with by a panel of 26 pediatric epileptologists and dietitians in 2006 (Kossoff 2008).

Results of this working group showed much agreement, and still some controversy. For example, 81% of the group agreed that the ketogenic diet should be tried after two unsuccessful AED treatment trials. It should be first-line treatment for seizures associated with pyruvate dehydrogenase deficiency and GLUT-1 deficiency. About 88% of the panel started the diet in a hospital setting. Obviously vitamin, etc., supplements should be initiated, and careful monitoring also started. The patients should consult a dietitian and neurologist every 3 months. The ketogenic diet should be given at least 3 months before being discontinued, and risks balanced against benefits.

Flexibility should be maintained in that the long-chain triglyceride and medium chain diet were similar so either could be used. Controversial subjects included the importance of fasting before initiation of the diet. For example, 31% said children should never be fasted, while 58% said they believed in a fasting period. About 42% did not think the ketogenic diet should be offered to good surgical candidates.

The panel of 26 believed their conclusions and recommendations provide an international consensus of how to approach the ketogenic diet for intractable epilepsy in children.

One question asked is at what stage of ketogenic treatment do results begin to be seen. In a retrospective study from two major hospitals, records were analyzed in order to address this question (Kossoff et al. 2008a, b). In a general way, results showed about equal numbers of male and female subjects. Of 118 patients, mean ketogenic diet onset was 3 years of age (range 0.3–15 years). The majority of patients had generalized seizures which included Lennox-Gastaut, myoclonic, infantile spasms, Dravet syndrome, partial epilepsy, etc.

Results showed that 84% (99 patients) had some level of improvement in seizures. The median time to the first noted improvement was 5 days (range 1–65 days). About 75% showed some level of improvement in the first two weeks, and 90% by day 23. Fifty-five patients had a greater than 90% seizure reduction at 3 months.

At 6 months, 71% were improved such that they had a greater than 50% reduction of seizures. Results were statistically similar between the two institutions contributing data. Of the 99 children with seizure reduction, 75 maintained or continued to improve on the diet. Patients who did not show benefit by 65 days after initiation of diet never did.

The authors note the first month of treatment is most critical. They feel an initial starvation period and close follow up and changes in the diet if necessary are distinct advantages. The authors note that while fasting is not absolutely necessary, it may lead to a more rapid improvement (Freeman and Vining 1999). The fasting period acts to "jump start" the process by raising ketone body levels in advance. The authors also state that since the parents maintained seizure calendars, there could have been an optimistic bias. However, much improvement was physician documented, and most parents seemed reliable as regards record keeping. This study speaks to the potential efficacy of the ketogenic diet, and defines the period of time in which to look for improvement in seizure control.

There is always a risk for long-term sequelae from anticonvulsant administration. In terms of the ketogenic diet, there might be metabolic changes due to high circulating levels of ketone bodies, or from the increased levels of lipid in the diet. There have been some reports of liver damage using the diet (Ballaban-Gil et al. 1998). To examine the risk, enzymes associated with liver damage (alkaline phosphatase-ALP, and aminotransferases- ALT and AST) were measured in blood in animals on a ketogenic diet regime, and in animals with status epilepticus plus the ketogenic diet (da Silveira et al. 2010).

Status epilepticus was induced by using a lithium-pilocarpine model. The status was characterized by chewing, then repetitive clonic activity in both the trunk and extremities. Rearing and falling with forelimb clonus were also features. Animals were fasted for 24 h before initiation of the ketogenic diet. The ketogenic diet contained 25% protein.

Results showed that with a level of 25% protein, rats remained healthy in appearance. The ketogenic diet by itself increased adenosine nucleotide (ATP, ADP, and AMP) hydrolysis. The hydrolysis was unchanged in animals made epileptic with lithium-pilocarpine and treated with the diet. Lithium-pilocarpine alone increased ATP hydrolysis, but decreased ADP hydrolysis. To evaluate possible liver involvement, the activation of ALP, ALT, and AST in blood showed increases in levels. Status epilepticus alone without the ketogenic diet also elevated ALP and AST levels.

The authors state their data show increased ATP hydrolysis and a decrease in ADP hydrolysis in lithium-pilocarpine induced status epilepticus. In addition, suggesting possible liver damage, ALT was elevated in blood of rats on the ketogenic diet.

The concerns for long-term health in children on the ketogenic diet have been reviewed (Vining 2008). While there have been some reports of serious sequelae in patients on the ketogenic diet, some of the patients were otherwise quite sick, before being placed on the diet (Kang et al. 2005), as seen above. Conversely, several patients have been on the diet for over 20 years without any major problems. Even in children starting therapy at an early age, growth and development was normal.

Yet, in a large study (Vining et al. 2002) children did not have a significant weight gain, and were at lower percentiles after 1 year on the diet.

Other studies looking at criteria such as cardiovascular features, renal stones, cholesterol levels, low-density lipoproteins, etc., are equivocal. Some studies even show that a ketogenic diet might be beneficial in terms of adult weight loss, and decreased BMI (Dashti et al. 2006). In seizure patients there has been the observation (see above) that behavior seems to be better in epileptic patients on the diet.

Clearly what is needed is controlled randomized studies using significant numbers of patients in order to clarify the long-term sequelae related to the ketogenic diet. The increased use of this diet warrants a proper assessment of risk factors.

Animal models can fulfill some of the boundaries of efficaciousness of the ketogenic diet treatment and safety issues, but what features are essential? A recent review has examined some of these desirable features (Holmes 2008). First should be the applicability of the model to the human disorder it is supposed to resemble. This means that symptoms, time course, histological results, and treatment efficacy should all be as similar to the human version as possible. Then when data are collected from focal brain areas, essentially impossible in human patients, it will have relevancy to human mechanisms. Another example is that of the anticonvulsant felbamate. This drug is a potent anticonvulsant (Faught et al. 1993) in which adverse effects were thought to be low based on rat studies. In humans, however, cases of aplastic anemia, and hepatotoxicity surfaced which were not predicted based on rat studies (Pellock and Brodie 1997).

In regards to efficacy, it is important to determine whether any given drug or treatment is effective in generalized or partial epilepsies. Common models of seizures are maximal electroshock (generalized tonic-clonic seizures), pentylenetetrazole (myoclonic seizures and absence seizures), and kindling models (partial and generalized seizures). Other multiple animal models are used including: transgenic models, knockout mice, hyperthermia, hypoxia, bilirubin encephalopathy, infections, etc.

In terms of models, the putative toxicity of both the method of seizure induction and the potential treatment is critical. The model is less effective if it produces some other unidentified toxicity which clouds interpretation of data. Similarly, if toxic reactions are produced by the treatment, then clarification is needed. Negative results in this area need confirmation. Tests for potential toxicity need to be behavioral, and neurological, as well as looking at development, and possible teratological effects.

As regards the ketogenic diet, more data are required on effects on behavior and memory. Some clinical studies suggest a failure to gain weight in younger children. This would be easy to determine in animal models, even in primates. Some rats treated with a ketogenic diet had lower brain growth rates, data which needs further confirmation. In terms of the diet, remember that most ketogenic diet studies are not exactly identical in composition. These deficiencies lead to differing data and conclusions. This problem is compounded since the "human model" is highly heterogeneous genetically as compared to Sprague-Dawley rats, inbred for hundreds of generations.

The author notes that potential antiepileptic drugs are usually tested in adult male rats. Time and money constraints often govern models of seizures. The ketogenic diet, as seizure therapy has been well tested, with few remaining questions for animal models. The underlying mechanisms of action of the diet in terms of the exact anti-seizure properties remain to be elucidated. The author comments that there may be many mechanisms which could impact the effectiveness of the diet. Conversely, there might be only one mechanism, basic enough to in turn affect other pathways and metabolites. Because of the ability to ask questions which cannot readily be answered in humans, especially in newborns and children, animal models will continue to have utility, even in studies of the ketogenic diet, and results translated to human patients.

The effects of the ketogenic diet on the expression of signaling pathway proteins AMP-activated protein kinase (AMPK) and heat shockprotein 70 have been investigated (Jeon et al. 2009). Mice were placed on a ketogenic diet for 6 weeks, then given kainic acid. Two days later they were sacrificed and AMPK, acetyl CoA carboxylase (ACC), and HSP70 analyzed.

Results showed that phosphorylation of AMPK and ACC were increased in mice treated with kainic acid. The ketogenic diet, however, served to lower the phosphorylation of AMPK and ACC. The diet also lowered HSP70, which had been elevated by kainic acid. These measurements were made in the hippocampus.

The authors note that the data show that the ketogenic diet acts to protect against the kainic acid-induced cell death in the hippocampus, through down regulation of the AMPK cascade. Further, the diet lowered the kainic acid induced increase in HSP70. This was present in the CA3 pyramidal neurons as shown by immunofluorescence. The authors state that ACC phosphorylation which is induced by the ketogenic diet, could be involved in neural protection.

An interesting study relating to the above comments has been published (Nylen et al. 2009). This study relates to a genetic defect called succinic semialdehyde dehydrogenase deficiency (SSADH), which is a rare autosomal recessive defect in GABA metabolism (Gibson et al. 1998). A mouse model, Aldh5a1 (Hogema et al. 2001) is available which develops absence like seizures. These seizures pass into a status epilepticus stage. The present study was undertaken to look at mechanisms, specifically those related to mitochondria and energy metabolism.

In the study, animals were fed either control diet, or ketogenic diet. Tissue was examined by electron microscopy. At sacrifice, mice were perfused and tissue harvested for EM. For ATP assay, tissue was quickly frozen following decapitation, and hippocampi tissue extracted for assay. The luciferin/luciferase assay was used to estimate ATP.

Results showed that the actual number of mitochondria are not lower in control diet mutant mice as compared to control mice. However, the present data show that the ketogenic diet does increase mitochondrial numbers. The size of the mitochondria in the hippocampi from ketogenic diet mutants was slightly larger than those in the hippocampi of control diet mice. When compared to control mice, the ketogenic diet mice had increased ATP levels. Wild-type control mice also had ATP levels higher than non-ketogenic diet mutants. So, the ketogenic diet acted to restore ATP and energy metabolism in the mutant mice. Most importantly, the ketogenic diet acted to decrease seizure activity in the SSADH-deficient mice.

The authors speculate that the SSADH-deficient mice are energy deficient due to a decreased ability to utilize glucose properly. The energy deficit begins to overtly materialize in the form of first absence seizures, then status epilepticus. Interestingly, weight loss and seizure onset correspond to the time of weaning in the Aldh5a1 mice. Subsequently, when on a ketogenic diet, alternative energy sources (ketone bodies) are available, and the seizures are suppressed, weight loss slows, and importantly, hippocampal ATP levels rise. This paper is significant because ATP was measured regionally, in the hippocampi, not in whole brain.

The authors further state that the restoration of ATP and normal mitochondria act together to restore normal function to the many energy processes which are secondarily affected by an ATP deficit. The problem, the authors note is that the "disease process" is not stopped, only slowed. But it must be remembered that any metabolic encephalopathy starts as a biochemical alteration, which invariably, if untreated, progresses to structural changes. In the case of the Aldh5a1 mice, more studies are warranted to look at when and how this progression occurs. The key finding is the reversibility of the ATP deficit at a time when symptoms are improving (slowing), and the highly focal (hippocampi) nature of these energy metabolite results.

In a similar vein, a paper describes studies on the ketogenic diet effects on succinic dehydrogenase activity and mitochondrial density, localized to the cerebellar Purkinje cells in rats (Balietti et al. 2010). The study was motivated by the potential use of the diet on neurodegenerative disorders which affect older patients. Accordingly, the effects of the ketogenic diet on succinic dehydrogenase activity and mitochondrial density, volume, etc., were examined in aged (19 months) rats. Studies focused on Purkinje cells of the cerebellar vermis. Results showed an increase in the mitochondrial density in ketogenic diet animals as compared to all controls (young, age-matched normal chow). These data show the ketogenic diet treatment counteracted age-related decrease in succinic dehydrogenase positive mitochondria, thereby enhancing the metabolic efficiency. The study, like the previous one carries more significance because of the highly regional focus of the findings.

A recent study looking at the efficacy of the ketogenic diet used a blinded, randomized, crossover method (Freeman et al. 2009). In this study, children with the Lennox-Gastaut syndrome were utilized. Criteria for inclusion were age 1–10, prior use of at least two anticonvulsants, EEG evidence of the Lennox-Gastaut syndrome, and evidence of at least 15 atonic myoclonic seizures per day. The ketogenic diet treatment for this group represented their first trial. Children enrolled (20) were admitted, and had an extensive EEG workup. Parents were employed to be certain to record seizure attacks. All 20 children received a ketogenic diet with 13 receiving a 4:1 ratio diet, and the other seven receiving a 3:1 ratio diet.

Results showed that by the end of the first six days, there was a significant decrease in the EEG-documented seizure events. There was a further decrease by day 12. The authors comment that their study achieved one goal – to have a controlled, blinded study. No one knew that the diet had acted to control seizures until the code had been broken at the end of the study.

The study employed a crossover feature in which patients were first given either glucose (to reduce ketone bodies) or saccharin as a placebo. A fasting period punctuated the crossover of glucose or saccharin between patients. All patients enjoyed seizure reduction, reflecting the efficacy of the ketogenic diet.

Another study (Villeneuve et al. 2009) looked at the ketogenic diet effects on intractable focal seizures which were worsening. This was a study of 70 ketogenic diet patients from which 22 children with focal seizures were selected. These patients had significant worsening of symptoms in recent weeks of study. Seizure frequencies 1 week before the treatment regime were compared to frequencies 1 month after the initiation of the diet.

Results showed that of the initial 22 patients, 11 had at least a 50% or greater reduction in seizure frequency, and seven were still seizure free after 6 months on the ketogenic diet. Results were best in those with more significant worsening of seizures. Tolerability was excellent in ten of the patients, although five quit the study due to adverse side effects. The authors conclude the ketogenic diet is efficacious in focal seizures, and in those with worsening symptoms.

Another paper (Beniczky et al. 2010) examined ketogenic diet efficacy in a variety of seizure types. The study included 50 consecutive patients with severe pharmacoresistant epilepsy. After 3 months on the diet, 2/3 of the patients enjoyed a 50% or more seizure reduction. Unfavorable responses correlated significantly with temporal lobe EEG discharges, and complex partial seizures just missed significance as regards less favorable outcomes. All other types of seizures examined benefited significantly from the ketogenic diet.

Chapter 31
Gene Therapy for Epilepsy*

Epilepsy is a devastating disorder which affects about 1.5–3 million people in the United States (about 0.5–1.0% of the population), and as many as 50 million worldwide. Antiepileptic drugs (AEDs) can successfully treat about 2/3 of epileptic patients (Nadkarni et al. 2005). Thus, many patients are left with intractable seizures, with surgery as a possible last resort.

The source of epilepsy is sometimes thought to be an intracerebral imbalance between excitation and inhibitory neurotransmission. A major excitatory neurotransmitter is glutamate, and a key inhibitory neurotransmitter is GABA. Most AEDs have a mode of action directed at one or the other of these neural processes. A more generalized effect on neural excitation and inhibition is caused by neuropeptide compounds (Hokfelt 1991). In the human, precursors of neuropeptides are encoded by several dozen genes. Consequently, one potential mechanism for modulating epileptic seizures is gene therapy, in which genes are inserted directly into the cells. A common method for implementing gene therapy is through viral vectors that deliver modified genes to produce proteins and enzymes. In the case of epilepsy, the goal of viral vectors is to increase the intracerebral production of inhibitory products, which may be particularly beneficial because of the large variety of etiologic epilepsies. With adeno-associated viral vectors (AAVs), the approach is to transfer the appropriate modified gene into the infected cell, resulting in the expression of gene products. AAVs have been used frequently for treatment of numerous conditions. This vector has low immunogenicity with success in delivery of potential treatment forms to non-dividing neurons. In addition, inflammation is low, and AAV vectors are long acting.

The approach of gene therapy via viral vectors has been fairly well researched. For example, this method has been used to attempt treatment of certain muscle and eye diseases in humans (Boison 2007). In animals, various successful treatments have been demonstrated in Gunn rats (a model of the Crigler-Najjar Syndrome, with hyperbilirubinemia). In particular, gene therapy has served to provide Gunn rats with enough bilirubin conjugation enzymes to essentially eliminate jaundice.

---

*Author David McCandless was deceased.

D.W. McCandless, *Epilepsy: Animal and Human Correlations*,
DOI 10.1007/978-1-4614-0361-6_31,

This effect has been achieved in one treatment which served the life span of the rat (2 years). This treatment should be soon ready for clinical trials (McCandless 2011).

Key treatment mechanisms with viral vectors involve GABA, adenosine, galanin, and neuropeptide Y (NPY). The associated forms of treatment are endogenous anticonvulsants which have different modes of action. Of these four therapeutic approaches, GABA has been used specifically for epilepsy. In particular, the GABA approach has been used to treat a temporal lobe epilepsy rat model which uses pilocarpine (Raol et al. 2006). In this treatment, adeno-associated viral Z vector was prepared with an alpha 1 gene driver by an alpha 4 promoter. Results showed that in rats with pilocarpine-induced status epilepticus, the GABA treatment resulted in a threefold increase in seizure-free time, and a 60% decrease in the number of rats developing epilepsy. However, because some GABA-increasing AEDs have poor efficacy in treating epilepsy, this method for gene therapy is not fully resolved.

Results with adenosine have been more successful. With this treatment, the approach has been to engineer fibroblasts and myoblasts to have a decrease in adenosine kinase. These cells, which release adenosine, are then placed in semipermeable membranes and implanted. Results showed that the adenosine-releasing implants protected rats from seizures as well as epileptiform after discharges in a hippocampal kindling model. The anti-seizure effect lasted 8 weeks (Guttinger et al. 2005; Huber et al. 2001).

A third treatment option is galanin. Galanin is a 29/30 amino acid neuropeptide (Tatemoto et al. 1983), which is widely distributed in the CNS (Melander et al. 1986). The action of galanin is to antagonize glutamate excitatory neurotransmission in the hippocampus (Zini et al. 1993). Not only does galanin reduce seizure activity in several epilepsy models, galanin antagonists also act to increase seizure activity. In terms of gene therapy, the emphasis has been on AAV-regulated cerebral expression and secretion of galanin. In one study (Lin et al. 2003), kainic acid, a kindling agent, was administered 2.5 months after gene therapy with AAV vectors programmed to increase galanin release. Results showed that even 10 weeks later, the gene therapy decreased the number of seizures, with a decrease in seizure time in those rats that seized. The efficacy of intra hippocampal injection of vector demonstrated a significant anti-seizure outcome (Fig. 31.1).

In another study with galanin and viral vectors, galanin injected into the piriform cortex was found to reduce kainic acid produced epileptic seizures. Over 90% of rats studied showed no limbic seizures. In the same study, additional rats were electrically kindled, then given the galanin expressing viral vector. Results showed that 1 week after the gene therapy, additional stimulation was required to produce seizures. In other words, the kindling effect was attenuated by galanin-stimulating AAV.

A fourth treatment option is NPY (36 amino acid polypeptide), first discovered by Tatemoto (1982). NPY is the protein most studied as a possible agent for successful gene therapy in epilepsy. While NPY has several physiological functions, the possible role in seizure modulation has already been described (Vezzani et al. 1999; Vezzani and Sperk 2004). At least six NPY subtypes have been analyzed, and Y1 and Y2 are important in the hippocampus (Redrobe et al. 1999). Knockout mice

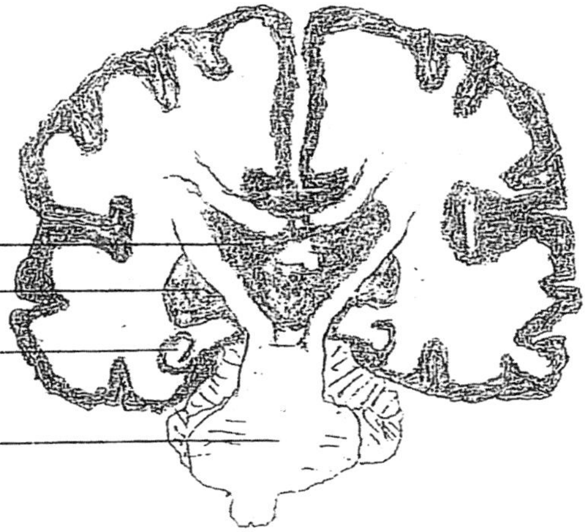

Fig. 31.1 Schematic drawing of brain slice showing thalamus, globus pallidus, pons, and hippocampus

with NPY are more prone to seizures than control mice. The first description of in vivo efficacy of NPY was in 1997 (Woldbye et al. 1997). Anti-NPY immunoglobulin was demonstrated to increase seizure incidence in rats.

In terms of viral vectors, AAV vectors have been widely used in studying NPY administration. AAV vectors are capable of packaging DNA as large as 4.5 kDa (Dong et al. 1996). These vectors do not replicate, and a tissue specific promoter is used to limit gene expression to brain. AAVs have a relatively low capacity for reversion to wild types, or insertional mutagenesis, but high doses may induce tumorigenesis in rodents (Donsante et al. 2001).

The direct injection into the hippocampus of AAV vectors-neuronyl-specific enolase, NPY has the potential to attenuate seizures (Richichi et al. 2004). This finding was based in part on studies showing that in patients with pharmacologically intractable seizure disorders, there is an over expression of hippocampal GABA containing interneurons. This may lead to an increase in inhibitory activity (Milner et al. 1997).

Results from the Richichi et al. (2004) landmark study showed that rats injected as above had NPY gene over expression limited to the hippocampus, and none in any other cerebral region. Rats injected with "empty" vector had little tissue damage, limited to the injection site. NPY tissue over expression extended for about 1.5 mm around the injection site, and was restricted to neurons. Over expression was at a maximum 3 weeks after injection, and lasted at least 3 months. Over expression of NPY had no effect on any endogenous peptides. The over expression was seen in the dentate gyrus (injection site), and outer molecular layers. Results on seizures induced by kainite were reduced by 50–75%, and seizure onset delayed. In a subset of rats injected in the hippocampus with the chimeric serotype 1/2 viral vector, status epilepticus was eliminated, and kindling significantly delayed. The authors state that the mechanism of NPY activity is likely the suppression of depolarization-induced glutamate release from mossy fibers and Schaffer collaterals. Recent work by Foti et al. (2007) essentially confirms this NPY study. It is likely that NPY transfer of inhibitory capacity may provide a novel methodology for human epilepsy treatment.

Although the successful in vivo treatment of seizures using AAVs expressing NPY appears promising, this approach is not without potential problems. For example, a recent study by Sorensen et al. (2009) found that both synaptic plasticity and transmission are affected by kindling and NPY over expression. Fortunately, results showed that in contrast to previous findings (Sorensen et al. 2008), the AAV viral vector treatment expressing NPY does not worsen whatever memory loss was already present. In a related study, Sorensen et al. (2009) found that long-term potentiation was not affected by NPY in gene therapy.

From a pragmatic standpoint, a fundamental goal of epileptic treatment is to resolve epilepsy that has already been established. Accordingly, a recent study by Noe et al. (2008) examined the effect of NPY in an established (16 weeks) model of status epilepticus. In this study, an electrically produced status epilepticus rat model was achieved via electrodes implanted directly into the hippocampi. These also allowed monitoring of seizure activity. Well after status was established, an AAV expressing human NPY gene was injected into the dorsal and ventral aspects of both hippocampi.

Results showed that status epilepticus was severe, and video EEG verified this in the experimental rats. Rats were injected bilaterally in the hippocampi 16 weeks after induction of status epilepticus. Control rats received "empty" viral vector. The delivery of NPY occurred during the characteristic progressive phase of seizure activity. The administration of NPY to the status rats resulted in a dramatic decrease in the progressive nature of the epilepsy, as well as a reduction in the frequency of seizures in 40% of the affected rats.

This finding demonstrates the efficacy of NPY treatment in a pathologically established seizure model. The effect of cessation of the progression of the status as well as a decrease in frequency in about one half of the experimental animals is noteworthy. The authors comment that the viral vector treatment resulting in over expression of NPY was long lasting, and had a positive effect on seizures. This study offers the first proof of the principle of the efficacy of gene therapy in epilepsy. This could provide a new therapeutic alternative to major surgery in intractable epilepsy once future studies determine efficacy, long-term benefits, and safety in humans.

The above study used stereotaxic-controlled placement of the vectors expressing NPY directly into the hippocampi. This technique has only minimum risk. A proof of principle method has been published (Laing et al. 2006) in which intranasal delivery of a herpes simplex viral vector with an anti-apoptotic gene transduced hippocampal neurons. The level of transgene expression was limited, but the intra nasal route to the brain certainly needs further investigation.

Clearly, the evolving science of gene therapy for epilepsy is compelling and promising. The use of NPY to suppress seizures, and an effective viral vector for delivery have been proven effective in rodents by several investigators. Long-term efficacy has been demonstrated in animals pretreated prior to seizure induction. Speculation is that successful gene therapy is possible in most seizure types, even though several etiologies contribute to epilepsy. As has been stated by various authors above, this could portend a cure for most epilepsy cases.

With millions of people coping with epilepsy, the stakes are high. Of the cases that are refractory to AED treatment, over half are poor candidates for surgery, leaving many in need of new and better treatments. From the standpoint of translation, studies and trials in primates should be an important priority. In addition, more research on specific treatments need further study. For example, the effects of NPY treatment on traits such as behavior, psychology, intelligence, and memory should be evaluated. Also, risks associated with bilateral hippocampal injection (such as the possibility of immunoreactivity problems) need evaluation in primates to supplement the extensive examinations already assessed with rodents. Furthermore, the potential ability to use less invasive routes for vector administration (such as a nasal route) needs further study. The next goal should be to further assess gene therapy for seizure correction in primate models, with the ultimate aim to ameliorate human seizures.

Part VI
Surgical Epilepsy Treatment

Chapter 32
Neuroanesthsia in Epilepsy Patients

As of 2008, epilepsy was one of the three most common treatment areas in functional neurosurgery. Functional neurosurgery is often reserved for cases of intractable epilepsy. The diagnosis of intractable epilepsy is frequently made after appropriate trials of anticonvulsant medications along with documentation of therapeutic levels of medications (NIH Consensus Development Conference 1990). Functional neurosurgery is defined as a surgical subspecialty that aims to alter the current function of the nervous system along with improving the functional status of the patient (Heit et al. 2009).

Of those patients suffering from epilepsy, research shows that patients with partial epilepsy secondary to a lesion most benefit from surgery. Functional neurosurgery consists of surgical resection or vagal nerve stimulation. This chapter will discuss the anesthesia for the resection of the epileptogenic focus. Vagal nerve stimulation is typically reserved for those patients who are not candidates for surgical resection. Epileptogenic lesions are most often noted in the temporal lobe of the brain, and the most common operation for this type of lesion is a temporal lobectomy. This surgery often consists of a partial resection (resection of the focus and surrounding area, cortical resection, or resection of cortical structures) or a complete resection of the anterior temporal lobe along with the posterior temporal lobe. This resection is dependant on cerebral dominance.

There are several approaches to surgical resection. A corpus callosotomy consists of sectioning the corpus callosum. This approach is utilized when a patient suffers from atonic seizures or partial seizures with secondary generalization (Shuer et al. 2009). This entails dividing the corpus callosum in the midline with sectioning including the anterior two-thirds or the entire corpus callosum. The next procedure includes a resection of an epileptogenic focus that is the consequence of a structural lesion. This type of resection can include a frontal, parietal, or occipital craniotomy and is facilitated by stereotaxic localization. With use of stereotaxic localization, the craniotomy may be performed utilizing image guidance or in a stereotaxic head frame. Yet, another type of resection is a selective amygdalohippocampectomy. This is a variation on an anterior temporal lobe resection. This specific procedure is utilized when the focus is located on the dominant side of the

D.W. McCandless, *Epilepsy: Animal and Human Correlations*,
DOI 10.1007/978-1-4614-0361-6_32,

Table 32.1 Types of functional neurosurgery for epilepsy

Types of functional neurosurgery for epilepsy
Vagal nerve stimulation
Temporal lobectomy
Partial frontal lobectomy
Partial parietal lobectomy
Partial occipital lobectomy
Corpus callosotomy
Hemispherectomy

brain. This is in an attempt to lower the risk of postoperative speech and language dysfunction. The last category of resection is known as a hemispherectomy. This resection involves the removal or neuronal disconnection of an entire cerebral hemisphere, as the epileptic focus is broadly localized to the resected or disconnected hemisphere of the brain. This form of surgical resection is generally reserved for extreme cases of epilepsy that are refractory to medication and less invasive procedure (see Table 32.1).

A preoperative evaluation is absolutely imperative prior to anesthetic induction and surgical treatment. When conducted in a thorough manner, the preoperative evaluation can alert the anesthetist to key considerations when developing the anesthetic plan, which includes considering possible intraoperative and postoperative complications, avoidance of those complications, and key treatments for unavoidable complications. The preoperative evaluation begins with a review of the patient's chart, which includes a review of the history, physical and any laboratory, cardiac, and neurological evaluations on the patient's chart. These evaluations include, but are not limited to, laboratory results, electrocardiograms (EKGs), and electroencephalograms (EEGs). It is also vital to obtain a current height and weight from the patient's chart, as many anesthetic medications and interventions are based on patient's height and weight. It also bears worth mentioning that if the patient has undergone anesthesia for other surgical procedures, the anesthetist maybe able to view the past anesthetic records for a specific anesthetic history.

Next, a thorough patient interview should take place. During the interview, a full neurologic exam is utilized to establish a baseline for the patient's neurologic functioning (Morgan et al. 2006). This baseline exam is especially key for comparison to postoperative evaluation. Throughout the patient's interview, it is imperative that key information is elicited. If the preoperative interview is occurring immediately prior to surgery, it is of utmost importance that the anesthetist asks the patient when the last time he or she had anything to eat or drink. This information is important, as there are safety guidelines in regards to the time that has elapsed since the last oral consumption and induction of anesthesia. This crucial information that is obtained guides the anesthetist's plan of induction for the safety of the patient. This is in an effort to prevent pulmonary aspiration of gastric contents during induction of anesthesia. Information such as the patient's drug and food allergies aids the anesthetist in the avoidance, selection, and utilization of medications during the preoperative, induction, intraoperative, and postoperative periods.

A discussion of the patient's current medications must also take place. The list of current medications alerts the anesthetist to possible side effects that can occur with anesthesia along with guiding the choices of medications that can work synergistically or in an inhibitory fashion to employ the most effective anesthetic possible for the patient. There are specific medications that patients are advised to continue up to the time of surgery. These medications include antiepileptic drugs (AEDs) and antihypertensive medications, with the exception of angiotensin converting enzyme (ACE) inhibitors, and angiotension-2 receptor blockers (ARBs). It has been well documented that those patients who continue ARBs and ACE inhibitors until the morning of surgery have experienced profound hypotension intraoperatively (Coriat et al. 1994). Both ACE inhibitors and ARBs should be held 24 h prior to surgery. Patients are advised to cease taking anticoagulants one to 14 days prior to surgery. This advisory comes with the knowledge that anticoagulation at the time of surgery puts the patient at risk for excessive intraoperative and postoperative bleeding or hemorrhage. Patients who consume prescription diuretics are advised to not take the diuretic the day of surgery, as diuretics promote intravascular dehydration and decrease the blood pressure. This scenario often translates into intraoperative hypovolemia and hypotension. These are two situations that the anesthetist strives to avoid.

As mentioned earlier, laboratory values should be obtained from the patient's chart. If during the patient interview, the anesthetist elicits information regarding medications or medical history that implies that further laboratory workup is necessary, the anesthetist may proceed by ordering further labs. It can be expected that the patient who will be undergoing a neurosurgical procedure will have a complete blood count (CBC) and a comprehensive metabolic panel (CMP). These labs are utilized, respectively, to rule out anemia, thrombocytopenia, infection, metabolic derangements, and hepatic dysfunction. If the patient is currently on an anticoagulant, there should be a coagulation panel completed, which includes a prothrombin time (PT), an international normalized ratio (INR), and partial thromboplastin time (PTT). These laboratory values indicate whether or not enough time has elapsed for the patient's coagulation factors to normalize in an attempt to prevent excessive perioperative bleeding or hemorrhage. If, in fact, these laboratory values are not within the expected range, the surgeon or anesthetist may order further testing, administration of medications, or the administration of blood products to normalize the coagulation factors. There are certain medications that require laboratory evaluation of current drug levels to ensure maximum efficacy of drugs. AEDs are one of several classes of drugs that require laboratory evaluation of drug levels prior to surgery. Patients who are undergoing surgical resection of epileptogenic foci can be expected to be taking AEDs preoperatively. AEDs have been associated with bone marrow suppression and hepatotoxicity. Along with the above-mentioned CBC, the anesthetist should expect that a hepatic profile evaluation be available on the patient's chart (Stoelling and Hilling 2006). Women of childbearing age should have a human chorionic gonadotropin (HCG) test performed to rule out pregnancy prior to surgery.

A past medical history (PMH) is another key piece of information to be obtained from the patient. A PMH includes information from all body systems. As mentioned previously, a neurologic baseline is established in an attempt to compare to the patient's preoperative and postoperative neurologic status. It is also of importance to question the patient regarding a history of cerebral vascular accidents (CVA), transient ischemic attacks (TIA), spinal cord injuries that may or may not have resulted in paraplegia or quadriplegia, and developmental delays. A respiratory history includes considerations such as smoking status, emphysema, asthma, bronchitis, recent cough, cold or fever, and obstructive sleep apnea. The cardiovascular history includes a history of myocardial infarction, angina, chest pain, exercise intolerance, and heart valve abnormalities. For those patients who have a positive cardiovascular history, the anesthetist should expect an EKG, stress test results, and transthoracic (TTE) or transesophageal echocardiogram (TEE) to rule out acute changes, cardiac ischemia, valvular dysfunction, and heart wall abnormalities, respectively. The patient is asked about his or her gastrointestinal history that includes a history of gastroesophageal reflux disease (GERD), nausea, vomiting, diarrhea, or constipation. The history also includes any genitourinary problems; of particular interest to the anesthetist is renal dysfunction or renal failure. Endocrine history includes a history of diabetes, adrenal dysfunction, thyroid derangements, and morbid obesity. The patient is also questioned regarding a history of arthritis (rheumatoid and osteoarthritis), fractures, stiff joint syndrome, and spinal fusions. These musculoskeletal pathologies exhibit a great effect on intraoperative positioning considerations. A psychiatric history that includes anxiety, depression, bipolar disorder, schizophrenia, or other psychiatric issues should also be asked of the patient. This history is of particular importance, as anesthetic medications may have interactions with psychiatric medications. The patient is asked about oncologic history that includes previous exposure to chemotherapy and/or radiation. Finally, a history includes any genetic disorders that the patient may suffer from. The PMH is of such great importance, as it also alerts and guides the anesthetist in the formulation of her/his anesthetic plan for the perioperative care of the patient.

During the patient interview, it is of great importance that the anesthetist and patient discuss the patient's past surgical history (PSH), past anesthesia history (PAH), and family history of anesthetic complications. These three histories tend to blend within each other during the preoperative interview. However, each history is of equal importance and provides the anesthetist with information regarding the patient's past surgical and anesthetic experiences, and provides the anesthetist with insight in regards to the avoidance of repeat complications. PSH includes all surgical procedures, dates, and any complications. PAH should include previous difficult airway management that the patient is aware of, intraoperative awareness, delayed emergence, postoperative nausea and vomiting (PONV), postoperative cognitive dysfunction (POCD), severe postoperative pain, and malignant hyperthermia. Family anesthetic complications are of significant interest, as anesthetic complications such as malignant hyperthermia have genetic associations. If the patient has a family history of malignant hyperthermia, this alerts the anesthetist and operating room staff that there needs to be special preparation for the patient in the operating room as well as particular intraoperative anesthetic considerations that need to be employed to maintain patient safety. A thorough discussion of these histories and complications guides the anesthetist in the anesthetic plan in an attempt to provide the patient with the most effective anesthetic while maintaining patient safety.

Finally, the patient interview should include a social history. The social history includes smoking, which is often mentioned in the respiratory history, alcohol use or abuse, and the use of illicit street drugs. Smoking includes cigarette, cigar, pipe, and illicit street drug smoking. Smoking is of interest to the anesthetist due to the impact of smoking on the patient's lung status and performance as well as airway irritability that so often occurs with smoking. Alcohol consumption, in regards to amount and frequency, is of substantial interest, as this information guides the anesthetist during the titration of anesthesia (Wolfson and Freed 1980). The use of illicit street drugs poses a challenge to the anesthetist in regards to medications used to manage a patient throughout the anesthetic setting. This is the reason the anesthetist must be able to rapidly establish rapport with his or her patient. The patient must feel that she or he can openly divulge information, as this information is used to effectively manage the patient's anesthetic and prevent serious, often fatal, interactions between anesthetic medications and illicit drugs.

Following the patient interview, the anesthetist conducts a head-to-toe physical exam. This exam includes current vital signs including temperature, heart rate, respiratory rate (RR), blood pressure, and pulse oximetry. A thorough airway evaluation provides the anesthetist with priceless information. This evaluation includes cervical spine range of motion, Mallampati score, interincisor gap, thyromental distance, presence of an overbite or underbite, visual inspection of the neck, and inquiring about temporomandibular joint instability. The airway evaluation information alerts the anesthetist whether or not the patient could possibly be a difficult intubation. This valuable information allows the anesthetist the opportunity to obtain pertinent equipment to aid in difficult intubation scenarios. The physical exam includes auscultation of lung fields, as the auscultation of adventitious breath sounds alerts the anesthetist to the possibility of reactive airway disease, pneumonia, and pulmonary edema. While auscultating lung fields, the anesthetist will also auscultate heart tones in an attempt to rule out cardiac defects/pathologies such as heart murmurs. The physical assessment can provide the anesthetist with cues to the patient's health status and invaluable information regarding the anesthetic plan.

After the patient interview and assessment, the anesthetist must use the elicited information to guide their anesthetic plan. This includes the use of preoperative anxiolytic medications. These medications are often used in the preoperative setting in an attempt to allay patient's anxiety immediately prior to the surgical procedure. However, the anesthetist must keep in mind that these medications will often alter the patient's neurological status, which maybe undesirable in this particular neurosurgical population. This warrants a discussion between the anesthetist and the surgeon. The patient interview may also provide the anesthetist with information that may prompt the anesthetist to order further laboratory evaluation, cardiology consultation, or medication administration. Cardiology consultation is elicited at times for those patients who have acute onset of chest pain, shortness of breath, or adventitious heart sounds that have not been evaluated prior to the day of surgery. Medications that the anesthetist may be prompted to order would include medications to help prevent gastroesophageal reflux and pulmonary aspiration. These medications may include metoclopramide, famotidine, and sodium bicitrate. These three medications decrease gastric volume and increase pH of gastric contents, thus decreasing the risk of pulmonary aspiration. Also of consideration is the type of intravenous fluids the anesthetist anticipates using throughout the perioperative period.

Anesthesia is based on the four tenets of amnesia (loss of memory), analgesia (loss of pain), akinesia (loss of movement), and autonomic blunting (preventing the sympathetic surges produced by pain and surgical stimulation). The intraoperative period is often divided into four distinct time periods: preinduction, induction and intubation, maintenance, and emergence from anesthesia. An analogy is often drawn between anesthesia and commercial aviation. The preinduction, induction, maintenance, and emergence periods parallel taxiing from the gate to the runway, take-off, cruising at altitude, and landing, respectively. As with take-off and landing, induction and emergence are considered the most dangerous portions of the anesthetic period.

The primary goals of neuroanesthesia include decreasing cerebral metabolic rate of oxygen consumption ($CMRO_2$), minimizing increases in intracranial pressure (ICP), and ensuring adequate cerebral perfusion pressure (CPP). Cerebral metabolic rate of oxygen consumption is directly proportional to electrical activity of the brain. As metabolic activity increases, $CMRO_2$ increases and cerebral arteries dilate to increase cerebral blood flow (CBF), allowing more glucose and oxygen to be delivered. By decreasing the electrical activity of the brain, the $CMRO_2$ decreases as well. The cranial vault is a fixed space comprised of the brain (80%), blood (12%), and cerebrospinal fluid (8%). If any of these components increase, the others must decrease to prevent drastic elevations in ICP. Anesthetic agents alter $CMRO_2$, CBF, and ICP directly. CPP is calculated as mean arterial pressure (MAP) minus ICP or central venous pressure (CVP), whichever is greater. Anesthetic agents alter CPP indirectly, by their effects on MAP.

Once in the operating room, the preinduction period begins. The preinduction period serves as a "final check" to ensure that all necessary personnel are available and that equipment is functional prior to induction of anesthesia. The patient is placed on the operating room table and the anesthetist connects the patient to several standard monitors: electrocardiogram, pulse oximeter, and blood pressure cuff. Vital signs are obtained and a facemask is placed over the nose and mouth to provide 100% oxygen. This "denitrogenation-preoxygenation" step helps purge the lungs of excess nitrogen found in room air, while filling the lungs with a reservoir of oxygen prior to induction and intubation. A time-out procedure is then performed to ensure the correct patient, surgical site, surgeon, antibiotic administration, blood product availability, special equipment and positioning, as well as verifying any patient allergies to medications and/or latex. The anesthetist, surgeon, and surgical staff must participate in the time-out procedure and voice any concerns during this period.

Once the anesthetist has preoxygenated the patient, a current set of vital signs is obtained and the induction period commences. A smooth, controlled induction and intubation are crucial to neuroanesthesia. A "standard" induction sequence generally includes the administration of three to five intravenous medications to ensure that sufficient amnesia, analgesia, akinesia, and autonomic blunting are obtained prior to surgical stimulation. When preparing medications for each case, the anesthetist takes into consideration the patient's age, weight, coexisting diseases, laboratory data, and surgical requirements. The anesthetist administers the induction sequence in a particular order, based on the purpose of each medication, as well as the onset and peak of the medications.

Midazolam, a benzodiazepine, is generally administered first in the sequence to allay patient anxiety. Midazolam is considered the drug of choice because it has a rapid onset, does not produce pain on injection, and produces anterograde amnesia, ensuring that the patient remembers very few events of the preinduction period (Dershwitz and Rosow 2008). Benzodiazepines decrease $CMRO_2$, effectively decreasing ICP. Intravenous administration reliably calms most patients, facilitating the induction process.

A potent narcotic, such as fentanyl, is then administered to produce analgesia. Fentanyl and its derivatives, remifentanil and sufentanil, are commonly used for neuroanesthesia because they produce profound analgesia, while maintaining hemodynamic stability. Morphine and meperidine are generally avoided because they are less potent and trigger histamine release, which results in vasodilation, increasing CBF and ICP, which may be detrimental. Meperidine also has the potential for lowering the seizure threshold.

Lidocaine, a local anesthetic, is often administered prior to the induction agent to minimize pain caused by vascular irritation resulting from additives in the induction agent solution. In addition, lidocaine has the ability to blunt sympathetic response to tracheal intubation (Abou-Madi et al. 1977). Intravenous lidocaine should be administered 3–5 min prior to the induction dose to have maximal effect.

The induction agent, most commonly propofol or thiopental, is then administered to produce unconsciousness within seconds. Unconsciousness is assessed by a loss of eyelid reflex in response to tactile stimulation. At this time, the eyelids are taped shut to prevent corneal injury during surgery. Thiopental, once the agent of choice, has been shown to consistently reduce $CMRO_2$, CBF, and ICP (Kassell et al. 1979). Propofol has effects similar to those of thiopental (Vandesteene et al. 1988). Propofol has the added benefit of having antiemetic properties. The induction agent ketamine is generally avoided because of its propensity to increase ICP and sympathetic nervous system stimulation. Etomidate, another induction agent, is generally used with caution because of its ability to lower the seizure threshold, as well as an increased risk of nausea and vomiting (Modica et al. 1990). However, etomidate has an excellent cardiovascular profile and may prove beneficial when hemodynamic stability is a concern (Dershwitz and Rosow 2008).

Shortly after unconscious, the patient's respirations cease, and the anesthetist manually ventilates the patient to determine ease of ventilation. If ventilation is easy, a nondepolarizing neuromuscular blocking agent (NMBA), such as rocuronium or vecuronium, is administered to facilitate tracheal intubation and prevent movement during the case. It should be noted that chronic use of AEDs induce metabolism of NMBAs, and the dose should be increased accordingly. That being said, cis-Atracurium doses remain unaffected, possibly due to the method of plasma metabolism (Hofmann elimination), which is pH- and temperature-dependent. Atracurium, another nondepolarizing NMBA, is relatively contraindicated because it produces histamine release, increasing CBF and ICP. Controversy exists regarding the use of succinylcholine, a depolarizing NMBA. Succinylcholine provides the most rapid onset of akinesia and optimal intubating conditions, but is relatively contraindicated for neuroanesthesia due to the transient rise in ICP (Minton et al. 1986). More recent data suggest that elevations in ICP may be minimized by pretreatment with a defasciculating dose of a nondepolarizing NMBA prior to administration of succinylcholine (Kovarik et al. 1994). Following administration of the NMBA, the patient is manually ventilated until the NMBA has taken effect (generally 1–2 min). If not performed preoperatively, an assistant may place an arterial catheter at this time for continuous monitoring of arterial blood pressure. A urinary catheter is also placed at this time.

Endotracheal intubation requires placement of a breathing tube into the trachea to provide oxygenation and ventilation during surgery. When akinesis is deemed adequate, a laryngoscope is inserted into the mouth to reposition the tongue and provide direct exposure of the vocal cords. The endotracheal tube is passed through the vocal cords under direct visualization, the stylet is removed, and the cuff is inflated to produce an adequate seal. The endotracheal tube is connected to the ventilator breathing circuit, which contains a sampling port for measurement of oxygen, carbon dioxide, and anesthetic gases. The continuous presence of end-tidal (exhaled) carbon dioxide ($ETCO_2$) is considered the gold standard for correct placement of the endotracheal tube and identification of inadvertent esophageal intubation (Sum Ping et al. 1991). Bilateral breath sounds are assessed to ensure that the tube has not migrated into a primary bronchus, then the endotracheal tube is secured to prevent movement during surgery. At this time, an orogastric tube may be placed to decompress the stomach, and an esophageal temperature probe and stethoscope may also be placed for continuous monitoring of core body temperature and breath sounds. A peripheral nerve stimulator is used to determine the level of paralysis throughout the remainder of the surgery. Postinduction medications may include dexamethasone 10 mg IV, an intravenous steroid used to decrease inflammation, and mannitol 0.5–1 g/kg, an intravenous osmotic diuretic used to decrease cerebral edema.

A balanced combination of inhalation and intravenous anesthetics are used to ensure adequate depth of anesthesia throughout the maintenance period. As mentioned previously, anesthetic agents are chosen to suppress $CMRO_2$, minimize elevations in ICP, and facilitate the surgical process. The most commonly used inhalation anesthetic is isoflurane, because of its ability to rapidly suppress electrical activity of the brain while maintaining hemodynamic stability. The minimum alveolar concentration (MAC) of an inhalation anesthetic is the concentration required to prevent movement in response to surgical stimulation in 50% of patients, and is equivalent to the effective dose (ED_{50}) of other medications. The MAC of isoflurane is approximately 1.1 volume-percent. Sevoflurane, another inhalation anesthetic, had been shown to be safe for use in neuroanesthesia (Baker 1997). However, recent data suggest that sevoflurane at surgical levels may produce epileptogenic activity in healthy subjects (Jaaskelainen et al. 2003) and should be used with caution. The use of desflurane, the least potent of the volatile anesthetics, may be advantageous due to its rapid onset and rapid awakening, allowing postoperative neurologic assessment prior to leaving the operating room. However, desflurane has been shown to cause airway irritability, which may lead to coughing, bronchospasm, and laryngospasm, all of which cause unwanted increases in ICP. Nitrous oxide (N_2O) is relatively contraindicated in neurosurgery, due to its rapid diffusion into air-filled cavities, increasing the risk of pneumocephalus, pneumothorax, and venous air embolism (VAE). Nitrous oxide use is absolutely contraindicated in patients who have undergone craniotomy within the previous 6 weeks (Reasoner et al. 1994).

Propofol along with fentanyl or remifentanil, are the agents of choice for maintenance of amnesia and analgesia. A low-to-intermediate propofol infusion serves as an excellent adjunct to the volatile anesthetic, reducing the concentration of volatile required to maintain amnesia and anesthesia. Fentanyl may be administered as intermittent intravenous boluses, based on surgical stimulation and hemodynamic parameters. Remifentanil infusions are commonly employed to maintain a steady plasma concentration, rather than the fluctuations associated with IV bolus dosing (with fentanyl). Remifentanil has the advantage of rapid plasma metabolism allowing prompt awakening postoperatively, but with the subsequent disadvantage of providing no analgesia postoperatively. Therefore, postoperative pain must be anticipated and a long-acting narcotic must be titrated in during the case to provide optimal postoperative pain control. Neuromuscular blocking agents may be redosed intermittently to maintain surgical relaxation. For patients undergoing evoked potential monitoring, the effects of NMBAs are allowed to subside and inhalation agents are kept to half the standard concentration delivered (0.5 MAC).

During any surgical procedure, the anesthetist constantly scans the operating room and surgical field to guide anesthetic maintenance. Aspects of the maintenance period include infusion of intravenous fluids to maintain intravascular volume, constant monitoring of vital signs, maintaining hemodynamic stability and adequate perfusion, ensuring proper oxygenation and ventilation, providing surgical relaxation and monitoring laboratory data.

The preferred intravenous fluid for administration during neurosurgical cases is 0.9% normal saline. 0.9% normal saline is an isotonic crystalloid solution that minimizes loss of fluid into the cerebral tissue. Intravenous fluids routinely are kept to a minimum in an effort to prevent cerebral and pulmonary edema. Dextrose-containing fluids (e.g., 5% dextrose or D_5W) shift into cerebral cells and exacerbate cerebral edema, increasing ICP. Lactated ringer's solution is relatively contraindicated because of its lactate content, which is converted to lactic acid and bicarbonate. An increase in serum lactate levels predisposes the patient to acidosis and subsequent vasodilation, resulting in increased CBF and ICP. Blood loss should be replaced with colloid solutions such as 5% albumin or hetastarch. Colloid solutions increase plasma oncotic pressure, drawing fluid into the intravascular space, and have been shown to remain within the intravascular space considerably longer than crystalloid solutions. Although blood loss during epilepsy surgery is generally minimal, considerable blood loss may be replaced by packed red blood cells.

Temperature regulation is of key concern, as intracranial procedures may last for several hours and temperature may fluctuate throughout the surgical course. The operating room is a cooler than normal environment (generally 65°F), and loss of body heat via radiation, evaporation, conduction, and convection occurs rapidly. A forced air warming blanket is placed over the patient's chest, abdomen, and extremities to minimize heat loss during the operation. Ideally, the patient should be kept normothermic. Increases in temperature may produce substantial increases in $CMRO_2$, CBF, and ICP.

Heart rate may be controlled via intermittent intravenous boluses of medications such as glycopyrrolate or esmolol to raise or lower the heart rate, respectively. In the presence of normal or near-normal ICP, MAP is the primary determinant of CPP. Thus, ensuring adequate MAP is paramount to ensuring perfusion to the brain. An arterial catheter, placed preoperatively or postinduction, allows continuous beat-to-beat monitoring of arterial blood pressure for rapid control. The transducer is aligned with the tragus (external auditory canal) to estimate CPP, and blood pressure may be pharmacologically increased or decreased using infusions of phenylephrine and nicardipine, respectively. Assessment of renal perfusion is guided by urinary output. Urine production should equal 0.5 ml/kg/h for pediatric patients or a minimum of 30 ml/h for adult patients.

Maintenance of oxygenation and ventilation is an ongoing process during surgery. The presence of an arterial catheter allows blood sampling for blood gas analysis and other laboratory tests. The arterial blood gas (ABG) provides the anesthetist with information regarding the blood pH, partial pressures of oxygen (PaO_2) and carbon dioxide ($PaCO_2$), serum bicarbonate (HCO_3^-) levels, oxyhemoglobin saturation (SaO_2), electrolytes, and fluid-volume status. The presence of an endotracheal tube and manipulation of ventilator parameters allow rapid control of oxygenation and ventilation. The fraction of inspired oxygen (FiO_2) may be increased to optimize oxygen saturation, whereas tidal volume (V_T), RR, and positive-end expiratory pressure (PEEP) may be utilized to improve alveolar gas exchange (ventilation). For most neurosurgical procedures, a slightly decreased exhaled carbon dioxide level is preferred, with a goal of 27–30 mmHg. Hypoxia and hypercarbia are avoided, as these potentiate increased CBF and ICP.

In order to perform surgery within the cranial vault, paralysis is commonly used to create an immobile surgical field. A peripheral nerve stimulator is used to assess the level of neuromuscular blockade as the procedure progresses. Two electrodes are placed over the ulnar nerve, and neuromuscular blockade is assessed by the generation of four electrical impulses in rapid succession. This process is known as a train-of-four. Stimulation of the ulnar nerve results in opposition of the thumb. The absence of thumb movement indicates that >90% of nicotinic neuromuscular receptors are occupied by the NMBA. Thumb opposition gradually returns as the NMBA diffuses away from the nicotinic receptors. The anesthetist assess the train-of-four every 5–15 min and may redose the NMBA, if necessary.

As mentioned previously, the presence of an arterial catheter allows frequent sampling of blood for laboratory analysis. Due to the diuretic effect of mannitol, hemoconcentration occurs, and electrolyte values may be falsely elevated.

Decreased electrolyte values may be corrected by intravenous administration of the particular electrolytes (e.g., potassium or calcium). CBC and coagulation panel may be used to determine if a coagulopathy exists. The cause of the coagulopathy must be identified and treated accordingly. Coagulopathies may be corrected by transfusion of packed red blood cells, platelets, fresh frozen plasma and/or cryoprecipitate.

Communication with the surgeon is imperative during the surgical procedure. The surgeon may ask for a deeper level of anesthesia prior to certain surgical procedures, such as placing the patient's head in a head frame using metal pins that pierce the scalp. Likewise, the surgeon may verbalize concerns, such as bleeding or cerebral edema, to the anesthesia provider. As the surgery progresses, the surgeon notifies the anesthetist of the approximate time of completion, allowing the anesthetist to titrate inhalation and intravenous anesthetics accordingly.

The risk of complications during neurosurgery necessitates close attention to detail. Complications may include bleeding, VAE, paradoxical embolism, infection, and postoperative neurologic deficits. Constant monitoring and recognition of complications, as well as prompt treatment, are crucial to reduce morbidity and mortality.

The scalp is highly vascular and, during most neurosurgical procedures, blood loss occurs during incision of the scalp. Damaged to any portion of the cerebral venous plexus or cerebral arteries may require rapid infusion of intravenous fluid and administration of blood products to maintain hemodynamic stability.

The risk of VAE is increased during any procedure in which the surgical site is 5 cm (2″) above the level of the heart. If a large vein is damaged, gravity allows entrainment of room air into the venous system, which carries the air to the heart. A small air embolus may produce minimal effects, however, large air emboli may produce respiratory distress and rapid cardiovascular collapse. A precordial Doppler may be used to detect the presence of entrained air within the heart. A characteristic millwheel murmur indicates turbulent flow within the heart, caused by the presence of air within the heart chambers. The presence of end-tidal nitrogen (ETN_2) is also indicative of a VAE. Rapid desaturation and hypoxemia may result, increasing the risk of end-organ damage. Treatment of VAE includes notifying the surgeon and having the surgeon apply bone wax to the free skull edges, flooding the site with saline, placing the patient in left lateral Trendelenburg (head-down) position to prevent movement of the air embolus through the pulmonary or systemic vasculature. If a multi-lumen central venous line is in place, an attempt may be made to withdraw the entrained air through the proximal port of the catheter. Historically, VAE was a major risk during neurosurgical procedures due to utilization of the sitting position. In contemporary practice, many surgeons tend to perform these procedures in a modified lateral position, in an attempt to decrease the risk of VAE.

Paradoxical embolism refers to an embolism (air, fat or otherwise) that enters the heart and bypasses the pulmonary circulation via a patent foramen ovale or other cardiac defect, allowing the embolus to enter the systemic circulation. The embolus may become lodged at any point in the systemic arterial circulation, prevent delivery of oxygen, blood and essential nutrients to points distal to that point. Paradoxical emboli may result in TIAs, CVAs (strokes), or other morbidities.

Infection is a concern for any patient undergoing a surgical procedure. Once the skin barrier is opened, bacteria have an open port of entry into the body. Infections within the central nervous system (brain and spinal cord) are often difficult to treat, and may prove fatal. Prophylactic antibiotics are administered to the majority of surgical patients to prevent iatrogenic (hospital-acquired) infections. Preoperative antibiotics are generally administered within one hour of incision to allow peak plasma levels of the antibiotics. As previously mentioned, the administration of antibiotics is confirmed during the time-out procedure.

Postoperative neurologic deficits are a considerable risk due to the invasive nature of neurosurgery. The preoperative evaluation establishes a baseline for postoperative comparison. The temporal lobe houses centers for hearing, speech, memory and certain emotional responses. Damage to the temporal lobe may result in hearing and speech deficits, altered emotional responses, and impaired short- or long-term memory. Following the conclusion of surgery and emergence from anesthesia, a neurologic exam is performed by the surgeon to establish the presence or absence of neurologic deficits.

Toward the end of the surgery, the anesthetist often administers a long-acting analgesic such as hydromorphone. Inadequate analgesia increases blood pressure and ICP. An antiemetic is administered to reduce the risk of PONV. Nausea and vomiting increase intrathoracic pressure, which is transmitted to the head, causing an unwanted increase in ICP. If paralysis is maintained throughout the procedure, the anesthetist evaluates the train-of-four for the presence of at least one twitch. A single twitch must be present prior to the administration of reversal agents. Glycopyrrolate, an antimuscarinic, is administered intravenously to counteract the unwanted effects of mass-activation of muscarinic receptors, namely, bradycardia and increased salivation. Neostigmine, a cholinesterase inhibitor, is then administered intravenously to allow accumulation of acetylcholine within the neuromuscular junction. The acetylcholine competitively displaces the NMBA from the nicotinic receptor, allowing the gradual return of muscular function.

Prior to extubation, the anesthetist suctions the patients oropharynx and ensures that the patient is breathing spontaneously with a regular rate and adequate tidal volumes with each breath. The patient should open the eyes spontaneously and should be able to follow simple commands, such as squeezing the fingers or raising the head. An active head lift for 5 s is considered the gold standard for extubation.

Following extubation, the surgeon performs a brief neurologic exam to ensure that there are no postsurgical complications, such as speech or sensorimotor deficits. The patient is transferred to the intensive care unit, where he or she will be monitored closely for 24–48 h. Postoperative concerns include adequate analgesia, monitoring for residual seizures, avoidance of events that increase ICP (e.g., hypertension, nausea, and vomiting), and monitoring for signs and symptoms of an intracranial bleed.

In conclusion, neuroanesthesia is a complex and fascinating subspecialty requiring keen attention to detail and ability to think critically and respond to intraoperative events rapidly and accurately. The surgical team and anesthetist must work in concert to provide optimal outcomes for patients undergoing surgery for epilepsy. Fortunately, many patients have excellent postoperative results and are able to enjoy a higher quality of life.

Chapter 33
Vagus Nerve Stimulation

In a recent paper (Nei et al. 2006), vagus nerve stimulation was assessed in terms of efficacy and safety as compared to corpus callosotomy. Results showed that although corpus callosotomy was more effective for generalized tonic clonic seizures and for atonic seizures, the risks and complications excelled those of vagus nerve stimulation by a significant margin. The much lesser invasiveness of vagus nerve stimulation would suggest that in cases in which invasive surgery is not an option, vagus nerve stimulation might be indicated. Complications in the above paper listed infections and a defective battery for vagus nerve stimulation. Complications of corpus callosotomy included death, ataxia, hemiparesis, infection, etc.

Probably in excess of 60,000 vagus nerve stimulators have been inserted into the upper left anterior thorax of patients with intractable epilepsy. Leads are attached to the vagus nerve at the bifurcation of the left common carotid artery. The devices then apply bursts of electrical current to the left vagus nerve.

The vagus nerve sends afferent fibers from the G.I. tract, heart, aorta, and lungs to medullary nuclei. Motor and visceral efferents and somatic and visceral sensation fibers are also vagal. The vagus nerve also projects to the frontal lobes. It also projects to the limbic system, the amygdala, and to the locus coeruleus (Schachter and Saper 1998). In a primate study in which topical brain convulsants were used to initiate seizures, vagus nerve stimulation attenuated ictal and interictal EEG alterations (Lockard et al. 1990).

The study by Lockard showed the safety of the vagus nerve stimulation procedure. A later study (Koo 2001) showed a "normalization" of the frequency of spikes, and a reduction in slowing in responders. Kuba et al. (2002) have shown that vagus nerve stimulation decreases interictal discharges, and this early result can be a predictor of further vagal nerve stimulation success. The animal seizure model of maximal electroshock probably has tonic seizure activity originating in the brainstem, and vagal nerve stimulation is effective in this model, as well as the limbic onset convulsant pentylenetetrazole (Woodbury and Woodbury 1991).

Of interest is that afferent vagal input acts to modulate neural systems in ways that have implications for other neurological disorders. Vagal nerve stimulation,

D.W. McCandless, *Epilepsy: Animal and Human Correlations*,
DOI 10.1007/978-1-4614-0361-6_33, © Springer Science+Business Media, LLC 2012

for example, can attenuate nociceptive pain in experimental animals (Randich and Gebhart 1992). The vagus nerve also may play a role in modulating memory.

When human testing began in humans using vagal nerve stimulation, the safety and success in animal models had already been established with MES and pentylenetetrazole models. However, the effect on other animal seizure models was less clear. Careful and extensive testing using blinded, controlled, randomized trials were performed in epileptic patients. When establishing stimulation characteristics, patient safety and comfort were key considerations. High frequencies were not used.

Early patients' results showed a reduction in seizure frequency and a reduction in seizure severity (Penry and Dean 1990). In later studies, the fraction of patients achieving at least a 50% reduction in seizures was one third. Other studies are in general agreement (DeGiorgio et al. 2000). Studies of vagal nerve stimulation suggested the procedure might also be effective in partial complex seizures and in the Lennox–Gastaut syndrome (Hosain et al. 2000). The treatment is effective in children and also in patients who have less-intractable epilepsy.

There are complications associated with vagus nerve stimulation, including laryngeal tingling and hoarseness, probably due to stimulation of the recurrent laryngeal nerve. Other adverse effects include drooling, the Horner syndrome, psychosis, etc.

Mechanisms of action are not clear, but it has been shown that vagus nerve stimulation has a stimulating effect on GABA levels in the brain. In an experiment looking at 2-deoxy glucose uptake in animals with vagal nerve stimulation, no brain area showed any alteration in labeled glucose uptake. Other studies using fMRI and PET have been equivocal. Results from vagal nerve stimulation may act even after the stimulation has ceased. Even after cessation of vagal nerve stimulation, kindling current required to institute a seizure remained higher in the vagal-stimulated animals. Vagal nerve stimulation may also alter neurogenesis in areas where brain-derived neurotrophic factor and nerve growth factor are able to reverse pathology (Follesa et al. 2007).

While AED treatment carries a high success rate, many patients achieve less than optimal results either because of poor efficacy of AEDs or due to adverse effects of AEDs. For those patients, a variety of possibilities exist, including the ketogenic diet, electrical stimulation of central cerebral areas such as the thalamus and cerebellum, the Atkins diet, dietary supplements like creatine, resective surgery of large areas such as the temporal lobe and vagus nerve stimulation. All of these are subjects for various chapters in this book.

In terms of vagal nerve stimulation, the procedure has been detailed before (Terry et al. 1990). The essential components of the system consist of a pulse generator, a bipolar VNS lead, a programming wand with software, a tunneling tool, and handheld magnets (see Fig. 33.1). The generator is implanted, as stated above, in the left anterior chest wall, and leads are attached to the vagus nerve a manor, which does cause damage to the nerve.

The surgery is performed as day surgery in some settings, the procedure lasting about 90 min. The process is usually performed under a general anesthetic to eliminate any possibility for a seizure to occur during surgery. The procedure for vagus

Fig. 33.1 Schematic representation of changes with vagal nerve stimulation. Tracings from an anesthetized cat receiving vagal nerve stimulation. *Left arrow* = on, *right arrow* = off. Adapted from Chase, M., et al. Brain Res. (1967) 5:236

nerve exposure is similar to that used for carotid endarterectomy (Reid 1990). Two incisions are made, one over the anterior border of the sternocleidomastoid muscle for the helical lead coils, and the second is about 8 cm below the left clavical for placement of the generator. The tunneling tool moves the electrode ends from the vagus nerve inferiorly to the generator. The output current is gradually increased with patient comfort in mind.

Although similar, the left and right vagus nerves have a somewhat different functioning. The right vagus carries many afferent and efferent fibers to the atria, whereas the left vagus serves the ventricles. The vagal innervation of the ventricle is less dense than that of the atria, thereby decreasing the chance for cardiac problems. The vagus is a mixed nerve, having a parasympathetic function, and carrying sensory afferents. The nodose ganglia contain cell bodies of afferent axons. Associated vagal pathways include those of the reticular formation and fibers to the thalamus.

Vagal stimulation has been shown to produce an EEG slowing (Magnes et al. 1961). High-frequency stimulation (over 30 Hz) results in desynchronization, whereas low-frequency stimulation (below 18 Hz) acts to synchronize the EEG. The anticonvulsant effect of vagal nerve stimulation is based on "C" fibers (large myelinated fibers) stimulation. Thus, experiments showed that "C" fiber stimulation in rats attenuated the seizure activity produced by pentylenetetrazole. There was a direct relation between the number of fibers stimulated and the quantitated efficacy (see Fig. 33.1).

In terms of human trials, hundreds of patients were ultimately involved in initial double blind controlled studies in multiple centers. The primary efficacy evaluation was based on percent improvement in seizure frequency as compared to baseline. There were high-stimulation patients and low-stimulation groups. Results in low-stimulation patients showed between a 6 and 15% improvement in the low group, and a 24.5–28% significant decrease in the high-stimulation group. Efficacy in a small group of children showed better results in terms of frequency decrease. From the safety standpoint, some patients experienced hoarseness, and there were reports of a couple of patients experiencing Cheyne-Stokes respiration.

These very first studies indicated that vagal nerve stimulation was safe. There were no deaths, and relatively minor sequelae. The efficacy of the procedure was encouraging given the highly refractive nature of the seizures of the selected patients. The vagal stimulation method can be used with AEDs, and is a relatively noninvasive well-tolerated method of seizure control (Ben-Menachem 1998).

An interesting aside is that vagus nerve stimulation has been shown to suppress experimentally induced pain (Kirchner et al. 2000). The idea was to test whether or not an antinociception effect was associated with vagus nerve stimulation. Ten patients with AED intractable seizures were tested before and after implantation of a vagus nerve stimulator. Stimuli included heat, impact, and tonic pressure.

Results showed that pain associated with impact stimuli and heat were unaffected by vagus nerve stimuli. However, pain from consecutive stimuli and pain from tonic pressure was reduced significantly (0.001) as compared to control patients. The authors note that the results showed pain was reduced by procedures in which pain magnitude was amplified by CNS processing. The results showed a potential role for vagus nerve stimulation for pain, and possible risk for epilepsy patients not recognizing some pain types.

Another study examined possible mechanisms of action of vagus nerve stimulation in patients with refractory complex partial seizures (van Laere et al. 2000). In this study, 12 patients, mean age 33 years old with a duration of complex partial seizures of 20 years, were treated with vagal nerve stimulation. Patients were evaluated with regards to regional cerebral blood flow using 99 m TC-ethyl cysteinate dimer. Each patient had two SPECT scans.

Results showed that the 12 patients were all unsuited for resective surgery, because of multiple foci. Four patients had a greater than 50% improvement from vagal stimulation. Three more patients showed moderate (30–50%) reduction of seizures. Regional CBF showed a decrease in the left thalamus and the left parietal cortex. No areas showed increased blood flow.

The authors note that the results of vagal nerve stimulation are similar to other results showing an efficacy of 30–40% (Lesser 1999). This is added support for the concept that vagal nerve stimulation is an effective treatment especially in light of the fact that the group was refractory to AEDs, and poor candidates for resective surgery.

The mechanism of action of vagal nerve stimulation is unclear. Two mechanisms are suggested, the first being the hypothesis that vagal nerve stimulation raises the threshold for seizure propagation in the nucleus of the salitory tract and other structures. The second hypothesis states that the vagal nerve stimulation serves to increase inhibitory neurotransmitters and/or lower excitatory neurotransmitters (Ben-Menachem 1996).

Results in the present study have shown that 2–3 min after left vagal stimulation, the left thalamus, hippocampus, and brainstem demonstrated reduced regional CBF. The authors state that this suggests decreased perfusion could be involved in the mechanism of action of vagal nerve stimulation. Earlier studies have shown that vagal nerve stimulation affects neuronyl networks outside of the brainstem.

The concept is that vagal nerve stimulation alters several sites causing an increase in transsynaptic neurotransmission (Henry et al. 1998). The authors state that their results tend to support the direct connection theory, but do not rule out changes in neurotransmitters in brain due to vagal nerve stimulation. Further studies seem warranted.

Another paper describes studies looking at the ability to obtain fMRI images from vagal-stimulated patients (Bohning et al. 2001). The goal was to develop a method to obtain fMRI images, then to determine where intracerebral activity was occurring immediately after stimulation. PET images do not have sufficient resolution for an accurate description.

In the study, nine patients who had a vagal nerve stimulator implanted for treatment of depression were utilized in the study. The initial implantation was done with this study in mind and was MRI compatible. A 1.5 T clinical MR scanner was used for multislice gradient echo, and echoplanar for obtaining MRIs. This allowed viewing the entire brain except for the inferior cerebellum. Vagus nerve stimulation-fMRI methodology resulted in activation bilaterally in the orbitofrontal cortex, parietal cortex, hypothalamus, left temporal cortex, and left amygdala.

The authors comment that this pilot study shows that synchronized vagal stimulation and fMRI imaging is possible. Areas of brain affected can be determined, and there appears to be an overall diffuse increase in brain activity. Other imaging attempts are in agreement that the thalamus is a structure which is involved (van Laere et al. 2000; Ring et al. 2000). Differences in study results probably exist because of methodology differences, and more work is warranted. This study shows that fMRI can be used to elucidate the mechanisms of vagus nerve stimulation.

A paper describing a six-center retrospective study on the efficacy and safety of vagus nerve stimulation in patients with the Lennox–Gastaut syndrome has been published (Frost et al. 2001). This study involved 50 patients, median age 13 years old, and a median age of onset of 1.4 years. Twenty one of the patients were under the age of 12. Sixty percent had IQs of less than 70. In terms of AEDs, a median of nine had been tried, and one third of the patients had been on the ketogenic diet.

Results showed that after 1 month of vagus nerve stimulation, seizures had decreased by over 75% in 7 patients, and by 50% in 20 patients. None were seizure free. After 3 months of vagus nerve stimulation, seizure reductions of 75% or more involved one third of patients, and 56% of patients had a 50% or more reduction in seizures. Seizures increased by over 50% in one patient. Results were unchanged at 6 months after implantation.

Drop attacks before vagus nerve stimulation, occurred in 2/3 of the patients, but had decreased by 47, 55, and 88% after 1, 3, and 6 months, respectively, after treatment. All decreases in drop attacks were statistically significant. Age of the patients was not correlated with outcomes, and those over 12 responded similarly to those under12 years of age. In terms of quality of life, alertness improved in over half of the patients 1 month after implantation. At 6 months, verbal communication and school work improved in 25% of patients.

Adverse effects were nearly nonexistent. Two patients had brief infections at the site of implantation. Forty five percent of patients had voice alterations or hoarseness. Other changes including neck tingling, shortness of breath, hiccups, insomnia, etc. resolved over time. There were no serious adverse effects as a result of vagus nerve stimulation.

The authors comment that vagus nerve stimulation showed improvements in all seizure types seen in Lennox–Gastaut syndrome patients, including atonic seizure attacks. Absence seizures responded favorably to stimulation. The authors note that corpus callosotomy may have a better seizure reduction rate (Sorensen et al. 1997).

However, the sequelae are notably worse, so the recommendation is to first try vagus nerve stimulation.

In terms of potential adverse affects of vagus nerve stimulation, a clear effect of stimulation on sleep apnea hypopnea index has been reported (Marzec et al. 2003). This was a prospective study of 16 patients aged 21–58. Patients with previous sleep disorders were excluded. Results showed using polysomnograms before and at 3 months after vagus nerve stimulation, that 5/16 had an elevated apnea–hypopnea index. These data show an adverse effect of vagus nerve stimulation on sleep apnea. While the effect may be mild, two patients had moderate effects.

The issue of children under 12 and over 12 years old was once again examined (Murphy et al. 2003). This was a retrospective study of 100 children who underwent vagus nerve stimulation for refractory epilepsy. The average age was 10.5, average number of years with seizures was 8.4, and median seizure frequency was 120 per month. Results showed 45% of patients had a 50% reduction in seizures at follow-up. Differences between under 12 and over 12 in terms of results were minimal. The reduction rate compares to that previously reported (Morris and Mueller 1999).

Adverse effects were the usual ones reported in other studies, plus three patients developed abscesses associated with the generator, and they were removed. No deaths were reported in the vagus nerve-stimulated group, but in those who terminated stimulation (24 patients) one was found dead in bed, and another died from head trauma during a seizure. Five patients had increases in seizure frequency. Patients with more than 7 years seizure duration were compared with those of less than 7 years of seizures and no differences were found.

In one half of patients who had the ability to use the hand magnet for on demand current, there was a reduction in duration and intensity of the seizure. Nearly 2/3 of patients felt an improvement in the feeling of well being. Except for the Lennox–Gastaut syndrome, it is difficult to predict which patient might benefit from vagus nerve stimulation.

The authors note that since AED success is unlikely after three or more failures of drugs (Kwan and Brodie 2000), vagus nerve stimulation could be considered. The authors also state that vagal nerve stimulation should be available for children under 12 because it is just as efficacious as in children and has no additional adverse effects.

Yet another study reporting the efficacy of vagus nerve stimulation has been published (Uthman et al. 2004). This study shows results from a 12-year period. The average age of onset was 9, and first vagus nerve implantation was at age 32. Frequency of seizures was 35 total seizures per month. Seizure type included simple partial seizures, complex partial seizures, and generalized tonic clonic seizures.

Adverse effects of vagus nerve stimulation were hoarseness, parasthesia, dyspnea, throat pain, etc. Many of these types of adverse effects are related to the recurrent laryngeal nerve. Two deaths occurred, but not related to the implantation. Sixty percent of patients enjoyed a greater than 50% seizure reduction. Three patients remained seizure free after one year with the stimulator. There was no difference between those with a greater than 50% success and those without such success as regards age of implant, age of onset, duration of epilepsy, or etiology of the seizures.

The authors comment that their data show continued seizure reduction after long term adjunctive vagus nerve stimulation. Patients with unilateral epileptiform EEG have better outcomes than those with bilateral epileptiform EEGs. The authors state that patients did not develop tolerance over time. The observation is made that, based on this 12-year study, vagus nerve stimulation is a safe and effective adjunctive therapy for partial epilepsy treatment.

Another paper describes studies aimed at identifying factors that might predict seizure outcome from vagus nerve stimulation (Jansky et al. 2005). Vagus nerve stimulation has been shown to be efficacious, with at least 1/3 of patients showing a better than 50% reduction in seizure frequency (see above). The actual success in terms of seizure freedom shows a large variation ranging from 0 to 24% (Labar 2004). The results looking at duration and outcomes are equivocal.

Results from this study showed only two variables correlated with outcome. One was the presence or absence of bilateral interictal epileptiform discharges, and the other was the presence of malformation of cortical development. More sophisticated statistical analysis (logistic regression analysis) did not show a correlation of cortical development malformation with ability to predict vagus nerve stimulation outcome. Overall, 13% of patients in this study became seizure free. This is a higher percentage than most other studies.

One reason for the 13% cited by the authors was that there were no restrictions on AEDs, or changing them during the evaluation period. This places a limitation on the study, and the authors say, and is an asset. The effectiveness of vagus nerve stimulation has been shown to improve with time (Boon et al. 2001), so if patients who needed AED changes were eliminated, the study would have been clinically compromised. On the other hand, if more than one variable is manipulated, then what caused the benefit? More results should be gleaned from controlled studies with more patient numbers.

Epilepsy may occur in 50 to 90% of Rett syndrome patients. Rett syndrome is a genetic disorder characterized by cognitive impairment, seizures, growth impairment, ataxias, etc. The present paper describes vagus nerve stimulation in seven female patients with Rett syndrome and refractory epilepsy (Wilfong and Schultz 2006). This study examines vagus nerve stimulation results in five patients with classic Rett syndrome and two with atypical Rett syndrome, and all were refractory to AEDs early in the course of their disease.

Results showed that the patients' ages ranged from 1 to 14 years, mean 9 years old, and had seizures for 6 years, failed at least two AED treatments, and averaged 150 seizures per month. After 12 months of follow-up after vagus nerve stimulation, six patients had a seizure reduction of 50% or greater. No patients had any significant complications. Increased alertness was reported by all seven patients. There were no difficulties in communication skills.

The authors suggest that vagus nerve stimulation was a successful therapy for Rett syndrome patients with refractory epilepsy. The key was no alterations in mood or communication.

In an animal study (Smith et al. 2006), rats were subjected to lateral fluid percussion brain injury, then the efficacy of vagus nerve stimulation was evaluated in terms of recovery of function. Previous work showed that recovery was accelerated when vagus nerve stimulation was initiated 2 h after brain injury. In the present study, stimulation was delayed 24 h posttrauma, and recovery evaluated.

Recovery evaluation of the rats using the beam walk, skilled forelimb reaching, water maze, etc., provided data that confirmed that at 2 weeks, they were better in recovery than trauma rats without vagal nerve stimulation. In addition, they were indistinguishable from sham-operated rats. The recovery data at 2 weeks post-trauma, reinforce data collected from rats in which stimulation was initiated 2 h posttrauma. Mechanisms, while unknown, may involve changes in neurotransmitters similar to changes in seizure patients similarly treated.

Another interesting paper describes a treatment regime involving a combination of the ketogenic diet and vagus nerve stimulation (Kossoff et al. 2007). In this retrospective study, a total of 30 children were treated over a 6 year period. The median age was 10 years. Analytic differences were not apparent between the six centers involved in the study. The goal was to see if "polytherapy" consisting of nonpharmacological treatments might work together to improve the efficacy of seizure treatment in the group of AED-intractable children.

Results showed that of the 30 epileptic patients, half had no known etiology, whereas the other half had various etiologies, including encephalitis, tuberous sclerosis, cortical dysplasia, stroke, etc. Some patients had an initial polytherapy started with the ketogenic diet, others had an initial start with vagus nerve stimulation. Results showed that more than 2/3 of patients reported additional benefits when the second therapy was started. Some patients in the diet group were on a modified Atkins diet. They were grouped together because the differences were insignificant.

The authors comment that the delay in trying these two epilepsy treatments together may relate to the "work intensive" nature of each treatment, and that not each is available at all medical centers. Other problems are that the long-term effects of the diets are not known; therefore, they are often discontinued after 2–3 years. Low vagus nerve stimulation cycles were effective in seizure control (DeGiorgio et al. 2005), so this would serve to increase battery lifespan. Further studies looking at various vagus nerve stimulation parameters with diets would be valuable.

The above question regarding stimulation parameters has been examined in a retrospective study addressing stimulation settings following generator replacement (Labar et al. 2008). In this study of 28 patients, seven patients had end of battery life (EOBL), or anticipated EOBL (21 patients). These patients who had previously responded to vagus nerve stimulation participated in the study.

Results showed the mean duration of generator (stimulator) was 41 months. Patients whose batteries had expired did experience an increase in seizures and a disappearance of side effects. There were 11 same battery changes, and 17 patients received different batteries. In 21 of the 28 patients, the reprogramming exercise was not tolerated. After 1 year, the baseline currents were still not tolerable in 17 patients.

The authors note that due to the potential symptoms of battery replacement after failure, projected battery replacement might be preferable to waiting for battery failure.

They also recommend simultaneous replacement of the generator. Various reasons for the change in tolerability could result from different programmers, different battery/generators used for replacement, chronic vagus nerve stimulation modifying the sensitivity of the vagus nerve, etc.

The key concept, states the author is that after battery/generator replacement, the patients struggle to tolerate a return to previous current intensity levels. It is important to note that this does not lead to worsening of the patient's seizures.

One problem with vagus nerve stimulation occurs if there is reason that the left side cannot be used. This can occur if serious infection requires removal of the left vagus nerve stimulating system, thereby rendering the left side unusable. A right sided approach has been tried with mixed results (Spuck et al. 2008). In this case, a 16-year-old boy with intractable seizures underwent a vagus nerve stimulation surgery implant of the left side. A deep infection required removal of the vagal nerve stimulator several weeks later.

Results showed that the resultant left nerve damage rendered it no longer a candidate for reimplantation of the vagal nerve stimulator, so the right side was utilized. The right side stimulator did provide adequate seizure suppression; however, cardiac symptoms were induced by the right vagus stimulation.

The authors comment that the right sided vagal nerve stimulator seemed to work, and can be advantageous in select patients. The current adjustments should be performed carefully and under surveillance using electrocardiogram control. The right vagus nerve innervates the SA node. Obviously further studies with more patients are necessary.

Another approach to the problem of a damaged left vagus nerve preventing a reattachment of vagus nerve stimulation electrodes has been presented (Smith et al. 2008). In this study, a transcutaneous electrical vagus nerve stimulation method was developed and tested. Functional MRI was used to evaluate noninvasive access to the vagus nerve. The approach was to stimulate the vagus nerve at the left tragus of the left external ear.

The vagus nerve has two named branches in the jugular fossa: the meningeal branch and the auricular branch. The auricular branch passes posterior to the internal jugular vein and communicates with the facial nerve. The auricular branch of the vagus nerve is an afferent somatic nerve and reaches skin around the auricle and external acoustic meatus (Gray 1973). The significance of this is that it represents a superficial approach to the ability to stimulate the vagus nerve on the left side above the carotid canal location, frequently damaged in complications from initial surgery. This greatly reduces the chance of cardiac involvement associated with right vagus surgical implantation.

In this study, four normal male subjects were monitored with a 1.5 T MRI scanner during transcutaneous vagal nerve stimulation. Both cortical and brainstem regions were evaluated as compared to baseline blood oxygen level dependent fMRI.

Results showed a positive BOLD response in relay nuclei of vagus nerve afferent pathways. These included those of the locus coeruleus, thalamus, prefrontal cortex, postcentral gyri, and insula. Deactivations were noted in the right nucleus accumbens and right cerebellum.

The authors state that the transcutaneous approach to the vagus nerve was successful and represents an alternative to the more invasive method in which vagal nerve stimulation has traditionally been used. Further studies using patients with seizures will show its efficacy.

Anatomically, afferent vagal fibers ascend, then pass through the nucleus of the solitary tract in the brainstem. From this site, afferents go to several brain regions including pontine/midbrain nuclei, cerebellum, thalamus, and cerebral cortex. Other sites receiving vagal communication include the frontal cortex, locus coeruleus, amygdala, and cerebellum. (McLachlan 1993; Naritoku et al. 1995).

Mechanisms of action of vagus nerve stimulation are still unclear, but may have one or multiple modes of effectiveness including an alteration in neurotransmitter release including glycine and GABA (Marrosu et al. 2003). Vagal nerve stimulation may be associated with a reduction in interictal epileptiform discharges (Koo 2001).

Studies on efficacy of vagus nerve stimulation show results on several hundred patients with complex partial seizures (see above). These studies show up to 45% of patients benefiting from vagal nerve stimulations as regards frequency and severity of partial seizures. Adverse effects related to the recurrent laryngeal nerve were a problem for some patients. Notable is the same success rate among children under 12 as in those over 12 years of age (Lundgren et al. 1998; Patwardhan et al. 2000).

In conclusion, vagus nerve stimulation is a reasonable approach for patients with AED refractory epilepsies. It offers an easier surgical approach than, for example, temporal lobe resection, and can be discontinued if adverse effects are intolerable. It can also be used with AEDs and even other nonpharmacological treatments such as the ketogenic and/of modified Atkins diet. Other sites of brain stimulation include deep brain structures such as the hippocampus and thalamus (see next chapter).

Chapter 34
Deep Brain Stimulation

Electrical brain stimulation has been attempted for several decades in order to try to treat a variety of neurological disorders, including epilepsy. Many cerebral areas have been stimulated, including the cerebellum, hippocampus, thalamus, caudate nucleus, subthalamic nuclei, and locus ceruleus among others. In general, electrodes are positioned on or in the target, and electrical current is delivered. The current is generated by internalized pulse generators (IPGs), which are battery powered. The batteries are placed under the skin in various places, or more recently in the cranial cavity.

The stimulating electrode is placed through skull burr holes into the selected cerebral area using stereotaxic techniques. The process is called deep brain stimulation. There are of course several variations of this surgical technique. The stimulation is variable in terms of frequency, duration, and amplitude. Stimulation can be set to be cycling or continuous.

The correct placement is a critical process, so much so that a team of neurophysiologists, engineers, neurologists, neurosurgeons, etc., all participate. Serious side effects may result if the target is missed. Transoperative confirmation of placement is monitored by evoked responses, electrode recording, and deep brain stimulation mapping.

In considering epilepsy patients, neural stimulation is a candidate treatment paradigm in cases where the seizures are serious, are intractable to AEDs, and when focal resective surgery is not considered appropriate. The first attempt of deep brain stimulation was undertaken by Cooper et al. (1973). The area selected was the thalamus. The trial was based on animal studies showing that low frequency stimulation (10 Hz) of the paravermian cerebellar cortex decreased paroxysmal EEG discharges and seizures in a maximal electroshock model.

In kindling, low amplitude, low duration of electrical stimulation is delivered such that after several episodes, the animal seizes at a lower level of stimulation than before. This increased sensitivity is termed kindling. If during kindling, a low amplitude direct current is also delivered, kindling is disrupted. This effect is termed quenching (Weiss et al. 1995).

D.W. McCandless, *Epilepsy: Animal and Human Correlations*,
DOI 10.1007/978-1-4614-0361-6_34,

In a review of brain stimulation for epilepsy (Litt and Baltuch 2001), both animal and human results of deep brain stimulation are examined. The authors note that efforts by researchers and industry have developed a variety of implantable electrical devices for stimulating the heart to treat dysrhythmias, and devices for stimulating the brain for treatment of Parkinson's disease, tremors, and epilepsy, among other disorders.

Several animal seizure models have been used to investigate efficacy of deep brain stimulation. The idea was developed that various animal models of different seizure types were originating and propagating from discrete sites, often different. Identified stimulation targets included the cerebellum, basal ganglia, subthalamic nucleus, locus ceruleus, etc. The vagus nerve was also a target (see Chap. 33).

Human trials were often based on animal models showing efficacy in reducing or preventing seizures. Studies include those in which the target was the cerebellum (Cooper and Upton 1978), the centromedian thalamus (Velasco 2000), and hippocampus. Results are somewhat spotty in that some show excellent efficacy, others are equivocal.

Stimulating devices can be simple in that they can be "on or off," or they can respond to physiological reactions (Fisher et al. 1997). The concept is that over time, the effects of therapeutic stimulation can build up resulting in a period in which continued stimulation is no longer needed. "Intelligent" stimulators capable of determining physiological parameters and able to predict seizure will have additional value.

The authors conclude saying that development of smart brain stimulating devices should prove increasingly effective in seizure modulation. Major challenges are to properly select the site(s) for deep brain stimulation, and determine the proper parameters for stimulation. Technical issues such as having a power conserving implantable device are also impatient. These developments will hopefully result in better targeted therapies aimed at control of seizures.

Another paper (Richardson et al. 2003) looks at the modulation of hippocampal epileptoform activity using radial electric fields. The purpose of this study was to use electric field stimulation to interact in a subthreshold fashion with brain activity in order to control seizures. A large electric field was created between two electrodes in contact with the tissue. The subjects were Sprague-Dawley rats with an average weight of 254 g, and they received kainic acid to induce seizures.

Experimentally, kainic acid was infused into the right hippocampal CA1 area at 20 min intervals until epileptiform activity appeared. Electric field stimulation was applied with sinusoidal or multiple phase square waves. The wave form had equal duration consecutive plateaus of amplitude.

Results showed that in five of six animals, root mean squared (rms) analysis showed a statistically significant increase in activity at the positive and/or negative phase of the sinusoidal field. Similar results were seen in phasic stimulation. In addition, in experiments, characteristic changes were seen in the neural response in advance of electrographic seizures.

The authors note that their experiments show that radial electrofields are effective in the modification of hippocampal epileptiform activity. Variability of effects of positive or negative fields may have been due to electrode placement vis-a-vis the hippocampal layers. The authors noted that bilateral modulation resulted from unilateral stimulation, and it might be interesting to determine if the same concept could apply to the temporal lobe. This would then indicate bilateral seizures could be controlled with unilateral modulation.

Data also suggest that in a preseizure state, before clear seizure onset, periodic field stimulation demonstrates distinctive modulation prior to seizure activity. This finding has important implications for the design of "smart" equipment that would be responsive to physiological measures and preseizure indicators.

The authors conclude saying that their method is suitable for seizure control. The ability to modulate the hippocampus using minimally invasive surgical strategies is a very favorable situation. The modulation is graded, which could decrease adverse cognitive effects.

In another review paper (Oommen et al. 2005), effects of thalamic brain stimulation in humans has been described. Some reservations concerning electrical stimulation (modulation) centers on the extent to which the stimulation will actually activate or inhibit neurons at increasing distances from the stimulator. The exact anatomical placement of electrodes, and the features of stimulation (intensity, frequency, and duration) are critical in determining outcomes.

The effect of stimulation is not completely clear, but is thought to involve 1 of 2 mechanisms. The first states that stimulation causes a preferential release of inhibitory neurotransmitters. The second hypothesis states that the stimulation inactivates neurons near the electrodes via overdepolarization and failure of sodium channels.

Thalamic brain stimulation has focused on stimulation in the thalamic centromedian nucleus. The centromedian nucleus is thought to be part of the thalamic reticular formation, but is also related to the basal ganglia and motor systems. An early study of centromedian stimulation showed that seizure frequency improved by 60–100% (Velasco et al. 1987). In the study, stimulation was via externalized electrodes for 2 h per day, up to 3 months, in patients with complex partial seizures. In a larger series, similar results were obtained (50% of patients showed a 50% reduction). Patients with generalized seizures had a more favorable response than those with complex partial seizures.

Another paper (Tellez-Zenteno et al. 2006) examined hippocampal electrical stimulation in four patients with mesial temporal lobe epilepsy. A rationale was to look for potential reversible, adjustable treatment paradigms for drug refractive epilepsies. Hippocampal stimulation may be one such treatment. In this study, all four patients with mesial temporal lobe epilepsy were implanted with a chronic depth electrode along the axis of the left hippocampus (see Fig. 34.1).

The treatment paradigm called for continuous subthreshold, electrical stimuli (90 ms, 190 Hz). The study used a double blind, multiple crossover, randomized controlled design. Outcomes were assessed at monthly meetings. The authors compared results between on/off and baseline periods.

Results of the small number study showed that hippocampal stimulation yields a mean reduction of seizures of 15%. Three of four patients achieved seizure improvements. There were no significant differences in other outcomes or any adverse effects.

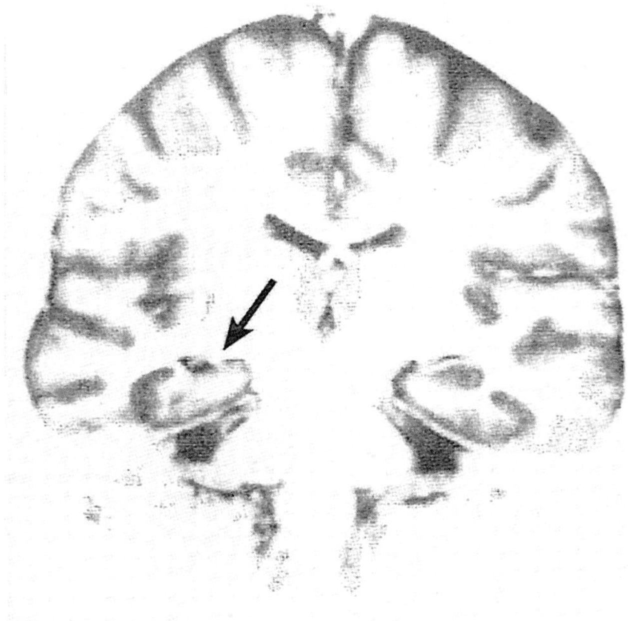

Fig. 34.1 Left hippocampus site of hippocampal stimulation. With kind permission of Springer Science+Business Media: Epileptic Syndromes and their Treatment, 2010, p. 449, Panaylotopoulos, C., Fig. 15.5

One patient has had long-term (4 years) seizure improvement with continued treatment. While this study shows positive effects, they are less than previously reported.

A recent paper from the Belgium group (Van Roost et al. 2007) examines the neurological aspects of temporal lobe deep brain stimulation for seizures. In temporal lobe epilepsy (complex partial epilepsy), resective surgery produces short-term results up to 85% and long-term cure in about 60% of cases. On the down side, resective surgery can be associated with visual field defects (temporary) and cognitive deficits. Plus, one third of patients do not benefit.

Deep brain stimulation has the feature of being able to be turned off. Deep brain stimulation may exact either excitatory or inhibitory effects. The concept is to try to restore intracerebral balance in excitation and inhibition. Deep brain stimuli was first applied to a hippocampus patient prior to resection, and found to terminate clinical seizures (Velasco et al. 2000).

Technique wise, hippocampal stimulation with electrodes can be inserted by a posterior/anterior approach. They are placed using a stereotactic frame. MRI imaging (1.5 T) is used, acquiring 1.2-mm thick slices. The position of the leads is verified by MRI. The finding of a unilateral or bilateral focal or regional medial temporal lobe seizure onset is reason to consider hippocampal deep brain stimulation. The stimulation consists of biphasic square pulses.

The authors note that deep brain stimulation at the microstimulation level could theoretically offer a more selective effect than is seen in macrostimulation. As a macrostimulation, the current influences many cell types in addition to neurons.

The previous paper was a methods paper for this paper, which is a pilot study to evaluate long-term deep brain stimulation (Boon et al. 2007). In this study, twelve patients with intractable temporal lobe epilepsy (complex partial epilepsy) underwent surgical procedures. Ten had medial temporal lobe deep brain stimulation, while two had selective amygdalohippocampectomy.

The methodology regarding electrode implantation is detailed in the preceding manuscript. Following video EEG monitoring, AEDs were decreased until habitual seizures were recorded. When the focus was located, the patients could select deep brain stimulation or return to AEDs. In patients with unilateral medial temporal lobe seizures, implantation was ipsilateral, when the seizure was bilateral, implantation was bilateral.

Results showed implantation of electrodes was uneventful, except that some bleeding occurred in one patient along the site of the electrode path. The mean follow up was 31 months (range 12–52 months). Follow up showed one of ten was seizure free, one had greater than 90% reduction in seizure frequency, five patients showed a greater than 50% reduction, two patients had a frequency reduction from 30 to 50%, and one had no change. Reported side effects were zero. The two patients who underwent amygdalohippocampectomy were seizure free.

The authors note that resective surgery for medial temporal lobe elipsy has a high rate of long-term seizure freedom rates of 60–75%. However, side effects such as verbal memory decline are worrisome. The apparent advantages of this reported treatment is that it is minimally invasive, and in this pilot study, there were no side effects.

The efficacy of the deep brain stimulus paradigm further validates previous studies showing that stimulation in the zone of seizure onset reduces seizures (Boon and Vandekerchove 1996). One group claims that the stimulus itself is not necessary; the act of implanting the electrode and the electrodes presence is sufficient to reduce seizure activity (Hodaie et al. 2002). The result of a lack of any side effects was based on MRI results, patient interviews, and neurologic and neuropsychological testing.

The authors state that the precise mechanism of action of electrical stimulation efficacy is not known, but local inhibition is likely a feature. This provides a "reversible functional lesion" in which the areas involved in the epileptic discharge are inhibited. Targeting of the actual seizure focus may have a similar effect and outcome. The authors conclude saying that this method is effective in both lesioned and nonlesioned patients. The preferred treatment for lesional patients remains resective surgery, but deep brain stimuli appears to be an alternative treatment method. Studies involving multicenter need to be done to confirm these findings.

In order to better assess possible intracerebral mechanisms of action of stimulating electrodes, Sprague-Dawley rats (180–200 g) were anesthetized, placed in a stereotaxic apparatus, and electrostimulus applied to the thalamic anterior nucleus (Toda et al. 2008). Stimulation lasted 1 h, and consisted of 2.5 V, 90 ms of pulse width, and variable sequences (10, 50, and 130 Hz).

The animals were subsequently injected with 5-bromo-2-deoxyuridine (BrdU) 1–7 days postsurgery. The rats were killed 1–28 days later, and brain prepared for microscopic examination. Some animals received daily injections of corticosterone several days before and after the stimulation of the thalamic anterior nucleus. The thalamic anterior nucleus was selected, because it is a site being used in epileptic patients.

Results showed that rats given thalamic anterior nucleus high-frequency stimulation had an increase in immunoreactivity for the immediate early gene protein Zif/268. Rats receiving thalamic anterior nucleus high-frequency stimulation had a 2–3-fold increase in cells showing BrdU as opposed to controls.

The maximum numbers of BrdU-labeled cells were seen 3 and 5 days after stimulation. Stimulation at 50 Hz and 130 Hz were most effective in producing the most highly BrdU-positive cells. There was still significant label of BrdU at 28 days after stimulation 85% of the cells positive for BrdU were inactive neurons, and only 2% were astrocytes. Finally treatment with corticosterone reduced BrdU cells, and this effect was reduced by thalamic anterior nucleus stimulation.

The authors comment that their study shows in rats that high-frequency thalamic anterior nucleus stimulator seems to significantly increase hippocampal neurogenesis. The exact mechanism for the neural stimulation of neurogenesis is unclear. One theoretical idea is that excitatory stimuli cause neural progenator cells to change their differentiation programs to more neurons instead of glia.

The effect of the change is not certain, but conditions such as stress decrease hippocampal neurons thereby producing a deterioration in performance (Bain et al. 2004). Conversely, enriched environments have the effect of increasing hippocampal neurons and enhancing performance (Gaulke et al. 2005). The authors suggest that increasing use of the efficacious deep brain stimuli may provide answers to these questions and more in human patients.

In another review (Ellis and Stevens 2008), effects of deep brain stimulation of various brain regions were examined. Results from stimulation of the substantia nigra pars reticulata in rats showed a complete blockage of amygdala-kindled seizures in 43% of rats (Shi et al. 2006). In this study, electrodes were implanted in both the amygdala and substantia nigra. Kindling was achieved by electrical stimulation of the amygdala, and current was sent to the substantia nigra one second after kindling stopped. Results showed that stimulation of the substantia nigra pars reticulata had the effect of stopping seizures in 60% of rats in subsequent trials when the substantia nigra was not stimulated.

In another study (Vercueil et al. 1998), using the GAERs genetic rat model, high frequency subthalamic nucleus deep brain stimulation showed it was capable of suppressing the seizures. Stimulation was bilateral at 130 Hz for 60 ms.

In humans, the subthalamic nuclear stimulation is used for treatment of Parkinson's disease, and the substantia nigra is involved in propagation of seizures due to its GABA transmission to the superior colliculus (Gale 1986). One study

(Benabid et al. 2001) showed that bilateral deep brain stimulus of the subthalamic nucleus led to an 83% reduction in seizure frequency.

In another study (Loddenkemper et al. 2001), five patients underwent deep brain stimulation in the subthalamic nuclei in order to treat intractable seizures. Stimuli was at 100 Hz for 60 ms. Two patients had an 80% reduction in seizures at 10 months and a 60% reduction at 16 months after the stimulation. The authors speculated that the superior colliculus is under inhibitory control of the efferent projections from the substantia nigra. This in turn may reduce the inhibitory effect of the substantia nigra on the dorsal midbrain anticonvulsant zone. This would act to raise the seizure threshold. The belief is expressed that many patients with intractable seizures (to AEDs) are not good surgical candidates and are in need of additional options. The deep brain stimuli (as well as vagus nerve stimulation) offer excellent noninvasive, safe, and efficacious alternatives. These methods have good grounding in animal epilepsy models, and represent an increasingly used epilepsy treatment paradigm.

While deep brain stimulation is often referred to as "noninvasive" compared to resective surgery, it is invasive in that electrodes are inserted into cerebral tissue. The use of a method such as repetitive transcranial magnetic stimulation (RTMS) is actually completely noninvasive. This method has been used previously to measure parameters such as motor cortex output. The RTMS method is thought to have some level of therapeutic use (Tassinari et al. 2003). Indeed, RTMS has been shown to have anticonvulsive efficacy against pentylenetetrazole induced seizures in rats (Akamatsu et al. 2001), and RTMS may suppress kainic acid induced seizures.

In humans, low-frequency RTMS decreases motor cortex excitability. In one epilepsy patient study, a round coil was placed over the vertex such that global excitability was decreased. Patients were exposed to magnetic stimulation for 5 days (0.33 Hz) after which they had an almost 40% reduction in seizures (Tergau et al. 1999). Some other studies failed to show efficacy for seizures, but a study looking at the effects of RTMS on seizures in patients with cortical malformations showed significant seizure reduction attributable to RTMS. The suggestion is that the RTMS method may be best suited to patients with obvious foci in the cortical convexity. Proximity to the surface seems advantageous (Saillet et al. 2009).

Brain stimulation, whether deep or superficial, using RTMS has the clear advantage that cerebral tissue is not removed. This serves to reduce operative risk, plus, the process is reversible, and postoperative complications avoided. Open loop stimulation refers to deep brain stimulation 24 h per day, whereas closed loop stimulation consists of stimulation timed to needs. The stimulator therefore may be off or on, in response to an impending seizure (Anderson et al. 2009).

Thalamic stimulation was initially proposed in 1945, and early attempts were made by the 1970s. Before long, several targets had been tried with success. Many early studies were not well designed or controlled, and so a large 17-site study of 110 patients was designed and executed. The results of this 5-year study were presented at the 2008 meeting of the American Epilepsy Society by the principal investigator (Fisher 2008). The results were overall quite impressive in that deep brain stimulation produced a dimunation in seizure activity greater than controls.

In a closed loop system, the exact site of seizures onset is determined by recording paradigms. Subsequently, the "responsive neurostimulator" (RNS) is placed on or in the focus. The device records electrical features that signal an impending seizure, then responds in a way that is designed to control the seizure.

Results of stimulation using the device were presented at the same meeting (2008) as above (Morrell et al. 2008). The results on 65 patients all with partial onset seizures was that patients with generalized tonic clonic seizures enjoyed a greater than 50% (65% of patients) in seizures, where the plus 50% reduction consisted of only 27% of complex partial seizure patients. The key seemed to be the recognition of specific sites of onset of seizures prior to implantation of the device. There was an increase of positives over time.

The conclusion from the RNS study was that the method is safe and efficacious for generalized tonic clonic seizures, and also (not as much) efficacious for complex partial seizures. With time, there was a significant increase in responder rates. Another study with a larger number of participants will further knowledge of this method of deep brain stimulation.

In some previous studies, the suggestion was made that when treating intractable epilepsy with deep brain stimulation, the actual act of inserting the probe into the region could cause a reduction in frequency. The speculation was that a small amount of target tissue is damaged or removed from function. In the case of the hippocampus, this might be termed a "microhippocampectomy". The presented paper is a case report of such a case in which an intra hippocampus electrode by itself caused seizure reduction (Schulze-Bonhage et al. 2009).

The case was of a 37-year-old woman with complex partial seizures of 10-year duration. Seizure frequency was about one seizure per 5–7 days. The seizures were characterized by an aura, complex partial seizures, behavioral arrest, and automatisms. A variety of AEDs were ineffective in seizure control. At that time the patient was evaluated as a possible candidate for resective surgery.

To that end, the patient had extensive presurgical workup. Evaluation included high-resolution MRI, fluorodeoxyglucose-positron emission tomography, video EEG monitoring, etc. The patient next had an electrode implanted into the hippocampus in order to record hippocampal seizure activity. The electrode was 1.0 mm in diameter, and placed using stereotaxic equipment, and the patient was under general anesthesia.

Following hippocampal implantation, no more clinically obvious seizures were seen over a 22-day period of video EEG monitoring, even with discontinuation of AEDs. There were no sequelae associated with the surgery.

The authors note that the actual act of electrode implantation ended the patient's epilepsy. Postimplantation MRI showed no evidence of damage or swelling from the implantation process. The authors state that a discrete destruction of allocortical tissue or network connectivity should be an interesting treatment trial. It would be easily done simply by waiting awhile after routine implantation procedures to see if there was any effect in seizure frequency or duration.

Another recent paper looks at the characteristics of scheduled or responsive stimulation (Jobst et al. 2010). Scheduled stimulation is the stimulation of deep brain structures based only on time. Most scheduled stimulation is aimed at the anterior nucleus of the thalamus or the hippocampus. Stimulation refers to stimulation in response to an impending seizure. This type of stimulus requires "on line" detection of epileptiform activity. During functional mapping, seizure activity can be interrupted by brief pulses of cortical stimulation.

Scheduled stimulation can be delivered constantly or at specific intervals. The pulse generator is usually implanted in the chest, and is connected to the target by four leads. The target is frequently the thalamic anterior nucleus, which is closely linked to the medial temporal lobe. Scheduled stimulation of the hippocampus has yielded very good responder rates as proof of principle (Velasco et al. 2007). Future controlled studies will be more definitive.

Responsive neurostimulation requires an exact knowledge of the seizure onset zone. This requirement is based on being able to predict exactly when an overt seizure is impending, then firing a stimulation in order to arrest the seizure immediately. Proof of principle was shown (Morrell et al. 2008) using an external stimulator that detected electrographic seizures and fires responsive stimulation. This requires the implantation of depth electrodes into the onset zone. The stimulating device is located in the skull, detects early epileptoform activity, and responds.

Current technology allows two zones of onset to be implanted, even including in areas of the eloquent cortex. The methodology has been demonstrated to be safe and efficacious (Morrell et al. 2008). There were no serious adverse effects. The median responder rate increased over the time of follow-up. Further studies in progress will increase understanding of this interesting epilepsy treatment paradigm.

In a recent report looking at the site-related efficacy of deep brain stimulation (Rahman et al. 2010), emphasis was placed on the thalamus and related nuclei (subthalamic nuclei, centromedian thalamic nucleus, and the thalamic anterior nucleus). The overriding concept is that the target is optimal if it has broad connections. Deep nuclei usually fill this criterion.

The subthalamic nucleus has garnered interest because it is frequently a target for deep brain stimulation in Parkinson's disease patients. The subthalamic nuclei contain sensorimotor, limbic, and association areas, thus rendering it an obvious target for deep brain stimulation for seizure modulation.

Animal models of epilepsy provided initial data showing that injection of NMDA into the subthalamic nucleus decreased seizures in GAER models of seizures. Early human studies of seizures in which deep brain stimulation of the subthalamic nucleus were performed by Benabid (Benabid et al. 2001). This study showed a 50–80% decrease in seizure frequency using subthalamic nucleus stimulation.

Another study (Lee et al. 2006) compared deep brain stimulation in the subthalamic nucleus to the anterior nucleus of the thalamus as regards seizure suppression efficacy. Results showed that the anterior thalamic nucleus group had better success (75 vs. 50%). These results were based on small patient numbers, and need reevaluation based on higher patient numbers.

The anterior nucleus of the thalamus is a frequent target of deep brain stimulation, because of its gate keeping role in seizures. The nucleus connects to the hippocampus, orbitofrontal cortex, amygdala, cingulated cortex, and caudate nucleus, as well as other sites. Animal models of deep brain stimulus studies are not consistent, some showing seizure depression, others showing the opposite. One attraction of the anterior nucleus is that it is easy to surgically approach due to its size.

The first human study (Upton et al. 1985) showed a 66% decrease in seizures with stimulation. Similar subsequent studies followed, each with similar results, but the patient numbers were small.

A large 110 patient study (see above) was undertaken (Fisher et al. 2010). In this study, patients received 5 V, 90 ms pulses for 1 min, +5 min without stimulation. Results showed a seizure frequency decrease of 40.5% in the experimental group, and the result was relatively longlasting. More than 6 patients were seizure free for at least 6 month at the 2-year follow up. Only patients with temporal lobe foci showed improvement, not those with frontal, parietal, or occipital lobe foci. The centromedian thalamic nucleus is a key structural area in the so-called reticulocortical system, playing a role in wakefulness, and it is a relay station. Some animal studies implicate the centromedian nucleus as having a role in excitability of the cerebral cortex in seizures (Meeren et al. 2002). The centromedian nucleus is also a gate-keeper in terms of electrical activity. This nucleus was first stimulated by Valasco (Velasco et al. 1987). Results showed that pulses led to an 80–100% decrease in generalized seizures, and 60–100% decrease in complex partial seizures after 3 months of treatment.

In a controlled study, (Fisher et al. 1992) in a small group of patients (7), the on/off paradigm was used to evaluate deep stimulation of the centromedian thalamic nucleus. At 3 months, generalized tonic clonic seizures were reduced by 30%. This was with a 2 h/day routine; when the stimulations were on longer, the reduction increased. This foci of stimulation was efficacious and encouraging, especially in generalized tonic clonic patients.

An important concept in deep brain stimulation, is that, in addition to the site for stimulation, the patient selection is critical. Many early studies can be criticized for small patient numbers, and for having only one site and multiple seizure types. For example, subthalamic nucleus deep brain stimulation seems best suited for patients with frontal lobe lesions. One other comment by the author is that virtually all patients in all studies have been involved with several AEDs, plus other attempts at seizure control, such as diets, failed resective surgery, etc. All of these act to initiate "rogue" areas of secondary epilepsy foci, confounding in unknown ways, the results of deep brain stimulation in epilepsy patients. This is yet another area in which translation of animal data to human treatment has been significant.

The mechanism of action of deep brain stimulation is not completely understood. A large animal (sheep) model has certain advantages over rats in that assessment is easier. In this study (Stypulkowski et al. 2011), the authors used sheep to implant anterior thalamic leads, and hippocampal depth electrodes and catheters using a stereotaxic apparatus.

Results using MRI-guiding techniques permitted correct placement of electrodes and leads in the targeted sites. Robust evoked potentials were produced dependent on stimulus location and parameters. Thalamus deep brain stimulation produced an obvious inhibition of spontaneous and also of penicillin-induced ictal discharges in the hippocampus. The ictal activity lasted longer than the stimulation. The authors conclude saying the sheep model is a useful model of epileptic seizures, and that high frequency deep brain stimulation suppressed excitability. The deep brain stimulation appears to have the ability to be a therapeutic mechanism. These results have a clear translational property.

Chapter 35
Surgical Resection

The issue of surgery for patients with intractable epilepsy is always a difficult situation. Patients, family, and physicians must always try to balance possible adverse effects of surgery against the potential for life style improvement with successful surgical treatment. One major consideration is the impact on the patient of his/her seizures. A seizure frequency of two per month might be entirely acceptable for one patient, while being totally unsatisfactory to another. Much depends on life style/expectations, career, etc. Even after surgery, there is a necessity for continued AED treatment in most cases. Factors in decision making include age of patient, length and severity of existing seizures, compliance and ability of the patient to cope with seizures, expectations of postseizure changes in seizures, etc.

The question of seizure intractability is perhaps the initial question to be addressed. In some measures, it is important to be sure of the initial diagnosis, since AED treatment is highly dependent on the diagnosis. For example, absence and complex partial seizures can be confused. If this leads to an inappropriate diagnosis and treatment, then the seizure state could be viewed as intractable. Proper diagnosis using video EEG, MRI, fMRI, PET, single photon emission computerized tomography (SPECT) are all essential to determine seizure type and lesion location before surgery. Even intracranial EEG (ECoG), an invasive technique may be necessary in order to completely localize the lesion. The obvious goal of seizure treatment is to eliminate seizure activity with monotherapy. Most postoperative patients continue to require AED treatment, but following surgery, the frequency and severity of seizure episodes may be dramatically reduced (see Table 35.1).

Other than intractable seizures, considerations regarding surgical intervention in epileptic patients include: increased mortality rates of epileptics, accidental injuries from falls, etc., memory loss, cognitive decline, depression, anxiety, reduced fertility, lack of employment, etc. Candidates for resection surgery must not have attained accepted AED control of seizures, and have disabling complex partial seizures with or without second generalization of seizures.

Surgical outcomes are variable, and contributing factors to variability include duration of epilepsy, completeness of surgical resection, extent or lack thereof of

D.W. McCandless, *Epilepsy: Animal and Human Correlations*,
DOI 10.1007/978-1-4614-0361-6_35, © Springer Science+Business Media, LLC 2012

Table 35.1 Factors suggesting intractable epilepsy

Factor	Comment
Age	Less than 1 year old
Epilepsy type	Rasmussen syndrome
	Infantile spasms
	Dravet syndrome
	Progressive myoclonic epilepsy
Seizure frequency	Subjective; up to several per year
AED failure	More than two AED failures
	Duration of unresponsiveness over 2 years

Adapted from Go, C., and Snead, O., 2008 Neurosurg Focus. DOI: 10.3171/foc/2008/25/9/E2

pathology in the resectioned tissue, more than one focus, etc. In addition, there is greater success in both adults and children in temporal resection surgery versus extratemporal resection. In children, a higher degree of co morbidities leads to a decreased incidence of seizure freedom following seizures (McLellan et al. 2005).

Another variable with surgical treatment is that of continued AED treatment following surgery. In one study (Schmidt et al. 2004), about 35% of postsurgical patients had a seizure by 5 years following cessation of AED treatment. In another study (Berg et al. 2006), the incidence of postoperative seizures in patients stopping AEDs was the same as that in patients still taking AEDs after surgery.

In a 10-year follow up of surgeries to correct cerebral malformations, over 70% of patients were still on AEDs. The usual outcome from resective surgery is good seizure control for the first postoperative year, followed by a slow deterioration over time leading to seizure recurrence in about 25% of cases (Spencer et al. 2005).

Not all patients, especially children, are candidates for temporal lobe resection. These include patients with Landau–Kleffner syndrome, and those with Lennox–Gastaut syndrome. These patients are better served by palliative surgery such as multiple subpial transections and corpus callosotomy. The success rate in children undergoing corpus callosotomy is relatively good, with from 60 to 80% showing some improvement, and at least 50% showing a frequency drop (Turanli et al. 2006). As many as 20% of patients following surgery may be seizure free.

Finally are both pediatric and adult patients who elect to not have surgery, or are a poor risk. The rate of seizure freedom in follow up in these patients is in the range of 11–20% in unoperated localized epilepsy. An additional highly significant statistic is the morbidity rate associated with seizures. Due to accidents, drownings, etc., the mortality rate even after surgery in patients still seizing is nearly sixfold that of seizure-free patients. In the case of patients rendered seizure free following surgery, the reduction of the mortality rate is a significant benefit.

The functional outcomes of surgery as related to language, motor, behavioral, and psychiatric results are sometimes hard to predict. Much depends on the extent of the epileptic focus, and its successful removal. Behavioral problems in epileptic patients may be significant, and can be a key element in epileptic patients.

How much of this is AED related is hard to assess. Behavioral problems are clearly an adverse effect of AEDs. Some patients may use epilepsy as an excuse to act in an inappropriate manner. Following surgery for intractable epilepsy, some children may develop new behavioral symptoms (Szabo et al. 1999).

Thorough neuropsychological testing before and after surgery is important. This technique can be helpful in localizing and assessing brain function. Memory deficits may also be assessed before and after mesial temporal resection (Scoville and Milner 1957).

There are of course potential complications of neurosurgery. These include infections, blood loss, infarct with hemiparesis, and surgery-resultant language/motor deficits. These potential complications are reduced with continued improvements in surgical methodologies, and in increasing familiarity with the procedures by a surgical team. In any case, the operative morbidity and mortality is less than that associated with unoperated patients who would be candidates for surgery. Further complications of neurosurgery in infants and children relates to blood loss. Since blood loss in temporal lobe resection may range from 250 to 500 ml, and the total blood volume of a 6 month old may be no more than 750 ml, over one half of the total blood volume could need replacement. Neurosurgical morbidity and mortality are usually under 5%, but are somewhat higher in infants. Waiting until an infant is 3–4 years old lowers the surgical risk, but increases the risk for increased developmental and behavioral complications of the surgery (Wyllie et al. 1996). A team approach to planning resective surgery acts to "customize" the surgery to each patient's needs, and reduce potential complications.

Finally, in children undergoing focal temporal lobe resection, the outcomes may not be deleterious as regards cognitive results. In one large study, over 80% of post-operative children having temporal lobe resection had no change in cognition as regards I.Q. (Westerveld et al. 2000). Some few studies (Smith et al. 2006) did show a decline in verbal memory, however this was a reversible feature, again stressing the "plastic" nature of the infant developing brain. This is further evidence in favor of early surgery in infants and children.

Thus overall, early surgery in infants and children who have AED-resistant seizures appear to be well advised to explore surgery provided preoperative studies assure the correct diagnosis, and a reasonable attempt was made at drug therapy. Once this is ascertained in children and adults, delay seems associated with a poorer prognosis. The goal of being seizure free is paramount. With this comes an overall reduction of morbidity and mortality, and a significant improvement in life quality.

The above comments pertain to temporal lobe resection for complex partial seizures, however there are other less-performed neurosurgeries for other types of epileptic seizures. Corpus callosotomy is a surgical technique developed before W.W.2 as a treatment for generalized seizures. The rationale for corpus callosotomy is to sever major CNS pathways between the right/left side of the brain. This in turn acts to decrease the severity and frequency of generalized seizures such as tonic, clonic, atonic, and clonic-tonic seizures.

This procedure is performed much less often than anteromesial temporal lobe resection. Pre surgery work up is essentially the same as for temporal lobe resection.

Criteria for selection includes intractable seizures, and generalized seizures not amenable to focal resection. Neuropsychological testing is indicated if specific problems such as memory integrity are important. Inhalation anesthesia is recommended in order to facilitate arousal and neurological evaluation. A major complication is air emboli produced by the inadvertent tearing of the superior sagittal sinus. Results show that this surgical procedure acts to decrease the frequency of seizures, but does not stop them.

Another surgical technique is called multiple subpial transection. This method is preferred to temporal lobectomy if the focus of epileptic activity lies in an important cortical area such as that associated with speech. Results from this procedure have been spotty. One study (Morrell et al. 1989), reported a 55% complete seizure control, and no major complications involving the precentral gyrus, but there were some complications involving other areas. This surgical method has also been successfully used in the Landau–Kleffner syndrome.

One area of significant concern is the occurrence of acute postoperative seizures after anterior temporal lobectomy for partial seizures. The appearance of acute postoperative seizures raises questions regarding the long-term prognosis of complex partial seizure patients. This occurrence can be particularly disturbing to relatives and staff. In a study examining this phenomenon (Malla et al. 1999), 160 patients were examined retrospectively following temporal lobectomy. All selected patients also had an amygdalohippocampectomy. Median age of surgical patients was 32 years. All patients were followed up for at least 1 year, and were evaluated via the Engel scale (Engel et al. 1993).

Results showed the mean preoperative duration of epilepsy was 23 years. Mean length of follow up was 37 months. Of the 160 patients, 32 had acute postoperative seizures, with 15 having acute early seizures, and 17 late postoperative seizures. Analysis of potential risk factors showed no significant differences in acute postoperative seizure when analyzed against side of surgery, presence of hippocampal atrophy, presence of spikes on EEGs, or presence of pathological findings in resected cerebral tissue. There was also no correlation of acute postoperative seizures with age, gender, duration of epilepsy, or extent of resection.

The significant finding was that those 32 patients with an acute postoperative seizure (early or late) had a statistically significant higher rate of unfavorable outcome. For example, those with acute postoperative seizures had a 62% favorable outcome at 1 year after surgery, compared to a rate of 83% for those without acute postoperative seizures. There was no significant difference between early and late acute postoperative seizure patients.

The authors note that about 20–40% of patients who undergo temporal lobectomy (Engel et al. 1993; So 1997) have seizure recurrence later, which may occur in the first 6–12 months after surgery (Penfield and Paine 1955). The authors note that acute seizures following surgery have been largely ignored, but their results show a significant correlation between these acute postoperative seizures and less favorable long-term outcomes. The authors conclude that acute postoperative seizures of the same type as were occurring preoperatively portend a worse prognosis. In acute

Fig. 35.1 The Engle postoperative seizure reduction scale

class I = seizure freedom

class II = rare seizure (1 - 3 per year)

class III = worthwhile seizure reduction (more than 90%)

class IV = not a worthwhile seizure reduction (less than 90%)

postoperative seizure patients who have different seizure types than were occurring before surgery, the prognosis is no worse as patients not having acute postoperative seizures.

In another paper (Boling et al. 2001), temporal lobe epilepsy surgery was examined in patients 50 years old or older. In this study, 218 patients ranging from 10 to 64 years of age were compared. Patients were chosen based on intractable temporal lobe epilepsy, absence of tumors, and a two or more year follow up. All surgeries were performed by one surgeon. There were 18 patients in the over 50 year group, and 150 in the under 50-year-old group. The group of patients were compared using the Engel postoperative seizure reduction scale (see Fig. 35.1).

The authors conclude that older patients with temporal lobe epilepsy do achieve favorable results following surgery as compared to other age groups. There was a "trend" toward lower success in older patients in Class I and II, but overall results were similar. There was an observation that older patients were less likely to return to work following temporal lobe resection.

Wiebe et al. (2001) published a paper which represents the first randomized, controlled trial of surgery for temporal lobe epilepsy. The authors correctly state that randomized studies have not previously been done due to the inherent difficulties associated with both the design and implementation of such studies.

In this study, 80 patients were randomly assigned to one of two groups (1) anteromesial temporal lobe resection, or (2) AED management. When assessed at 1 year following randomization, 58% of the surgical group, and 8% of the AED group were free of seizures which impaired awareness (consciousness). The patients in the surgical group had fewer seizures and a better life quality than those in the medical treatment group. Those in the surgical group had a trend toward more favorable employment or school status. Adverse effects in the surgical group included one case of thalamus infarct two patients with a decline in verbal memory, and one case of postoperative infection. In the medical (AED) group, there was one case of sudden unexplained death.

The authors note that their study is the first of its kind. They show that randomized, controlled clinical trials are possible to evaluate the efficacy of temporal lobe resection for epilepsy. Follow up greater than 1 year would provide additional data for assessment of this surgical approach.

In another retrospective study (Clusman et al. 2002), 321 patients who were operated on for temporal lobe epilepsy, were examined by uni and multi factorial analysis of a variety of factors such as imaging, neuropsychological, and surgical factors to determine predictors of surgical outcomes. Results of stepwise logistic regression exposed a predictive model which contains five factors which predict good seizure control:

(1) a clear abnormality as shown by MRI; (2) an absence of status epilepticus; (3) MRI evidence of a ganglioglioma or dysembryoplastic neuroepithelial tumor; (4) concordant lateralizing memory defect; and (5) absence of dysplasia as reflected by MRI. Seizure outcome correlated with the diagnosis and clinical features as opposed to various resection methods. Results from neuropsychological tests showed better results from limited focused resection as opposed to the anterior temporal lobectomy.

In this study, the most critical neuropsychological result was verbal memory. This criterion was worse postsurgery for patients with older age, left-sided surgery, and standard anterior temporal lobectomy. It is noted that future imaging techniques such as functional MRI may further improve the determination of the cognitive risk of surgery for temporal lobe epilepsy. Since epilepsy refractory to AEDs can be determined rather early (Kwan and Brodie 2000), given the apparent efficacy of neurosurgical approaches to temporal lobe epilepsy, the average duration of seizures prior to surgery (15–20 years) seems excessive. The demonstration of successful surgical treatments of successful surgical treatments of from 55% before 1985 to 71% in the present study argues in favor of increased awareness and increased earlier referral to epilepsy centers of cases refractory to AEDs.

As regards prognostic values, a paper (Suhy et al. 2002) has explored the prognostic differences between temporal epilepsy patients who do and who do not have increased hippocampal T2 signal, or atrophy on structural MRI. There are a variety of functional imaging modalities, such as fMRI, but these are complex and expensive procedures. This study examined patients with normal structural MRI. Results showed that patients becoming seizure free after surgery had N-acetylasparate/creatine values about normal. The N-acetylasparate/creatine ratios were normal both in the ipsilateral and contralateral foci, whereas patients with less successful outcomes had low ratios in foci bilaterally. The authors suggest that contralateral low ratios may indicate that there is epileptic tissue outside of the proposed lobe resection boundary.

Another paper examines outcomes of temporal lobe epilepsy surgery in adult patients from a national epilepsy center, and also examines preoperative factors predicting surgical success. This study was a retrospective longitudinal follow up of 140 consecutive adult patients (Jutila et al. 2002).

Patients were selected who had AED-resistant temporal lobe epilepsy, and who at surgery had amygdalohippocampectomies. Preoperative evaluation included MRI, video EEG recording, neuropsychological evaluation, psychiatric evaluation, and the WADA test. Postoperative follow up was done at 3 months, 1 year, and 3 years. Complications which impacted daily life were deemed "major."

Results of patient analysis showed a median age of 32 years. The median age of seizure onset was 12 years of age, and the median duration of epilepsy was 19 years. The majority of patients (113) underwent anterior temporal resection and amygdalohippocampectomy. Preoperative MRI showed a unilateral temporal lobe lesion in 53% of patients. This included some patients with hippocampal atrophy. MRI was normal in 34% of patients with unilateral temporal lobe epilepsy.

Results as regards seizures depend to some extent on seizure type. Unilateral temporal lobe epilepsy showed 45% of patients were seizure free 1 year postoperatively, and another 12% had only auras. In the cases in which there were seizure relapses, 86% occurred within the first year. Seizure relapses greater than 2 years represented only 5% of patients. Those operated on for temporal lobe seizures had better outcomes than a subgroup operated on palliatively. Overall, an Engel I–III seizure result occurred in nearly 70% of patients operated on for temporal lobe epilepsy, and this result was generally consistent with later (3 year) follow ups.

Microscopic examination of resected temporal lobe tissue revealed isolated cortical microdysgenesis. This was characterized by gray matter heterotopias, increased cellularity of white matter, neuronyl clustering, and rows of perivascular glia (Armstrong 1993). The significance of these changes is not clear (Kasper et al. 1999).

The authors conclude saying that surgical attempts to reduce seizure severity is better than long-term pharmacological treatments. Early onset of seizures, seizure type, volume of hippocampal reduction, and abnormalities in the temporal lobe demonstrated by MRI, all were predictors of Engel I–II outcomes. Most (86%) of postoperative seizure recurrence occurred in the first year. The first year results were consistent with results at 3 years or more, so were good predictors of long-term success. These data suggest the importance of establishing the correct time line for surgical intervention.

In another long-term seizure outcome study after selective amygdalohippocampectomy, 369 consective patients were retrospectively analyzed (Wieser et al. 2003). The surgery was performed using a specific procedure, selective amygdalohippocampsectomy which removes the amygdala, uncus, hippocampus, and anterior parahippocampal gyrus. Included were patients with a discrete unilateral seizure origination from one locus, and patients with more widespread multifocal temporal loci.

Results of follow up showed 67% of patients were free from disabling seizures (Engel I) after surgery. Patient's outcomes were somewhat better when surgery was performed early in the course of the disease. Overall, 70% of patients enjoyed reductions in AED treatment at the time of last follow up. This is significant because 200 patients participated in 5-year follow ups, and 100 participated in 10-year follow up assessment. Twelve percent of patients only achieved seizure freedom 2.7 years after surgery.

Another study (Tigaran et al. 2003) looked at the possible association of acute postoperative seizures and a poor postoperative outcome. The idea had been prevalent that acute postoperative seizures might portend a poor long-term outcome of surgery for intractable frontal lobe seizures. In this study, 65 patients who had surgery consisting of frontal lobe cortical resection were included for study. All patients were followed up for at least 1 year.

Results showed that 26% had acute postoperative seizures. Several potential causative factors were shown not to be related to the acute postoperative seizures: gender, length of time of seizures, etiology of seizures, use of subdural electrodes, or surgical pathology. A positive correlation was noted for age at surgery, and age of seizure onset. Most importantly, follow up has shown that there is no correlation between acute postoperative seizures and frequency of surgery success. Thus, the occurrence of a seizure state is similar to those with and without acute postoperative seizures. These data suggest, as have other studies, that acute postoperative seizures do not preclude a seizure-free outcome following frontal lobe epilepsy surgery.

One concern relating to temporal lobe surgery for epilepsy is older patients undergoing the procedure (Helmstaedter et al. 2002). One hundred eighty-seven patients before and after temporal lobe surgery were examined for verbal memory. Eighty patients had amygdalohippocampectomy, and over 107 had anterior 2/3 temporal lobectomy. Results showed that both procedures – left anterior 2/3 temporal lobectomy and amygdalohippocampectomy – age regression of verbal learning became steeper than that of younger controls. The older patients already had learning and memory deficits, which worsened following surgery. The authors argue in favor of earlier surgery for temporal lobe epilepsies.

The idea of providing prophylactic anticonvulsants following any neurological surgery is not new. A recent review examines this question (Temkin 2002). This review states that five studies evaluating the prophylactic use of phenytoin show a 44% decrease in the seizure risk in the first postoperative week. Other AEDs such as carbamazepine and valproate have not been evaluated, sufficiently to reach a conclusion. These authors summarize saying that there is a significant reduction in postoperative seizures following brain surgery by administering phenytoin. Prophylactic treatment with AEDs is not recommended after the first postoperative week of seizure-free results.

Not all temporal lobe surgeries are successful. One question is why does this occur, and is a second surgical attempt warranted. Several studies have examined this problem with interesting results (Salanova et al. 2005; Siegel et al. 2004). In the first of these two studies, from a total of 262 patients, following temporal lobe seizures, 16% continued to have Engel class III–IV seizures. Of these 41 cases, 29% had head trauma, 12% had febrile seizures, and 7% had encephalitis. In addition, over one half had abnormal MRIs showing residual posterior mesial temporal lobe spiking.

Postsurgical MRI showed posterior mesial temporal lesions in over 85% of cases. Of these 41 cases, 21 were operated on again with no morbidity or mortality. Of these, 57% became seizure free. There was a correlation between the necessity for re-operation and head trauma, encephalitis, and indications of a larger focus of epileptic activity.

The second study looked at 64 patients undergoing a second surgery for intractable seizures. Following a second surgery, 39% became seizure free, and 12% had some improvement. In this study, predictors of success of the second surgery were duration of seizures (less than 5 years) and onset of seizures (older than 15 years of age).

These studies are important since there is little data on the efficacy of a second attempt to surgically treat epilepsy when the first surgery fails. Factors which predict surgical failure were normal imaging, and a predisposition to widespread epileptic activity such as head trauma and encephalitis. The positive results on the second surgical attempt stress the importance of evaluation of failed first attempt cases, and consideration of a second surgical attempt.

The issue of multiple auras in temporal lobe seizures is of interest since the aura can have a localizing and lateralizing function. Visual auras, for example, may serve to localize the seizure origin to the occipital lobe; an abdominal aura may suggest a mesial temporal origin, etc. Multiple auras reported by a patient may signal multiple and different sites of origin. A recent study has examined some features of this problem (Widdess-Walsh et al. 2007).

In this study, 31 patients over a 16-year period reported multiple auras described as visual, olfactory, auditory, gustatory, etc. Results showed that all patients had EEG seizures which were restricted. Twenty of the 31 patients had surgery, and 53% were rendered seizure free. Subdural EEG recordings were interesting in that six patients had a march of sequential auras, or several ictal onset zones which resulted in several auras. The success rate of surgery for intractable epilepsy in these patients indicates that they are good surgical candidates.

The occurrence of somatosensory and visual auras in the same patient gave initial concern as regards multiple origin sites, but these almost always followed abdominal, psychic, olfactory, etc., auras, typical of temporal lobe origin. Interestingly, most patients remembering auras have a restricted seizure, and less likely to remember if the seizure is more sensitive and widespread (Schulz et al. 1995). Most patients with multiple auras have right hemisphere foci, and restricted to one hemisphere epilepsies.

A recent paper looks at medically intractable epilepsy in children, and discusses diagnosis and preoperative evaluation with an eye for early diagnosis and surgical attempt, which is associated with a greater success rate (Go and Snead 2008). These authors note that of children with epilepsy, 10–40% will continue to have seizures into adulthood despite optimal AED treatment. The adverse effects of AEDs and seizures as regards their effects on developing brain must be kept in mind. To achieve a complete cessation of seizures as early as possible is a key consideration. Therefore, early surgery (when children's brains are most "plastic") is a goal in the event of intractable seizures.

The authors of this paper define pharmacological intractability as inadequate seizure control despite at least two AEDs at maximal doses for 1.5–2 years, or control with unacceptable drug side effects. It is widely accepted that if two AEDs fail, only a 5–10% chance exists that a third AED will achieve success. The notion of minimal seizure frequency is variable among investigators, and ranges from 1 per month to 1 per 6 months to 1 in 3 years. The minimum acceptable frequency is also dependent on age-teenagers being less tolerant of seizures than are 6–8-year-old children. The duration of seizures before intractability can be definitely defined, varies from up to 2 years to more, depending on patient circumstances. Of course, parent and patient play an important role in decision making, as one family might be better equipped to handle the burden of seizures than another.

Some childhood epilepsies are associated with a relatively good prognosis for remission. These include absence seizures and febrile seizures. Other types are not associated with good outcomes, including progressive myoclonic epilepsy, Landau–Kleffner syndrome, West syndrome, etc. Certain etiologies of epilepsy reflect a less favorable outcome, including cortical malformations, microcephaly, abnormal glial cell migration, inborn errors of metabolism, etc. For the most part, these represent rare teratology of the CNS. The age of onset in a patient and possible seizure control is not clear. Results from a variety of studies are unclear.

Some childhood surgeries are palliative, such as corpus callosotomy for atonic epilepsy (drop attacks), while other seizure types are curative, such as extratemporal cortical resection in children, and temporal lobectomy in adults. Children are at an advantage if operated early for intractable seizures in that the length of time of seizures is relatively short, adverse effects of AEDs is minimized, and the plasticity of childhood cerebral tissue is greater than in the adult. The stated goal of surgical intervention is to render the patient free from seizures, without significant neurological sequelae, and hopefully free of AEDs.

In terms of pre surgical evaluation, a complete and detailed history regarding seizure semiological features is essential. Also, careful patient examination is key to determining the intractable nature of the patients AEDs. Intractability may be due to wrong diagnosis, wrong drug, poor compliance, etc. These must be ruled out, or corrected before embarking on a surgical plan. Video EEG monitoring is an important diagnostic tool for lateralizing and localizing seizure activity. It also serves to characterize inter-ictal activity during awakefulness and sleep. This also can show effects of AEDs when administered.

Imaging techniques are evolving rapidly, and the latest procedures should be used for evaluation. These include fMRI, and the use of fluorine-18-labeled fluorodeoxyglucose PET to assess regional metabolic rate. Ictal SPECT scans can show ictal hyperperfusion in the epileptic zone. New MRI methods such as diffusion tensor imaging enable better resolution in order to look for subtle changes in white matter, for example. Also important in the pre surgical evaluation is the lateralization of language function.

Neuropsychological testing assesses overall cognitive function, and can help lateralize the dominant hemisphere for not only language, but for cognitive function. Neuropsychological data can help in analyzing semiological and EEG data. The WADA test involves injecting amobarbitol into the carotid arteries. This serves to determine language dominance, and to assess risk for postoperative amnesia and motor deficits. The WADA test, being invasive, carries its own risk factor. In some cases another invasive test, intracranial EEG, obtained via an intracranial subdural grid and/or depth electrodes. This test is especially valuable in extra temporal lobe epilepsy, common in the pediatric population, and when lateralization is in some doubt. In cases in which the epilepsy seems to be non lesional, or other data conflicts with lesional data, or any other inconsistencies are present in any pre surgical work up, subdural EEG results are warranted, and may be definitive.

The authors of this excellent review on pre surgical considerations state that treatment of refractory epilepsy in children is both a significant challenge, and can be quite rewarding. As many as 30% of epileptic children fit the definition of having intractable epilepsy. The authors state that surgery does have the potential to alter the natural history of epilepsy in children. The authors correctly note that many factors may bear on the intractability of seizures to AEDs, and all of these must be carefully examined prior to the neurosurgical approach.

The majority of epilepsy surgeries are those involving the temporal lobe. Still, 15–20% of cases are extra temporal. A recent paper examines the long-term outcome of extra temporal epilepsy surgery in adults and preoperative prognostic factors (Elsharkawy et al. 2008). This was a retrospective study examining 154 consecutive patients, all operated on at the same site. Results were classified according to three criteria (1) general outcome, (2) outcome in relation to pathologic findings, and (3) outcomes related to resected areas.

Patients were included in the study based on an extra temporal diagnosis, surgery to correct an extra temporal problem, and patient availability for a minimum of a 5-year follow up. This resulted in 154 patients over a 10-year period. Of these, 94% had lesions, and 6% had non lesioned disease. Patient characteristics were that 91 were males, and 63 were females. Mean age of seizure onset was 12.6 years of age. Mean age of first surgery was 28.6 years, and of second surgery was 30.4 years.

Preoperative evaluation included neuroimaging by MRI, PET, and SPECT. The WADA test was performed in most patients, as well as functional MRI. Subdural grids were used to localize and lateralize lesions in cases in which the lesion was not well defined. Somatosensory recordings were made during surgery in some cases. Intra operative electrocorticography was usually performed. Patients were referred to a clinic after discharge for EEG evaluations and follow up.

Results from this study showed that 50% were seizure free after 2 years. This represents the percentage in the Engel I class. About 10–15% of patients were determined to be in each of the other Engel classes (II–IV). There was a significant difference postoperatively between patients in the lesioned group versus those in the non lesioned group (lesioned group = better outcome). Long-term follow up showed the difference was no longer significant. In terms of extra temporal resection, 40% had frontal lobe resection, 51% posterior cortex resection, and nearly 10% had multilobe resection. Seizure-free outcomes were similar in follow up between frontal and posterior lobe resection, but less favorable in patients receiving multilobe resection.

Pathological results showed that one third had neoplasms, another one third had cortical dysplasia, and 20% had gliomas and/or vascular anomalies. Only 2% had hippocampal sclerosis. Eight percent had more than one pathological condition. Most patients had Engel class I results at long-term follow up (in the 50% range) regardless of site of cerebral resection. About 31% of Engel I patients underwent AED withdrawal and maintained seizure freedom for over 2 years. If there were recurring seizures, most were in the first year after drug withdrawal. Patients suffering seizure recurrence (76 patients) usually had seizures in the first 6 months after surgery.

Prognostic factors for successful surgery outcomes included preoperative invasive monitoring, single operation, and surgery being performed within 5 years of seizure onset. Poor outcome correlated with auditory auras, previous surgery, and tonic-clonic seizures. Postoperative complications included infection, paresis, visual defects, and subdural hemorrhage. Total postoperative complications amounted to 13 patients.

The authors note that there are only a very few previous studies looking at surgical outcomes in extra temporal surgeries. This current study was well done in that the patients were not heterogeneous. In this study, pre operative work up, surgery, and postoperative follow up were all performed by the same team. In previous studies of frontal lobe resection (Rasmussen 1991), 56% of patients had either freedom from seizures, or a marked reduction of seizures. In another study (Tellez-Zenteno et al. 2005), the freedom from seizure rate for parietal and occipital lobe resection was 46%. Thus, previous studies had results which were at least consistent with the present study. Long-term outcome in this study was better in patients with well-defined lesions.

As regards prognostic factors, the present study supports the concept that surgical intervention is more successful the shorter the duration of seizures prior to surgery. In terms of AEDs, relapses following AED cessation was about 20%. All regained the seizure-free state following AED resumption. The present data also confirms that there is no difference in risk of long-term seizure relapse between temporal and extra temporal lobe resection.

The authors conclude by stating that their study showed that surgery is effective in both lesioned and non lesioned patients. They also note that surgery results in a better outcome than continued AED treatment. In addition, the outcome by 2 years postsurgery effectively predicts outcomes up to 14 years postsurgery.

In terms of predictive factors for success in seizure surgery, I.Q. has been a controversial subject. Many believe that low I.Q. predicts a lower success rate, so are reluctant to operate on such patients. A recent study (Malmgren et al. 2008) examines this concept. In the study, 325 patients with temporal lobe resections, and extra temporal lobe resection patients were included. The patient population was a mix of children and adults. Results showed at a 2-year follow up, 56% were seizure free: 22% in the group with I.Q.s below 50, 37% in the 50–70 I.Q. group, and 61% seizure free in the over 70 I.Q. group. This was a statistically significant difference between groups. Thus, I.Q. was an independent predictor of surgical outcomes in epilepsy lobectomy. However, the authors note that even the lowest I.Q. group enjoyed some freedom from seizures, and the authors suggest that a low I.Q. should not be an a priori reason to deny surgery to patients who otherwise might benefit.

The problem of clinically differentiating between pseudoseizures and true seizures has been examined in terms of breathing patterns (Azar et al. 2008). Previously, the differentiation between true seizures of a tonic-clonic nature and pseudoseizures has been difficult. Characteristics of pseudoseizures such as pelvic thrusting and bicycle pedaling leg movements are characteristic, and once thought to be diagnostic. Incontinence, physical injury, and a positive response to initial AED trials were thought of as being a good indicator of true seizures.

The present study examined breathing frequency and characteristics as suggestive of whether a patient was having true seizures or pseudoseizures. Results showed that breathing after tonic-clonic seizures had deep and prolonged inspiratory and expiratory phases, whereas pseudoseizure breathing was noted to have irregular frequency, and short inspiratory and expiratory phases. The two groups showed differences in the amplitude of breathing, with tonic-clonic post-ictal breathing being noticeably louder than pseudoseizure breathing. The authors conclude that the breathing pattern following an episode can serve to predict whether a patient is experiencing a generalized tonic-clonic seizure or not.

Continuous simple partial seizures may occur in an epilepsy patient, and carries the name epilepsia partialis continua, and is a form of status epilepticus. These seizures frequently originate near or within the motor cortex, rendering surgical correction difficult. A recent paper examines this problem and surgical approaches (Lega et al. 2009).

In this paper, two cases are described with epilepsia partialis continua, and the surgical approach including ictal recording obtained during awake craniotomy surgery. This is done in conjunction with direct cortical stimulation mapping. This results in a customized surgical approach for each patient. Both patients had refractory nearly continuous focal seizure activity. This feature permits the recording of the seizure activity during awake craniotomy. Thus, the exact anatomical focus of the seizure activity can be determined. This in turn allows a sparing of immediately adjacent non affected yet very important cortical motor areas.

The authors note that in the literature there are only five other reported cases with epilepsia partialis continua who were operated on. In these five cases, intraoperative electrocorticography was also used, but the patients were under full general anesthesia. In two of the other cases, no motor mapping was performed. The authors note that their method using awake craniotomy allows a precise mapping of the ictal zone. This permits an effective resection, at the same time decreasing risk of a postoperative deficit.

The authors achieved this by first recording ictal discharges and mapping critical functional areas, and last, resecting the ictal zone while preserving the functional motor areas. Then, the authors were able to see cessation of seizures both electrographically and clinically. It is the unique feature of epilepsia partialis continua which allows for the surgeon to record the locus of seizure activity intraoperatively. Having the patient awake in the operating room allows direct visual confirmation of the elimination of the epilepsia partialis continua. The ability to actually observe the cessation of seizure activity facilitates knowing when the resection is complete. It is the uniqueness of epilepsia partialis continua which permits a continuous intra operative ability to have a constant ability to determine the anatomical limits of the seizure focus. The authors also suggest that direct cortical recordings have advantages in localizing seizure foci than do subdural grids.

Many studies exist regarding surgery in temporal lobe epilepsies, but not near as many exist for seizures originating in the posterior cortex. Posterior cortex in this context refers to the occipital lobe, or parietal/occipital border of the temporal lobe. The present study (Elsharkawy et al. 2009) aimed to examine long-term outcomes in adult patients who had posterior cortical epilepsy, treated by resective surgery. The study included 80 patients operated on over a 15 year period, and followed up for more than 2 years, and up to 10 years.

In this retrospective study, 44 patients were male, 36 female. Mean age of epilepsy onset was 11.9 years, mean age of epilepsy surgery was 29.5 years. The mean duration of seizure activity was 17.5 years. Auras consisted of gustatory, visual, clonic, hypermotor, and absence alike. EEG results show abnormal recordings most likely over the posterior cortical areas. In 11 patients, no inter-ictal epileptiform activity was detected. Following surgery, all patients showed pathological findings in the resected tissue, and MRI showed that the resection was complete in about 60% of postsurgery patients.

In terms of preoperative work up, patients had video EEG monitoring, high-resolution MRI, neuropsychological testing, and when necessary, PET and SPECT. Over one-third of patients had subdural grid recordings to aid localization and lateralization. In some patients, the WADA test was used. Lesions were classed as well circumscribed and less well circumscribed.

The most common surgical procedure was lesionectomy (81%), with some patients having cortical resection, and lobectomy – with or without subpial transections. Results showed that recurring seizures took place in the first 2 years after surgery. Over 50% of patients were seizure free at the 2-year follow up. About 70–90% of patients were classed as either Engel I or II up to 10 years after surgery. If a patient was seizure free at 2 years postsurgery, the chances of seizure freedom at 10 years was 70%. In 15 patients AEDs were stopped, and two had recurring seizures, which stopped with the reinstitution of AEDs.

Positive predictors for surgical outcome included childhood onset, short epilepsy duration, well-localized lesion, and ipsilateral spikes. Unfavorable outcomes correlated with widespread spikes, tonic-clonic seizures during video EEG, and somatosensory auras. Complications of surgery which become permanent included paresis (two patients) and visual field deficit (one patient). Other complications such as infection were successfully treated.

The authors note that their study is the first looking at long-term outcomes following surgical treatment of posterior cortical epilepsy. The authors note that their study reconfirms the concept that a 2-year follow up accurately predicts long-term (up to 10 years) outcomes. The positive correlation of outcomes with early surgery, localized lesion, and short duration of epilepsy were demonstrated. The most important long-term predictor of outcome in this study was the absence of generalized tonic-clonic seizures. Additionally, a complete resection of the lesioned area was associated with a better outcome. The authors suggest that high-resolution MRI may not clearly differentiate the actual borders of the lesion. Six patients underwent additional multiple subpial transactions, without effect. The reasons for possible incomplete resections include the possibility that a lesioned area is not epilepsy related, misleading information from scalp EEGs, and/or existent very small (microscopic) areas of epileptic tissue which is outside the lesioned areas. A lesion area may, with time, shrink and disappear, or undergo a progressive expansion.

Post operative EEG showing inter-ictal epileptiform discharges is a poor predictor for a seizure-free outcome. This indicates the lesioned area may recover and increase, or that a focal area was missed. As has been shown in frontal lobe epilepsy

(Spencer et al. 2005) the absence of tonic-clonic seizures is a powerful predictor of a good surgical outcome.

Another recent paper (Cheung et al. 2009) examined the ability of functional MRI to accurately judge memory processing before and after temporal lobe surgery for epilepsy. Previous studies show that patients with left temporal lobe epilepsy show verbal memory impairment (Helmstaedter et al. 1997), whereas those with right temporal lobe epilepsy tend to show non verbal memory deficits (Gleissner et al. 1998). In this study, functional MRI was used to evaluate verbal and non verbal memory both pre- and postoperatively in patients with temporal lobe epilepsy.

In this study, 17 right-handed preoperative patients and eight controls were utilized. Pre surgical evaluation included video EEG, neuropsychological testing, the WADA test, and functional MRI. For functional MRI testing, the patient was evaluated following a 20 s picture encoding task and a 20 s visual fixation task. The same test was administered 1 year after surgery. High-resolution functional MRI was used and 16 contiguous slices (7 mm) were reconstructed to evaluate the right and left mesial temporal lobes.

Results showed that preoperatively, left temporal lobe epilepsy patients had a worse verbal memory than controls, whereas right temporal lobe epilepsy patients had impaired preoperative visual memory performance. The two temporal lobe epilepsy groups were similar to each other as regards verbal and non verbal memory performance. Results from postoperative memory performance showed that overall, the two groups performed worse than controls. Thus, left temporal lobe resection patients were worse than controls in verbal memory, whereas right resected patients were worse as regards non verbal (visual) memory. Hippocampal volumes were decreased after surgery, but there was no significant correlation between volumes in either right or left surgically treated patients.

The authors note that only left temporal lobe epileptic patients had a verbal memory deficit, while both left and right patients had a decrease in functional activation in the ipsilateral mesial temporal lobe. Memory changes after surgery could be due to the capacity of contralateral mesial temporal lobe, or to a certain functional reserve and hippocampal adequacy. Which of these is actually occurring is not clear, but the authors suggest their data support the idea that the contralateral mesial temporal lobe plays an important role in support of postoperative ipsilateral memory. Further studies with larger numbers of patients seems justified.

In terms of outcomes, a paper describes results from corpus callosotomy surgery for generalized tonic-clonic epilepsy and atonic (drop attack) seizures. In this study, 95 patients were examined retrospectively as regards seizure type and outcome. These seizures were refractory to AED treatment, and judged to be candidates for corpus callosotomy. This procedure is believed to slow/stop the rapid spread of seizure activity from one side of the brain to another, thereby stopping tonic-clonic and atonic seizures.

One question relates to the optimal extent of callosotomy sectioning. Results from this study demonstrate that when comparing outcomes, patient undergoing a corpus callosotomy involving only the anterior half as compared to those having a 2/3s callosotomy had a poorer outcome. A complete callosotomy is associated with a higher rate of complications, such as the occurrence of disconnection syndromes. This study also determined that callosotomy had no adverse effect on I.Q., nor was there any effect of extent of callosotomy and I.Q.

Overall, corpus callosotomy should be used in patients who suffer severe epilepsy and are prone to injuries from falls.

Vagus nerve stimulation has been approved for refractory partial epilepsies, and has even been used palliatively for generalized epilepsy (Ng and Devinsky 2004). Many patients who are candidates for corpus callosotomy undergo vagus nerve stimulation before callosotomy, since it is a much less invasive treatment, requiring only a very short time to implant, and is performed as day surgery.

In another study of surgery for cortical dysplasia, outcomes were compared to completeness of resection. Cortical dysplasia makes up about 15% of all epilepsy surgeries, and the vast majority of seizures in children under two. Cortical dysplasia exists in two classes: the first is mild, the second severe, based on the presence of abnormal (balloon) cells in type 2 (Lerner et al. 2009). Recent studies show that inter-ictal ECoG has a unique pattern in cortical dysplasia, and also that high intensity MRI may be indicative of damaged areas in 50–90% of patients (Cohen-Gadol et al. 2004a, b). This means that surgery for cortical dysplasia is, and will continue to be increasing.

Results demonstrated that the only significant predictor of a cohort of 149 patients was the completeness of the surgical resection. This was confirmed by MRI before and after surgery. The cortical region showing ictal and inter-ictal abnormalities on intracranial EEG represents the target for resection. All other factors such as age, duration of seizures, seizure frequency, other pathologies including hippocampal sclerosis, extent of EEG findings, etc., were not predictors of outcome. Twenty-five percent of patients had a significant improvement in that they had a higher classification of Engel's model after the second postoperative year. The authors conclude that this surgery can improve seizures associated with cortical dysplasia, and the extent of the resection is a positive predictor of outcome.

It is well known that the suicide rate is several times higher in epilepsy patients than in the non epileptic population. The effect of surgical control of seizures and suicide is not as clear, but has recently been studied (Hamid et al. 2011). In this study, a standardized assessment was administered which detailed the patient's history, diagnostic testing, and neuropsychiatric condition. The Beck Depression Inventory and Beck Anxiety Inventory were both also administered.

Results showed that of 396 enrolled patients, four had deaths attributed to suicide. The standardized suicide rate, corrected for age and gender was 13.3. Only one suicide patient had a Beck Depression Inventory showing severe depression. Three others showed no depression, or only mild depression. In addition, two showed moderate to severe anxiety on the Beck Anxiety assessment. The authors conclude that even in cases in which the patients report highly successful seizure control following resective surgery or epilepsy, suicide can occur.

Depression is a key factor in suicide, and its treatment is essential for health and a satisfactory quality of life. Patients with depression often show a more severe seizure disorder. It is possible that suicide, depression, and epilepsy may show a common pathophysiology. The precise mechanisms of depression and suicide as related to epilepsy AEDs are still vague (Wen et al. 2010).

Part VII
Other Aspects of Epilepsy

Chapter 36
Pediatric Concerns

Somewhere between 25,000 and 40,000 children will have a first non-febrile seizure each year in the US. The mortality rate is low in these children, but it is important to have a thorough evaluation of these patients. Several common conditions can be mistaken for seizures, including apnea, breath holding, ADHD, and sleep disorders. Defining the etiology, if possible, is important. If, for example, they are due to metabolic or toxic insults, then correction of these conditions will likely ameliorate the seizures. In childhood seizures, the type seizure should be determined according to ILAE criteria. Several types of seizures predominate in infancy and very early life. These include infantile spasms, childhood absence seizures, febrile seizures, etc. Children are most vulnerable in the first year of life.

In a history, preceding events are important to inquire about, and postictal events are also important. Children themselves are not completely reliable until age 3–4 as regards auras. Seizures associated with awakening are usually generalized tonic clonic events. The witnessed event should be described by the patient or caregiver in great detail to the physician. Forced eye closure may signal a pseudoseizure. The length of time of the seizure could indicate status epilepticus or not. The nature of postictal events are critical.

The developmental history is important to ascertain since as many as 30% of children with cerebral palsy have epilepsy, and 10% or more of autistic children also have epilepsy (Shinnar and Pellock 2002). Developmental mile stones may afford clues about postnatal development.

The family history is important both in terms of genetic history and history of seizures, as well as the pregnancy history. Obviously, a neurological examination can be indicative of seizure activity. Laboratory tests should be performed as indicated. Suggested tests should include CSF creatine, glucose transport, and neurotransmission tested (GABA and glutamate) if possible.

The EEG is the gold standard in seizure evaluation, particularly if a seizure episode is captured. The EEG can help define the seizure type, help distinguish if the episode is a pseudoseizure, and aid in determining appropriate AED treatment. One study (Yoshinaga et al. 2001) showed a 20% error in diagnosis of seizure type in cases in which the EEG was not used.

D.W. McCandless, *Epilepsy: Animal and Human Correlations*,
DOI 10.1007/978-1-4614-0361-6_36, © Springer Science+Business Media, LLC 2012

Although controversial, imaging studies such as MRI are recommended for seizure evaluation (Hirtz et al. 2000). Prolonged seizures and an abnormal neurological exam are the usual indication that an MRI would be beneficial. Even with normal examination and normal EEG, 10% of cases show abnormal MRI (Shinnar et al. 2001a, b).

Seizures are highly upsetting for parents, especially at first, and a careful evaluation of the affected child is essential. Imaging alleviates worries about structural brain anomalies as well as stroke. MRI is a preferred imaging method.

Development does not stop at birth, but continues in several organ systems including pulmonary, G.I. and liver, cerebral-myelination, renal, etc. Studies have shown that AEDs are affected by this (as compared to adults) for the first 6 months during which time metabolism and elimination are slow. Later, metabolism may be faster in adults. These considerations are discussed in a paper recently published (Gilman et al. 2003).

When children enter a hypermetabolic stage, blood levels of AEDs should be monitored and the doses adjusted accordingly. Low serum concentrations can indicate hypermetabolism or non-compliance, and the physician must distinguish between the two. The maturation of the G.I. System is important. Intestinal transit time and surface area are low in young children. These differences may offset each other. Liver metabolism may be slow in young children as well. Finally, the brain of young children may not respond as quickly as in an older patient. The blood brain barrier may not be as mature for transport of various substances as in more mature patients; these differences in pediatric patients bears careful observation.

Children with epilepsy have special needs especially as regards educational needs, generally greater than those needed by adult epilepsy patients (Berg et al. 2005). This study looked at the timing of service needs in relation to the onset and progression of the patients' epilepsy.

In this study, 613 children were recruited at the time of epilepsy diagnosis. Mean age at first seizure was 5 years 11 months (range 1 month to 15 years 8 months of age). The parents had an initial interview 5 years after the first seizure. Seizures were classified as being idiopathic, cryptogenic, or remote symptomatic.

Symptomatic and encephalopathic children received available services more frequent by 88% versus the other epileptic patients (49%). Interestingly, services were sought by 15% of children before the first seizure. This could signal that these patients were showing behavioral or cognitive changes before seizures started. Several other reasons could have also influenced this statistic.

The authors note that many epileptic children take advantage of special educational programs and services. Since epileptic children may frequently have adverse social and cognitive outcomes, the educational services available should be utilized whenever possible.

Another study looks at the possible factors involved in educational delay in epileptic children (Aldenkamp et al. 1999). Many factors could have an impact on the development of learning problems in epileptic children. These include epilepsy type, seizure frequency, age of event, side effects of AEDs, etc. Several studies support an increased risk for educational problem in epileptic children compared to the generalized population.

The present study was designed to look at the two issues: the educational delay in epileptic children, and the type of learning problems in these children. In both cases, the epileptic children were compared to children with learning problems, but without epilepsy.

Results showed age of seizure onset in the epilepsy group was 24 months. Seizure types included complex partial (4), absence seizures (16), tonic clonic (1), other (3). Seizure frequency ranged from seizure free for the last 6 months (6) to a high seizure frequency (several per month). Full-scale I.Q. was 90 for epilepsy children, and 94 for the control group.

The effects of epilepsy on school achievement could not be properly assessed due to the recent epilepsy diagnosis in almost one half of the patients. School achievement scores were lower in epileptic children with localized epilepsies.

The authors of this study showed that the educational delay for epileptic children was about one half a grade. This is in keeping with other studies (Seidenberg et al. 1987). This level of underachievement was not statistically different from the learning-disabled group without epilepsy. The epilepsy group was not referred more often than the group without epilepsy. Analysis of separate academic skills in the epileptic group showed similar results.

The authors state that girls were over represented in this study which may have skewed the data. This study also shows, say the authors, that epileptic children with less AED-controlled seizures have poorer results than those whose seizures are better controlled. This conclusion is somewhat speculative, and should be further studied.

A review paper (Sanchez-Carpintero and Neville 2003) examines aspects of attention ability in children with epilepsy. Epilepsy is associated with a variety of other problems including ADHD, autism, learning problems, cognitive impairments, etc. Attention is thought to be a skill necessary for cognitive and behavioral function. Sustained attention serves to maintain focus and alertness.

Children with benign childhood epilepsy with centro temporal spikes have sustained attentional problems, and in these patients, right-sided interictal electrical activity interferes with sustained attention. This can also occur during sleep. Patients with complex partial seizures may have sustained attention deficits, but selective or divided attention are normal. Early age onset of seizures have more frequent cognitive difficulties.

Prospective studies which examine different aspects of attention, both visual and auditory modalities, would be helpful. The actual financial cost of epilepsy has been examined (Beghi et al. 2005). The authors note that epilepsy treatment has a continuing cost basis consisting of hospital care + new AEDs and new non pharmacological tools. Many of these are expensive. Economic costs are different between children and adults.

Costs are usually high in secondary and tertiary facilities due to the fact that they usually have more severe patients, and necessary equipment to handle these cases.

The diagnostic procedure process is usually most expensive, then costs drop. The tertiary care centers tend to use the latest treatment modalities, thus raising individual patient costs. Most studies focus on epilepsy cost on adults, rarely on children. In one study (Mirza et al. 1993), there was a reduction in health care requirements which could be attributed to a reduction in polytherapy to 1 or 2 AEDs from 4 to 5, 1/3 of patients became seizure free, and most of the rest had a 50% reduction in seizures. This led to a significant reduction in cost.

It is also clear that the type of epilepsy, age of patient, and disease severity all can alter epilepsy costs. New onset epilepsy costs are highest, possibly four times as much as the second year's cost (De Zelicourt et al. 2000).

The authors conclude saying that overall, childhood epilepsy costs are higher than those of adults. Seizure frequency was correlated with higher costs, and treatment with AEDs was also directly correlated to cost of epilepsy. The two longest components of total expenditure were hospital admissions and drug treatment.

Future studies should stress consistency. Comorbidity costs should be eliminated from consideration. The study population should be well defined. All costs should be available. Studies should be conducted in the same country. The health care system should be documented. Authors should describe limits of the studies.

Another problem facing patients and epilepsy patients alike relates to compliance as regards AED prescriptions. As many as 75% of epilepsy patients miss taking their medication at some point, which can have serious adverse effects, even life-threatening consequences (Mitchell et al. 2000). The present study aims to examine features of successful drug taking, and suggest ways to increase compliance (Wilmot-Lee 2008).

Various factors influence the taking and/or administration of AEDs. These include uncertainty about the diagnosis, fear of addiction, worry about side effects, etc. For these reasons and others, as many as 20% of newly diagnosed epilepsy children were not taking AEDs as prescribed within 1 month of diagnosis.

Suggestions for compliance include better education for parents and children. The epileptic children should be taught by health care personnel as to why taking AEDs regularly and not missing any days is highly important. Children should receive basic information about medicine and the general importance of compliance.

Studies in asthmatic pediatric patients have shown that by age 7, children know about their disorder, and its treatment (Sanz 2003). The adolescent period can be a challenge since children are facing several additional problems, and may rebel against parental/adult suggestions. Education is a key to gaining compliance in this age group.

The author concludes saying that the promotion of long-term AED taking can be a daunting task. It involves continual attention to make sure there are no lapses. The patients experience with medication for epilepsy is at the core of successful treatment.

In another recent paper (Arthur et al. 2008) the risk of seizure recurrence after the first seizure is examined. Previous studies on seizure recurrence are variable because of different definitions. The practice parameter of evaluating first seizures stress the utility of MRI. The conclusion is that routine time neuroimaging is questionable since results rarely influenced initial treatment (Hirtz et al. 2000). The correct study further evaluates the utility of both EEG and MRI in predicting seizure recurrence after an initial seizure.

In this study, 1,540 children were used in the EEG portion of the study and 125 in the MRI study. Each child had extensive history and neurological examinations, as well as EEGs and MRIs. Each patient was classified as to seizure type. Parents were interviewed every 9 months for 27 months to determine seizure recurrence.

Results showed that 35% of patients had generalized seizures, and 63% had partial onset seizures. The mean age of the patients was 9.7 years of age, and mean I.Q. was 104. A recurrent unprovoked seizure did occur by 9 months after the initial seizure in 58% of children. Results showed no association between seizure recurrence and EEG findings.

Results comparing recurrence of seizures after the initial seizure and MRI results showed an association between early (less than 9 months) recurrence and abnormal MRIs. This significant recurrence association was not present after 9 months up to 27 months. Among patients with abnormal MRIs, 100% had a recurrence in seizures by 27 months.

The authors state their purpose was to examine seizure recurrence in a normal well-defined group of epileptic children to see if EEG and/or predicted seizure recurrence.

Recurrence rates in the present study compared favorably with those of Martinovic (Martinovic and Jovic 1997). Somewhat lower rates have been described by other workers (Shinnar et al. 1996).

The authors state their EEG studies did not show any risk factors as regards recurrence. The present study was of normal subjects; other studies have examined epileptic children with behavioral and/or cognitive problems and found a correlation between abnormal EEG results and recurrence.

The authors question whether the abnormalities in MRI can predict whether these children have a risk for more severe epilepsy and intractability. Further long-term studies are needed to answer this important question. If this question could be correctly answered, and it could be shown that the abnormal MRI could predict intractability, this could suggest early surgery should be considered. Early MRI (for new onset epilepsy) may be warranted at least to better classify the seizure activity.

Another paper looks at seizure recurrence, and the issue of the initiation of AED therapy (Arts and Geerts 2009). Much data has been collected previously regarding the course and progression of seizures, which can serve as guidelines for treatment.

The authors note that recurrence of seizures after the first seizure is around 50%. Outcomes are comparable regardless if treatment is started after the first or second seizure. Similarly, outcome after status epilepticus is similar to that of patients with shorter seizures. This implies treatment after the first seizure, or the first status is not essential. Patients with benign partial seizures may also not need immediate treatment.

However, the authors state that the actual danger of the seizures themselves, the risk for intellectual decay argue in favor of timely AED treatment. Some types of epilepsy (generalized idiopathic absence epilepsy) have risks for learning problems and accidents encourage early AED treatment. Other epilepsies such as Dravet's syndrome, the Lennox–Gastaut, and the West syndrome all call for aggressive early treatment, even including surgical procedures.

The authors conclude saying that postponing early treatment after a single seizure, or even one status episode probably does not worsen prognosis. Certain types of epilepsies, seizures with progressive features, and certain etiologies all suggest early aggressive AED treatment, progressing to other treatments if AEDs are not efficacious.

The issue of cognitive deficits is an important feature of childhood epilepsy. The present study (Hermann et al. 2008) examines the cognitive development in children with new onset epilepsy over a 2-year period. In adults with epilepsy, cognition is relatively stable, but in children, there is a changing level of cognition associated with brain development, making cognition more vulnerable.

Several epilepsy factors can cause cognition problems including epilepsy etiology, level of seizure control, and AEDs. In this present study, two groups of children had comorbid condition which adversely affected cognition: ADHD and children with academic problems.

In this study, 100 children between ages 8 and 18 were utilized. Fifty-two children were with new/recent onset epilepsy, and 48 were healthy first cousin controls. The epilepsy patients had idiopathic epilepsy, no abnormal findings on MRI, no other neurological abnormalities and diagnosis of epilepsy within 12 months of inclusion in the study.

Results showed that 38 control subjects had no comorbidities, while three had ADHD and four had academic problems. Comorbidities in the epilepsy groups included 15 with ADHD, and 13 with academic problems. This was significantly higher than the first cousin controls ($p=0.001$).

The effect of no comorbidity/epilepsy, and + comorbidity epilepsy) had significant influence on outcomes. The + comorbidity epilepsy group was significant for 5 of 6 domains, notably: intelligence, academic achievement, language, executive function, and psychomotor speed. There were no significant differences between controls and no comorbidity epileptic patients.

In most cognitive domains, the differences between groups were unchanged overtime. Group-by-time changes were noted in the executive function domain.

The authors note that the presence or absence of the two comorbidities during epilepsy predicts abnormal cognitive development, and further predicts similar results 2 years later. Those epileptic children without comorbidities have normal neuropsychological scores at baseline, and 2 years later, as compared to controls. Lastly, the presence/absence of comorbidities in epileptic children were not associated with epilepsy type, age of onset, AEDs, or time course.

It is well recognized that comorbidities may be recognizable at or before the initial seizure onset (Austin et al. 2001). It has also been noted that neurobehavioral signs/symptoms are evident at seizure onset in some children (Berg 2007). The present study shows the two comorbidities result in a distinct pattern of cognition.

Another set of interesting findings is that epileptic children without comorbidities were otherwise indistinguishable from healthy controls in every area tested including intelligence. The authors state this finding might be of special importance. Many previous studies show significant problems of epileptic patients with psychosocial issues, quality of life, psychiatric status, etc. This pattern is even seen in benign, mild epilepsies, and patients who are seizure free. It has been shown that these comorbidities in the general (non-epileptic) population can produce the same abnormal psychosocial outcomes (Spencer et al. 2007; McKay and Halperin 2001). This data could indicate a "subclass" of epileptic children who could benefit greatly from early, aggressive cognitive/social progress aimed at improving outcomes.

The authors conclude saying their study shows the presence or absence of important comorbidities (ADHD and academic problems) can greatly influence cognition and subsequent social parameters. Long-term psychosocial outcomes may be amenable to normalization depending on how these comorbidities are treated. This paper is highly significant.

In a recent study (Lewis and Parsons 2008) the actual understanding of epilepsy by children who have epilepsy, was examined. Some other studies have suggested that the school experience has been a key determinant of future quality of life. Underachievement in school could result from seizures, AEDs, or school factors. Stigma may play an important role. The present study includes two overlapping data sets – one was a survey of 44 epileptic children and interviews with 22 separate epileptic children. Age of all participants ranged from 3 to 23 years of age.

Results showed that primary school children could describe seizures, and were willing to talk openly about the disorder. A lack of discussion about seizures in school may have led to same misconceptions such as that epilepsy was contagious, or that pale skin could transmit epilepsy. Half of the primary school students thought medication would help the epilepsy go away.

Students with epilepsy aged 16–17 complained of AED side effects such as weight gain. Many complained of feeling tired. The tiredness seemed more permanent in younger patients, this could be misinterpreted by teachers who do not realize it is a part of epilepsy.

There was reluctance for epileptic children to tell friends about the seizures. Having control over who knew about the patient's epilepsy was important to children. Some children commented on events in which they could not attend, such as sleep overs at a friend's house. Some epileptic children had few friends. Conversely, some patients had supportive and understanding friends who helped significantly with positive attitudes.

Many children expressed worry and concern about seizures at school, and how the teachers viewed epilepsy. Children who were most "relaxed" at school were also confident in the teachers' response to their seizure state. In older children, the idea was that epilepsy "got in the way" of plans and activities. Not being able to drive was a serious problem for some students with epilepsy. Still, most epilepsy patients had optimistic attitudes. There was no major feeling by the students, that epilepsy would hinder their future plans.

The authors comment that there was a certain "invisibility" to the epilepsy in the students in that they could in some ways conceal the disorder. It did not permeate daily life. Many patients acted to conceal epilepsy because they were unsure of other peoples' responses. Many patients felt "not normal," which in actuality reflected others perceptions of epilepsy rather than the patients intrinsic impairments.

The authors conclude saying perhaps epilepsy is "over medicalised" at school. For example, medications at school are to be "administered to" children rather than available for the children to take for their epilepsy, too much emphasis may be placed on what the child cannot do, rather on activities the epileptic school child can do. This is seen easily in the children's overall perception of how epilepsy is accepted at home versus how epilepsy is accepted at school.

In another similar paper regarding epilepsy perceptions, the impact of the disorder is not limited to children, but can affect other family members (Ferro and Speechley 2009). Mothers of epileptic children are at particular risk for clinical depression. This paper purports to examine maternal clinical depression and its effects on child progression and outcome in children with epilepsy.

Results from this study show as many as 50% of mothers of epileptic children are candidates for significant clinical depression at some time after the diagnosis of epilepsy in their offspring. Specific risk factors correlate with the clinical depression including ambiguity of the maternal role, significant worry and maternal satisfaction with interpersonal relations.

These depressive symptoms are worthy of treatment not only because of the maternal mental health, but due to the negative effect on the epileptic child. These maternal depressive symptoms can adversely affect their child's behavior, and health-related quality of life index. The authors note that their studies warrant further more extensive clinical assessments of maternal depression.

An I.L.A.E. subcommittee for pediatric neuroimaging has examined the features of imaging in the evaluation of children with newly diagnosed epilepsy. This was a retrospective time and prospective study. Over 30 studies with several thousand patients provided data.

Imaging results are most often significantly abnormal in either localization-related seizures, or remote symptomatic epilepsy. As many as 50% of imaging studies in localized new onset epilepsy are abnormal, and as many as 20% provide significant information. Information from a few patients will prove critical (King et al. 1998; Chang et al. 2002).

The authors note that neuroimaging is a key component in the evaluation and treatment of childhood epilepsy. Children with remote symptomatic localization-related epilepsy have a less favorable prognosis, and may require additional treatment. Imaging serves to identify lesions, finding chronic processes which might require immediate attention (tumor), and identifying acute process requiring immediate attention (stroke). Neuroimaging may not be needed in cases of idiopathic local or generalized epilepsy due to their "benign" nature (Stroink et al. 1998; Sillanpaa et al. 1998). The authors note that MRI is the imaging mechanism of choice because of excellent anatomical resolution. MRI can resolve abnormalities better than CT scans. Since immature myelination can obscure some lesions in children under 2, special sequences are required. Skilled pediatric imaging professionals should interpret results.

The authors conclude saying that neuroimaging is recommended for all new cases which do not have features characteristic of classical idiopathic local or generalized epilepsy, and any child under 2. MRI is superior due to increased resolution. MRI permits visualization of structures which cannot be resolved with C.T. the idea is to be able to identify lesions requiring an alteration in medical/surgical management.

Another recent study evaluated the effect of AED treatment on vitamin D metabolism and markers of bone turnover in 38 epileptic children on AEDs and 44 healthy control children (Nettekoren et al. 2008). In this study, patients were between 5 and 12 years of age and used AEDs for a minimum of 3 months. Exclusion included any condition which could not affect bone metabolism, or taking drugs or vitamins which could affect bone metabolism.

Results showed that 75% of the epilepsy patients had 25-hydroxy vitamin D (25-OHD) levels below normal levels, indicating vitamin D deficiency. By contrast, 23% of the controls were also vitamin D deficient. Polytherapy AED patients had a greater incidence of vitamin deficiency than seen in monotherapy patients. No statistically significant correlation was found between vitamin D and bone turnover in epileptic patients.

The authors comment that their study is in general agreement with other studies showing that AED treatment is associated with lower serum 25-OHD levels as compared to control subjects (Baer et al. 1997). AED-treated patients may be at risk for osteopenia and osteoporosis. Vitamin D also plays a role in decreasing the importance of certain autoimmune diseases.

There were no correlations between bone markers and 25-OHD levels, so while AEDs may alter the skeleton, mechanisms remain obscure. Acute/chronic exercise affect bone metabolism, and this was not assessed in the present study, the authors state that given the effects of AEDs on bone, children taking AED treatment for epilepsy should be regularly monitored, and supplemental vitamin D administered if necessary.

A major concept regarding early brain damage has been that with early damage comes an improved recovery (Kennard 1936). These concepts have changed somewhat in that the cerebral damage is probably more extensive than previously thought (Anderson et al. 2005). The present study (Max et al. 2010) looks at brain plasticity following stroke. Vulnerability and age of lesion are examined in terms of outcomes of overall cognition and other domains.

The present study was designed to determine the effect of plasticity and vulnerability based on age of lesion onset and on cognitive and psychiatric outcomes in stroke patients. The results from this study could have relevance for epilepsy children as they too have early focal brain lesions. Control patients were children with club-foot and scoliosis.

Results showed differences between stroke patients and controls were largest in the early groups as regards the neurocognitive domains of language, visuospatial, and memory, lens affected were domain of reading and spelling, as well as academic function. The rate of psychiatric disorder was not significantly different between early and late stroke groups. The mean number of psychiatric diagnosis was higher in both stroke groups than in the orthopedic group.

The authors state the most significant finding of their study was that prenatal, or up to 1 year postnatal stroke was associated with a low to high degree of lesser performance on a variety of neurocognitive and/or psychiatric measures. The affected domains included those of intellectual function, visuospatial function, and psychiatric domains.

The authors note that the later vulnerability for certain executive function deficits suggests changes are not an indication of a global impairment. The findings are consistent with a relation of executive function and age in patients with local frontal lesions form trauma, penetrating injuries, cerebral dysplasia, demyelination, stroke (and seizures).

The authors comment saying these data suggest that even in the presence of local cerebral lesions, vulnerability rather than plasticity is the characteristic norm. Vulnerability after early lesions could be due to metabolic demands arising from neural networks rapidly developing skills. This is coupled to myelination demands, which require increased synthesis and energy requirements. Repair and reconnecting would also increase metabolic demand, in turn increasing vulnerability.

Another idea concerning the difference between early and late damage and outcome states that the early insult serves to disrupt subsequent brain development (Hebb 1942) and would limit the acquisition of knowledge and skills. This could have a "domino" effect such that the outcomes would worsen over time. Some skills undergoing rapid development at insult could be more vulnerable, plus some domains might not show effects until later in development.

The authors conclude saying longer studies (more patients) would allow a better assessment of effects of lesion size, location, and laterality. Increased size of the study would permit better assessment of windows of critical periods in these studies. This interesting study has a bearing on features of seizures and their outcomes on cerebral development because of similarities of epilepsy to stroke in terms of timing, lesion size, neurocognitive and psychiatric outcomes, plasticity, etc., mechanisms of brain damage, and are informative, and valuable.

Table 36.1 Epileptic syndromes and inborn errors of metabolism

Epileptic Syndromes	Inborn errors of metabolism
Neonatal seizures (ILAE 3.1)	Urea cycle defects: argininosuccinic acidemia, ornithine transcarbamylase, carbamylophosphate synthetase Organic acidurias: maple syrup urine disease Disorder of biotin metabolism: early-onset multiple carboxylase deficiency (holocarboxylase synthetase deficiency) Peroxisomal disorders: Zellweger syndrome, acyl-CoA oxidase deficiency
Cryptogenic myeloclonic epilepsies (ILAE 2.2) other than infantile spasms or Lennox–Gestaut syndrome	GM1 gangliosidosis GM2 gangliosidosis Infantile neuroaxonal dystrophy Neuronal ceroid lipofuscinosis Glucose transporter defect 1 deficiency Late-onset multiple carboxylase deficiency Disorders of folate metabolism, methylenetetrahydrofo late reductase deficiency Arginase deficiency (urea cycle defect) Tyrosinemia type I
West syndrome, generalized 2.2	Phenylketonuria/hyperphenylalaninemia Pyruvate dehydrogenase deficiency Pyruvate carboxylase deficiency Carbohydrate-deficient glycoprotein syndrome (type III) Organic acidurias Amino acidurias
Early myeloclonic encephalopathy/ early infantile epileptic encephalopathy (ILAE 2.3.1)	Nonketotic hyperglycinemia Propionic acidemia D-Glyceric academia Leigh disease

Adapted from Pellock, J. 2009, Chap 11, Jain, S., and Morton, L.

Finally, several seizure types are associated with a variety of inborn errors of metabolism. Most inborn errors are quite rare, and many will not be encountered in a lifetime of clinical practice. They, however, are frequently associated with more severe epilepsies, and should be tested for if there is any suspicion (see Table 36.1).

Chapter 37
Pseudoseizures

Pseudoseizures (nonepileptic seizures) appear overtly to be genuine epileptic seizures, but are unaccompanied by EEG changes. These pseudoseizures are uncontrolled and those with pseudoseizures do not consciously produce them. Pseudoseizures are unresponsive to AEDs. There are always some patients with pseudoseizures who also have real seizures.

Symptoms include many which also appear in true epileptic seizure patients, and some real symptoms can distinguish pseudoseizures. After the pseudoseizure builds over several minutes, some features resemble panic attacks. The ability to use video EEG in order to record overt seizure-like movements, at the same time confirming no EEG activity, is an important diagnostic technique.

Seizure comparison may be useful, although true seizures and pseudoseizure symptoms overlap nearly 100%. Automatisms are usually not present in pseudoseizure patients, and also highly infrequent, are incontinence, and postictal depression. In some cases pseudoseizure patients may yell and use loud language, where as true seizure patients may groan and mumble.

In a sense, pseudoseizures fill a gap between psychiatry and neurology. An adverse feature of the disorder not really being psychiatric, and not really a neurologic disorder, has resulted in little or no treatment. AEDs have no significant effect beyond a placebo effect. As stated above the advent of video EEG has eased the diagnosis aspect and prompted additional studies (Cragar et al. 2002).

In spite of improvements in diagnosis, no meaningful advances in treatment regimens have been developed. There may be etiologies which will suggest treatments. The phenotypic manifestations of pseudoseizures are varied, and several possible etiologies, including physical abuse, trauma, comorbid psychiatric problems, and reinforced behavioral problems are described.

There is certainly a correlation between pseudoseizures and hysteria. The idea that family dysfunction may prompt pseudoseizures has suggested the treatment of leaving home. Few if any scientific controlled randomized studies have been preformed in pseudoseizure patients. In a prospective study (McDade and Brown 1992), one half of the patients (16) showed cessation of seizures following supportive psychotherapy, physiotherapy, operant conditioning, and occupational therapy.

D.W. McCandless, *Epilepsy: Animal and Human Correlations*,

DOI 10.1007/978-1-4614-0361-6_37,

Table 37.1 Unique features of pseudoseizures

Unique features of pseudoseizures
1. Screaming, roaring, shouting
2. Strange movements (e.g., pedaling leg movements)
3. High frequency of episodes
4. No response to AEDs
5. Continued awareness
6. Abrupt onset and termination
7. Dramatic nature

Adapted from Stores, G. (1999) J. Child Psychology and Psychiatr 40: p. 851

Another patient group had 75% success of the receiving group therapy, family therapy, and personal therapy (Kim et al. 1998).

Just as in epilepsy, there are several or more subtypes in pseudoseizures, and these may predict various treatment regimens. Delineation of these subtypes is critical. Interview methods could lead to the successful diagnosis of pseudoseizure subtypes, using risk factor analysis. Defining risk factors for the cessation of pseudoseizures could aid in preventative treatment.

One interesting treatment finding was that telling the patient that he/she has pseudoseizures, compiled with explanations, served to terminate the pseudoseizures (Buchanan and Snars 1993). Since psychiatric comorbidities such as depression and anxiety are often present in pseudoseizure patients, their successful treatment may also reduce pseudoseizures.

Ultimately, a coordinated effort on the part of both neurologists and psychiatrists may be needed for successful treatment. Unlike true epilepsy, early studies may indicate a high treatment success rate. Psychotherapy may be a key therapeutic device in efficacious treatment.

The issue of the recognition of pseudoseizures in children and adolescents is a problem. Many of the "classic" signs and symptoms are similar to regular epilepsy making the diagnosis difficult. It is important to correctly diagnose the condition in order to avoid AEDs when not necessary. The present paper aims to clarity features of each in order to avoid misdiagnosis (Stores 1999).

There is a broad spectrum of symptoms displayed by pseudoseizure patients, including opisthotonic posturing, loss of bone, loss of awareness, unresponsiveness, and complicated behavior. Unusual diagnostic features of pseudoseizures include bizarre and dramatic characteristics. The problem arises when regular seizures have bizarre phenotypes such as those of frontal lobe epilepsy in which symptoms such as pedaling leg movements, screaming, shouting, hitting, rocking, etc., may occur. Misdiagnosis here could lead to no AED prescriptions and further brain damage.

Some features are thought of as being those of pseudoseizures: elevated frequency of attacks with the AED effect, no loss of awareness, many hospital visits without any findings, and abrupt onset and termination. But these above symptoms are also sometimes seen in frontal lobe epilepsy. Further, cases have been described with both frontal lobe epilepsy and pseudoseizures in the same patient (see Table 37.1).

Other clinical features of pseudoseizures may include: attacks are never witnessed by any independent observes, readily induced by (verbal) suggestion, especially in children, fast evolving attacks of a wide variety.

The EEG findings are important. Pseudoseizures are not associated with any alteration in EEG. Care must be taken, because similar EEGs are seen in mesial frontal epilepsy (Stores et al. 1991). However, a normal EEG during a convulsive attack is highly significant. Measuring plasma prolactin and estimating creatine kinase may help distinguish pseudoseizures from epilepsy (Wyllie et al. 1985).

The authors note that confusion surrounds the differential diagnosis between pseudoseizures and true epilepsy. The distinction is important from the treatment (mistreatment) standpoint. Because of the psychological natures of pseudoseizures, early therapeutic intervention is essential. Psychological prognosticators include anxiety, depressive states, dissociative disorders, and somatoform disorders. Successful treatment should involve several medical specialties.

The issue of AED treatment in patients with pseudoseizures has been examined (Benbadis 1999). In this study, 31 consecutive patients with pseudoseizures were examined, and the administration of AEDs to these patients was quantified. The pseudoseizure diagnosis required a minimum of one seizure without any EEG alteration.

Results showed 5 men and 26 women patients. The age range was 18–75 years. Twenty-four patients were on at least one AED at the time of diagnosis. The delay between seizure onset and proper diagnosis was from 5 months to 9 years.

The author notes that the suggestion has been made that many patients with pseudoseizures have been on AEDs. This study shows an AED percentage administered to the 31 patients was 77%. These results are generally in keeping with those in other studies. This, of course, exposes nonepilepsy patients to the potential adverse effects of AED. This suggests AED treatment is initiated without totally convincing evidence of true seizures. This emphasizes the importance of video EEG confirmation of pseudoseizures.

As many as 25–30% of refractory seizure cases are ultimately found to have pseudoseizures (Holmes et al. 1980; Sutula et al. 1981). Numerous studies have shown that patients with pseudoseizures also have an impaired score in the Halstead Impairment Index of more than half of the subtests in the Halstead Reitan Battery, tactile performance tests – location, etc.

This represents widespread neuropsychological impairment. There were also indications of impaired motor function and handedness. The purpose of the study (Sackellares and Sackellares 2001) was to evaluate motor and grip strength in both right- and left-handed pseudoseizure patients.

Results showed a group of 40 patients with well-documented psychogenic pseudoseizures plus 40 normal controls matched for handedness, sex, and age. Mean age for controls was 33.2 years, and for the pseudoseizure group, 32.8 years. Thirty percent of the pseudoseizure groups were left-handed, significantly higher than the 10% estimated general population rate of left-handed subjects.

Results further showed a reduction in motor speed and grip strength in both the dominant hands in pseudoseizure vs. control subject. The authors comment that these findings indicate an impairment in the frontal lobe of pseudoseizure patients. The incidence of left-handed pseudoseizure patients is three times the rate of the general population.

The authors postulate that pathologic left-handed people develop left handedness due to a left cerebral hemisphere insult. Evidence from some left-handed patients in this study show evidence of brain damage associated with birth. Other similar cases were found (three). No such history was found in eight other left-handed patients.

In this study, pseudoseizure patients were found to have diminished motor speed and grip strength. The normal dominant hand advantage was attenuated in pseudoseizure patients. The results of motor strength in the nondominant hand was also less in pseudoseizure patients than in controls.

In the past, stated by the authors, psychogenic pseudoseizures were considered to be a conversion disorder. These are thought to be a class of somatoform disorders. This is frequently a diagnosis of exclusion. The authors suggest the pseudoseizure patients may be reflecting evidence of impairment of bilateral motor pathways. Supporting evidence shows 82.5% of the patients in this study had some level of trauma, including head trauma. Physical abuse occurred in 27% of patients. Automobile accidents occurred in 42% of patients and alcohol/drug abuse occurred in 17.5% of pseudoseizure patients. Other studies support their findings (Binder et al. 1998). The authors suggest that neurological disturbances may play a role in pseudoseizures and more studies are needed.

In a similar study, Fleisher et al. (2002) examined 31 adult patients with pseudoseizures, and looked at the difference in trauma-related measures between pseudoseizure patients and those with refractory epilepsy.

Diagnoses were confirmed using video EEG. Each patient took a variety of test looking at trauma distress, sleep quality, impact of event scale, or seizure history, etc.

Results showed that pseudoseizure patients had statistically significantly higher scores on the Davidson trauma scale, impact event scale, Mississippi scale for combat-related PTSD, and the sleep quality index when compared to intractable epilepsy patients.

The authors note that the patients with pseudoseizures had a significantly higher level of trauma indicator that did the other epilepsy group. These traumatic events have been shown in the present study as well as other studies as being physical abuse and trauma, as well as emotional stress and sexual abuse. The association between pseudoseizures and abuse/trauma seems consistent, and further studies are warranted, including those related to interventions for trauma-related issues.

Another study (Prueter et al. 2002) have looked at the incidence and significance of dissociation in patients with pseudoseizures. The dissociative experience scale was used in this study. This test has been previously used, but results were somewhat mixed (Alper et al. 1997; Bowman 1993).

In this study, 60 patients underwent the dissociative test. Patients were classed into three groups: those with pseudoepilepsy, those with epilepsy, and those with both epileptic seizures and pseudoseizures.

Results showed that the Dissociative Experience Scale was highest in pseudoseizure patients, followed by the class with both seizures and pseudoseizures. In addition, general psychopathologic symptoms were associated with pseudoseizure patients. The increases were statistically significant in each incidence.

The authors note that their study shows a higher (statistically significant) incidence of dissociative symptoms in pseudoseizure patients. This study is the first to look at dissociative symptoms in patients with both epilepsy and pseudoseizures.

Comorbidity of a psychiatric nature exists in pseudoepilepsy patients. The subscales showing psychopathologic symptoms included those of phobic anxiety, somatization, anxiety, and depression. Important indicators suggest that dissociation is a critical, but not the only psychopathic syndrome involved. The authors state they agree with Harden (1997), in that dissociation is key in pseudoseizures with episodic disturbance of memory and perception. If present, there is a history of severe trauma, especially physical and sexual, and, there is a diagnosis of personality disorder, post-traumatic stress disorder, or affective disorder (Kuyk et al. 1996). They note an increase in dissociative amnesia in pseudoseizure patients with a history of regular sexual or physical abuse.

The authors point out that their conclusions are not absolute, and not all pseudoseizure patients showed these exact results. Therefore, these psychopathologic features are likely not the whole story. More studies are needed. Treatments must take into consideration the above findings.

Seizures occur in 1–1.5% of all children by age 14, and account for as many as one half of initial in-patient epilepsy visits. Pseudoseizures are episodes which closely resemble true epilepsy, but for which there can be found no cerebral organic cause. The distinction between pseudoseizures and epileptic seizures can be complicated, especially in newborns and infants. The key to diagnosis is video EEG in which overt seizures and the absence of any abnormal cerebral electrical activity can be seen.

The current paper (Paolicchi 2002) looks at "problem" diagnostic features in patients (especially children) who have pseudoseizures. Neonatal behavioral features such as oral sucking movements, blinking, or pedaling movements can be a sign of pseudoseizures, but also occur in normal infants, making correct diagnosis very difficult. Ictal EEGs are very important and can rule out pseudoseizures. The Sandifer syndrome mimics tonic seizures, and is often misdiagnosed. The presence of the Sandifer syndrome is an example of a nonepileptic event in children which is often mistaken for tonic seizures (Mandel et al. 1989).

Other examples of pseudoseizures which are not true seizures include benign paroxysmal vertigo. This disorder has no postictal period, and no loss of consciousness can differentiate it from complex partial seizures. Another problem is a child who stares. Staring can be associated with complex partial seizures or absence seizures. This action can also be simply day dreaming. The gold standard is video EEG.

Pseudoseizures usually start in adolescents (Wyllie et al. 1999). Key features suggesting pseudoseizures are absence of incontinence, atypical clinical features, absence of postictal depression, and absence of any AED effect. Difficulties can arise when trying to diagnose pseudoseizure in retarded patients and cognitive deficits. The author stresses the importance of a correct early diagnosis in order to prevent possible adverse effects of AEDs.

Another paper stresses the importance of video EEG (see above), and the views of neurologists and psychiatrists (Harden et al. 2003). There are overlaps is pseudoseizures as regards who made the diagnosis and which dates are used. Many view pseudoseizures as a psychiatric condition. This paper explores this phenomenon.

Results showed that neurologists and psychiatrists differed on the usefulness of video monitoring EEG. While the role of video EEG is widely accepted and established, psychiatrists are reluctant to accept the method. This may be a general reluctance for psychiatrist to accept mechanisms such as video EEG to rule out organic brain disease.

Frequently pseudoseizure patients are thought of as "falling through the cracks," whereas psychiatrists see this as a failure of doctors to properly diagnose and treat pseudoseizure patients.

What is needed say the authors is for terms to be similar, and using the best available methodology to diagnosis and treat pseudoseizure patients. Each specialist should support the treatment regardless of which discipline is primarily responsible for the successful outcome.

The incidence of pseudoseizures is sometimes difficult to assess because many such patients are not properly diagnosed for many years. Unfortunately, during this period, many patients have unnecessary drug treatment, which can be deleterious. The present paper (Davis 2004) evaluates EEG and lack of AED response as predictors of pseudoseizures.

There were 13 patients in the study, all of whom had at least two pseudoseizures per week, video EEGs showing no epileptic form activity, and no effect of AEDs. Eighty percent of 10 patients who met all criteria had pseudoseizures. In these ten patients, the median number of years with symptoms until proper diagnosis was 3 years.

The authors note the success rate of 80% was excellent, and might be lowered by expanding the criteria. Lowering the threshold would improve the validity. The author notes the numbers in this study are low. The potential difficulties other than small numbers are the reliability of reports of previous EEG results, number of AEDs tried, and pseudoseizure frequency. The use of video EEG should be tried as quickly as possible in order to reduce inordinate time loss before accurate diagnosis.

A recent case study (Haines 2005) describes a 45-year-old woman who had been hospitalized for "stress seizures." She experienced a sudden episode of dizziness, loss of balance, and could not stand. She was taken to the hospital, and brain scan and blood work were negative.

She returned to the hospital 2 days later and at that time, MRI and EEG were normal. At physical examination, the patient grimaced and arched her back, then had a thrashing movement of arms and legs. The patient appeared unresponsive during the several minute long episode. She was given phenylalanine and lorazepam.

A video EEG during episodes as above revealed no aberrant electrical brain activity, and the diagnosis was pseudoseizures. She refused further treatment.

The author comments the pseudoseizures closely resembled epileptic episodes, yet the episode is psychologic unaccompanied by any abnormal paroxysmal electrical cerebral discharges. Interestingly, Gowers (1885) original reports described the back arching (similar to opisthotonus) after which the arms and legs trash about randomly.

Mechanisms responsible for pseudoseizures are unclear. A dissociative component serves to "remove oneself" from conflict. The motor component serves to reduce tension and anxiety. Secondary gain is achieved by the patient in that he/she receives increased attention, lowering anxiety. Psychotherapy is the usual treatment of choice, the goal of which is to lower emotional stress.

In another paper, Bowman and Markand (2005) looks at various features of pseudoseizures. Patients with pseudoseizures can mimic real seizures, but are psychogenic, and have no evidence of cerebral epileptiform activity. They also note that the "epileptic features" such as atonic–clonic movements reflect what the patient thinks seizures look like. Rarely do these psychogenic movements bear a close correlation to true seizures.

Movements during a pseudoseizure may be uncontrolled, even chaotic. Arms and legs may even move in opposite directions. While crying/sobbing are rare in epilepsy, they are common in those with pseudoseizures. Autonomic features do exist in pseudoseizure patients including flushed face, dilated pupils, tachycardia, coughing, etc., but not incontinence. Other changes which are common in pseudoseizures and rare in epilepsy do occasionally occur in epilepsy, confounding diagnosis.

Many reasons exist for a correct diagnosis, as a diagnosis of epilepsy means no driving, no swimming, employment problems, insurance problems, and stigma. This is addition to, as mentioned above, the adverse effects of AEDs, and delay an absence in psychiatric treatment. In one study, the misdiagnosis rate approached 25% (Alsaadi et al. 2004).

EEGs alone cannot differentiate pseudoseizures from real seizures. For one thing, they may both occur in the same patient. The key is to have video EEG which can show visual proof of seizure while simultaneously recording abnormal cerebral electrical activity. Negative conclusions could be drawn if the electrical activity is deep in the brain and is missed by scalp electrodes.

In terms of psychiatric treatment of pseudoseizures, the diagnosis should be explained in an educational manner in order to motivate the patient to undertake treatment. Presentation should reduce anxiety. If the family is also present, all hear the same definitions. Conversions are unconscious emotions and conflict, and patients have difficulty recognizing and expressing these emotions.

Many have noted the association between pseudoseizures and abuse, usually sexual trauma such as incest and rape, which is reported at an incidence rate as high as 85% (Arnold and Privitera 1996). A common problem with pseudoseizures and dissociated severe childhood abuse occurs when a reminder brings the abuse to consciousness, or near consciousness. Detailed histories can often elucidate amnesic periods which may correlate with trauma.

Another key life event can be when anger by the patient is suppressed by punishment of verbal abuse by the parents. These patients, when asked about anger state "I don't get angry," yet their facial expressions show anger. Patients may express anxiety about this.

Some pseudoseizure events appear as staring events in which the patient effectively dissociates him/herself from themselves and the environment. The subjects of conversation which elicit the dissociative behavior is an indicator of the cause of the pseudoseizures. Patients with pseudoseizures need careful evaluation to determine childhood as adolescent trauma.

Psychiatric treatment of pseudoseizures should be by psychotherapy in a non-judgmental way. Associated comorbidities should be identified and treated in order to stabilize the patient. Psychoanalysis seems to be not as efficacious. The patients should be taught, if possible, to not enter into dissociative states. Ignoring trauma does not resolve it, and therefore the pseudoseizure patient should address their trauma in a measured way. Childhood abuse should be discussed, but not too quickly or show individual "customized" therapy works best. The authors conclude saying that pseudoseizures are treatable if correctly diagnosed and carefully treated.

The vast majority of cases of pseudoseizures occur in adolescents and adults. There are however cases occurring in childhood (Said et al. 2006). This retrospective study showed 12 children under the age of 12 who had psychogenic pseudoseizures, as shown by video EEG.

Results showed six boys and six girls with a mean of 8 years of age, and age of onset range from 3.5 to 11 years of age. The main feature in younger patient was staring and unresponsive to verbal questions. Less frequent semiology included myoclonic jerks, shaking, focal right arm waving, and pelvic thrusting.

There was concomitant pseudoseizures and epilepsy shown in one patient, and possibly in two others. Highly significant psychosocial stress factors were identified in almost all children, and two had recent sexual abuse, and three in family situations with parental separation/divorce. One was under child protection custody.

While most pseudoseizure disorder patients are adolescents and adults, children as young as 3–7 are susceptible to stress significant enough to produce pseudoseizures. The presence in pseudoseizures in very young children are thought to be linked to secondary gain or with an attentional component. It has been previously noted that the semiology of pseudoseizures in older patients might represent what the patient thinks seizures look like. Therefore, not surprisingly, very young patients exhibit staring spells and unresponsiveness.

Always on the look out for signs and symptoms which might easily characterize pseudoseizures, a recent paper evaluates postictal breathing. In this study, postictal breathing patterns were carefully evaluated in 23 cases of generalized tonic–clonic epilepsy patients, and compared to postictal patterns in 24 pseudoseizure patients (Azar et al. 2008).

Results showed breathing in generalized tonic–clonic seizure patients after the seizures to be deep and prolonged in both inspiration and expiration phases. The breathing was regular and loud. In patients with pseudoseizures, breathing was characterized by hyperpnea with short inspiratory and expiratory phases, similar to what is seen following exercise. The two groups diffused in loudness, with the generalized tonic–clonic seizure group showing postictal snoring. The pseudoseizure groups had irregular breathing with pauses.

The authors conclude saying the divergent breathing characteristic between the two groups can help differentiate between them. The authors note that the diagnosis of pseudoseizures is often difficult, and this rather simple feature can aid a correct diagnosis.

Another paper (Syed et al. 2008), looking at easily distinguished features of pseudoseizures which might prove to be reliable indicators of pseudoseizures, examined eye closure. A cogent reason for this is that there can be as long a period as 7.2 years on average between onset and correct diagnosis (Reuber and Elger 2003). This is actually a shocking statistic given the length of time on "failed" AEDs, length of time in which appropriate therapy is not used, and the suffering from the stigma of epilepsy.

There is an obvious benefit from the use of video EEG for diagnosing pseudoseizures, but the limited availability and high cost can limit access. Eye closure has been described in pseudoseizure patients (Detoledo and Ramsay 1996), and it is associated with pseudoseizures (Chung et al. 2006). The present study examines whether the observer or self-reporting can predict pseudoseizures.

Results showed 43 patients with pseudoseizures and 69 epileptic patients without pseudoseizures. Age of onset was 29.9 years for the pseudoseizure group and 20.1 years for the non-pseudoseizure group. Video EEGs were performed on all patients and the eye closure recorded. This showed that eye closure occurred in 64% of pseudoseizure and 26% of non-pseudoseizure patients.

The authors note that neither observer nor self-reporting of eye closure was able to predict pseudoseizures. This is in contrast to previous reports suggesting eye closure could indicate pseudoseizures. The authors note that some pseudoseizure patients do report eye closure during their history interview, which suggests awareness during the episode. The differences between the present study and other may reflect methodological differences (Bounds 2007).

The authors note limits of their study include the exclusion from the study of patients with associated cognitive impairments. This could have underestimated the reliability of observer reports. The patients were those who had been referred to a tertiary epilepsy center. The authors conclude saying video EEG remains the key component of accurate diagnosis of pseudoseizure episodes.

Dozen of papers have appeared relating pseudoseizures to childhood trauma as a certain cause. In a recent report (Selkirk et al. 2008), clinical differences were noted between pseudoseizure patients who reported antecedent sexual abuse and patients who do not. Reports of sexual abuse in pseudoseizure patients can be as high as two-thirds of patients.

Results of this study showed that of 176 patients with pseudoseizures, 130 (74%) were women, and of these 59 women (45%) and 5 men (11%) reported antecedent sexual abuse. Of the 64 patients (59 + 5), 32 (50%) also reported physical abuse. Of the 112 patients not reporting sexual abuse, only 17 reported physical abuse.

Associations showed that the patient's age at onset of pseudoseizures was lower in sexually abused patients than those not abused. It took 2 years longer to diagnose pseudoseizures in those with sexual abuse than in those with no sexual abuse. Those with sexual abuse had more AEDs prescribed, more often receiving state benefits and more often living alone. The pseudoseizures of those with sexual abuse were more severe, more likely to have more triggers, flashbacks, nocturnal spells, self-injury, and incontinence.

The authors comment that these data are in keeping with other studies emphasizing the significance of sexual abuse in pseudoseizures. Whereas the average delay of diagnosis is about 7 years, the delay in sexual abuse cases is longer. The problem of the underreporting of those sensitive can lead to incorrect information, but the authors note they tried to minimize this by having each patient see the same doctor, and also having consults with a female psychologist. This emphasizes, as stated earlier in this chapter, the importance of a "team" approach to pseudoseizure patients.

This paper stresses the idea that there are subgroups within the clinical entity of pseudoseizures. The subgroups of pseudoseizure patients who were sexually abused is a clear and obvious case. These patients have a poorer prognosis in part because of a longer delay to diagnosis which prolongs much needed psychotherapy. These patients have poorer psychological health and worse (more dramatic) spells. Studies are much needed looking at the long-term outcomes of treatment of sexually abused pseudoseizure patients.

Chapter 38
Psychosocial Aspects of Epilepsy

It has long been known that epileptic patients suffer various levels and types of psychosocial deficits. These include intelligence levels, academic deficits, social interactions both at home and in the public realm, and overall quality and satisfaction with life. Contributing to these aspects of existence are types, severity, and frequency of seizures, treatment regimes such as effects of AEDs, and the view by the patient of his or her disorder.

These issues have been well studied in a variety of disease states, but epilepsy has not been as well studied as regards these issues. It is important for the patient to have as complete a life experience as is possible given the presence of epilepsy. Obviously, patients who were mildly epileptic at an early age, then had seizure frequency completely controlled, will do better than a patient with severe progressive intractable seizures. Each individual case is unique, and this should be considered by health care workers and family members alike.

Most epilepsy cases have an onset in childhood, and therefore one of the first "hurdles" faced by patient and family is school. It is essential to derive as much as possible from school, both in terms of academic achievement and social interaction with both classmates and teachers. The day in day out socialization cannot be underestimated, especially in epileptic patients. As adults, epileptic patients often struggle with social interaction and lack of education. Both impact employment. The support (or lack thereof) of family and the socioeconomic status of the family affect achievement (Fastenau et al. 2004).

A variety of testable variables are also important for academic success. These include language achievements, memory, and reading and spelling capabilities. There is currently a lack of understanding as how to intervene to help epilepsy patients with academic deficiencies. As stated above, there are a plethora of causes which interact to produce academic under achievement, and these are not all known. This has in some sense limited academic intervention in epileptic children.

There are always parental concerns about the "reception" on the part of the school/teacher with regards to an epileptic student. Again, much depends on the degree of the epilepsy and ability of the child to deal with the situation. Clearly, epileptic children are at risk for academic failure and tend to show behavioral and

D.W. McCandless, *Epilepsy: Animal and Human Correlations*,
DOI 10.1007/978-1-4614-0361-6_38,

social difficulties. Some teachers are uneasy or feel inadequate in being able to cope with an epileptic child in their classroom. Even now, some teachers hold older notions of epilepsy, and what it means (Bishop and Boag 2006).

Some level of language disorder is common in epileptic patients. This is especially prevalent in patients with complex partial (temporal lobe) seizures (Schoenfeld et al. 1999). Language deficits in complex partial seizure patients are related to age of onset, seizure frequency, and AEDs. In severe epileptic patients being considered for surgery, cognitive skills are below age expectations (Smith et al. 2002). Any type of language intervention aimed at improving language deficits should be coordinated with a speech pathologist and relevant to the school curriculum. The goal is to improve academic success.

Reading skills represent another area of critical importance for success in academics and postschool job opportunities. Several key components ensure reading success and the acquisition of these is essential for skilled reading. If the exact nature of deficient reading can be gleaned, a customized intervention can be formulated. Once again, complex partial (temporal lobe) epilepsy patients are identified as those needing help in this area more than other seizure-type patients.

Phonological deficits may be central to many with reading disorders. The phonological defects limit the patient's ability to decode single words, to learn to spell based on word sounds, etc. There are phonic programs to help circumvent these reading deficits and should be used in any intervention.

It is important to initially do a complete neuropsychological evaluation of any epilepsy patient who is a good candidate for psychosocial intervention. This provides a focus for action and baseline information for further analysis of efficacy of treatment. This should be performed by qualified personnel. There should be a team approach including teachers, physicians, health care providers, parents, and especially the patient when appropriate. A health care worker or even a guidance counselor can facilitate the process due to familiarity with steps to be taken, etc. Educating school teachers and school administrators about epileptic's needs is very important whether for young students in grade school or adolescents in high school.

Also related to psychosocial aspects of epilepsy, and stated above, is the additive effect of some AEDs on intellectual and behavioral effects. Over the years, several AEDs have been shown to have a propensity to produce adverse cognitive side effects. In addition, in early treatments of epilepsy patients the main focus (100%) was eliminating seizures and adverse effects were barely considered. By the 1970s, there was an increase in concern regarding AED side effects.

Unfortunately in the case of epilepsy, new drugs have little information available regarding possible adverse cognitive effects. Even then, the presence of seizures may contribute to cognitive effects irrespective of which AED is used. When an AED is effective in suppressing seizure frequency, cognitive abnormalities may abate, but the mechanism could be dual. The type epilepsy is of significance, for example, cognitive effects are greater in complex partial seizures as compared to generalized tonic–clonic seizures (Mandelbaum and Burack 1997).

Another potential problem is the "control" groups in some studies. This is a problem in many pediatric studies of all types in which placebo treatments are not

Table 38.1 Some psychiatric side effects of AEDs

AED	Side effect
Barbiturates	Depression, irritability, and ADHD
Carbamazepine	Possible mania/depression
Ethosuximide	Psychosis – greater change in adults
Phenytoin	Dosage-related schizophrenia, possible encephalopathy
Valproate	Possible encephalopathy

Adapted from Hirsch, E., et al. (2003) Acta Neurol Scand 108: p. 23

usually done. Here too, in the case of AEDs, controls vary from study to study. Indeed, many definitions, criteria, methods of selecting patients, etc., render study comparisons problematic. Even statistics are often inappropriate, parametric statistics, for example, being used on small groups as small as 3–5, etc.

Of interest is the finding that socioeconomic status of the family is a strong predictor of intelligence in epileptic children. Another group (Oostrom et al. 2005) found that cognitive/behavioral outcomes were related to early learning history and parental ability to cope with the epilepsy diagnosis. Disorganized home environment foretold a less than satisfactory outcome (Fastenau et al. 2004). It would be beneficial to have EEG monitoring at the same time as cognitive/intelligence testing for correlative reasons.

In contrast to intellectual and cognitive effects of epilepsy and AEDs in children, there are similar problems in elderly patients (Hirsch et al. 2003). All patients with epilepsy are at risk for confounding deficits from either the seizures or treatments or both. Risk factors include cerebral lesions, genetics, and severe epileptic encephalopathies.

As the number of people over 65 years of age continues to increase, so will the number of epileptic patients. In addition, there are increasing numbers of elderly patients developing reactive epilepsy from stroke, renal disease, liver failure, etc. A major problem with treating elderly epilepsy patients are comorbidities. An example is determining proper doses in a patient with liver disease. Aging patients are more likely to develop cognitive side effects from AEDs than younger patients. Some cognitive problems already may exist in elderly patients, so the cause of problem is hard to determine.

Results have shown that elderly patients have a greater risk for negative side effects from phenobarbital and benzodiazepines compared to younger patients (Smith et al. 1987). Other AEDs such as carbamazepines and valproate seem to have less chance of adverse effects in elderly patients and little negative effect could be discerned (Prevey et al. 1996). Gabapentin is another AED which has little risk for changing cognition in elderly patients (see Table 38.1).

In the treatment of seizures in elderly patients, extra care needs to be used in terms of explaining to the patient aspects of epilepsy. Anything which minimizes falls should be taken. Broken bones, especially hips, are hard to heal in the elderly. Low-risk drugs should be chosen, remembering an efficacious dose might be only half of a dose for younger patients.

It is estimated that as many as 30 million children have epilepsy worldwide, and 80% are without treatment due to gaining access to health care in Third World developing countries. The psychosocial impact on these children can be as deleterious as the epilepsy itself (Pal 2003).

Low self-esteem and behavioral problems can significantly worsen the overall effect of epilepsy. The stigma associated with seizures is also an important negative factor for the patient. This is largely due to lack of education and harks back to medieval times when prejudice and pseudoscience were rampant. This translates to a significant decrease in educational opportunities in Third World countries for epileptic children. This is due to expulsion or withdrawal from school by epileptic children by as many as 85%.

Parents may not be aware of a child's potential and overprotect them. Epileptic children are frequently shunned by peers and excluded from ordinary childhood activity. Uneducated parents may express alarm at seeing their children seize. Parents may have feelings of low self-esteem and be depressed. Feelings of guilt, etc. often accompany the mother of an epileptic child.

Access to health care is often difficult, for example, there are only 400 neurologists in India. Social costs are as high as doctor fees and drug costs in deciding to try to access health care (Pal et al. 2002). Attempts to solve some of these problems have been undertaken. One example is that in one area of India, disabled workers meet with teachers and children in order to help the children enter school.

Education and the realization that epileptic children can often lead satisfying, productive lives holds promise. Social support intervention for mothers of epileptic patients can offer better understanding and hope. Care should be taken, however, to take into consideration the cultural values. The above type of intervention should be carried out if possible by peers. Failure to educate epileptic children is indefensible. This has a hidden impact in families with epileptic members.

The variables which may affect academic underachievement have not all been proven in well-designed studies. The present study (Fastenau et al. 2004) examines the relation between neuropsychological function and academic achievement in epileptic children. The authors point out that many studies are not consistent in identifying seizure variables such as age of onset, seizure type, frequency, and severity. Even something like age of onset and cognition is controversial – some studies show strong correlations, others none.

The present study involves the recruitment of a large number of school age patients with various seizure types and uses a comprehensive battery of neuropsychological tests, and examines academic outcomes. In this study, 165 children were enrolled and completed the neurophysiological and achievement tests. The age range of patients was 8–15 years, and all were on AEDs when enrolled.

As regards seizure types, one-third had complex partial seizures, 20% had generalized tonic–clonic seizures, and 18% had absence seizures. Other types included complex partial seizures with generalization (17%), absence seizures (17%), atonic, myoclonic, etc. Age of onset was 6.9 years of age (range 0.0–13.9) and duration was 4.6 years (range 0.3–14.4).

Results showed that three latent factors explained 56% of the variance. They were verbal/memory/executive (V.M.E.), rapid naming/work memory (RN/WM), and psychomotor tasks (PM). It was the hypothesis that all three factors would serve to predict academic success. Each factor was thought to be a predictor to varying degrees. This was substantiated by the goodness-of-fit indices.

The authors note their model compared favorably with previous results (Seidenberg et al. 1987). The authors state the study identified another factor (RN/WM) which was not described previously. That was in part due to more recent emergence of these factors as being of importance.

The authors comment their study shows that an epileptic group is at significant risk for less than favorable academic outcomes. These include patients with neuropsychological deficits, and with disorganized, less functional home environments. The possibility exists that these problems might be lessened given intervention to increase structure and support in the home for the patient.

The use of resective surgery for AED intractable epilepsy is frequently used, and a study was undertaken in order to ascertain outcomes (Bjornaes et al. 2004). In this study, 31 patients, including eight children under the age of 18 were examined. All had I.Q.'s less than 70, and all underwent neuropsychological testing before and 2 years after resective surgery. Scant few studies exist in which seizure outcomes are analyzed in patients with I.Q.s under 70.

Results showed an age of onset of seizures as a man of 6.7 years of age (range 0–32.0) and duration of seizures was 12.2 (range 6.0–46.8). Forty-five percent of epilepsy was reactive from head trauma. Complex partial seizures were high in frequency and most had secondary generalizations

After surgery, 48% were seizure-free. Forty-five percent had no significant improvement in seizure frequency. Comparison of characteristics of seizure-free vs. not seizure-free patients showed that the shorter period of time with epilepsy was a good predictor of success. Significance as regards predictors did not occur with respect to age, sex, etiology, CT/MRI findings memory, laterality, handedness, etc. Test scores showed no difference between seizure-free patients after surgery and those not seizure-free.

When seizure duration was subdivided into three groups (under 12 years, 12–27 years, and over 27 years), there was a difference in that those with a shorter duration of seizures had a significantly greater chance of a seizure-free result after surgery. Even with a duration of less than 27 years, the chance was greater for seizure freedom than in the longest group.

Mean test scores before and after surgery showed no significant differences. When statistically analyzed before and after surgery, results in terms of seizure freedom showed no difference in seven children, but improvement after surgery in adults had improved in test scores. There were no significant pre/postsurgical outcomes in terms of schooling or employment.

The authors comment that their study shows 52% of temporal lobe-resected patients and 38% of those with extratemporal lobe resection were seizure-free 2 years later. The average I.Q. was less than 70. This shows that low I.Q. does not represent an initial reason to deny surgery in refractory cases. The additional finding that a higher success correlated with a shorter duration of epilepsy was an important observation. From the studies, it seems prudent to consider surgery in less time than 20 years.

Another paper published at about the same time (Smith et al. 2004) reported a study which looked at the benefit of seizure control on cognitive, psychosocial, and family functioning following pediatric surgery for epilepsy. The study was based on the concept that successful epilepsy surgery in children will also show an improvement in mental, behavioral, and social functioning.

Results were based in part by a comparison of 30 epileptic surgical cases with 21 refractory epileptic patients on AEDs. A level of 0.01 was chosen as a measure of statistical significance. Baseline I.Q., age, and duration of epilepsy were covariates. Cognitive findings showed that the only difference was seen in the "story recall" test in which the comparison (control) group had higher scores. Predictors of change in the "perceptual organization index" were that patients with multilobe lesions had decreasing scores as compared to single lobe lesion patients.

Results from psychosocial testing showed that analysis of covariance yielded no effects or interactions. Seizure outcome and site of surgery (temporal vs. extratemporal) were not associated with any variable in psychosocial outcomes. Results further showed that younger patients were more likely to improve over time. This seemed to correlate with surgery. Self-concept improved more in older surgical children than in younger patients. In terms of family functioning, results showed an increase in independence in the family in the surgical group. Patients in the nonsurgical group showed a decline.

In terms of seizure control, after 1 year, 57% of children were seizure-free (Engel class I) and 17% had rare seizures (Engel class II). This compares favorably to other series (Gillam et al. 1997; Snead 2001). All children were still taking AED treatment at the 1 year time point.

The authors note that there were only a few psychosocial improved outcomes. This raises questions about the widely held view that the elimination of seizures will improve cognitive, psychosocial, and family functioning outcomes. The authors state that they found few changes, thereby casting doubt for patients and parents about improved psychosocial performances after epilepsy surgery. The success for seizure control was good. Some improvements in other areas may be seen, but parents and patients should not be overly optimistic.

Attention has recently been paid to the problem of the awareness of epilepsy by epileptics themselves (May and Pfafflin 2005). Epilepsy is still stigmatized, with fears and myths, and a general lack of knowledge regarding this disorder. This even applies to those who have epilepsy. Patients with long-standing seizure disorders have a surprising lack of epilepsy knowledge. This probably is based on overhearing comments by others based on misinformation. This in turn could serve to further patients' depression and worsen the medical condition.

Psychoeducational programs have evolved in order to address this problem. Intervention and programs include those offered by psychologists, specialized nurses, neurologists, etc. These professionals provide information and advice for patients with epilepsy. The idea is to raise self-awareness regarding the disorder. This in turn gives confidence, raises self-esteem, and improves social interaction. This improves the ability of the patient to manage his/her seizure disorder in terms of medication, keeping records of seizures, etc.

Several areas have been identified in which increased knowledge proves helpful for epileptic patients. First among these is epilepsy knowledge possessed by the patient. All educational programs, whether for adults or children should improve epilepsy knowledge. Relatively simple pretests can be administered in order to assess preprogram patient levels of epilepsy knowledge. Any course needs to be customized for the age and intellectual level of the patients.

Epilepsy-related fears are another area of concern. Patients probably have fears and worries which are ill-founded. Frequently, patients' fears are not even verbalized. An education program with this component foremost can ease tensions. These programs also can aid in teaching patients how to cope with epilepsy.

Noncompliance regarding medication is an ongoing problem, especially in adolescents and young adults. This can have disastrous results. Programs and interventions regarding AED medications and treatment and its importance can be improved. This is an example where a short-term program can have significant results. In concert with this, some programs emphasize the avoidance of things/situations which could serve as seizure triggers. Results from studies on these aspects show a lowering of seizure frequency (Lewis et al. 1991).

The authors conclude saying that many if not all epileptic patients have deficits in their knowledge concerning their own malady. The described programs and interventions have the potential to greatly decrease fears, worries, and outright misinformation about epilepsy. Problems include a lack of information or reviews of the efficacy of such programs.

However, many programs are valuable, and do not even reduce seizure frequency. They certainly have the potential to reduce psychosocial problems, if nothing else, by raising awareness. Health care workers should pass along to patients sources of information regarding the location of these programs. Feedback between epilepsy patients themselves would be invaluable.

In another study, self-reporting of epilepsy and depression was studied (Kobau et al. 2006). Previous studies have shown that epileptic patients have more days with depression, anxiety, and recent psychological distress as compared to nonepileptics. There is also a higher level of psychiatric comorbidity in the epilepsy patients as compared to nonepilepsy controls (Hermann et al. 2000).

In the present study, questions to determine the level of incidence of seizure disorders were included in a 2004 Health Styles survey. The survey is designed to determine attitudes and beliefs about chronic disease and details of seizures as well as to ask about depression and anxiety.

Results show that 2.9% of adults had been informed that they had epilepsy. Interestingly, those in households with less than $25,000 annual income were 1.9 times more likely to have self-reported epilepsy. Of self-reporters, only 54% of adults were taking AEDs and only 58% reported they were seizure-free.

The authors note this study is the first to look at the relationship between epilepsy and self-reported anxiety and depression. Actual results showed Hispanics and people in the lowest income category were more likely to self-report seizures and depression. When the data were corrected for the association of the two features, the significance disappeared. More studies are needed in this area.

The epilepsy incidence derived from the survey of over 4,300 people was 1.6%, higher than previous estimates. Adults with self-reported epilepsy were twice more likely to also report depression or anxiety. This is in keeping with previous studies (Strine et al. 2005).

Another interesting finding was that those with epilepsy saw a primary care physician more frequently in the previous year than those without epilepsy, but of the epilepsy group, only 45% saw a neurologist in the preceding year. Limitations of the study, noted by the authors, included no validation of epilepsy in those self-reporting epilepsy. Also, the depression/anxiety questions were not validated for possible mental illness in the patients. Another limitation was that severely epileptic patients might not have been able to complete the survey, thereby underestimating the frequency of seizures and depression. The key finding was that those self-reporting epilepsy patients had a higher incidence of depression/anxiety than nonepileptics.

Another paper has been published describing the long-term (7–17 years) follow-up based on 110 completed questionnaires from previous mesial temporal lobe epilepsy surgery for refractory epilepsy (Dupont et al. 2006). The present study was designed to include an evaluation not only of seizures at follow-up, but psychosocial outcomes. Many epileptic patients with refractory seizures feel stigmatized and psychosocial problems may continue beyond corrective epilepsy surgery.

The population target base of patients consisted of all who underwent surgery for temporal lobe epilepsy with hippocampal sclerosis from 1988 to 2004. Four categories of patients were defined: (1) completely free of seizures after surgery, (2) patients free of seizures for at least 1 year, (3) patients with rare seizures (Engel's class II), and (4) patients with disabling seizures (Engel's class III and IV). Follow-up patients were grouped as 1–5, 6–19, and 11–17 years after surgery follow-up.

Results showed 110 patients completed the questionnaire. Mean age was 42, mean age at surgery was 35 years, and age at onset was 11 years. Mean duration of seizures was 24 years. Fifty patients had left temporal lobe epilepsy, 60 had right-sided epilepsy.

At the time of follow-up, 71% of patients had Engel's class I and 19% had Engel's class II. Seventy-three percent had been seizure-free for at least 1 year before assessment. Side and type of resection were not predictors of outcome. All patients took AEDs for at least 1 year after surgery, after which tapering was customized to each patient's needs. This process was not rushed.

At the time of assessment, 62% had driver licenses. Sixty-one percent reported a change in employment after surgery and almost half of these had an improved job. Sixty-eight percent of patients reported an improvement in nonfamily social relationships. Forty-two percent of patients reported improvement in marital relations. Sixty percent reported an improvement in relations in their families. Overall, 91% of patients were satisfied with the surgical outcome. Of those dissatisfied, half were seizure-free, but had some deficit such as lateral hemianopsia, motor deficits, memory changes, etc.

The authors conclude that their study details the long-term advantages in terms of psychosocial outcomes. The results may be biased since only 60% of those patients eligible actually participated. Even patients who did not achieve full seizure freedom reported improvement in psychosocial outcomes. There were still a significant number of patients who enjoyed seizure improvement, but still had some level of psychosocial difficulties.

A similar study looked at cognitive and psychosocial outcomes in pediatric patients undergoing surgery for epilepsy (Houghton and Welham 2006). In this study, 27 children were evaluated both before and after epilepsy surgery. Mean age was 9 years (range 1.0–14.7).

Results showed only a few psychosocial or cognitive functional changes after surgery. An I.Q. test showed no difference in presurgery and postsurgery test scores, and language and memory scores were similarly not different between pre/post-surgery. This result can certainly be viewed as positive, showing no diminution in scores after surgery. Some positive meaningful changes in scores were shown in the differences in the verbal scale and the performance scale before and after surgery. These data represent a positive improvement resultant from the surgery. The authors conclude saying their study on a small number of cases suggests alternative ways of clarifying scores in order to better evaluate psychosocial changes (or lack thereof) following epilepsy brain surgery.

Another paper (Freilinger et al. 2006) describes a patient study aimed at assessing both behavioral and emotional issues in children with epilepsy. Analysis included seeing if specific behavioral and emotional problems related to specific epilepsy syndromes. It has been known for sometime that behavioral/emotional problems were associated with seizures. In the general population, behavioral/emotional problems exist at a rate of 6.6%. In epileptic children, the rate is 28.6% (Rutter et al. 1970).

In this study, 108 consecutive patients age 5–18 were evaluated. All had long-standing epilepsy for over 1 year. The study included parental participation in that they completed a child behavior checklist while their children were having an EEG. The child behavior checklist has 113 items for the parents to comment on.

Results showed the mean age to be 11.6 years with 62 boys and 46 girls; age of onset was 5.9 years of age. The behavior checklist was completed by 92 mothers and 16 fathers. Cutoff points were chosen based on previous studies and showed a moderately disturbed group (scores 67–70), and a group with scores over 70, labeled as having severe behavioral and emotional problems. This group constituted 11.1% of the total 108 patients. Symptoms included clinging to adults, acting young for age, feeling unliked, being teased, and often overweight.

The following domains had high scores: the withdrawn scale – ten patients, severely disturbed group – seven patients. Seven patients had high scores on the somatic complaints scale. These symptoms consisted of dizziness, overtired, and aches and pains.

A high score on the social problems scale was correlated with symptomatic epilepsies and an earlier onset of seizures. High scores on four scales (withdrawn, social problems, attention problems, and aggressive behavior) were associated with polytherapy. No other correlations between child behavior checklist scores and factors associated with epilepsy were found.

Children taking polytherapy AEDs scored higher on social problems, attention problems, and aggressive behaviors. Children with focal seizures scored better than those with generalized seizures. Significantly, higher scores in the category were associated with idiopathetic epilepsies.

The authors state their study showed that 22.2% of the 108 patients had moderate to severe behavioral or emotional problems. The results showed no gender differences, which differs from some other studies showing higher scores in boys, especially in the aggressive category.

The scores do show that children with symptomatic seizures have more risk for aberrant behavior/emotional features. Also, results show that polytherapy carries an increased risk of aggressive behavior and have increased social problems. It has been shown that externalizing behavior predicts higher social competence scores.

The authors conclude saying that childhood epilepsy has a high prevalence of associated behavioral and emotional problems. Etiology, age of onset, and polytherapy are risk factors. Evaluation of an epileptic child's potential psychosocial problems should be part of treatment evaluation, and furthermore, should be done early.

Another paper looks at various aspects of behavioral and psychiatric comorbidities in children with epilepsy (Austin and Caplan 2007). In this study, factors associated with behavioral changes in epileptic children were classified into two groups. The first group relates to epilepsy itself and consists of age of onset, duration, type of seizure, AEDs used, etc. The second group involves stress factors, coping, family adjustment, etc.

Three large previous studies have shown that there is an association between seizure control and parent-rated behavioral problems (Hermann et al. 1988). Children with chronic epilepsy tend to internalize problems, and the influence of gender is unclear. Other studies (Rodenburg et al. 2006) were unable to demonstrate any effects of seizure severity on parent-reported behavioral problems. In addition, another study showed the parent-reported behavioral problems were more likely related to negative parental perceptions as opposed to actual seizure-related problems. Other studies fail to find a significant relationship between actual seizures and behavior.

Studies looking at the relation between seizure types and behavior have also been mixed as regards results of seizure type and behavior. Some report no difference between complex partial seizures and generalized epilepsy (Ott et al. 2001). There was no demonstrable difference in psychiatric diagnosis when compared to seizure types. Although 10% of complex partial seizure patients had schizophrenia-like psychoses compared to generalized epilepsy patients, the types of epilepsies examined seem to be those which are relatively mild to moderate, and excluded more serious seizures such as those in Landau–Kleffner epilepsy which would certainly show different results.

Age of onset and duration of epilepsy are two features for which there are also mixed results. Age of onset is complex due to the fact that early age of onset (under 3 years of age) is more serious due to continued brain development postnatally. Children with epilepsy and intellectual disabilities seem to have more behavioral problems than seen in epilepsy alone.

Studies of AEDs and behavior have been retrospective studies and only a few have been prospective. These have shown only a minimal effect of AEDs on behavior (Stores et al. 1992). The authors of this review do note that in retrospective studies, patients with intellectual disability, refractory seizures, and polytherapy do show cognitive/behavioral effects.

In terms of psychosocial variables and behavior, findings have been consistent in showing that family-related variables are associated with the patient's behavior. The child's abnormal behavior can be associated with preexisting family characteristics such as low family esteem, fewer adoptive resources, negative perceptions of epilepsy (stigma), poor family adjustment, etc. Interpretations of these results are obviously highly subjective, and as stated in a previous study (see above), effort must be provided by health care workers to educate both family and patient about epilepsy.

Another recent study looks at temporal lobe epilepsy (complex partial epilepsy) and social cognition and emotional intelligence has been reported (Walpole et al. 2008). The authors note that despite speculation, the exact impact of psychosocial difficulties on patients with chronic epilepsy is unclear. Emotional intelligence is a relatively new concept and relates to perceptions of ones own feelings and those of others, and to an appropriate response (Salovey and Mayer 1990).

Participants in this study included 16 temporal lobe epilepsy patients with an EEG, MRI, and confirmed seizure type. Controls were 14 participants matched to the epilepsy patients, but without seizures.

Results showed no significant differences between epileptic patients and controls regarding cognitive I.Q., but the epilepsy group scored lower on emotional intelligence. The epilepsy group had more errors on the facial expression recognition test than controls. Additionally, the epilepsy group scored high on the anxiety/depression correlation test than the control group. There was no significant difference between right and left temporal lobe epilepsy. Other negative associations included sex, number of seizures, and onset of seizures.

The authors comment their study shows that a group of temporal lobe epileptics were impaired on emotional intelligence as compared to controls. These results are consistent with other previous studies (Shaw et al. 2007). While emotional intelligence was not significantly related to quality of life, there was a "trend" in this direction. The authors state that further studies should be undertaken to examine this more completely to determine if a decrease in emotional intelligence is specific to any other seizure types besides temporal lobe epilepsy (complex partial seizures).

Long-term follow-up of psychosocial outcomes in children 2 years after seizure surgery has been published (Elliott et al. 2008). This was a follow-up to a previous study which builds on the previous study by the authors (Smith et al. 2004).

Some studies find improvement in domains of psychosocial function, whereas others did not see positive changes. The usual problems of divergent definitions and methodologies and statistical methods render results somewhat difficult to compare. Many have no control groups. This current study circumvents these problems in that the same investigators reexamine the same group of postsurgery patients 1 year later.

The current study compared a subset of epilepsy children 2 years after surgery who had data from the child behavior checklist 1 year earlier. The child behavior checklist (Achenbach 1991) is a widely used test in order to gather information regarding behavior. In this current measure, data were obtained from 22-year post-surgical children vs. 12 nonsurgical controls.

Results showed that parents reported significant improvement over time on total behavior, externalizing, aggression, and delinquent behavior. Significant group × time interaction was observed for social and social problem subscales. Parents of children in the surgical group reported improvements; those in the nonsurgical group reported a decline.

The authors note the baseline results showed a high prevalence of behavioral/social problems, the current 2-year study showed significant improvement in social and social problem subscale scores in the surgery group. The same scores were negative in the nonsurgery (but epileptic) group. The authors note the inclusion of a nonsurgery epileptic group allowed a better comparison for the surgical group.

The positive responses in the surgical group included improved social outcomes. In some of the other previous studies, it is unclear exactly which social changes improved. There are also differences in the time frames for change, and more longitudinal studies are suggested by the authors. The length of time (2 years) for recovery of some but not all factors suggests the possibility of continued brain dysfunction even after surgery and seizure control.

The authors state another important factor relates to the possibility the surgical child's home/school environment may not change following seizure control. This could effectively "mask" positive changes in the child's psychosocial improvement, producing less favorable outcomes not related directly to changes in seizure outcomes.

Chapter 39
Quality of Life for Epileptics

Several years ago, the benchmark of epilepsy treatment was viewed as effects on seizures – frequency, severity, etc. As inevitable changes in health care occur, more emphasis is placed on the overall well-being of the patient, especially in pediatric cases. The outcomes of treatment over many years are of concern, and physicians, health care workers, and parents are all concerned with "quality of life."

The increased focus on life quality has led to involving health care professionals, such as clinical psychologists, social workers, and health care workers to become involved in defining quality of life, developing questionnaires, assessment tools, etc., in order to even more formally answer the question "how are you?".

The evaluations and assessments are quite subjective, and have geographical overtones. Patients from some areas and from some socioeconomic backgrounds might respond different to the "how are you" question, than a patient with endless resources. The subjective issue is sometimes hard to overcome, but the inherent personality of an epilepsy patient will surely influence the perception of the health care worker.

Overall, the issue of quality of life may be largely determined by the patient's own perception of his/her state of function. Of course the evaluation should be customized to the type of epilepsy the patient has. It is important to evaluate quality of life in terms of the boundaries imposed by epilepsy. To that end, a multidomain approach is used. One worker in this field proposed four key domains: physical, psychological, social, and school/work (Austin et al. 1996). Others have suggested more domains to include home life, how seizures feel, life fulfillment, and medication/treatment. Most early studies have centered on one domain.

Many papers have been written on school/work-related problems for epileptic patients. Academic achievement has been uniformly described as somewhat delayed in the epilepsy field. One study (Yule 1980) reports on average a delay in reading of about 1 year, and 20% of patients had severe deficits. In terms of jobs, epileptic patients often report they have jobs which are beneath their capabilities. The exact relationship between epilepsy and school/job successes is not completely known.

D.W. McCandless, *Epilepsy: Animal and Human Correlations*,
DOI 10.1007/978-1-4614-0361-6_39, © Springer Science+Business Media, LLC 2012

In terms of the family of epileptic patients, an organized and supportive family fares better on quality of life for the epileptic patient than an unsupportive family. Most parental concerns center around fears about treatments and medications, causes of epilepsy, injury, and effects of seizures on intelligence, and further brain damage. There was concern about the patient's future life, and how to prevent mental health problems.

The health-related quality of life when applied has a goal of selecting the best treatment for the illness at hand. Structured interviews and self-report questionnaires are two most widely used instruments. Interviews also act as teaching mechanisms because the interviewed person can ask questions.

Important features are reliability and validity. Reliability translates to consistency, and validity means the instrument actually measures what it is supposed to measure. Responsiveness indicates an instrument's ability to analyze in "3D" – detect change over time. Other important considerations include length of time to complete the assessment, complexity of the scoring, staff cooperation is important, etc. Finally, the use of collected data could be important in improvement of treatment in both children and adults. When the physician is in tune to quality of life issues in patients, more thoughtful decisions can be made.

An important study was done (Austin et al. 1994) which compared the quality of life in children having epilepsy or asthma. The authors state their study represents the first of its kind as regards longitudinal results comparing quality of life between epilepsy and asthma in children. This study included the four domains listed above. The authors compared 136 epileptic children with 134 children with asthma. Data were collected from children, mothers, and school teachers. School records and questionnaires were also accessed for data.

Results showed the epilepsy children did worse on quality of life in psychological, social, and school domains. The epilepsy group did better than the asthma group on physical domain. In a 4-year follow-up, the results remained essentially the same, and the epilepsy group was experiencing a more compromised quality of life than the asthma group. The authors conclude saying that it is important for health care workers to focus on more than just seizure control, and to also focus on the full range of quality of life in epilepsy patients.

In an interesting paper (Breier et al. 1998), a comparison was made between 43 patients with complex partial seizures and 25 patients with pseudoseizures, as regards quality of life. A recently developed instrument, the quality of life epilepsy inventory-89 (QOLIE) was used in this study to measure quality of life in epileptics.

In this study, two groups were analyzed – those with intractable seizures originating from the temporal lobe (right or left) and those (25) with pseudoseizures. Pseudoseizures were identified by an accompanying observer, as being typical for that patient. Objective features such as EEG and duration were typical of pseudoseizures, or the patient was excluded.

Results showed overall a reduced verbal memory/language ability for patients with left temporal lobe epilepsy and decreased nonverbal memory for right temporal lobe epileptics as compared to patients with pseudoseizure. From the 17 scales on the QOLIE, four domains were identified as: seizure-specific effects, cognition, physical health, and mental health.

Results of the QOLIE domain responses show no statistical difference in the seizure-specific and cognitive domains. In the physical health domain, there was a clear difference in that the pseudoseizure group consistently reported decreased physical health as compared to the complex partial epilepsy group. In addition, the pseudoseizure group rated lower in the mental health domain, with fatigue and low energy being most significant.

The authors comment that their study compared epilepsy patients with severe intractable temporal lobe epilepsy (complex partial seizures) to patients having pseudoseizures. The data support the concept that patients with pseudoseizures focus on physical features rather than psychological alterations. The authors also note that some of the measures on the QOLIE were below a normal range, they were not as low as might have been suspected.

In keeping with the above study, another group looking at health-related quality of life in patients with epilepsy were compared to others suffering serious medical problems. These included angina pectoris, $n=785$; rheumatoid arthritis, $n=1,030$; asthma, $n=117$; and COPD, $n=221$. The epilepsy group consisted of 397 patients. Results showed the epilepsy group reported the highest ranking in health-related quality of life when compared to the results of the other four groups. The rheumatoid arthritis and COPD ranked lowest.

In a comprehensive review, major features of quality of life and epilepsy are examined (Berto 2002). The paper examines quality of life not only just as regards epilepsy, but includes AED treatment and surgical treatment. The authors point out that lack of standardization of definition, research methodologies, etc., makes it difficult to compare studies and results.

The idea of developing valid tests designed to assess psychosocial issues and quality of life evolved in the 1980s. This was done in addition to research on mechanisms of epilepsies and new treatment modalities. Different health care professionals such as psychologists, teachers, family practitioners, etc., became involved in the well-being of the epilepsy patient. Tests and measurements also evolved in order to assess quality of life questions.

Two types of instruments were designed to examine quality of life issues. The first generic instruments measured a broad range of well-being, and disability and distress. They are not epilepsy-specific, but can be applied to (and compared to) different diseases. They are instruments having multiple scales. The second type instrument for examining quality of life is designed for specific diseases, such as epilepsy. One such test looked at effects of adult epilepsy on family interactions, vocational adjustments, etc.

Several conclusions can be gleaned from previous studies on epilepsy and quality of life. One is that the quality of life is worse in epileptic patients than in those without epilepsy. Another finding is that the quality may be equal to or worse than that in patients with other serious disorders. Frequency of seizures is one feature of epilepsy deemed a feature which lowers quality of life. Life goals can actually be close to normal if the seizures are well controlled. Depression is a significant comorbidity in many epilepsy patients.

Assessment of drugs on life quality indicates that AEDs may have both positive effects and negative effects on quality of life. Surgical treatment is viewed as positive in part because of the positive outcomes after perhaps years of failed alternative (AEDs) treatments. Early surgery (childhood) was associated with improved quality of life scores.

The author comments that the overall picture of epilepsy and quality of life is fragmentary. Concerns and fears play an important role in the comorbidity of depression. The impact of epilepsy on the quality depends in some measure on the type of epilepsy and initial treatment success. It is clear that the diagnosis of epilepsy greatly affects quality of life. The author notes that much more data are needed using standardized methods.

A paper from British Columbia looked at the quality of life in a group of patients who had been referred for a standard neuropsychological assessment as part of a presurgery work up. The patients were children and adolescents with severe epilepsy. Data were collected for 121 cases and analyzed. The mean age was 6.7 years (range 0.2–17.9).

Results showed that 73% of patients had localized symptomatic epilepsy, and 8.7% had localized cryptogenic seizure types. Generalized seizures contributed 11.1% of the total seizure types. A large number (45%) of children had clinical level executive impairment as indicated by the Global Executive Composite. There was a correlation between executive dysfunction and decreasing quality of life. Results from the Behavior Rating Inventory of Executive Function (BRIEF) and quality of life shared moderate correlations in most clinical scales. Seizure frequency was not associated with health-related quality of life, and age of onset and duration of seizures were only minimally associated with quality of life.

The authors note that the BRIEF scales were a good predictor of health-related quality of life. Data suggested the executive dysfunction is a barrier to quality of life in pediatric patients and their families, and further influence several components of quality of life. The finding of only mild effects of seizure frequency on quality of life may relate to the absence of the degree of sensitivity in the formula. The sample of patients was from those considering surgery and therefore biased the sample toward more severe cases.

The authors conclude that executive dysfunction must now be added to other features in predicting risk for poor health-related quality of life outcomes in children with epilepsy. The other factors include low adaptive behavior, low I.Q., low family income, intractable seizures, and psychosocial problems.

The evaluation of quality of life has been extended to epilepsy patients who have intractable epilepsy (Szaflarski et al. 2006). This study looks at age, age of onset, and seizure duration on quality of life in the above-stated group of patients. Epileptics have a decreased health-related quality of life, and interest in this area has spurred development of instruments for measuring treatment responses and for defining risk factors.

In this study, 99 patients were selected, and all completed the QOLIE-89 test as well as the Profile of Moos States (POMS), and an adverse events profile instrument. Epilepsy diagnosis was based on video EEG monitoring and neurological examination. Patients were 18 years of age or older and had no mental handicap. The main independent variables were age, age of onset, and duration of epilepsy.

Results were based on a final patient number of 99. Sixty-three percent of the patients were female and the average age was 37 years old. The average age of onset was 19 years and the duration of epilepsy was 18 years. Correlations were noted between age and age of onset (positive correlation), and also between age of onset and duration (negative correlation). The number of comorbidities was correlated positively with age of onset and negatively with duration of epilepsy. The correlation of results from the QOLIE-89 assessment and age of onset, and duration of epilepsy were modest and significant. Bivariate association between the QOLIE-89 instrument and TOMS and adverse effects profile was statistically significant.

The authors comment that this study did not show a correlation between age and quality of life and duration was a significant factor in the patient's quality of life. The present study showed increasing age of onset resulted in a decrease in quality of life. The present study showed an increase in health-related quality of life with a longer duration of epilepsy. This is attributable to the idea that adults can more readily cope, and since their life is more established, adjustments are less upsetting.

The authors comment that their results are not directly applicable to all epileptics since the population was a select one having intractable seizures. The patients' group also had many members with diverse severe epilepsy types. The authors suggest that in patients with adult onset of epilepsy, attention to minimizing AED numbers and treating mood disorders could decrease quality of life problems.

Mood disorders, as might be suspected, are a frequent psychiatric comorbidity in epileptic patients. The prevalence of this comorbidity may range from 20 to 50%, depending on the study. Major depressive episodes may be a feature of mood disorders, but need careful diagnosis and classification since several types exist, and require different treatments. For example, some antidepressants (drugs of choice in depressive disorders) may worsen bipolar disorders.

Several previous studies have shown depression to be a reliable predictor of health-related quality of life and a predictor of decreased disability scores (Cramer et al. 2004). There is a relation between epilepsy and mood disorders in that those with depression have a much higher risk for epilepsy. There are common biochemical pathways which are changed and associated with both disorders. These include glutamate and GABA neurotransmitter systems. Serotonin and norepinephrine are also altered in both epilepsy and mood disorders.

The authors conclude saying it is important to refer patients with major depression disorders to a psychiatrist if the disorders continue after two antidepressant drug trials. Also any patient with bipolar disorders or psychiatric depression episodes or mood disorders with suicidal ideation should be referred at once. Indeed, any patient problems or indication of pharmaco-resistance should be referred to a psychiatrist with experience in this field.

Comorbidities are a common feature in patients with epilepsy. Patients with epilepsy may have a two- to fivefold increase in comorbidities such as migraine, cardiovascular and GI disorders, mood disorders, anxieties, etc. (Wiebe and Hesdorffer 2007). The type and prevalence of comorbidities in epileptic patients generally follows that of the general public, but of a higher frequency. Thus, asthma might be seen in pediatric patients and cardiovascular problems in elderly patients (Tellez-Zenteno et al. 2007). It is not unusual for patients falling into any disease category to have comorbidities.

Comorbidities have been evaluated as regards quality of life, and other criteria such as health resource usage and survival. Comorbidities measure the burden of the patient's mix of conditions, and no single comorbidity can predict outcomes. A key factor of any comorbidity is the severity of the disorder. Severity of comorbidity such as major depression and suicide ideation represent the extreme.

The existence of a comorbidity should result in added burden and decreasing quality of life, but that does not always occur. Comorbidities in Third World countries with limited medical resources are even more impacted by these conditions, as well as making access to proper care even more difficult for the patient. Even in developed countries, low socioeconomic people may influence the number, type, severity, and impact of comorbidities (Schrijvers et al. 1997).

The authors conclude saying that few studies exist looking at treating or preventing epilepsy comorbidities. Clinical guidelines are focused on single conditions and rarely deal with co-treatments for comorbidities. Studies are mixed in saying if patients with comorbidities receive better or worse care. Elderly patients may have several comorbidities. Often patients with comorbidities are excluded from clinical trials. Future studies need to be directed to features of comorbidity, especially in epileptic patients.

A recent paper looks at health-related quality of life in two groups – patients with epileptic seizures and patients with psychogenic nonepileptic seizures (Testa et al. 2007). The idea that patients with psychogenic seizures have a self-defined quality of life worse than epileptic patients was previously reported (Breier et al. 1998). The current study was designed to further explore the effects of mood on health-related quality of life between epilepsy patients and patients with psychogenic seizures.

Data were collected from 219 patients with either definite epilepsy (95) or definite psychogenic seizure (77), plus "possible" epileptic or psychogenic seizure (57). Instruments related to mood were administered to all participants. The quality of life in epilepsy inventory-89 (QOLIE) was administered. The profile of mood states (POM) plus the MMPI were also used to evaluate patients' moods.

Results showed the mean age of participants was 36 years (10.6 age of seizure S.D.) and age of onset 18 years. Seizure duration was 15.6 years (12.9 S.D.). Statistical analysis showed the seizure diagnosis was significant in the results of the POMS instrument. Seizure diagnosis also had an effect on the MMPI. Personality variables such as psychological distress and somatization were two factors in psychogenic seizure patients which influenced MMPI as opposed to epileptic patients. The MMPI subscales, hysteria, and hypochondriasis were elevated in psychogenic seizure patients.

The authors note their study is unique in that the relationship between personality variables and health-related quality of life in epilepsy patients and those with psychogenic seizures are compared. Problems relate to the lack of validation of the QOLIE-89 in terms of psychogenic seizures. This is because individual experience with either actual epilepsy or psychogenic seizures may not be the same. In addition, the authors note that there were some rather large variances in data, and some differences barely caught statistical significance.

In a pilot study, the specific items suitable for parental evaluation of quality of life are examined in their children with epilepsy (Soria et al. 2007). While all agree that several domains influence health-related quality of life, there is disagreement as to which are most important. Most include four domains as listed above: physical, psychological, social, and cognitive (Austin and Dunn 2000). Even in light of the importance of cognition, some studies of quality of life exclude patients with severe cognitive problems.

The pilot study described here looks to compensate for a lack of adapted quality of life instruments and examine parental concerns. Several factors may impact the quality of life in epileptic children, such as frequency of seizures, onset of seizures, types of seizures, comorbidities such as behavioral problems, socioeconomic status of the family, etc.

In their pilot study, 35 subjects were used. All had confirmed epilepsy, special educational needs, presence of cognitive deficits, and age of 3–11 years. Questionnaires were provided to the parents covering general items such as the history of the disorder, quality of life scales, impact of illness scale, and a questionnaire relating to behavior.

Results showed the 35 patients had a mean age of 8.2 years (range 4.8–10.7), age of onset of 1.6 years (range 0–6.3). Forty percent of parents felt the information provided was insufficient. All parents reported their children had learning difficulties. Overall quality of life was judged bad or very bad in only 12% of parents, and one-third judged the quality of life as good or very good. The most cited need was for physical supervision. Concern was expressed in the cognition section and most parents thought epilepsy in their child "frightened" other people.

In the "impact" questionnaire, concern on the part of parents centered on the need for supervision (to prevent accidents), lack of autonomy, and learning difficulties. The behavioral questionnaire showed concern about attention–concentration and hyperactivity questions. Depression, lack of energy, etc., were not strong areas for parental concerns.

The authors state the parental questionnaire concept did not show particular difficulties in answering questions, except in behavioral items, in which the percentage of "not applicable" response was given many times. Questions regarding complex social interaction may not have been appropriate for parents of young children.

The impact of epilepsy in children had a significant effect on parents, in keeping with earlier observations (Sabaz et al. 2001). Parents reported major change in their everyday life. However, only a few percent of parents said their child's quality of life was poor. The authors conclude saying this study details questions which should be included (or not included) in quality of life parental questionnaires. Further study is necessary to ensure validity. The quality of life questionnaire is a suitable tool for parents to complete in evaluation of their children and epilepsy.

In a somewhat similar study, incorporating self-reported questionnaires regarding quality of life in epileptics was done on adult patients over 18 years of age ($n=41{,}494$ patients) who participated in a California Health Interview Survey, 2003 (CHIS).

The CHIS is a telephone survey conducted biannually to evaluate answers related to health behavior, insurance concerns, etc. Initial results indicated that of 41,494 interviewed people, 550 stated they had been told they have epilepsy or a seizure disorder (1.2%). Of these, 1.8% stated they no longer had epilepsy and 1.7% were not taking AEDs. Twenty-six percent of those with recent seizures reported not taking any medicine to control seizures.

Results showed age, sex, race, etc., did not have statistical significance in relation to incidence of epilepsy. There were differences in terms of fair or poor health. About 20% of new epileptics reported fair to poor health, whereas 47% of those with epilepsy reported fair to poor health.

The authors note this in the first study examining lifetime prevalence of epilepsy using self-reports. Findings include the substantial financial burden. The findings show the burden of impaired quality of life is similar to that of other disorders such as asthma, diabetes, etc. Some burden may be associated with compromised lifestyle (job opportunities, lower socioeconomic status) based on other features besides epilepsy. Quality of life in this study was not related to use of medications. It was a "striking" finding that one-fourth of epilepsy patients were not taking AEDs. Most were satisfied with availability of medical care.

The authors conclude saying their study was large and evaluated the assessment of validity and verification of validity in the questionnaires. Limits include consideration that self-reporting is subject to memory, it was not possible to verify types of seizures, and seizure prevalence could be underreported because of patient reluctance to disclose the presence of a neurological disorder. The authors conclude saying this study shows the potential importance of this type survey in that it allows guidance of quality of life intervention. It also suggests looking for and analyzing the importance of comorbidities and injury in this group of patients.

Another recent study (Kwan et al. 2009) assessed the effects of subjective anxiety, depression, and sleep disturbance on quality of life in adult epileptics. The study was performed on 247 patients who completed a variety of questionnaires including the Quality of Life in Epilepsy-31 (QOLIE-31) questionnaire. The authors note that anxiety, depression, and sleep disorders are prevalent among epileptic adult patients and can influence quality of life.

This cross-sectional study employed consecutive patients visiting an epilepsy clinic. Acceptable patients were 18 years of age or older, had epilepsy for at least 6 months, and stable AED treatment for at least 1 month. Seizure types included generalized (23%), partial onset (61%), and unclassified (16%). Mean age was 39 years (range 18–76). Average duration of seizures was 16.4 years. The mean QOLIE-31 score was 64.3 (range 5–98).

Linear regression showed the number of AEDs, seizure freedom in the last month, medical outcome score (MOS sleep score), and the HADS anxiety subscore were all associated with the QOLIE-31 score. The numbers of AEDs and presence of psychiatric diagnoses also correlated with QOLIE-31 scores.

The authors note that subjective symptoms of anxiety, depression, and sleep deprivation were independent predictors of decreased QOL scores in adult epileptics. This has previously been reported except for the relation to sleep disturbances (Boylan et al. 2004).

The contribution of current numbers of AEDs and seizures in the previous month made only mild contributions to quality of life. Long-term seizure freedom correlates with seizure improvement (Spencer et al. 2007). Self-reporting methods offer economy of administration and can be repeated easily in order to record change. Disadvantages relate to recall bias in that results may be overemphasized. Other standardized methods were not employed due to their time-consuming nature.

The authors conclude their study saying that it has clinical practice and research implications. They state that practicing physicians should look at psychopathology as a comorbidity. The authors also relate a decrease in reporting of psychiatric problems reflects a relative neglect of psychosocial aspects of patient care.

In another quality of life study, patients who had a corticoamygdalohippocampectomy were analyzed using the Epilepsy Surgery Inventory (ESI). The questionnaire was administered before and 3 years after surgery. Employment is an important predictor of integration of an epilepsy patient into society. Many variables influence employment, including education, family support, fear, and possible employer discrimination. This study aimed to evaluate employment before and after surgery for mesial temporal lobe epilepsy.

Results showed mean age of the 57 patients was 31.4 years, and 39 patients had left surgery, 19 had right-sided surgery. After surgery, a three-year survey showed 50% were seizure-free. Employment status was unchanged at the 3-year time. Fifty-one patients had no employment change. Features such as age, age of onset, duration, etc., had no significant effect on employment status.

The authors note there was a modest employment gain after surgery for mesia, temporal lobe epilepsy. The gain consisted of employment rates moving from 29 to 38%. In the employment group, 46% were seizure-free and 32% had only auras. Health perceptions, emotional well-being, and quality of life were all associated with employment status.

A fair definition of quality of life is one's own perception of their position in life, in their own culture, and this in relation to their goals. The present paper looks at quality of life as described by children with epilepsy, and their parents' perception (Verhey 2009). In this study, 375 children and 378 parents completed the CHEQOL-25, a condition-specific quality of life measure.

Results showed the mean age of onset of seizures in this study was 6.9 years and the duration mean for seizures was 4.6 years. Two hundred ten of 375 subjects had seizures described as having a high severity. In terms of quality of life, the parents rated health-related quality of life lower than did the patients. The patient/child level of agreement was greatest when children were enrolled in regular classes in school.

The authors comment that agreement between parents and their epileptic children is in agreement in domains which reflect externally visible perception of epilepsy. Internal expressions in life, more easily kept "secrets," were less likely to be in agreement between parents and their epileptic children. These are private feelings not readily shared. Different life experiences between parents and children also no doubt contribute to discrepancies.

As the child's age increases, there is more likelihood of the parents rating their child's quality of life lower. Age of onset, duration of seizures, and severity of seizures did not correlate with level of discrepancy between parent's and children's assessment of quality of life.

The authors conclude that the discrepancies between parent's and children's responses to quality of life questions likely also apply to other moderate to severe neurological disorders. The study also shows that parental responses alone are insufficient to determine quality of life in epileptic children. The CHEQOL-25 is a practical tool for quality of life evaluations.

The effects of epilepsy on quality of life feature of marital adjustment has been examined (Vibha et al. 2010). In this study, the goal was to evaluate the interrelationships between epilepsy and marital problems, and to compare those results with similar results in chronic psychiatric illness.

Three groups for study were selected. These were patients with bipolar disorder, schizophrenia, or epilepsy. The first two were collectively called the psychiatric illness group. A final group of spouses was used. Quality of life was evaluated using the WHO-Quality of Life-BRES scale. It is a self-administered scale.

Results showed that 30 patients with epilepsy and 60 with psychiatric illness were selected. The mean age of epilepsy patients was 34 years, and the mean epilepsy duration was 6.9 years. The couples had been married for a mean time of 14.1 years. The majority were from urban areas and about one half of the women were housewives. In terms of marital adjustment, differences were noted, but about 60% showed good adjustment as compared with those in the psychiatric illness group.

The patients with epilepsy had a higher income and lower unemployment than those in the psychiatric group. Good marital status was seen in those with epilepsy. These results may have been due to the fact that 70% of patients in the epilepsy group were seizure-free. The perception of epilepsy patients and their families showed a better outlook as regards marital adjustment. Correlating marital adjustment with quality of life showed improvement of quality of life perceptions.

The authors note future studies should use larger numbers of subjects and better match the epilepsy group to the comparative group.

In another very recent study (Mathiak et al. 2010), the quality of life was evaluated in children with lateralized epileptic foci. One rationale for this study was the finding that in adults, left-sided foci were associated with a lower quality of life compared to patients with right-sided foci.

In this study, 51 children were enrolled in the study. These patients all had an age range of 6–15 years, I.Q. over 85, unilateral epileptic seizure focus, and no other chronic diseases. Following some exclusions, 16 patients with right temporal lobe epilepsy and 15 patients with left foci epilepsy were selected (total 31). Parents of epilepsy children completed the health-related quality of life questionnaire. While the parents were filling out the questionnaire, the children were undergoing neurological examinations, EEG procedures, and imaging procedures. CT/MRIs were performed on 23 of 31 children.

Results showed that children with right-sided epileptiform discharges had a statistically significant lower overall quality of life. Specifically, the anxiety subscale, quality of life, stigma, and general health subscale were more compromised in children with right-side foci.

The authors comment that their study shows for the first time that lateralization of seizure foci plays a major factor in quality of life. The authors correctly note that the childhood period is a critical period for socialization, and therefore a disease process during this period could disrupt acquisition of social skills, perhaps forever. The earlier the deficits starts, the worse the outcome. One study (Camfield et al. 1993) showed that clinical features such as treatment and seizure control did not predict social functioning. Mood was a strong predictor of quality of life in terms of socialization.

The authors also note that their study contributes to the understanding of developmental aspects of a correlation between psychosocial problems and cerebral pathophysiology. The problem of childhood depression could have been overlooked because parents might have missed it in their children.

Based on clinical, EEG, and imaging, no anatomical malformations were suspected. The authors point out that more studies with higher numbers are needed. The authors summarize saying that lateralization of temporal lobe epilepsy affects quality of life. Right-sided foci lower quality of life, especially in social functioning and emotions. This knowledge could impact treatment strategies. Health care workers must consider cerebral development (postnatal) in the course of childhood epilepsy.

A recent paper (Ciechanowski et al. 2011) examines depression a factor in the lowering of quality of life in epilepsy patients. A long-term intervention, PEARLS (problem solving, behavioral activation, and psychiatric evaluation), which is a home-based intervention technique was evaluated for its effectiveness. Results showed the PEARLS program is effective in significantly reducing depression in adult epilepsy patients, and is maintained for more than 1 year after cessation of the program.

Table 39.1 Quality of life in children with epilepsy

Factor	Quality of life issues
Epilepsy and treatment	Neurologic functioning
	Cognitive functioning (attention, memory, abstract reasoning, psychomotor functioning)
	Antiepilepsy medication effects (physical, cognitive, and behavioral side effects)
Psychologic	Emotional status (happiness and satisfaction; anxiety, depression, behavioral problems, and psychiatric disturbance; self-esteem)
	Feelings about epilepsy (concerns and fears)
Social	Completion of age-appropriate psychosocial developmental tasks
	Satisfaction with family relationships
	Peer relationships
	Engagement in activities (sports, clubs, hobbies, teams, organizations)
School	Academic achievement
	Learning problems
	Adaptive characteristics
Family	Seizure-management skills
	Psychologic adjustment to epilepsy (concerns and fears)

Adapted from Pellock, J. Austin, J., and Santilli, N. 2009

Overall, factors influencing quality of life include the disorder itself, treatment, psychologic, social, school/work, and family. Each factor has issues which have a significant impact on individual epilepsy cases. Sometimes various of the "subfactors" are not pertinent in selective individual cases (see Table 39.1).

Chapter 40
Epilogue: Comments and Future Directions

The preceding 39 chapters concern a vast amount of information in a synoptic fashion. Obviously, much has been omitted or reduced in order to have brevity. For those needing more, 1,500 references are in the back which will in turn lead to other papers. The studies quoted provide at least some level of detail in order to be initially evaluated for validity and reliability. What follows in the epilogue is a synopsis of some important features and key points which were stressed in the full text. Promising directions and concepts are indicated.

The issue of world health concerns cannot be overstated. The World Health Organization estimates that there are 50 million epileptics worldwide, with 75% living in developing countries. About 50% are children, and in Zimbabwe for example, there are no neurologists. Reporting of health care statistics from third world countries is often manipulated for political reasons, therefore incidence, treatment, etc., are difficult to ascertain.

Some AEDs are very inexpensive. There should be a framework by which the incidence of epilepsy can be determined in developing countries, and inexpensive effective AEDs distributed along with educational materials to indigenous peoples to insure compliance. Regular visits to clinics should ensure continued taking of prescribed AEDs. Some measure of training and regular contact with on-site health care workers is essential for any measure of success.

This is a prerequisite for successful control of not only epilepsy, but also malaria, AIDS, cholera, etc. The effort for control should be not only one or two communicative diseases, but for all in which good treatment paradigms exist. Because of cultural bias, indigenous peoples are most effective in running training programs to reduce morbidity and mortality of those afflicted by those with potentially catastrophic syndromes.

In terms of epilepsy treatment, AEDs are the usual first attempt at controlling seizures. This is due to ease of administration, low cost of some early AED choices, and ease of patient compliance. On the other hand, the health care provider should be aware and mindful of multiple possible AEDs, and be watchful for evidence of toxicity or adverse effects. Once again, a certain level of recognition of the importance of proper training for compliance is necessary for both good outcomes and

D.W. McCandless, *Epilepsy: Animal and Human Correlations*,
DOI 10.1007/978-1-4614-0361-6_40,

level of success compared to cost. There are named foundations which fund these programs, and are highly interested in their success.

Even as education and training of both health care workers and patients is occurring, there should be an attempt to have consistency in terminology, methods of treatment, methods of measuring success, methods of reporting and evaluating results, etc. – these must be standardized. One can hardly read a manuscript without seeing a variety of differences in the materials and methods of both animal and clinical studies. Each author has his/her research idiosynchranies which render the majority of papers not suited for meaningful comparisons. Even simple definitions of, for example, febrile seizures are not all similar. This is unacceptable. Energy is being expended on how to define a particular type of seizure instead of defining mechanisms and determining treatment. The above is true in most aspects of the study of epilepsy. One more example: most experimental seizures affect focal brain areas, at least initially. What neurochemical results can be interpreted following use of whole brain when comparing differences between experimental and control animal brain?

In addition to AEDs, many other treatment paradigms exist, including electrical stimulation and dietary means. AED trials involving two or more AEDs usually serve to decrease the mean chance of success in intractable seizure patients, and suggest alternative treatment attempts. Frequently several AEDs (over five) are tried in hope of success in modulating either frequency or severity, or both. Statistics show such patients have a success of less than 5%. These patients are refractory to treatment, but often many years are expended trying various AEDs and combinations thereof, before other more "drastic" methods are tried.

The ketogenic diet (and modified Atkins diet) is often attempted in intractable cases of epilepsy. This diet has a high lipid/protein and low carbohydrate content. It is judged unpalatable by many patients and compliance (especially in teenagers) is difficult. New mixtures and "shakes" have recently been commercially introduced to overcome this distinct disadvantage. The exact mechanism of action of the ketogenic diet is unclear, but probably involves energy metabolism. The ketone bodies can "feed" the TCA cycle at various sites thereby increasing ATP levels. This energy "supercharging" of ATP seems to have a protective brain effect. Of interest is that creatine, precursor of phosphocreatine, appears to have a similar mechanism and efficacy. Some studies of these nutritional treatments report a greater than 75% success in lowering frequency by over 50%, and these studies are in cases refractory to AEDs.

The ketogenic diet was not taken seriously for many years, but has recently been more recognized. Creatine as an antiepileptic still seems little used. More studies are needed in this area, and additional studies on mechanisms of raising ATP intracerebrally are warranted. Combinations should be studied as AEDs are usually still taken in patients successfully treated with the ketogenic diet. This is an area in which animal studies should result in readily translatable data.

Gene therapy is another area of study which has been investigated in animals for over 10 years, with positive results. Several studies have shown adeno-associated viral vectors can carry neuropeptide Y into cerebral areas with rather impressive efficacy. This modality has been used successfully in several other metabolic encephalopathies, including bilirubin encephalopathy. The data derived from animal studies would seem to be more than adequate to justify human trials. Translation should be a high priority for gene therapy.

Vagus nerve stimulation is another modality for epilepsy treatment. This treatment has enjoyed at least 60,000 implantations, with many patients (over half) showing a 50% or greater seizure reduction. Similar or better animal results preceded the human trials. A major advantage of vagus nerve stimulation is that it is relatively minor surgery. A stimulator is implanted, and a wire is guided to the bifurcation of the left common carotid artery.

Side effects (hoarseness) are minor and the stimulator can be cycled off and on when necessary. Recent studies have shown that the tragus of the external ear can also serve as a site for the stimulator rather that the carotid artery. This reduces the surgical manipulation, and is overall more convenient. This surgical epilepsy treatment usually only requires 1 day of hospitalization. Remember, the 50% of patients showing improvement are from a group of intractable to AEDs.

Similar to vagus nerve stimulation is deep brain stimulation as a modality for AED intractable epilepsy. Several deep brain sites have been employed, however, hippocampal stimulation is perhaps most frequently used. Basically, electrode(s) are stereotaxically implanted into or adjacent to the hippocampus. Trains of stimuli then reduce seizure frequency and severity up to 40%. Interestingly, some studies show that a sham surgical procedure (placement of the electrodes) may be as efficacious as implantation + stimulation. The concept is that placing of the electrode is a mini hippocampectomy and by itself can reduce seizures. This needs more study.

Adverse effects of surgery include infection, missing the target, etc. The mechanism, like that of vagus nerve stimulation, seems to be that small subtle electrical discharges at the site acts to decrease sensitivity to intrinsic stimulation (seizure activity). The surgical procedure itself is somewhat more difficult than vagal nerve stimulation.

Generally, resective surgery or lobectomy is considered as a final attempt to control seizures. These procedures involve considerable presurgery investigation in order to pinpoint the foci. In addition, extensive study is directed at making sure the initial diagnosis is correct, as, remarkably, some reported as many as 20% of patients being evaluated for surgery have been incorrectly diagnosed (and treated?). Two to three hospital days may be necessary to complete this surgical workup.

Even after the surgical procedure, AED treatment is continued. Surgical success can be judged by imaging techniques, EEGs, and by histological methods of the excised tissues. Overall success varies patient to patient and by surgical approaches. For example, corpus callosotomy may show a 60–80$ improvement in lowering frequency and severity of seizures. Proponents of surgery to correct seizures argue in favor of performing this process at an early stage in order to improve outcome. The current mean delay until surgery is about 23 years post epilepsy diagnosis.

Pediatric considerations are important since at least one half of epilepsy cases develop during pediatric years. This is seen by some as a disadvantage because of vulnerability of the brain to interruptions, which obviously is continuous postnatally.

Conversely, there is an element of plasticity shown by children which can serve to benefit recovery. Compliance in AED administration can be monitored by parents, which is more difficult in teen years. Overall, many studies indicate a better outcome when surgery is performed earlier.

Other pediatric considerations include AED dosages. A dose suitable for an adult is likely too high for children. Comorbidities are important in the pediatric population. For example, children with liver disease must be evaluated regarding the ability of a compromised liver to metabolize drugs. Careful monitoring of children with epilepsy is essential.

Comorbidities in all epilepsy patients must be carefully evaluated as regards both epilepsy effects and drug administration. Examples include patients with hypertension so that the pressure-raising effects of the seizure do not precipitate further intracerebral damage. Patients with compromised liver function must be evaluated as to the correct AED dose to provide efficacy. Patients with GI disorders affecting absorption need additional study. As more and more primary care physicians and interprofessional care takers treat epilepsy patients, these considerations must be kept in mind.

Surgical procedures are often attempted in animals prior to human trials. In many ways, translation of surgical results to humans from animal studies do not take as long as other treatment types. This is partly because the surgical treatments represent a last hope for efficacious treatment and a satisfactory outcome. Diagnostic modalities prior to and during surgery have experienced major improvements over the years, and these developments have reached patients in a timely manner.

Another concept which should be considered relates to the probable outcomes of various epilepsies. This of course involves several features of any individual's seizures, including age of onset, genetic background, type of seizures, etc. The upshot of this is that patients with seizures such as infantile spasms have a much worse outcome than absence seizure patients. These huge diversities act to modulate treatment modalities. Fast action is necessary to correctly make the appropriate diagnosis and implement the appropriate treatment.

The concern is that fast action might result in an inaccurate diagnosis and incorrect treatment, which could have a lifelong adverse effect. Remember, in surgical workups, 20% of patients have an incorrect diagnosis. So, while early diagnosis and aggressive treatment are conducive to outcome success, the diagnosis needs to be accurate. Medical facilities should have an organized predetermined clear-cut protocol to follow in order to reach a correct diagnosis.

Significant psychosocial issues face epileptic patients which should always be considered by those in contact with epileptic patients. Stigma is one major problem facing especially epileptic children. This can be a continuing problem, at the patient's school, neighborhood, and even family members. This often comes across as a feeling of inferiority on the patient's part.

Attempts to correct stigma should revolve around education and a better understanding of epilepsy on the part of parents, teachers, classmates, neighborhood children, etc. Many years ago, the feeling was that epileptic people were "possessed" and were inferior to normal people. Even in the twenty-first century, this erroneous bias still exists and needs to be eliminated. This takes education.

Epileptic patients benefit from having a social worker advocate who can help patients find jobs. This involves explaining to potential employers exactly what is involved in employing an epileptic, and a social worker advocate is an ideal health care worker to assist.

In many cases, depression is a significant outcome of the stigma epileptic patients endure. There should be a psychological component in the long-term treatment of epileptic patients, especially children. This may become critical during teen years. This is also a time when treatment compliance is at risk, which can in turn increase stigma. Vigilance must always be maintained regarding both psychological and medical sequelae.

Psychological concerns can arise over effects of seizures, effects of AEDs, and effects of peer pressures. An example might be aggressive behavior of an epileptic school child directed toward a classmate. This behavior could be an adverse effect of an AED, but actually only worsens the view of classmates toward an epileptic patient. This type problem needs immediate attention on the part of school administrators and teachers, to include consultation with health care workers. Every attempt must be made to reduce/eliminate stigma, and to encourage the meeting of realistic goals established by each individual epileptic patient.

Quality of life is a highly subjective consideration. A patient with complex partial seizures consisting of three 30 s seizures per week may appear to be too much to many patients. Yet parents of a child with progressive myoclonic seizures might trade in a heart beat. The idea here is that all concerned need to be realistic regarding any particular patient's epilepsy.

It is unfair to try to convince parents and patients alike that "you can do anything you want." Uncontrolled epileptics cannot drive, swim, even be alone lest a fall prove disabling. This of course depends on the frequency and severity of individual's seizures. Quality of life therefore is subjective and must be reality-based. This facilitates a balanced realistic optimistic outlook on the outcome. This in turn reduces disappointment, depression, and possible psychiatric sequelae.

Based on frequency numbers in worldwide epilepsy cases as well as new cases in the United States, the seemingly long period of translation of animal study results to human treatment paradigms needs attention. A mechanism should be developed in which the process could be shortened. This could be achieved without any sacrifice in safety. There are previous examples in which hasty decisions have resulted in death. And yet months or years of delay also result in morbidity and mortality.

In previous times, the availability of volunteers for scientific clinical trials resulted in successful testing of various drugs and procedures in a timely manner. Such studies are not as plausible today in the pediatric setting in which over half of all epilepsy cases originate. Investigators must design studies in ways which render the data valid and reliable, thereby facilitating unequivocal interpretation of data. When this is not done, equivocal data cloud many issues regarding epilepsy. This does a distinct disservice to the study of seizures and to those patients in need of

Fig. 40.1 Schematic graph showing supposed temporal relationships between various theoretical initiators of seizures. These are energy metabolism (ATP), neurotransmitters, and channels. X, Y, and Z refer to other unspecified causes of seizures. The concept is that these alterations all occur in milliseconds of seizure induction, but one is the primary initiator of seizure activity

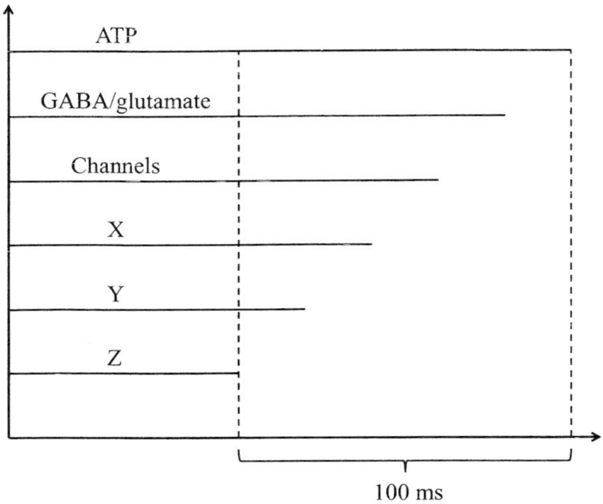

better more efficacious treatments. So, not only the time involved in translating data from animals to humans needs to be speeded up whenever possible, but the scientific quality of studies needs significant improvement.

There are several proposed mechanisms of epileptogenesis, any of which could be responsible for seizures in any particular patient. Among these are alterations in channels, changes in the balance of neurotransmitters GABA and glutamate, and changes in energy metabolites such as ATP. In various seizure types, one or another of these above-stated potential mechanisms may play a key role in seizures.

Although polytherapy as regards AEDs is common, using more than one of other types of antiepileptic methods has not been satisfactorily explored. For example, the ketogenic diet (or modified Atkins diet) acts to increase energy metabolism potential by producing ketone bodies which thereby can enter energy-producing cycles and generate excess ATP. The excess ATP in turn seems to act by lowering the sensitivity of neurons to have discharges. This protects against seizures.

Creatine supplementation has a similar effect in that it increases the production of phosphocreatine which is a high energy phosphate donor to ADP, forming ATP. Once again, the potential for an increase in intracellular energy compounds such as ATP is present. The idea is to try in animal models combinations of treatments, as is done with AEDs, in order to see if efficacy is increased. This could easily be established using the ketogenic diet supplemented with creatine.

Another treatment attempt is along similar lines. When a seizure occurs, it is initiated in a very short period of time. More than one neurochemical alteration occurs, and prevalent one may be determined by seizure type. The other changes also occur. If a seizure occurs due to an imbalance in GABA/glutamate, there is also an almost immediate change in intracellular energy metabolism (see Fig. 40.1).

One idea is to control the neurochemical milieu such that changes in energy metabolism are primary and most significant, they increase ATP metabolism/levels such that the seizures are modulated. Ideally, this might result in a large number and type of seizures all being controlled by one treatment modality – that of cerebral metabolism.

A similar situation exists in the use of gene therapy. Neuropeptide Y is administered using viral vectors to an extent whereby an inhibitory milieu is spread across the brain in a manner in which the inhibition is capable of control over both frequency and severity of seizures. Efficacy in animals is high and multiple seizure types are affected. This in turn increases efficacy of gene therapy.

The question is, can such an approach also work in terms of energy metabolism? Animals could be placed on a ketogenic diet, then supplemented with creatine, then tested for seizure activity using several seizure-producing paradigms. The hypothesis to be tested is that a "super charging" of brain energetics could modulate/reduce seizures from more than one source, producing a situation similar to that seen in gene therapy using neuropeptide Y. That is to say that the energetically supercharged brain might be chemically equipt to resist increased seizure activity regardless of source/cause. This could easily be tested in animals, then translation should be rapid since as in AED polytherapy, the ketogenic diet (or modified Atkins diet) and creatine are both currently utilized, or close to trials, in human epilepsy treatment.

In the case of AEDs, months to years are expended seeking efficacious drug treatments which are unsuccessful in about 30% of cases. During that time, adverse effects of AEDs and seizures may occur. If a less invasive and equally effective overall treatment with significant effectiveness could be defined, long-term sequelae could be reduced in epilepsy patients. This experimental suggestion and variations thereof are worthy of examination.

References

Abel, M., and McCandless, D.W. (1992) The kindling model of epilepsy. In: Animal models of neurological disease II, Bolton, A., Baker, G., and Butterworth, R., eds. Humana Press, Baltimore, pp 153–168

Abou-Madi, M., et al. (1977) Cardiovascular reaction to laryngoscopy and intubation following small and large intravenous doses of lidocaine. Canad. Anesthesia Soc J. 24:12–19

Achenbach, T. (1991) Manual for the child behavior checklist/4-18 and 1991 profile Univ Vermont Department of Psychiatr. Burlington, VT

Ackermann, R., Engel, J., and Baxter, L. (1986) Positron emission tomography and autoradiographic studies of glucose utilization following electroconvulsive seizures in humans and rats. Ann. N.Y. Acad Sci. 462:263–269

Adab, N., et al. (2001) Additional educational needs in children born to mothers with epilepsy. J Neurol. Neurosurg Psychiatr 70:15–21

Adab, N., et al. (2004) The longer term outcome of children born to mothers with epilepsy. J Neurol Neusurg Psychiatr 75:1575–1583

Adachi, N., et al. (2005) Intellectual prognosis of status epilepticus in adult epilepsy patients: analysis with Wechsler Adult Intelligence Scale-revised. Epilepsia 46:1502–1509

Adelow, C., et al. (2009) Newly diagnosed single unprovoked seizures and epilepsy in Stockholm, Sweden. Epilepsia 50:1095–1101

Aeby (2005) above, not 2009

Aeby, A., et al. (2009) Levetiracetam efficacy in epileptic syndromes with continuous spike and waves during slow sleep: experience with 12 cases. Epilepsia 46:1937–1942

Aghakhani, Y., et al. (2004) fMRI activation during spike and wave discharges in idiopathic generalized epilepsy. Brain 127:1127–1144

Aguglia, U. Tinuper, P., and Gastaut, H. (1984) Startle induced epileptic seizures. Epilepsia 25:712–720

Aicardi, J. (1988) Epileptic syndromes in childhood. Epilepsia 29:s1–s5

Aicardi, J., and Chevrie, J. (1987) Convulsive status epilepticus in infants and children: a study of 239 cases Epilepsia 11:187

Aihara, M., et al. (1997) Ictal EEG and single photon emission computed tomography in a patient with cortical dysplasia presenting with atonic seizures. 38:723–727

Akamatsu, N., et al. (2001) Decreased susceptibility to pentylenetetrazole induced seizures after low frequency transcranial magnetic stimulation in the rat. Neurosci Letters 310:153–156

Aker, R., et al. (2009) Localized cortical injections of ethosuximide suppress spike wave activity and reduce the resistnce to kindling in genetic absence epilepsy rats (GAERS). Epilepsy Res. Doi:10.1016j, 10.013

Alberman, E. (1984) Epidemiological aspects of severe mental retardation. In: Dobbing, J., et al. Scientific studies in mental retardation. London: Royal Soc. Med. pp 3–13

D.W. McCandless, *Epilepsy: Animal and Human Correlations*,
DOI 10.1007/978-1-4614-0361-6, © Springer Science+Business Media, LLC 2012

Aldenkamp, A., Over-Plandsoen, W., and Diepman, L. (1999) Factors involved in learning problems and educational delay in children with epilepsy Child Neurol 5:130–135
Aldred, C., Green, J., and Adams, C. 2004) A new social communication intervention for children with autism: pilot randomized controlled treatment study suggesting effectiveness. J Child Psychol Psychiatr 45:1420–1430
Alonso, N., et al. (2009) Employment and quality of life in mesial temporal lobe epilepsy with hippocampal sclerosis: is there a change after surgical treatment? J Epil Clin Neurophysiol doi: 10.1590/s1676-26492009000200008
Alper, K., et al. (1997) Dissociation in epilepsy and conversion nonepileptic seizures Epilepsia 38:991–997
Alsaadi, T., et al. (2004) Video telemetry can be crucial tool for neurologists experienced in epilepsy when diagnosing seizure disorder Seizure 3:32–34
Alsaadi, T., et al. (2005) Clinical experiences of levetiracetam monotherapy for adults with epilepsy. Seizure 14:139–142
Alsaadi, T., Morris, G., and Mueller, W. (2003) Lateralising valueof ictal features in partial seizures: effect on postsurgical outcome. Seizure 12:257–260
Amato, E., et al. (2000) Seizures induce widespread upregulation of cystatin B, the gene mutated in progressive myoclonus epilepsy, in rat forebrain neurons. Eur. J. Neurosci. 12:1687–1695
American Psychiatric Association (1994) Diagnostic and statistical manual of mental disorders, 4th ed. Amer Psychiatr. Assn.
Anderson, C., Davis, K., and Baltuch, G. (2009) An update on brain stimulation for epilepsy Curr. Neurol. Neuro Sci Reports 9:327–332
Anderson, V., and Hauser, W. (1997) Genetic counseling. In Engel, J and Pedley, T. eds. Epilepsy: A comprehensive textbook Phila. Lippincott-Raven
Anderson, V., et al. (2005) Functional plasticity or vulnerability after early brain injury? Pediatr 116:1374–1382
Andres, R., et al. (2008) Functions aand effects of creatine in the central nervous system Brain Res Bull 76:329–343
Annegers, J et al. (1980) Seizures after head trauma: a population study Neurology 30:683–689
Annegers, J., et al. (1987) Factors prognostic of unprovoked seizures after febrile convulsions. N. Eng. J. Med. 316:493–498
Annegers, J., et al. (1990) Recurrence of febrile convulsions in a population based cohort population. Epil. Res. 66:1009–1012
Annegers, J., et al. (1999) The incidence of epilepsy and unprovoked seizures in multiethnic urban health maintenance organization Epilepsia 40:502–506
Annegers, J., Hauser, W., and Anderson, V. (1982) The risks of seizure disorders among relatives of patients with childhood onset epilepsy Neurology 32:174–179
Annegers, J., Rocca, W., and Hauser, W. (1996) Causes of epilepsy: contributions of the Rochester epidemiology project. Mayo Clin Proc 71:570–575
Anoue, M., et al. (1990) Spontaneous occurrence of spike wave discharges in five inbred strains of rats. Physiol. Behav. 48:199–201
Appleton, D., and DeVivo, D. (1974) An animal model for the ketogenic diet. Epilepsia 15: 211–227
Aranda, A., et al. (2010) Generalized convulsive status epilepticus management in adults: a cohort study with evaluation of professional practice. Epilepsia DOI: 10. 1111/. 1528–1167. 2010/02688.x
Arboix, A., et al. (1997) Predictive factors of early seizures after acute cerebrovascular disease Stroke 28:1590–1594
Armstrong, D. (1993) The neuropathology of temporal lobe epilepsy. J. Neuropathol. Exp. Neurol. 52:433–443
Armstrong, M. D., and Tyler, F. H., (1955) Studies on phenylketonuria. 1. Restricted phenylalanine intake in phenylketonuria. J Clin Invest. 1955 Apr;34(4):565–80
Arnold, L., and Privitera, M. (1996) Psychopathy and trauma in epileptic and psychogenic seizure patients Psychosamat 37:438–443

Arroyo, S., et al. (2004) Pregabalin add-on treatment: a randomized double blind placebo controlled dose response study in adults with partial seizures. Epilepsia 45:20–27
Arthur, T., et al. (2008) Seizure recurrence risk following a first seizure in neurologically normal children Epilepsia 49:1950–1954
Arts, and Geerts, A. (2009) When to start drug treatment for childhood epilepsy: the clinical epidemiological evidence Eur J Pediatr Neurol 13:93–101
Arts, W., et al. (2009) Landau-Kleffner syndrome and CSWS syndrome: Treatment with intravenous immunoglobulins Epilepsia 50:55–58
Arzimanoglou, A., Guerrini, R., and Aicardi, J. (2004) Aicardi's epilepsy in children. Phila. Lippincott Williams & Wilkins pp 274–275
Astradsson, A., et al. (1998) Rolandic epilepsy: an incidence study in Iceland. Epilepsia 39:884–886
Atillo, A., et al. (1983) Pathogenesis of brain lesions caused by experimental epilepsy. Acta Neuropathol. 59:37–49
Au, K., et al. (2007) Genotype/phenotype correlation in 325 individuals referred for diagnosis of tuberous sclerosis complex in the United States. Genet. Med. 9:88–100
Austin, J., and Caplan, R. (2007) Behavioral and psychiatric comorbidities in pediatric epilepsy: toward an integrative model. Epilepsia 48:1639–1651
Austin, J., and Dunn, D. (2000) Children with epilepsy: Quality of life and psychosocial needs. Ann. Rev of Nurs Res 18:26–47
Austin, J., et al. (1994) Childhood epilepsy and asthma: Comparison of quality of life Epilepsia 35:608–615
Austin, J., et al. (1996) Adolescents with active or inactive epilepsy or asthma: a comparison of quality of life Epilepsia 37:1228–1238
Austin, J., et al. (2001) Behavioral issues involving children and adolescents with epilepsy and the impact of their families: recent research data Epil Behav 5:s33–s41
Avorn, J (2008) Drug warnings that can cause fits-communication risks in a data poor environment N Eng J Med 359:991–994
Azar, N., et al. (2008) Postictal breathing pattern distinguishes epileptic from nonepileptic convulsive seizures. Epilepsia 49:132–137
Bachevalier, J., and Merjanian, P. (1994) The contribution of medial temporal lobe structures in infantile autism in Bauman, M., et al. eds. The Nerobiology of autism. Johns Hopkins Univ. Press pp 119–145
Badawy, R., et al. (2010) Can changes in cortical excitability distinguish progressive from juvenile myoclonic epilepsy? Epilepsia 51:2084–2088
Baer, M., et al. (1997) Vitamin D, calcium, and bone status in children with developmental delay in relation to anticonvulsant use and ambulatory status Am J Clin Nutr 65:1042–1051
Bagdorf, R., and Lee, S. (1993) Midline spikes: is it another benign EEG pattern of childhood? Epilepsia 34:271–274
Baglietto, M., et al. (2009) Epilepsia partialis continua in type 1 diabetes: evolution into epileptic encephalopathy with continuous spike-waves during slow sleep. Neurol. Sci. 30:509–512
Bahi-Buisson, N., et al. (2008) The three stages of epilepsy in patients with CDKL5 mutations. Epilepsia 49:1027–1037
Bain, M., Dwyer, S., and Rusak, B. (2004) Restraint stress affects hippocampal cell proliferation differently in rats and mice Neurosci. Letters 368:7–10
Baker, K. (1997) Desflurane and sevoflurane are valuable additions to the practice of neuroanesthesiology. J. neurosurg Anesth 9:66–68
Balietti, M., et al. (2010) A ketogenic diet increases succinic dehydrogenase activity and recovers age related decreases in numeric density of SDH positive mitochondria in cerebellar Purkinje cells of late adult rats. Micron 41:143–148
Ballaban-Gil, K., et al. (1998) Complications of the ketogenic diet. Epilepsia 39:744–748
Balogou, A., et al. (2007) Management of epilepsy patients in Batamariba district, Togo Acta Neurol Scand 116:211–216
Bancaud, J., (1969) Physiopathogenesis of generalized epilepsies of organic nature. In: Gastaut, H., et al. eds Springfield Il: Charles Thomas Publ.

Bancaud, J., and Talairach, J. (1992) Clinical semiology of frontal lobe seizures Adv Neurol New York, pp 3–58

Baracskay, P., et al. (2008) Generalization of seizures parallels the formation of dark neurons in the hippocampus and pontine reticular formation . Brain Res. 1228:217–228

Baracskay, P., et al. (2009) Status epilepticus affects the gigantocellular network of the pontine reticular formation. BMC Neurosci. 10:133–144

Baram, T., Gerth, A., and Schultz, L. (1997) Febrile seizures: an appropriate aged model suitable for long term studies. Br. Res. Dev Br Res 98:265–270

Bauer, J. (1996) Seizure inducing effects of antiepileptic drugs: a review. Acta Neurol. Scand 94:367–277

Baulac, S., et al. (2001) First genetic evidence of GABA A receptor dysfunction in epilepsy: a mutation in the gamma 2 subunit gene. Nat Genet 28:46–48

Beghi, E., et al. (2003) Adjunctive therapy versus alternative monotherapy in patients with partial epilepsy failing on a single drug: a multicenter randomised, pragmatic controlled study. Epil Res 57:1–13

Beghi, E., et al. (2005) A review of the costs of managing childhood epilepsy Pharmacoecon 23:27–45

Ben-Menachem, E. (1996) Modern management of epilepsy: vagus nerve stimulation Baillieres Clin Neurol. 5:841–848

Ben-Menachem, E. (1998) Vagus nerve stimulation for treatment of seizures. Arch. Neurol. 55:231–232

Ben-Menachem, E., and Falter, U. (2000) Efficacy and tolerability of levetiracetam 3000mg/day in patients with refractory partial seizures: a multicenter double blind responder selected study evaluating monotherapy. Epilepsia 41:1276–1283

Ben-Menachem, E., et al. (2007) Efficacy and safety of oral lacosamide as adjunctive therapy in adults with partial onset seizures. Epilepsia 48:1308–1317

Benabid, A., et al. (2001) Deep brain stimulation of the corpus luysi (subthalamic nucleus) and other targets for bilateral Parkinson's Disease Appl Neurophys 50:344–346

Benbadis, S. (1999) How many patients with pseudoseizures receive antiepileptic drugs prior to diagnosis? Eurol. Neurol. 41:114–115

Benbadis, S. (1999) Observations on the misdiagnosis of generalized epilepsy as partial epilepsy Seizure 8:140–145

Benbadis, S., Tatum, W., and Gieron, M. (2003) Idiopathic generalized epilepsy and choice of antiepileptic drugs. Neurology 61:1793–1795

Benedict, A., Verdian, L., and Maclaine, G. (2010) The cost effectiveness of rufinamide in the treatment of Lennox-Gastaut syndrome in the U.K. Pharmacoecon. 28:185–199

Bengzon, J., et al. (1977) Apoptosis and proliferation of dentate gyrus neurons after single and intermittent limbic seizures. Proc Soc Natl Acad Sci. 94:10432–10437

Beni , L., et al. (1996) Subclinical status epilepticus in a child after closed head injury. J. Trauma 40:449–451

Beniczky, S., et al. (2010) Effectiveness of the ketogenic diet in a broad range of seizure types and EEG features for severe childhood epilepsies. Acta Neurol. Scand. 121:58–62

Benn, E., et al. (2008) Estimating the incidence of first unprovoked seizure and newly diagnosed epilepsy in the low income urban community of Northern Manhattan , New York City Epilepsia 49:1431–1439

Berg, A. (2007) Cognitive and psychological issues in epilepsy: the scope of the problem. Presented at NINDS Conference entitled Curing epilepsy 2007. Natcher Conf Ctr, NIH Bethesda Md

Berg, A., et al. (1992) Predictors of recurrent febrile seizures. New Eng J Med 327:1122–1127

Berg, A., et al. (2004) Mortality in childhood onset epilepsy. Arch. Pediatr. Adol. Med. 158:1147–1152

Berg, A., et al. (2005) Special education needs of children with newly diagnosed epilepsy Dev Med Child Neurol 47:749–753

Berg, A., et al. (2006) Multicenter study of epilepsy surgery. Epilepsia 47:64–71

Berkovic, S. (2008) Progressive myoclonic epilepsies. In Pediatric Epilepsy Pellock, J., et al., eds. Demos, NY, pp 367–375
Berkovic, S., et al. (1986) Progressive myoclonus epilepsies: specific causes and diagnosis. N. Eng J. Med. 315:296–305
Berkovic, S., et al. (1988) Hypothalamic hamartomas and ictal laughter: evolution of a characteristic epileptic syndrome and diagnostic value of magnetic resonance imaging Ann Neurol 23:429–439
Berkovic, S., et al. (1993) Twin birth is not a risk factor for seizures Neurology 43:2515–2519
Berkovic, S., et al. (1998) Epilepsies in twins: genetics of the major epilepsy syndromes Ann Neurol 43:435–445
Berkovic, S., et al. (2003) Hypothalamic hamartoma and seizures: a treatable epileptic encephalopathy Epilepsia 44:969–973
Berkovic, S., et al. (2006) Human epilepsies: interaction of genetic and acquired factors Trends Neurosci 29:391–397
Berkovic, S., Kuzniecky, and Andermann, F. (1997) Human epileptogenesis and hypothalamic hamartomas: new lessons from an experiment of nature Epilepsia 38:1–3
Bernasconi, A., et al. (2003) Magnetic resonance spectroscopy and imaging of the thalamus in idiopathic generalized epilepsy. Brain 126:2447–2454
Berry, H., et al. (1967) Treatment of phenylketonuria Amer J Dis Child 113:2–15
Berto, P. (2002) Quality of life in patients with epilepsy and impact of treatments Pharmacoecon. 20:1039–1059
Bertrand, D., et al. (2005) The CHRNAB2 mutation 1312M is associated with epilepsy and distinct memory deficits Neurobiol. Dis. 20:799–804
Besag, F. (2001) Tonic seizures are a particular risk factor for drowning in people with epilepsy. Brit Med J. 322:975–976
Besset, A. (1982) Influence of generalized seizures on sleep organization. In Sterman, M., Shouse, M., and Passouant, P., eds. Sleep and Epilepsy N.Y. Academic Press pp 339–346
Bessman, S., and Carpenter, C. (1985) The creatine-creatine phosphate energy shuttle. Ann Rev Biochem 54:831–862
Betting, L., et al. (2006) EEG features in idiopathic generalized epilepsy: clues to diagnosis. Epilepsia 47:523–528
Beydoun A. 2000. Safety and efficacy of oxcarbazepine: results of randomized, double-blind trails. Pharmacotherapy. 20(8 Pt 2):152S–158S.
Beydoun, A. (1999) Monotherapy trials with gabapentin for partial epilepsy. Epilepsia 40:S13–16
Beydoun, A., et al. (2000) Pregabalin add on trial: double blind multicenter study in patients with partial epilepsy. 41:253–254
Beydoun, A., et al. (2009) Lacosamide: pharmacology, mechanisms of action and pooled efficacy and safety data in partial onset seizures. Exp. Rev. Neurosci. 9:33–42
Beyer, B., et al. (2008) Absence seizures in C3H/HeJ and knockout mice caused by mutation of the AMPA receptor subunit Gria4. Human Mol. Gen. 17:1738–1749
Beyreuther, B., et al. (2007) Lacosamide: a review of preclinical properties. CNS Drug Rev, 13:21–42
Bharucha, N. (2003) Epidemiology of epilepsy in India Epilepsia 44:9–11
Bickel, H., Gerrard, J., and Hickmans, E. (1954) Influence of phenylalanine intake on chemistry and behavior of a phenylketonuric child ACTA Pediatr 43:64–70
Bidmon, H., et al. (2008) Glutamine synthetase becomes nitrated and its activity is reduced during repetitive seizure activity in the pentyleneterazole model of epilepsy Epilepsia 49:1733–1748
Bien, C., et al. (2002) The natural history of Rasmussen's encephalitis Brain 125:1751–1759
Biervert, C., et al. (1998) A potassium channel mutation in neonatal human epilepsy. Science 279:403–406
Binder, L., et al. (1998) Neuropsychologic impairment in patients with nonepileptic seizures Arch Clin Neuropsychol 13:513–522

Bindoff, L., et al. (1991) Multiple defects of the mitochondrial respiratory chain in a mitochondrial encephalopathy (MERRF). J Neurol Sci 102:17–24
Bindokas, V., et al. (1998) Changes in mitochondrial function resulting from synaptic activity in the rat hippocampal slice. J. Neurosci. 18:4570–4587
Binnie, C, and Prior, P. (1994) Electroencephalography. J. Neurol. Neurosurg. Psychiatr. 57:1308–1319
Bishop, D. (1985) Age of onset and outcome in acquired asphasia with convulsive disorder (Landau-Kleffner syndrome) Dev Med Child Neurol 27:705–712
Bishop, M., and Boag, E. (2006) Teachers' knowledge about epilepsy and attitudes toward students with epilepsy: results of a national survey Epil. Behav. 8:397–405
Bjornaes, H., et al. (2004) Resective surgery for intractable focal epilepsy in patients with low I.Q.: predictors for seizure control and outcome with respect to seizures and neuropsychological and psychosocial functioning Epilepsia 45:131–139
Bladin, C., et al. (2000) Seizures after stroke: a prospective multicenter study Arch Neurol 57:1617–1622
Blatt, G., et al. (2001) Density and distribution of hippocampal neurotransmitter receptors in autism: an autoradiography study J Autism Dev Disord 31:537–544
Blumenfeld, H., et al. (2003) Selective frontal, parietal, and temporal networks in generalized seizures. Neuroimage 19:1556–1566
Bodenant, M., et al. (2008) Interest in the ketogenic diet in a refractory status epilepticus in adults Rev Neurol 164:194–199
Boero, J., et al. (2003) Restricted neuronal expression of ubiquitous mitochondrial creatine kinase: changing patterns in development and with increased activity Mol Cell Biochem 244:69–76
Bohning, D., et al. (2001) Feasibility of vagus nerve stimulation-synchronized blood oxygenation level dependent functional MRI. Invest. Radiol. 36:470–479
Boison, D. (2007) Cell and gene therapies for refractory epilepsy. Curr. Neuropharm. 5:115–125
Boison, D. (2008a) Adenosine as a neuromodulator in neurological diseases Curr Opin Pharmacol 8:2–7
Boison, D. (2008b) The adenosine kinase hypothesis of epileptogenesis Prog Neurobiol 84: 249–262
Boling, W., et al. (2001) Surgery for temporal lobe epilepsy in older patients. J. Neurosurg. 95:242–248
Bolton, P., and Griffiths, P. (1997) Association of tuberous sclerosis of temporal lobes with autism and atypical autism Lancet 349:392–395
Bonnani, P., Parmeggiani, L., and Guerrini, R. (2002) Different neurophysiological patterns of myoclonus characterize Lennox-Gastaut syndrome and myoclonic astatic epilepsy. Epilepsia 43:609–615
Boon, P., and Vandekerchove, T. (1996) Relief from seizures after intracranial EEG recording in non operated patients with rfractory partial epilepsy Epilepsia 37:146–152
Boon, P., et al. (2001) Vagus nerve stimulation for refractive epilepsy. Science 10:448–455
Boon, P., et al. (2007) Deep brain stimulation in patients with refractory temporal lobe epilepsy Epilepsia 48:1551–1560
Borowicz, K., et al. (2002) Effect of gabapentin on the anticonvulsant activity of antiepileptic . Epilepsia 43:956–963
Boschert, U., et al. (1998) Regulated expression of dual specificity protein phosphatases in rat brain. Neurorep. 9:4081–4086
Bouchlariotou, S., Tsikouras, P., and Maroulis, G., (2009) Undiagnosedmaternal phenylketonuria: own clinical experience and literature review. J Matern Fetal Neonatal Med. 22(10):943–8
Bough, K. (2008) Energy metabolism as part of the anticonvulsant mechanism of the ketogenic diet Epilepsia 49:91–93
Bough, K. J., et al. (2006) Mitochondrial biogenesis in the anticonvulsant mechanism of the ketogenic diet. Ann Neurol 60:223–235

Bough, K., and Rho, J (2007) Anticonvulsant mechanisms of the ketogenic diet. Epilepsia 48:43–58
Bough, K., Yao, S. and Eagles, D (2000) Higher ketogenic diet ratios confer protection from seizures without neurotoxicity. Epil. Res. 38:15–25
Bounds, J. (2007) Ictal eye closure is a reliable indicator for psychogenic nonepileptic seizures Neurology 68:963–964
Bourgeois, and VanLente, F. (1994) Effect of clonazepam on antiepileptic potency, neurotoxicity and therapeutic index of valproate and ethosuxamide in mice Epilepsia 35:8
Bourgeois, B. (1988) Problems of combination drug therapy in children Epilepsia 29:S20–S24
Bourgeois, B., et al. (1987) Monotherapy with valproate in primary generalized epilepsies. Epilepsia 28:S8–S11
Bourgeous, B. (1987) Combination of valproate and ethosuximide: antiepileptic and neurotoxic interaction. J. Pharmacol Exp Ther 247:1128–1132
Bouron, A. (2001) Modulation of spontaneous quantal release of neurotransmitters in the hipppocampus. Prog. Neurobiol. 63:613–635
Bouwman, B., and van Rijn (2004) Effects of levetiracetam on spike and wave discharges in WAG/Rij rats Seizure 13:591–594
Bowman, E. (1993) Etiology and clinical course of pseudoseizures: relationship to trauma, depression, and dissociation Psychosomatics 34:333–342
Bowman, E., and Markand, O. (2005) Diagnosis and treatment of pseudoseizures Psychiatr Annals 35:306–316
Boylan, L., et al. (2004) Depression but not seizure frequency predicts quality of life in treatment resistant epilepsy Neurology 62:258–261
Brazil, C. (2004) Nocturnal seizures. Sem. Neurol. 24:293–3000
Breier, J., et al. (1998) Quality of life perception in patients with intractable epilepsy or pseudoseizures Arch. Neurol. 55:660–665
Brenton, D., and Pietz, J. (2000) Adult care in phenylketonuria and hyperphenylalaninemia: the relevance of neurological abnormalities Eur J Pediatr 159:s114–s120
Brewster, A., et al. (2005) Formation of heteromeric hyperpolarization activated cyclic nucleotide gated channels in the hippocampus is regulated by developmental seizures. Neurobiol. Dis. 19:200–207
Brodie, M. (2004) Pregabalin as adjunctive therapy for partial seizures. Epilepsia 45:19–27
Brodie, M., et al. (2007) Comparison of levetiracetam and controlled release carbamazepine in newly diagnosed epilepsy. Neurology 68:402–408
Brodtkorb, E., and Picard, F. (2006) Tobacco habits modulate autosomal dominate nocturnal frontal lobe epilepsy. Epil. Behav. 9:515–520
Brown, W. (1973) Structural substrate of seizure foci in the temporal lobe. In: Brazier, M., ed. Epilepsy: its phenomena in man. Acad. Press, N.Y. pp 339–374
Brown, W., and Babb, T. (1987) Central pathological considerations of complex partial seizures. In: Hopkins, A., ed. Epilepsy Chapman and Hall, London pp 261–277
Browne, T., et al. (1983) Clinical and EEG estimates of absence seizure frequency. Arch. Neurol. 40:469–472
Bruno, R., et al. (1995) Pathophysiology of a central cause of post polio fatigue. Ann NY Acad Sci 753:257–275
Brunson, K., et al. (2001) Corticotropin (ACTH) acts directly on amygdala neurons to down regulate coricotropin releasing hormone gene expression. Ann Neurol 49:304–312
Buchanan, N., and Snars, J. (1993) Pseudoseizures (non epileptic attack disorder)-clinical management and outcome in 50 patients Seizure 2:141–146
Burke, G., Fifer, S., and Yoder, J. (1992) Early detection of Rasmussen's syndrome by brain SPECT imaging Clin Nuc Med 17:730–731
Buzsaki, G. (1991) The thalamic clock: emergent network properties. Neurosci. 41:351–364

Cahalane, S. (1968) Phenylketonuria: mass screening of newborns in Ireland Arch Dis Child 43:985–991
Camfield, C., et al. (1993) Biologic factors as predictors of social outcome of epilepsy in intellectually normal children: a population based study J Pediatr 122:869–873
Camfield, C., et al. (1993) Outcome of childhood epilepsy: a population based study with a simple predictive scoring system for those treated with medication. J. Pediatr 122:861–868
Camfield, C., et al. (1996) Incidence of epilepsy in childhood and adolescence: a population based study in Nova Scotia from 1977–1985 Epilepsia 37:19–23
Canevini, M., et al. (2010) Relationship between adverse effects of antiepileptic drugs, number of coprescribed drugs and drug load in a large cohort of consecutive patients with drug refractory epilepsy. Epilepsia 51:797–804
Canitano, R. (2007) Epilepsy in autism spectrum disorders Eur Child Adol Psychiatry 16:61–66
Canitano, R., Luchetti, A., and Zappella, M. (2005) Epilepsy, electroencephalographic abnormalities, and regression in children with autism. J Child Neurol 20:27–31
Caplin, R., et al. (2005) Social competence in pediatric epilepsy: insights into underlying mechanisms Epil Bhav. 6:218–228
Capovilla, G., et al. (2004) Efficacy of levetiracetam in pharmacoresistant continuous spikes and waves during slow sleep Acta Neurol Scand 110:144–147
Carson, B., et al. (1996) Hemispherectomy: a hemidecortication approach and review of 52 cases J Neurosurg 84:903–911
Carter, J., et al. (2004) Increased prevalence of epilepsy associated with severe falciparum malaria in children Epilepsia 45:978–981
Cataltepe, O., et al. (1995) Effect of hypoxia/ischemia on bicuculline induced seizures in immature rats. Epilepsia 36:396–403
Cavalleri, G., et al. (2007) A multicenter study of BRD2 as a risk factor for juvenilemyoclonic epilepsy. Epilepsia 48:706–712
Cavazos, J., Das, I., and Sutula, T. (1994) Neuronal loss induced in limbic pathways by kindling: evidence for induction of hippocampal sclerosus by repeated brief seizures. J. Neurosci. 14:3106–3121
Cavazos, J., Golarai, G., and Sutula, T. (1991) Mossy fiber synaptic reorganization induced by kindling: time course of development, progression, and permanence. J. Neurosci. 11:2795–2803
Cavazos,J., and Spitz, M. (2009) Seizures and epilepsy, overview and classification. E-Med.: Seiz and Epil./1184846
Caviness, V., and Rakic, P. (1978) Mechanisms of cortical development: a view from mutations in mice. Ann. Rev. Neurosci. 1:297–326
Cendes, F., et al. (1993) Early childhood prolonged febrile convulsions, atrophy and sclerosis of mesial structures, and temporal lobe epilepsy: an MRI volumetric study. Neurology 43: 1083–1087
Chang, T., et al. (2002) Neuroimaging in children during the acute evaluation of new onset seizures Ann Neurol 52:s134–s135
Charlier, C., et al. (1998) A pore mutation in a novel KQT-like potassium channel gene in an idiopathic epilepsy family. Nat. Genet. 18:53–55
Chassagnon, S., et al. (2003) Restricted frontomesial epileptogenic focus generating dyskinetic behavior and laughter Epilepsia 44:859–863
Chen, C., et al. (2003) Olfactory auras in patients with temporal lobe epilepsy. Epilepsia 44:257–260
Chen, J., et al. (2010) Effects of lamotrigine and topiramate on hippocampal neurogenesis in experimental temporal lobe epilepsy. Brain Res. 1313:270–282
Chen, L., Chan, S., and Yung, W. (2002) Rotational behavior and electrophysiological effect of GABA B receptor activation in the rat globus pallidus Neurosci. 114:417–425
Chen, L., Chan, Y., and Yung, W. (2004) GABA B receptor activation in the rat globus pallidus potentially suppresses pentylenetetrazole induced tonic seizures. J. Biomed. Sci. 11:457–464

Cheung, M., Chan, A., and Lam, J. (2009) Pre and postoperative fMRI and clinical memory performance in temporal lobe epilepsy. J. Neurol. Neurosurg. Psychiatr. 80:1099–1106
Chevrie, J., and Aicardi, J. (1972) Childhood epileptic encephalopathy with slow spike wave. A study of 80 cases. Epilepsia 13:259–271
Chin, R., et al. (2006) Incidence, cause, and short term outcome of convulsive status epilepticus in childhood: prospective population based study Lancet 368:222–229
Chin, R., Neville, B., and Scott, R. (2004) A systematic review of the epidemiology of status epilepticus. Eur J Neurol. 11:800–810
Chinchilla, D., et al. (1994) Reappraisal of Ramussen's syndrome with special emphasis on treatment with high doses of steroids J Neurol Neurosurg Psych 57:1325–1333
Chisholm, D. (2005) Cost effectiveness of first line antiepileptic drug treatments in the developing world: a population level analysis Epilepsia 46:751–759
Cho, Y., et al. (2003) A Korean kindred with autosomal dominant nocturnal frontal lobe epilepsy and mental retardation. Arch. Neurol. 60:1625–1634
Christian-Herman, J., Emons, M., and George, D. (2004) Effects of generic only drug coverage in a medicare HMO Health Affairs 4:455–468
Chuang, Y., et al. (2009) Contribution of nitric oxide, superoxide anion, and peroxynitrate to activation of mitochondrial apoptotic signaling in hippocampal CA3 subfield following experimental temporal lobe status epilepticus. Epilepsia 50:731–746
Chugani, H., et al. (1987) The Lennox-Gastaut syndrome: metabolic subtypes determined by 2-deoxy-2 fluorodeoxyglucose positron emmision tomography. Ann Neurol 21:4–13
Chung, S., (2009) Lacosamide: pharmacology, action and efficacy in partial onset seizures. Expt Rev Neurother. 9:33–42
Chung, S., Gerber, P., and Kirlin, K. (2006) Ictal eye closure is a reliable indicator for psychogenic nonepileptic seizures. Neurology 66:1730–1731
Cicolin, A., et al. (2006) Effect of levitiracetam on nocturnal sleep and daytime vigilance in healthy volunteers Epilepsia 47:1
Ciechanowski, P., et al. (2011) Long term outcomes from the PEARLS randomized tial for the treatment of depression in patients with epilepsy Epil. Behav. PMID 21333607, Feb 16
Clarke, D., et al. (2005) The prevalence of autistic spectrum disorder in children surveyed in a tertiary care epilepsy clinic Epilepsia 46:1970–1977
Clarren, S., Alvord, E., and Hall, J. (1980) Congenital hypothalamic hamartoblastoma, hypopituitarism, imperforate anus, and postaxial polydactyly: a new syndrome? Am J Med Genet 7:75–83
Cloix, J., and Hevor, T. (1998) Methionine sulfoximine epileptogeny and related metabolic changes in the brain of rodents Biogenic Amines 14:624–644
Cloix, J., Ardourel, M., and Hevor, T (2008) Glycogen metabolism and brain pathologies Cent Nerv Sys agents Med Chem 8:187–197
Cloix, J.-F., and Hvéor, T., (2009) Epilepsy, Regulation of Brain Energy Metabolism and Neurotransmission. Curr. Med. Chem. 16:841–853.
Clusman, H., et al. (2002) Prognostic factors and outcome after different types of resection for temporal lobe epilepsy. J. Neurosurg. 97:1131–1141
Co, J. P., et al. (2007) Proposal of an algorithm for diagnosing and treatment of neonatal seizures in developing countries. Epilepsia 48:1158–1164
Cockerell, O., et al. (1996) Clinical and physiological features of epilepsia partialis continua: cases ascertained in the UK. Brain 119:393–378
Coenen, W., et al. (1992) Genetic models of absence epilepsy with emphasis on the WAG/Rij strain of rats. Epilepsy Res. 12:75–86
Cohen-Gadol, A., Deluna, M., and Spencer, D. (2004a) Partial epilepsy presenting as episodic dyspnea: a specific network involved in limbic seizure propagation. J. Neurosurg. 100: doi:10.3171/jns.2004.100.3.0565
Cohen-Gadol, A., et al. (2004b) Long term outcome after surgery for focal cortical dysplasia. J. Neurosurg. 101:55–65

Collins, R. C., et al. (1976) Metabolic anatomy of focal motor seizures. Arch. Neurology 33:536–542
Collins, R.C. (1976) Metabolic response to focal penicillin seizures in rat: spike discharges vs. afterdischarge. J. Neurochem. 27:1473–1482
Collins, R.C., Posner, J., and Plum, F. (1970) Cerebral energy metabolism during electroshock seizures in mice. Amer. J. Physiol. 218:943–950
Collins, R.C., Tearse, R., and Lothman, E. (1983) Functional anatomy of limbic seizures: focal discharges from medial entorhinal cortex in rat. Brain Res. 280:25–40
Colosimo, E., et al. (2007) Electroclinical features of a family with simple febrile seizures and temporal lobe epilepsy associated with SCN1A loss of function mutation. Epilepsia 48:1691–1696
Commission on Classification and Terminology of the International League Against Epilepsy (1989) Proposal for revised classification of epilepsies and pileptic syndromes. Epilepsia 30:389–399
Commission on classification and terminology of the international league against epilepsy. (1981) Proposal for revised classification of epilepsies and epileptic syndromes. Epilepsia 30:389–399
Commission on Epidemiology and Prognosis, International League Against Epilepsy. Guidelines for epidemiologic studies on epilepsy. Epilepsia (1993) 34:592–596
Cooper, I., and Upton, A. (1978) Effects of cerebellar stimulation on epilepsy, the EEG and cerebral palsy in man Electroen. Clin Neurophys Suppl 34:349–354
Cooper, I.S., et al. (1973) Clinical and physiological effects of stimulation of the paleocerebellum in humans. J Am Geriatr Soc. 21(1):40–43
Cooper, L., et al. (1976) Chronic cerebellar stimulation in epilepsy: clinical and anatomical studies Arch Neurol 33:559–570
Coriat, P., et al. (1994) Influence of chronic angiotensin converting enzyme inhibition on anesthetic induction. Aesth. 81:299–307
Cortez, M., Gibson, K., and Snead, O. (2004) Absence seizures in succinic semialdehyde dehydrogenase deficient mice: a model for juvenile absence epilepsy. Pharmacol. Biochem Behav 79:547–553
Cortez, M., McKerlie, C., and Snead, O. (2001) A model of atypical absence seizures: EEG, pharmacology, and developmental characterization. Neurol. 56:341–349
Coulter, D., Sombati, S., and DeLorenzo, R. (1993) Selective effects of topiramate on sustained repetitive firing and spontaneous bursting in cultured hippocampal neurons. Epilepsia 34:123
Courchesne, E., et al. (1988) Hypoplasia of cerebellar lobules VI and VII in infantile autism. N Eng J Med 318:1349–1354
Cragar, D., et al. (2002) A review of diagnostic techniques in the differential diagnosis of epileptic and nonepileptic seizures Neuropsych Rev 12:31–64
Craig, I., and Tallis, R. (1994) Impact of valproate and phenytoin on cognitive function in elderly patients: results of a single blind randomized camparative study. Epilepsia 35:381–390
Cramer, J., et al. (2004) The impact of comorbid depression on health resource utilization in a community sample of people with epilepsy Epil Behav 5:337–342
Crepin, S., et al. (2007) Link between epilepsy and malnutrition in a rural area of Benin Epilepsia 48:1926–1933
Crespel, A., Baldy-Moulinier, M., and Coubes, P. (1998) The relationship between sleep and epilepsy in frontal and temporal lobe epilepsies. Epilepsia 39:150–157
Crome, L., Tymms, V., and Woolf, L. (1962) A chemical investigation of the defects of myelination in phenylketonuria J Neurol Neurosurg Psychiatr 25:143–150
Cross, J., and Neal, E. (2008) The ketogenic diet-update on recent clinical trials. Epsilepsia 49: 6–10
Cross, J., and Neville, B. (2009) The surgical treament of Landau-Kleffner syndrome Epilepsia 50: 63–67
Cukiert, A., et al. (2009) Outcome after extended callosal section in patients with primary idiopathic generalized epilepsy. Epilepsia 50:1377–1380

Cuppens, K., et al. (2009) Detection of nocturnal frontal lobe seizures in pediatric patients by means of accelerometers: a first study. IEEE Eng. Med. Biol. Sci. 2009:6608–6611
Curatolo, P., and Cusmai, R. (1987) Autism and infantile spasms in children with tuberous sclerosis Dev Med Child Neurol 29:190–198
Curatolo, P., et al. (2004) Autism in tuberous sclerosis Eur J pediatr Neurology 8:327–332
D'Arcangelo G, D'Antuono M, Tancredi V, Avoli M. 2006. Neocortical hyperexcitibility in a genetic model of absence seizures and its reduction by levetiracetam. Epilepsia 47:1144–1152.
D'Arcangelo, G., et al. (2002) Thalamocortical oscillations in a genetic model of absence seizures. Eur. J. Neurosci. 16:2383–2393
da Silvera, V., et al. (2010) Effect of ketogenic diet on nucleotide hydrolysis and hepatic enzymes in blood serum of rats in a lithium pilocarpine induced status epilepticus. Metab. Brain Dis. 25:211–217
Dailey JW, Reigel CE, Mishra PK, Jobe PC. 1989. Neurobiology of seizure predisposition in the genetically epilepsy-prone rat. Epilepsy Res. 3:3–17.
Dailey, J., et al. (1990) Neurobiology of seizure predisposition in the genetically epilepsy prone rat. Epil. Res. 3:131–136
Dalsgaard, M., et al. (2007) High glycogen levels in the hippocampus of patients with epilepsy sy J Ceb Blood Flow Met 27:1137–1141
Dan, B., and Boyd, S. (2005) Nonconvulsive (dialeptic) status epilepticus in children. Curr Pediatr. Rev. 1:7–16
Dang-Vu, T, et al. (2008) Spontaneous neural activity during human slow wave sleep. Proc Natl Acad Sci U.S.A. 105:15160–15165
Danielsson, S., et al. (2005) Epilepsy in young adults with autism: A prospective population based follow up study of 120 individuals diagnosed in childhood. Epilepsia 46:918–923
Dashti, H., et al. (2006) Long term effects of ketogenic diet in obese subjects with high cholesterol level. Mol. Cell. Biochem. 286:1–9
Davis, B. (2004) Predicting nonepileptic seizures utilizing seizure frequency, EEG, and response to medication Euro Neurology 51:153–156
Dawson, R., Webster, J., and Gurdjian, E. (1951) Serial electroencephalography in acute head injury J Neurosurg 8:613–630
de Boer, H. (1994) Work and epilepsy training project. Epilepsia 40:77
de Boer, H. (2005) Overview and perspectives of employment in people with epilepsy Epilepsia 46:52–54
de Silva, M., et al. (1996) Randomized comparative monotherapy trial of phenobarbitone, phenytoin, carbamazepine, or sodium valproate. Lancet 347:709–713
De Zelicourt, M., et al. (2000) The contributing factors to medical cost of epilepsy: an estimation based on a French prospective cohort of patients with newly diagnosed epileptic seizures Seizure 9:88–95
Dean, J., and Penry, J. (1988) Carbamazepine/valproate therapy in 100 patients with partial seizure failing carbamazepine monotherapy: long term follow up. Epilepsia 29:687
Debiais, S., et al. (2007) Epilepsy and language development: the continuous spike-waves during slow sleep syndrome Epilepsia 48:1104–1110
Deckers, C. et al. (2001) Monotherapy versus polytherapy for epilepsy: a multicenter double blind randomized study Epilepsia 42:1387–1394
DeFrnce, J., and McCandless, D (1991) Energy metabolism in rat hippocampus during and following seizure activity Met Brain Dis 6:83–91
DeGiorgio, C., et al. (2000) Prospective long term study of vagus nerve stimulation for the treatment of refractory seizures. Epilepsia 41:1195–1120
DeGiorgio, C., et al. (2005) Vagus nerve stimulation for epilepsy: randomized comparison of 3 stimulation paradigms. Neurology 65:317–319
Delalande, O., and Fohlen, M. (2003) Disconnecting surgical treatment of hypothalamic hamartoma in children and adults with refractory epilepsy and proposal of a new classification Neurol. Med. Chir 43:408–412

Delgado-Eschear, S., et al. (2004) Mutations in EFHCI cause juvenile myoclonic epilepsy. Nat. Genet. 36:842–849
Delgado-Escueta, A. (2007) Advances in genetics of juvenile myoclonic epilepsies. Curr. Rev Clin Sci. 7:61–67
Delgado-Escueta, A., et al. (1981) The nature of aggression during epileptic seizures. N Eng J Med 305:711–716
Delgado-Escuetta, A., and Enrile-Bacsal, F. (1984) Juvenile myoclonic epilepsy of Janz. Neurol. 34:285–294
DeLorenzo, R. (2006) Epidemiology and clinical presentation of status epilepticus. Adv. Neurol. 97:199–215
DeLorenzo, R., et al. (1996) A prospective population based epidemiologic study of statur epilepticus in Richmond Virginia. Neurology 46:1029–1035
DeLorenzo, R., et al. (1998) Persistent nonconvulsive status epilepticus after the control of convulsive status epilepticus Epilepsia 39:833–840
DeLorenzo, R., et al. (1999) Comparison of status epilepticus with prolonged seizure episodes lasting from 10–29 minutes. Epilepsia 40:164–169
Denays, R., et al. (1988) Single photon emission computed tomography in seizure disorders Arch Dis Child 63:1184–1188
Deonna, T., et al. (1986) Partial epilepsy in neurologically normal children: clinical syndromes and prognosis. Epilepsia 27:241–247
Deonna, T., et al. (2009) Sign language in Landau-Kleffner syndrome Epilepsia 50:77–82
Depaulis, A., et al. (1997) Anxiogenic like consequences in animal models of complex partial seizures. Neurosci. And Biobehav Reviews 21:767–774
Depaulis, A., Vergnes, M., and Marescaux, C. (1994) Endogenous control of epilepsy: the nigral inhibitory system. Prog Neurobiol 42:33–52
Dericioglu, N., and Saygi, S., (2010) Generalized seizures aggravated by levetiracetam in an adult patient with phenylketonuria. Metab Brain Dis. 25(2):207–9
Derouiche, A., Heimrich, B., and Frotscher, M. (1993) Loss of layer specific astrocytic glutamine synthetase immunoreactivity in slice cultres of hippocampus Eur J Neurosci 5:122–127
Derry, C., et al. (2006) Distinguishing sleep disorders from seizures. Arch. Neurol. 63:705–709
Derry, C., et al. (2008) Severe autosomal dominant nocturnal frontal lobe epilepsy associated with psychiatric disorders and intellectual disability. Epilepsia 49:2125–21229
Dershwitz, M., and Rosow, C. E. (2008) Pharmacology of intravenous anesthetics, in Longnecker DE, Brown DL, Newman MF, et al. (eds): Anesthesiology. New York: McGraw-Hill, pp 849–868
DeToledo, J., and Ramsay, R. (1996) Patterns of involvement of facial muscles during epileptic and nonepileptic events: a review of 654 events Neurology 47:621–625
Devinsky, H., et al. (2011) Suicide outcomes after resective epilepsy surgery Epil. Behav. PMID 21334984 Feb 18
Devinsky, O. (1995) Cognitive and behavioral effects of antiepileptic drugs. Epilepsia 36:46–65
DeVivo, D., et al. (1978) Chronic ketosis and cerebral metabolism. Ann. Neurol. 3:331–337
Devlin, A., et al. (2003) Clinical outcomes of hemispherectomy for epilepsy in childhood nd adolescence Brain 126:556–566
Dhanuka, A., et al. (2001) Juvenile myoclonic epilepsy: a clinical and sleep EEG study. Seizure 10:374–378
Dietrich, S., et al. (2008) A novel transcutaneous vagus nerve stimulation leads to brainstem and cerebral activations measured by fMRI. Biomed Technik Biomed Engineering 53:22–26
Dimova, P., and Bojinova, V., (2004) Case of subacute sclerosing panencephalitis with atypical absences and myoclonic atonic seizures as a first symptom. J. child neurol. 19:548–552
Ding, D., et al. (2006) Assessing the disease burden due to epilepsy by disability adjusted life year in rural China Epilepsia 47:2032–2037
DiRocco, C., et al. (2001) A 4 year old boy with complex seizures. Pediatr. Neurosurg. 34:319–324
Dizdarer, G., et al. (2000) A comparison of carbamazepine and oxcarbazepine in partial epilepsies. Epilepsia 41: 67–68

Djukic, A., et al. (2006) Are early myoclonic encephalopathy (EME) and the Ohtahara syndrome (EIEE) independent of each other? Epil Res 70:S58–S67

Dobelis, P., et al. (2003) GABAergic systems modulate nicotinic receptor mediated seizures in mice. J PET 306:1150–1166

Donati, F., et al. (2007) The cognitive effects of oxcarbazepine versus carbamazepine or valproate in newly diagnosed children with partial seizures. Seizure 16:670–679

Dong, J., Fan, P., and Frizell, R. (1996) Quantitative analysis of the packaging capacity of recombinant adeno-associated virus. Hum Gene Ther 7:2101–2112

Donner, E., Smith, C., and Snead, O. (2001) Sudden unexplained death in children with epilepsy. Neurology 57:430–434

Donsante, A., et al. (2001) Observed incidence of tumorigenesis in long term rodent studies of rAAV vectors. Gene Ther 8:1343–1346

dos Santos, L., et al. (2006) The time has come: a new scene for PKU treatment Gen. Mol. Res. 5:33–44

Dravet, C., et al. (1992) Severe myoclonic epilepsy in infants. IN Roger, J., et al. Eds. Epileptic Syndromes in Infancy, Childhood, and Adolescence. London, J. Libby pp 75–88

Dreifuss, F. (1981) Proposal for a revised clinical and electrographic classification of epileptic seizures. Epilepsia 26:268–278

Du, W., et al. (2005) Calcium sensitive potassium channelopathy in human epilepsy and paroxysmal movement disorder. Nat. Genet. 37:733–738

Dua, T., et al. (2006) Epilepsy care in the world: results of an ILAE/IBE/WHO global campaign against epilepsy survey Epilepsia 47:1225–1231

Dube, C., Brewster, A., and Baram, T. (2009) Febrile seizures: mechanisms and relationship to epilepsy. Brain Dev 5:366–371

Dube, C., et al. (2000) Prolonged febrile seizures in the immature rat model enhance hippocampal excitability in limbic circuits. Nat. Med. 5:888–894

Dube, C., et al. (2002) Prolonged febrile seizures in the immature rat model enhance hippocampal excitability long term. Ann Neurol 47:336–344

Dube, C., et al. (2005) Interleukin 1 beta contributes to the generation of experimental febrile seizures. Ann. Neurol. 57:152–155

Dube, C., et al. (2006) Temporal lobe epilepsy after experimental prolonged febrile seizures. Brain 129:911–922

Dube, C., et al. (2007) Fever, febrile seizures and epilepsy. Trends Neurosci. 30:490–496

Duffy, T., Howse, D., and Plum, F (1975) Cerebral energy metabolism during status epilepticus. J. Neurochem. 24:925–934

Dulac, O., and Kaminska, A. (1997) Use of lamotrigine in Lennox-Gastaut and related epilepsy syndromes. J Child Neurol 12:S23-S28

Dulla, C., et al. (2005) Adenosine and ATP link Pco2 to cortical excitability via pH. Neuron 48:1011–1023

Dunlop, S., and Taitz, J. (2005) Retrospective review of the management of simple febrile convulsions at a tertiary pediatric institution. J. Pediatr. Child Health 41:647–651

Dupont, S., et al. (2006) Long term prognosis and psychological outcomes after surgery for MTLE Epilepsia 47:2115–2124

Dura-Tave, T., Yoldi-Petri, M., and Gallinas-Victoriano, F. (2008) Panayiotopoulos syndrome: epidemiological and clinical characteristics and outcome. Eur J neurol 15:336–341

Dworsky, S., and McCandless, D.W. (1987) Regional cerebral energy metabolism in bicuculline induced seizures. Neurochem. Res. 12:237–240

Dybdal, D., and Gale, K. (2000) Postural and anticonvulsant effects of inhibition of the rat subthalamic nucleus. J Neurosci 20:261–265

Eadie, M. (2002) The epileptology of Theodore Herpin (1799–1865). Epilepsia 10:1256–1261

Earnest, M., et al. (1992) The sudden unexplained death syndrome in epilepsy. Epilepsia 33:310–316

Edwards, M., et al. (2002) N-acetylcysteine and Unverricht-Lundborg disease: variable response and possible side effects. Neurology 59:1447–1449

Egli, M., et al. (1985) The axial spasm: the predominant type of drop seizure in patients with secondary generalized epilepsy. Epilepsia 26:401–416

Eid, T., et al. (2004) Loss of glutamine synthetase in the human epileptogenic hippocampus: Possible mechanism for raised extracellular glutamate in mesial temporal lobe epilepsy Lancet 363:28–37

Eid, T., et al. (2008) Glutamate and astrocytes-Key players in human mesial temporal lobe epilepsy Epilepsia 49:42–52

Eisermann, M., et al. (2003) Infantile spasms in Down's syndrome-effects of delayed anticonvulsant treatment Epil Res 55:21–27

Elger, C., Helmstaedter, C., and Kurthen, M. (2004) Chronic Epilepsy and cognition. Lancet Neurol. 3:663–672

Ellenberg, J., Hirtz, D., and Nelson, K. (1986) Do seizures in children cause intellectual deterioration? New Eng J Med 314:1985–1091

Elliott, I., et al. (2008) Psychosocial outcomes in children two years after epilepsy surgery: has anything changed? Epilepsia 49:634–641

Ellis, T., and Stevens, A. (2008) Deep brain stimulation for medically refractory epilepsy. J. Neurosurg 25: DOI:10.3171/FOC/2008/25/9/E11

Elsharkawy, A., El-Ghandour, N., and Oppel, F. (2009) Long term outcome of lesional posterior cortical epilepsy surgery in adults. J. Neurol. Neurosurg. Psychiatr. 80:773–780

Elsharkawy, A., et al. (2008) Long term outcome of extratemporal epilepsy surgery among 154 adult patients. J. Neurosurg. 108:676–686

Enev, M., et al. (2007) Imaging onset and propagation of ECT induced seizures. Epilepsia 48: 238–244

Engel, J. (2001) A proposed diagnostic scheme for people with epileptic seizures and with epilepsy: report of the ILAE task force on classification and terminology. Epilepsia 42:796–803

Engel, J., et al. (1985) Local cerebral metabolic rate for glucose during petit mal absences. Ann Neurol 17:121–128

Engel, J., et al. (1993) Outcome with respect to epileptic seizures. In; Engel, J., ed Surgical Treatment of the Epilepsies, 2nd Ed. N.Y., Raven Press, pp 609–622

English, R., et al. (1989) Five patients with Rasmussen's syndrome investigated by single photon emission computed tomography Nuc Med Com 10:5–14

Englot, D., and Blumenfeld, H. (2009) Consciousness and epilepsy: why are complex partial seizures complex? Prog Brain Res 177:147–170

Etholm L, Heggelund P. 2009. Seizure elements and seizure element transitions during tonicclonic seizure activity in the synapsin I/II double knockout mouse: a neuroethological description. Epilepsy Behav Apr 14(4): 582–590.

Falconer, M. (1971) Genetic and related etiologic factors in temporal lobe epilepsy: a review. Epilepsia 12:13–31

Falip, M., et al. (2009) Postictal psychosis: a retrospective study in patients with refractory temporal lobe epilepsy. Seizure 18:145–149

Farrell, K. (1986) Benzodiazepines in the treatment of children with epilepsy. 27:9–25

Fastenau, P., et al. (2004) Neuropsychological predictors of academic underachievement in pediatric epilepsy Epilepsia 45:1261–1272

Fatemi, S., et al. (2009) GABAa receptor downregulation in brains of subjects with autism J Autism Dev Disord 39:223–230

Faught, E. (1999) Epidemiology and drug treatment of epilepsy in elderly people Drugs Aging 15:255–269

Faught, E. (2004) Treatment of refractory primary generalized epilepsy. Rev. Neurol. Dis. 1:s34–s43

Faught, E., et al. (1993) Felbamate monotherapy for partial onset seizures: an active control diet. Neurology 43:688–692

Faught, E., et al. (1996) Topiramate placebo controlled dose ranging trial in refractory partial epilepsy using 200, 400, and 600 mg daily doses. Neurology 46:1684–1690

Fayad, M., Choueiri, R., and Mikati, M. (1997) Landao-Kleffner syndrome: consistent responses to repeated intravenous gamma globulin doses: a case report Epilepsia 38:489–494
Federico, P., et al. (2005) Functional MRI analysis of the preictal state. Brain 128:1811–1187
Feichtinger, M., et al. (2006) Efficacy and safety of radiosurgical callosotomy: a retrospective analysis Epilepsy 47:1184–1191
Fenoglio, K., et al. (2007) Hypothalamic hamartoma: basic mechanisms of intrinsic epileptogenesis Semin Pediatr Neurol 14:51–59
Ferraro, G., and Sardo, P. (2009) Cholecystokinin-8 sulfate modulates the anticonvulsant efficacy of vigabatrin in an experimental model of partial complex epilepsy. Epilepsia 50:721–730
Ferraro, T., et al. (1997) Mapping murine loci for seizure response to kainic acid. Mamm Genome 8:200–208
Ferraro, T., et al. (1999) Mapping loci for pentylenetetrazole induced seizure susceptibility in mice. J. Neurosci 19:6733–6739
Ferri, R., et al. (2004) Different EEG frequency band synchronization during nocturnal frontal lobe seizures. Clin. Neurophysiol. 115:1202–1211
Ferrie, C., Agathonikou, A., and Panayiotopoulos, C. (1998) Electroencephalography and video-electroencephalography in the classification of childhood epilepsy syndromes. J R Soc Med 91:251–259
Ferrie, C., et al. (1997) Early onset benign occipital seizure susceptibility syndrome. Epilepsia 38:285–293
Ferro, J., and Pinto, F. (2004) Poststroke epilepsy Drugs Aging 21:639–653
Ferro, M., and Speechley, K. (2009) Depressive symptoms among mothers of children with epilepsy: A review of prevalence, associated factors, and impact on children Epilepsia 50:2344–2354
Figl, A., et al. (1998) Two mutations linked to nocturnal frontal lobe epilepsy cause use dependent potentiation of the nicotinic Ach response. J. Physiol. 513:655–670
Filho, G., et al. (2009) Are personality traits of juvenile myoclonic epilepsy related to frontal lobe dysfunctions? A protein MRS study. Epilepsia 50:1201–1209
Filipek, P., et al. (1999) The screening and diagnosis of autistic spectrum disorders J. Autism Dev Disorders 29:437–482
Fiorella, D., et al. (2001) 18F-fluorodeoxyglucose positron emission tomography and MR imaging findings in Rasmussen's encephalitis Am J Neuradiol 22:1291–1299
Fisher, J., and Vorhees, C. (1992) Developmental toxicity of antiepileptic drugs: relationship to postnatal dysfunction. Pharmacol Res 26:207–221
Fisher, P., Sperber, E., and Moshe, S (1998) Hippocampal sclerosis revisited. Brain Dev. 20:563–573
Fisher, R. (1989) Animal models of the epilepsies. Brain Res. Rev. 14:245–278
Fisher, R. (2008) SANTE (Stimulation of the anterior nucleus of the thalamus for epilepsy) 62nd ann meeting Amer Epil Soc
Fisher, R., et al. (1992) Placebo controlled pilot study of centromedian thalamic stimulation in treatment of intractable seizures Epilepsia 33:841–851
Fisher, R., et al. (1997) Assessment of vagus nerve stimulation for epilepsy Amer Acad Neurology 49:293–297
Fisher, R., et al. (2010) Electrical stimulation of the anterior nucleus of the thalamus for treatment of refractory epilepsy Epilepsia, 51:899–908
Fittipaldi, F., et al. (2001) EEG discharges on awakening: a marker of idiopathic generalized epilepsy. Neurology 56:123–126
Fleischer, W., et al. (2002) Comparative study of trauma related phenomena in subjects with pseudoseizures and subjects with epilepsy Am J Psychiatr 159:660–663
Flury, T., Aebi, C., and Donati, F. (2001) Febrile seizures and parental anxiety: does information help? Swiss Med Wkly 131:556–560
Fogarasi, A., et al. (2005) Predictive clinical factors for the differential diagnosis of childhood extratemporal seizures. Epilepsia 46:1280–1285

Folbergrová J, Passonneau JV, Lowry OH, Schulz DW. 1969. Glycogen, ammonia and related metabolites in the brain during seizures evoked by methionine sulphoximine. J Neurochem 16:191–203.

Follesa, P., et al. (2007) Vagus nerve stimulation increases norepinephrine concentration and the gene expression of BDNF and BFGF in the rat brain. Brain Res. 1179:28–34

Folling, A. (1934) Uber ausscheidung von phenylbrenztraubensaure in den Harn als Stoffwechselanomalie in Verbindung mit Imbezzillitat Z. Physiol Chem 227:169–174

Fombonne, E. (2003) Epidemiological surveys of autism and other pervasive developmental disorders: an update J Autism Dev Disord 4:365–382

Fombonne, E., et al. (1997) Autism and associated medical disorders in a French epidemiological survey. J. amer Acad Child Adolesc Psychiatr 36:1561–1569

Fonck, C., et al. (2003) Increased sensitivity to agonist induced seizures, Straub tail, and hippocampal theta rhythm in knock in mice carrying hypersensitive alpha4 nicotinic receptors. J. Neurosci. 23:2582–2590

Fonck, C., et al. (2005) Novel seizure phenotype and sleep disruptions in knock in mice with hypersensitivity alpha 4 nicotinic receptors. 25:11396–11411

Fonck, C., et al. (2006) Spontaneous seizures and nicotine induced c-fos in knock in mice with hypersensitive alpha4 nicotinic receptors. Soc. Neurosci Abst. 326.316/c384

Fons, C., et al. (2009) Epilepsy spectrum in cerebral creatine transporter deficiency Epilepsia 50:2168–2170

Food and Agriculture Organization (2006) Available at http://www.fao.org/ag/agn/nutrition/ben-e.stm

Forsgren, L., et al. (1996) Influence of epilepsy on mortality in mental retardation: an epidemiological study Epilepsia 37:956–963

Forsgren, L., et al. (2005) The epidemiology of epilepsy in Europe-a systematic review Eur. J. Neurol. 12:245–253

Foti, S., et al. (2007) Adeno-associated virus mediated expression and constitutive secretion of NPY or NPY13-36 suppresses seizure activity in vivo Gene Ther 14:1534–1536

Franaszczuk, P., et al. (1998) Time frequency analysis using the matching pursuit algorithm applied to seizures originating from the mesial temporal cortex. Electro. Clin. Neurophys. 106:513–521

Francois, J., Boehrer, A., and Nehlig, A. (2008) Effects of carisbamate in two models of genetically determined generalized epilepsy, the GAERS and the audiogenic Wistar AS Epilepsia 49:393–399

Frank, L., et al. (1999) Lamictal monotherapy for typical absence seizures in children. Epilepsia 40:973–979

Frank, M., Issa, N., and Stryker, M. (2001) Sleep enhances plasticity in the developing visual cortex. Neuron 30:275–287

Fratelli, C., et al. (2001) Cognitive deficits in children with hypothalamic hamartoma Neurology 57:43–46

Frattali, C., et al. (2001) Cognitive deficits in children with gelastic seizures and hypothalamic hamartoma Neurology 10:43–46

Freeman, J. (2005) Rasmussen's syndrome: progressive autoimmune multi-focal encephalopathy Pediatr Neurol 32:295–299

Freeman, J., and Vining, E. (1999) Seizures decrease rapidly after fasting: preliminary studies of the ketogenic diet. Arch Pediatr Adolesc Med 153:946–949

Freeman, J., et al. (2003) Generalized epilepsy in hypothalamic hamartoma. Evolution and postoperative resolution Neurology 60:762–767

Freeman, J., et al. (2006) The ketogenic diet: from molecular mechanisms to clinical effects. Epil. Res. 68:145–180

Freeman, J., et al. (2009) A blinded, crossover study of the efficacy of the ketogenic diet. Epilepsia 50: 322–325

Freilinger, M., et al. (2006) Behavioral and emotional problems in children with epilepsy J Child Neurol 21:939–945

French, J., et al. (2003) Dose-response trial of pregabalin adjunctive therapy in patients with partial seizures. Neurology 60:1631–1637

French, J., et al. (2004) Efficacy and tolerability of the new antiepileptic drugs, I: treatment of new onset epilepsy: report of the TTA and QSS subcommittees of the American Academy of Neurology and the American Epilepsy Society Epilepsia 45:401–409

French. J., diNicola, S., and Arrigo, C. (2005) Fast and sustained efficacy of levetiracetam during titration and the first 3 months of treatment in refractory epilepsy. Epilepsia 46:8

Frost, M., et al. (2001) Vagus nerve stimulation in children with refractory seizures associated with Lennox-Gastaut syndrome. Epilepsia 42:1148–1152

Fukuma, G., et al. (2004) Mutations of neuronal voltage gated Na channel alpha 1 subunit gene SCN1A in core severs myoclonic epilepsy in infancy and borderline SMEI. Epilepsy 45: 140–148

Gámez-Leyva, G., et al. (2009) Experience with intravenous levetiracetam in status epilepticus: a retrospective case series. CNS Drugs 23(11):983–987

Gabis, L., Pomeroy, J., and Andriola, M. (2005) Autism and epilepsy: cause, consequence, comorbidity, or coincidence? Epil. Behav. 7:652–656

Gaillard, W., et al. (2009) Guidelines for imaging infants and children with recent onset epilepsy Epilepsia 50:2147–2153

Gaily, E., Jonsson, H., and Lappi, M. (2009) Visual fields at school age in children treated with vigabatrin in infancy. Epilepsia 50:206–216

Gale, K. (1986) Role of the substantia Nigra in GABA mediated anticonvulsant actions Adv Neurol 44:343–364

Gambardella, A. et al. (2008) Monotherapy for partial epilepsy: focus on levetiracetam Neuropsy. Dis. Treat. 4:33–38

Gambhir, S., et al. (1995) Public awareness understanding, and attitudes towards epilepsy. Ind. J Med Res 102:34–38

Gant, J., et al. (2009) Decreased number of interneurons and increased seizures in neuropilin 2 deficient mice: implications for autism and epilepsy Epilepsia 50:629–645

Garcia, P., et al. (1994) Phosphorus magnetic resonance spectroscopic imaging in patients with frontal lobe epilepsy. Ann. Neurol. 35:217–221

Gardocki, J., et al. (1986) Anticonvulsant activity of McN-4853-beta-D-fructopyranose sulfamate in mice and rats. Epilepsia 27:648–649

Garzia, M., et al. (1990) Epileptic activity following 10 minute cerebral ischemia in Mongolian gerbils Neurosci letters 112:48–53

Gascon, G., and Lombroso, C. (1971) Epileptic laughter. Epilepsia 12:63–76

Gastaut, H. (1981) Individualisation des epilepsies dites benignes ou fonctionnelles aux differents agesde la vie. Rev EEG Neuro. 11:346–366

Gastaut, H., et al. (1963) An electroclinical study of generalized epileptic seizures of tonic expression Epilepsia 4:15–44

Gastaut, H., Tassinari, C. and Buereau,-Paillas, M. (1966) Polygraphic and clinical study of epileptic atonic collapses. Riv. Neurol. 36:5–21

Gastaux, H. (1982) The Lennox-Gastaut syndrome: comments on the syndromes terminology and nosological position amongst the secondary generalized epilepsies of childhood. Electr. Clin Neurophys 35:71–84

Gates, J., et al. (1984) Corpus Callosotomy: clinical and electroencephalographic effects. Epilepsia 25:308–316

Gaulke, L. et al., (2005) Environmental enrichment increases progenitor cell survival in the dentate gyrus following lateral fluid percussion injury Br Res Mol Br Res 141:138–150

Gelisse, P., et al. (2001) Clinical factors of drug resistance in juvenile myoclonic epilepsy. J. Neurol. Neurosurg. Psychiat. 70:240–243

Geller, E., et al. (1998) Rasmussen encephalitis: complementary role of multitechnique neuroimaging Am J Neurorad. 19:445–449

Genton, P., et al. (2000) Do carbamazepine and phenytoin aggravate juvenile myoclonic epilepsy? Neurology 55:1106–1109

Genton, P., Ferlazzo, E., and Thomas, P. (2008) Absence status epilepsy: delineation of a distinct idiopathic generalized epilepsy syndrome. Epilepsia 49:642–649

George, A. (2005) Inherited disorders of voltage gated sodium channels. J Clin Invest 115:1990–1999

Geschwind, D., and Levitt, P. (2007) Autism spectrum disorders: developmental disconnection syndromes. Curr. Opin. Neurobiol 17:103–111

Geshwind, N. (1983) Interictal behavioral changes in epilepsy. Epilepsia 24: 523

Gharbi, T., et al. (2010) Role of lumbar puncture for febrile seizure among infants under one year old. Tunis. Med. 88:178–183

Gibbs, F. (1951) Ictal and non-ictal psychiatric disoders in temporal lobe epilepsy. J. Nerv. Ment. Dis. 113:522–528

Gibson, K., et al. (1998) 4-hydroxybutyric acid and the clinical phenotype of succinic semialdehyde dehydrogenase deficiency an inborn error of GABA metabolism. Neuroped. 29:14–22

Gill, D., Ramsay, S., and Tasker, R. (2010) Selective reductions in subpopulations of GABAergic neurons in a developmental rat model of epilepsy. Brain Res. Doi:10.1016j/2010.03.054

Gillam, F., et al. (1997) Epilepsy surgery outcome: comprehensive assessment in children Neurology 48:1368–1374

Gilliam, F., et al. (2003) A dose comparison trial of topiramate as monotherapy in recently diagnosed partial epilepsy. Neurology 60:196–202

Gilman, J., Duchowny, M., and Campo, A. (2003) Pharmacokinetic considerations in the treatment of childhood epilepsy Pediatr Drugs 5:267–277

Giroud, M., et al. (1994) Early seizures after acute stroke: a study of 1640 cases. Epilepsia 35:959–964

Glötzner, F., Planner, M. and Gaab, M., (1979) Creatine kinase in serum after grand mal seizures Euro. Neurol. 18:399–404

Glauser, T. (1997) Topiramate Semin Pediatr Neurol 4:42–43

Glauser, T., and Morita, D. (2005) Ethosuximide. In Wylie, E., et al. Eds The treatment of epilepsy. Phila. Lippincott

Glauser, T., et al. (2006) Double blind randomized placebo controlled trial of adjunctive levetiracetam in pediatric partial seizures. Neurology 66:1654–1660

Glauser, T., et al. (2006) ILAE treatment guidelines: Evidence based analysis of antiepileptic drug efficacy and effectiveness as initial monotherapy for epileptic seizures and syndromes. Epilepsia 47:1094–1120

Glauser, T., et al. (2007) Topiramate monotherapy in newly diagnosed epilepsy in children and adolescents. J. Child Neurology 22:693–699

Glauser, T., et al. (2008) Rufinamide for generalized seizures associated with Lennox-Gastaut syndrome Neurology 71:1950–1958

Glauser, T., et al. (2010) Ethosuximide, valproic acid, and lamotrigine in childhood absence epilepsy. New Eng J Med 362:790–799

Gleissner, U., Helmstaedter, C., and Elger, C. (1998) Right hippocampal contribution to visual memory. J. Neurol. Neurosurg. Psychiatr. 65:665–669

Gloor, P., et al. (1977) Pathophysiology of generalized penicillin epilepsy in the cat. Electro. Clin. Neuropfysiol. 43:79–94

Gloor, P., et al. (1982) The role of the limbic system in experimental phenomena of temporal lobe epilepsy. Ann Neurol. 12:129–144

Go, C., and Snead, O. (2008) Pharmacologically intractable epilepsy in children: diagnosis and preoperative evaluation. Neurosurg. Focus 25:1–7

Goddard, G. (1967) Development of epileptic seizures through brain stimulation at low intensity. Nature 214:1020–1021

Goh, S., et al. (2005) Infantile spasms and iltellectual outcomes in children with tuberous sclerosis complex. Neurology 65:235–238

Gong, B., et al. (1999) Type 1 and type 2 Na channel alpha subunit polypeptides exhibit distinct spatial and temporal patterning, and association with auxilliary subunits in rat brain. J. Comp. Neurol. 412:342–352

Gowers, W. (1881) Epilepsies and other chronic convulsive diseases: their causes, symptoms, and treatment. London: J A Churchill
Gowers, W. (1885) Epilepsy and other chronic disorders N.Y., William wood and Co., p189
Gray, H. (1973) Anatomy of the human body. Goss, C., ed. Philadelphia, Lea and Febiger
Gray, T. (1989) Autonomic neuropeptide connections of the amygdala. In Tache, Y., Morley, J., and Brown, M. eds., Neuropeptides and stress. Springer-Verlag, New York pp 92–106
Greenwood, R. (2000) Adverse effects of antiepileptic drugs Epilepsia 41:s42-s52
Guelpa, G., and Marie, A. (1911) La lutte contre le pilepsie par la de sintoxication et par la re education alimentaire. Rev. Ther. Medico-Chir. 78:8–13
Guerrini, R., et al. (2002) Early onset absence epilepsy and paroxysmal dyskinesia. Epilepsia 43:344–352
Gupta, A., et al. (1983) Aura in temporal lobe epilepsy: clinical and electrographic correlation. J. Neurol. Neurosurg. Psychiatr. 46:1079–1083
Gupta, S., et al. (1996) Selective interaction of JNK protein kinase isoforms with transcription factors. EMBO J. 15:2760–2770
Gur, R., et al. (1982) Positron emmision tomography in two cases of childhood epileptic encephalopathy. Neurology 32:1191–1194
Gurbanova, A., et al. (2006) Effect of systemic and intracortical administration of phenytoin in two genetic models of absence epilepsy. Br. J. Pharmacol. 148:1976–1082
Guttinger, M., et al. (2005) Suppression of kindled seizures by paracrine adenosine release from stem cell derived brain implants. Epilepsia 46:1–8
Hackett, R., Hackett, L., and Bhakta, P. (1997) Febrile seizures in a south Indian district: incidence and associations. Dev. Med. Child Neurol. 39:380–384
Hackett, R., Hackett, L., and Bhakta, P. (1997) The prevalence and associated factors of epilepsy in children in Calicut District, Kerala, India. Acta Paediatr. 86:1257–1260
Hackett, R., Hackett, L., and Bhakta, P. (2001) The prevalence and associated factors of epilepsy in children in Calicut District, Kerala, India. Acta Pediatr 86:1257–1260
Haines, J. (2005) A case of pseudoseizures Southern Med J 98:122–123
Halford, J., and Lapointe, M. (2009) Clinical perspectives on lacosamide. Epil. Cur. 9:1–9
Hallbook, T., Lundgren, J., and Rosen, I. (2007) Ketogenic diet improves sleep quality in children with therapy resistant epilepsy. Epilepsia 48:59–65
Hamid, H., et al. (2011) Suicide outcomes after resective epilepsy surgery. Epilepsy Behav. 20(3):462–464
Hamiwka, L., et al. (2005) Surgery for epilepsy due to cortical malformations: ten year follow up. Epilepsia 46:556–560
Hanaoka, S., Takashima, S., and Morooka, K. (1998) Study of the maturation of the child's brain using 31P-MRS. Ped. Neurol 18:305–310
Hanaya, R., et al. (2010) Hippocampal cell loss and propagation of abnormal discharges accompanied with the expression of tonic convulsion in the spontaneously epileptic rat. Brain Res. 1328:171–180
Harden, C. (1997) Pseudoseizures and dissociative disorders: a common mechanism involving traumatic experiences Seizure 6:151–155
Harden, C., Burgut, F., and Kanner, A. (2003) The diagnostic significance of video EEG monitoring findings on pseudoseizure patients differs between neurologists and psychiatrics Epilepsia 44:453–456
Hari, R., and Lounasmaa, O. (1989) Science e244:432–436
Harkin, L., et al. (2007) The spectrum of SCN1A related infantile epileptic encephalopathies. Brain 130: 843–852
Hart, Y., et al. (1997) Chronic encephalitis and epilepsy in adults and adolescents: a variant of Rasmussen's syndrome? Neurology 48:418–424
Hartman, A., and Vining, E. (2007) Clinical aspects of the ketogenic diet. Epilepsia 48:31–42
Hauser WA, Annegers JF, Kurland LT. 1993. Incidence of epilepsy and unprovoked seizures in Rochester, Minnesota, 1935–1984. Epilepsia 34:453–468.
Hauser, W. (1990) Status epilepticus: epidemiologic considerations Neurology 40:suppl:9–13

Hauser, W. (1994) The prevalence and incidence of convulsive disorders in children. Epilepsia 35:s1–s6

Hauser, W. (2001) Epidemiology of epilepsy in children. In Pellock, J. et al. Pediatric epilepsy diagnosis and treatment New York Demos pp 81–96

Hauser, W., Annegers, J., and Kurland, L. (1993) Incidence of epilepsy and unprovoked seizures in Rochester Minnesota Epilepsia 34:453–468

Hauser, W., Annegers, J., and Kurland, L. (1993) The incidence of epilepsy in Rochester Minnesota, 1935–1984 Epilepsia 34:453–459

Hauser, W., et al. (1990) Seizure recurrence after a first unprovoked seizure: an extended follow up. Neurology 40:1163–1164

Hayman, M., et al. (1997) Autosomal dominant nocturnal frontal lobe epilepsy. Neurology 49:969–975

Hebb, D. (1942) The effects of early and late injury upon test scores, and the nature of normal adult intellegence Proc Am Philos Soc 85:275–292

Heeger, D., and Rees, D. (2002) What does fMRI tell us about neuronal activity? Nat Rev Neurosci 3:142–151

Heggelund, E. (2009) Seizure elements and seizure element transitions during tonic-clonic seizure activity in the synapsin I/II double knockout mouse. Epil. Behav. 14:582–590

Heida, J., and Pittman, Q. (2005) Causal links between brain cytokines and experimental febrile convulsions in the rat. Epilepsia 46:1906–1913

Heit, G., et al. (2009) Functional neurosurgery-introduction: the surgical treatment of pain, movement disorders, and epilepsy. In R. Jaffe (Ed), Anesthesiologists Manual of Surgical Procedures (pp 70–72). Phila: Lippincott, Williams & Wilkins

Heller, A., et al. (1995) Phenobarbitone, phenytoin, carbamazepine, or sodium valproate for newly diagnosed epilepsy: a randomized comparative monotherapy trial. J. Neurol. Neurosurg. Psychiatr. 58:44–50

Helmstaedter, C., et al. (1997) Differential involvement of left temporolateral and temporomesial structures in verbal declarative learning and memory: evidence from temporal lobe epilepsy. Brain Cog. 35:110–131

Helmstaedter, C., et al. (2000) Temporal lobe epilepsy: longitudinal clinical, neuropsychological and psychosocial follow-up of surgically and conservatively managed patients. Nervenarzt 71:629–642

Helmstaedter, C., Reuber, M., and Elger, C. (2002) Interaction of cognitive aging and memory deficits related to epilepsy surgery. Ann. Neurol. 52:89–94

Hendriks, M., et al. (2004) Relationships between epilepsy related factors and memory impairment. Acta Neurol Scand 110:291–300

Henry, T., et al. (1998) Brain blood flow alterations induced by therapeutic vagus nerve stimulation in partial epilepsy. Acute effects at high and low levels of stimulation. Epilepsy 39:983–989

Hermann, B., et al. (1988) Multietiological determinants of psychopathology and social competence in children with epilepsy Epil Rev 2:51–60

Hermann, B., et al. (2008) Growing up with epilepsy: a two year investigation of cognitive development in children with new onset epilepsy Epilepsia 49:1847–1858

Hermann, B., Seidenberg, M., and Bell, B. (2000) Psychiatric comorbidity in chronic epilepsy: identification, consequences, and treatment of major depression Epilepsia 41:s31–s41

Hernandez, M., et al. (2003) Attention, memory, and behavior adjustment in children with frontal lobe epilepsy. Epil. Behav. 4:522–536

Hesdorffer DC, Hauser WA, Annegers JF, Kokmen E, Rocca WA. 1996. Dementia and adultonset unprovoked seizures. Neurology 46:727–730.

Hesdorffer, D., et al. (1996) Dementia and adult onset unprovoked seizures Neurology 46:727–730

Hesdorffer, D., et al. (1998) Risk of unprovoked seizure after acute symptomatic seizure: effect of status epilepticus. Ann Neurol. 44:908–912

Hessel, E., et al. (2009) Phenotyping mouse chromosome substitution strains reveal multiple QTLs for febrile seizure susceptibility. Genes, Brain, and Behav. 8:248–255

Hill, A. (1953) Observation and experiment. New Eng J Med 248:995–1001
Hill, R., et al. (1999) Hemodynamic and metabolic aspects of photosensitive epilepsy revealed by functional magnetic resonance imaging and magnetic resonance spectroscopy. Epilepsia 40:912–920
Hirano, Y., et al. (2009) Differentiation of myoclonic seizures in epileptic syndromes: a video polygraphic study of 26 patients. Epilepsia 50:1525–1535
Hirose, S., et al. (1999) A novel mutation of CHRNA4 responsible for autosomal dominant nocturnal frontal lobe epilepsy. Neurology 53:1749–1793
Hirose, S., et al. (2005) Genetics of idiopatic epilepsy. Epilepsia 46:38–43
Hirsch, E., Schmitz, B., and Carreno, M. (2003) Epilepsy, antiepileptic drugs (AEDs) and cognition Acta Neurol Scand 108:23–32
Hirtz, D., et al. (2000) Practice parameters: evaluating a first non febrile seizure in children Neurology 55:612–623
Hochi, T., et al. (1987) Electron microscopic study of mossy fiber endings of the hippocampal formation in EL mice. Acta. Med. Okayama 41:81–84
Hodaie, M., et al. (2002) Chronic anterior nucleus stimulation for intractable epilepsy Epilepsia 43:603–608
Hogema, B., et al. (2001) Pharmacologic rescue of lethal seizures in mice deficient in succinate semialdehyde dehydrogenase. Nat. Genet. 29:212–216
Hokfelt, T. (1991) Neuropeptides in perspective: the last ten years. Neuron 7:867–879
Hollo, A., et al. (2001) Ictal perfusion changes during occipital lobe seizures in infancy: report of two serial ictal observations. Epilepsia 42:275–279
Holmes, G. (1986) Partial seizures in children. Pediatr. 77:725–731
Holmes, G. (2008) What constitutes a relevant animal model of the ketogenic diet? Epilepsia 49:57–60
Holmes, G., and Riviello, J. (2001) Treatment of childhood idiopathetic language deterioration with valproate. Epil. Behav. 2:272–276
Holmes, G., et al. (1980) Evaluation of childhood pseudoseizures using EEG telemetry and video monitoring J Pediatr 97:554–558
Holmes, G., et al. (2008) Lamotrigine monotherapy for newly diagnosed typical absence seizures in children. Epil Res 82:124–132
Holmes, M., Quiring, J., and Tucker, D. (2010) Evidence that juvenile myoclonic epilepsy is a disorder of frontotemporal corticothalamic networks. Neuroimage 49:80–93
Holtzman, D., et al. (1991) Maturational increase in mouse brain creatine kinase reaction rates shown by phosphorus magnetic resonance Dev Brain Res 58:181–188
Holtzman, D., et al. (1998) Creatine increases survival and suppresses seizures in the hypoxic immature rat. Pediatr Res. 44(3):410–414
Holtzman, D., et al. (1999) In vivo development of brain phosphocreatine in normal and creatine treated rabbit pups. J. Neurochem. 73:2477–2484
Holtzman, D., Obana, K., and Olson, J. (1981) Hyperthermia induced seizures in the rat pup: a model for febrile convulsions in children. Science 213:1034–1036
Hosain, M., et al. (2000) Vagus nerve stimulation treatment for Lennox-Gastaut syndrome. J Child Neurol 15:509–512
Hosford, A., and Wang, Y. (1997) Utility of the lethargic (lh/lh) mouse model of absence seizures in predicting the effects of lamotrigine, vigabatrin, gabapentin, and topiramate against human absence seizures. Epilepsia 38:408–414
Hosford, D., et al. (1992) The role of GABA B receptor activation in absence seizures of lethargic (lh/lh) mice Science 257:3980401
Hosford, D., et al. (1999) Studies of the lethargic (lh/lh) mouse model of absence seizures: regulatory mec. Adv. Neurol. 79:239–252
Houghton, J., and Welham, A. (2006) Assessing cognitive and psychological outcomes in a small pediatric epilepsy surgery program-challenges and solutions Epilepsia 47:s4–25

Hrachovy, R. Glaze, D., and Frost, J. (1991) A retrospective study of spontaneous remission and long-term outcome in patients with infertile sporms. Eplepsia 32:212–214
Hrachovy, R., and Frost, J. (2003) Infantile epileptic encephalopathy with hypsarrhythmia. J Clin Neurophys 20:408–425
Hrachovy, R., Frost, J., and Glaze, (1994) High dose long duration vrs low dose short duration corticotropin therapy for infantile spasms. J Pediatr 124:803–806
Hrachovy, R., Frost, J., and Kellaway, P. (1984) Hypsarrhythmia: variations on the theme. Epilepsia 25:317–325
Huber, A., et al. (2001) Grafts of adenosine releasing cells suppress seizures in kindling epilepsy. Proc Natl Acad Sci 98:7611–7616
Huijbregts, S. C., et al. (2002) The neuropsychological profile of early and continuously treated phenylketonuria: orienting, vigilance, and maintenance versus manipulation-functions of working memory. Neurosci Biobehav Rev. 26(6):697–712
Hunt, A., and Dennis, J. (1987) Psychiatric disorder among children with tuberous sclerosis. Dev. Med. Child. Neurol. 29:190–198
Hunt, J. (1922) On the occurrence of static seizures in epilepsy. J Nerv Ment Dis 56:351–360
Iadarola, M., and Gale, K. (1982) Substantia nigra: site of anticonvulsant activity mediated by gamma-aminobutyric acid. Science 218:1237–1240
Iasemidis, L., et al. (2004) Dynamical resetting of the human brain at epileptic seizures. IEEE Trans Biomed Eng. 51:493–506
Imaizumi, K., et al. (1959) The epilepsy like abnormalities in a strain of mice. Jpn. J. Pharmacol. 29:503–507
Ishida, N., et al. (1993) Epileptic seizure of EL mouse initiates at the parietal cortex. Brain Res. 608:52–57
Ishihara, K., et al. (1993) Abnormal excitability of hippocampal CA3 pyramidal neurons of spontaneously epileptic rats (SER), a double mutant. Exp Neurol 119:287–290
Isnard, J., et al. (2004) Clinical manifestations of insular lobe seizures: a stereo-electroencephalographic study. Epilepsia 45:1079–1090
Itoh, K., and Watanabe, M. (2009) Paradoxical facilitation of pentylenetetrazole induced convulsion susceptibility in mice lacking neuronal nitric oxide synthetase. Neurosci. 159:735–743
Jaaskelainen, S., et al. (2003) Sevoflurane is epileptogenic in healthy subjects at surgical levels of anesthesia Neurology 61:1073–1078
Jackson, J., and Singer, H. (1902) Observations of a case of convulsions (trunk fit or lowest level fit?) Brain 25:122–132
Jambaque, I., et al. (1991) Neuropsychological aspects of tuberous sclerosis sclerosis in relation to epilepsy and MRI findings. Dev Med Child Neurol 33:698–704
Jambaque, I., et al. (2000) Mental and behavioral outcome of infantile epilepsy treated with vigabatrin in tuberous sclerosis patients. Epilepsy Res. 38:151–160
Jansky, J., et al. (2005) Vagus nerve stimulation: predictors of seizure freedom Neurol Neurosurg Psychiatr 76:384–389
Janz, D,and Christian, W. (1957) Impulsiv-Petit mal. Dtsch Nervenheilkd 176:346–386
Jasper, H., and Droogleever-Fortuyn, J. (1947) Experimental studies on the functional anatomy of petit mal epilepsy. Res. Publ. Assoc. Res. Nerv. Ment. Dis. 26:272–298
Jayakar, P., et al. (1992) Ictal head deviation: lateralizing significance of the pattern of head movement. Neurology 42:1989–1992
Jayawant, S. and Libretto, S. (2003) Topiramate in the treatment of myoclonic astatic epilepsy in children: a retrospective hospital audit J Postgr Med 49:202–206
Jeavons, P., Bower, B., and Dimitrakoundi, M. (1973) Long term prognosis of 150 cases of West syndrome. Epilepsia 14:153–164
Jeavons, P., Clark, J., and Maheshwari, M. (1977) Treatment of generalized epilepsies of childhood and adolescence with sodium valproate. Dev Med Child Neurol 19:9–25
Jeon, B., et al. (2009) Ketogenic diet attenuates kainic acid induced hippocampal cell death by decreasing AMPK/ACC pathway activity and HSP70. Neurosci. Letters 64:49–53
Jervis, G. (1939) The genetics of phenylpyruvic oligophrenia J Ment Sci 85:719–728

Jervis, G. (1953) Phenylpyruvic oligophrenia : deficiency of phenylalanine oxidizing system Proc Soc Exp Biol Med 82:514–5516
Jilek, W., and Jilek-Aall, L. (1970) The problem of epilepsy in a rural Tanzanian tribe Afr J Med Sci 1:305–307
Jobst, B., et al. (2010) Brain stimulation for the treatment of epilepsy Epilepsia 51:88–92
Johnston, J., Smyth, M., and McKinstry, R. (2008) Basics of Neuroimaging in Pediatric Epilepsy in Pellock, J., Bourgeois, B., and Dodson, W. eds Pediatric Epilepsy, 3rd ed., New York, Demos pp 213–226
Jokeit, H., and Ebner, A. (1999) Long term effects of refractory temporal lobe epilepsy on cognitive abilities: a cross sectional study. J. Neurol. Neurosurg. Psychiatr. 76: 44–50
Jokeit, H., and Ebner, A. (2002) Effects of chronic epilepsy on intellectual functions Prog. In Brain Res. 135:455–463
Joo E., Tae, W., and Hong, S. (2008) Cerebral blood flow abnormality in patients with idiopathic generalized epilepsy. J. Neurol. 255:520–525
Joo, E., Tae, W., and Hong, S. (2006) Regional effects of lamotrigine on cerebral glucose metabolism in idiopathic generalized epilepsy. Arch. Neurol. 63:1282–1286
Jordan, K. (1999) Nonconvulsive status epilepticus in acute brain injury. J. Clin Neurophysiol 10:445–475
Jordan, K., (1993) Continuous EEG and evoked potential monitoring in the neuroscience intensive care unit J Clin Neurophysiol 16:332–340
Jost, C., et al. (2002) Creatine kinase B-driven energy transfer in the brain is important for habituation and spatial learning behavior, mossy fiber field size and determination of seizure susceptibility Euro J Neuro 15:1692–1706
Jouny, C., et al. (2007) Intrinsic ictal dynamics at the seizure focus: effects of secondary generalization revealed by complexity measures. Epilepsia 48:297–304
Jutila, L., et al. (2001) MR Volumetry of the Entorhinal, Perirhinal, and Temporopolar Cortices in Drug-Refractory Temporal Lobe Epilepsy. American Journal of Neuroradiology 22: 1490–1501
Jutila, L., Immonen, A., and Mervaala, E. (2002) Long term outcome of temporal lobe epilepsy surgery: analysis of 140 consecutive patients. J Neurol. Neurosurg. Psychiatr. 73: 486–494
Juul-Jensen, P., and Denny-Brown, D. (1966) Epilepsia partialis continua. Arch Neurol. 15:563–578
Kahn, R., et al. (2003) Atonic seizures in survivors of childhood cancer. J Child Neurol 18:397–400
Kalinin, W., and Polyanskiy, D. (2005) Gender differences in risk factors of suicidal behavior in epilepsy Epil Behavior 6:424–429
Kalume, F., et al. (2007) Reduced sodium current in Purkinje neurons from Nav1.1 mutant mice: implications for ataxia in severe myoclonic epilepsy in infancy. J. Neurosci. 27:11065–11074
Kalviainen, R., and Nousiainen, I. (2001) Visual field defects with vigabatrin: epidemilogy and theraupeutic implications. CNS Drugs 15:217–230
Kaminska, A., et al. (1999) Delineation of cryptogenic Lennox-Gastaut syndrome and myoclonic astatic epilepsy using multiple corresponding analysis. Epilepsy Res. 36:15–29
Kandel, A.M et al. (1996) Lack of hippocampal involvement in a rat model of petit mal epilepsy. Epil. Res. 23:123–127
Kang, H. C., et al. (2005) Efficacy and safety of the ketogenic diet for intractable childhood epilepsy: Korean multicentric experience. Epilepsia. 46(2):272–279
Kang, H., et al. (2007) Safe and effective use of the ketogenic diet in children with epilepsy and mitochondrial respiratory chain complex defects. Epilepsia 48:82–88
Kann, O. et al. (2005) Metabolic dysfunction during neuronal activation in the ex vivo hippocampus from chronic epileptic rats and humans. Brain 128:2396–2407
Kanner, A. (2007) Epilepsy and mood disorders Epilepsia 48:20–22
Kanner, A., and Balabanov, A. (2005) The use of monotherapy in patients with epilepsy: an appraisal of the new antiepileptic drugs Curr Neurol Neurosci Reports 5:322–328
Kanner, L. (1943) Autistic disturbances of affective contact. Nerv. Child. 2:217–250
Karimzadeh, P, Alaei, R., and Rahimpour, F., (2010) Prevalence ofseizure in PKU: an analytic historical study. Iranian J. Childhood Neurology. 4:15–20

Kasper, B., et al. (1999) Temporal lobe microdysgenesis in epilepsy versus control brains. J Neuropath. Exp. Neurol. 58:22–28

Kassell, N., et al. (1979) Alterations in cerebral blood flow, oxygen metabolism, and electrical activity produced by high dose sodium thiopental. Neurosurg 7:598–603

Kedmi, M., and Orr-Urtreger, A. (2007) Expression changes in mouse brains following nicotine induced seizures: the modulation of transcription factor networks. Physiol. Genom. 30:242–252

Keen, J., and Lee, D. (1973) Sequelae of neonatal convulsions study of 112 infants. Arch Dis Child 48:542–546

Keith, H., (1963) Convulsive disorders in children. Boston, Little Brown

Kemp, A., and Sibert, J. (1993) Epilepsy in children and the risk of drowning. Arch Disease Child 68:684–685

Kennard, M. (1936) Age and other factors in motor recovery from precentral lesions in monkeys Am J Physiol 115:138–146

Kerby, M., et al. (1987) Serum prolactins in the diagnosis of epilepsy: sensitivity, specificity, and predictive value. Neurology 37:1224–1226

Kerrigan, J., et al. (2005) The hypothalamic hamartoma: a model of subcortical epileptogenesis and encephalopathy. Sem. In Pediatr. Neurology 12:119–131

Kerrigan, J., et al. (2007) Hypothalamic hamartoma and infantile spasms. Epilepsia 48:89–95

Kessler, K., et al. (2002) Reduced inhibition within primary motor cortex in patients with post-stroke focal motor seizures Neurology 59:1028–1033

Keusch, G., Wilentz, J., and Kleinman, A. (2006) Stigma and global health: developing a research agenda Lancet 367:525–527

Khan, Z., et al. (2004) Abnormal motor behavior and vestibular dysfunction in the stargazer mouse mutant. L. Neurol. Doi:10.1016

Khanna, N., et al. (2000) Modulatory effects of nifedipine and nimodipine in experimental convulsions. Ind J Pharmacol. 32:347–352

Kim, C., Barry, J., and Zeifert, P. (1998) The use of inpatient medical psychiatric treatment for nonepileptic events Epilepsia 39:242–243

Kim, D., et al. (2008) GABA A receptor mediated activation of L-type calcium channels induces neuronal excitation in surgically resected human hypothalamic hamartomas Epilepsia 49:861–871

Kim, J. et al. (2007) Regional grey matter abnormalities in juvenile myoclonic epilepsy: a voxel based morphometric study. Neuroimage 37:1132–1137

King, L., et al. (1967) Effects of convulsants on energy reserves in the cerebral cortex. J. Neurochem. 14:599–611

King, M., et al. (1998) Epileptology of the first seizure presentation: a clinical, electroencephalographic, and magnetic resonance imaging study of 300 consecutive patients Lancet 352:1007–1011

Kinsman, S., et al. (1992) Efficacy of the ketogenic diet for intractable seizure disorders. Epilepsia 33:1132–1136

Kirchner, A., et al. (2000) Left vagus nerve stimulation suppresses experimentally induced pain. Neurology 55:1167–1171

Kislinger, T., et al. (2006) Global survey of organ and organelle protein expression in mouse. Cell 125:173–186

Kissani, N., et al. (2001) Epileptic seizures and epilepsy in subacute sclerosing panencephalitis. Neurophys Clin 31:398–405

Klaassen, A., et al. (2006) Seizures and enhanced cortical GABAergic inhibition in two mouse models of human autosomal dominant nocturnal frontal lobe epilepsy. PNAS 103:19152–19157

Klein, J., et al. (2002) The harlequin mouse mutation downregulates apoptosis inducing factor. Nature 419:367–374

Kliegman, R., et al. (2006) Nelson's essentials of Pediatrics Phila Elsevier Saunders 5th ed., pp 834–835, 839–840

Kob., et al. (2007) Prevalence of active epilepsy and health related quality of life among adults with self reported epilepsy in California: California Health Survey, 2003 Epilepsia 48:1904–1913

Kobau, R., Gilliam, F., and Thurman, D. (2006) Prevalence of self reported epilepsy or seizure disorder and its associations with self-reported depression and anxiety: results from the 2004 healthstyles survey Epilepsia 47:1915–1921

Kobayashi, E., et al. (2001) Seizure outcome and hippocampal atrophy in famial mesial temporal lobe epilepsy. Neurol. 56:166–172

Kobayashi, K., et al. (2000) Autosomal dominant frontal lobe epilepsy in a Japanese family Epilepsia 41:57

Koch, M., and Polman, S. (2009) Oxcarbazepine versus carbamazepine monotherapy for partial onset seizures Cochrane Database of Systematic Reviews DOI 10.1002/14651858.CD006453

Koch, R., et al. (2003) Large neutral amino acid therapy and phenylketonuria: a promising. approach to treatment. Mol Genet Metab. 79(2):110–113

Kodama, M., Russell, D., and Duman, R. (2005) Electroconvulsive seizures increase the expression of MAP kinase phosphatases in limbic regions of rar brain. Neuropsychopharm. 30:360–371

Koepp, M. (2005) Juvenile myoclonic epilepsy-a generalized epilepsy syndrome? Acta Neurol Scand 112:57–62

Kohler, C., et al. (2001) Association of fear auras with mood and anxiety disorders after temporal lobectomy. Epilepsia 42:674–681

Kohsaka, S., et al. (1999) Brainstem triggers absence seizures in human generalized epilepsy. Brain Res. 837:277–288

Kollevold, T (1976) Immediate and early cerebral seizures after head injuries. I. J. Oslo City Hosp. 26:99–114

Konishi, T., et al. (1999) Partial seizures during the course in patients with absence epilepsy. N T Hattetsu 31:395–401

Koo, B. (2001) EEG changes with vagus nerve stimulation. J Clin Neurophysiol 18:434–441

Koskiniemi, M. (1990) Myoclonic Epilepsy. In Pediatric Epilepsy. Sillanpaa, M., et al. eds. Wrightson, Hampshire

Kossoff, E. (2008) International consensus statement on clinical implementation of the ketogenic diet: agreement flexibility, and controversy. Epilepsia 49:11–13

Kossoff, E., et al. (2007) Combined ketogenic diet and vagus nerve stimulation: rational polytherapy? Epilepsia 48:77–81

Kossoff, E., et al. (2008) A case control evaluation of the ketogenic diet versus ACTH for new onset infantile spasms. Epilepsia 49:1504–1509

Kossoff, E., et al. (2008) When do seizures usually improve with the ketogenic diet? Epilepsia 49:329–333

Kostov, H., Larsson, P., and Roste, G. (2007) Is vagus nerve stimulation a treatment option for patients with drug resistant idiopathetic generalized epilepsy? Acta Neurol Scand. 115:55–58

Kothare, S., et al. (2006) Efficacy and safety of zonisamide monotherapy in a cohort of children with epilepsy. Pediatr Neurol 34:351–354

Kothur, K., Ray, M., and Malhi, P. (2008) Correlation of autism with temporal tubers in tuberous sclerosis complex Neurol India 56:74–76

Kotsopoulos, I., et al. (2002) Systematic review and meta-analysis of incidence studies of epilepsy and unprovoked seizures Epilepsia 43:1402–1409

Koutroumanidis, M., Aggelakis, K., and Panayiotopoulos, C. (2008) Idiopathic epilepsy with generalized tonic-clonic seizures only versus idiopathic epilepsy with phantom absences and generalized tonic-clonic seizures: One or two sydromes? Epilepsia 49:2050–2062

Kovarik, W., et al. (1994) Succinylcholine does not change intracranial pressure, cerebral blood flow velocity, or the EEG in patients with neurologic injury. Anesthesia and Analgesia 78:469–473

Kowal, C., et al. (2006) Human lupus autoantibodies against NMDA receptors mediate cognitive impairment Proc Natl Acad Sci 103:19854–19859

Koyama, R., and Matsuki, N. (2010) Novel etiological therapeutic strategies for neurodiseases: mechanisms and consequences of febrile seizures: lessons from animal models. J. Pharm. Sci. doi:10.1254/jphs.o9r19fm

Kraemer, D., and Branding-Bennett, E. (2009) Epilepsy surgery eMedicine 1–34 (http://emedicine.medscape.com/article/251449

Kraev, I., et al. (2009) Partial kindling induces neurogenesis, activates astrocytes and alters synaptic morphology in the dentate gyrus of freely moving adult rats Neurosci. 18: 254–267

Kramer, U., et al. (1998) Epidemiology of epilepsy in children: a cohort of 440 consecutive patients. Pediatr. Neurology 18:46–50

Krauss, G., and Morrison, P. (2007) Understanding and treating a channelopathy: severe myoclonic epilepsy of infancy. Neurology 69:233–234

Krebs, H. (1961) The physiological role of the ketone bodies. Biochem. J. 80:225–233

Kroll-Seger, J., et al. (2007) Severe relapse of epilepsy after vigabatrin withdrawal: for how long should we treat symptomatic infantile spasms? Epilepsia 48:612–613

Krsek, P., et al. (2009) Incomplete resection of focal cortical dysplasia is the main predictor of poor postsurgical outcome. Neurol. 72:217–223

Krumholz, A., et al. (1995) Complex partial status epilepticus accompanied by serious morbidity and mortality. Neurology 45:1499–1504

Kuba, R., et al. (2002) Effect of vagal nerve stimulation on interictal epileptiform discharges Epilepsy 43:1181–1188

Kudin, A., et al. (2002) Seizure dependent modulation of mitochondrial oxidative phosphorylation in rat hippocampus. Euro. J. Neurosci. 16:1105–1114

Kudin, A., et al. (2009) Mitochondrial involvement in temporal lobe epilepsy. Exp. Neurol. 68:326–332

Kunz, W. (2002) The role of mitochondria in epileptogenesis. Curr. Op. in Neurol. 15:179–185

Kunz, W., et al. (2000) Mitochondrial complex I deficiency in the epileptic focus of patients with temporal lobe epilepsy. Ann. Neurol. 48:766–773

Kutluay, E., et al. (2001) Seizure semiology and neuroimaging findings in patients with midline spikes. Epilepsia 42:1563–1568

Kuyk, J., van Dyck, R., and Spinhoven, P. (1996) The case for a dissociative interpretation of pseudoepileptic seizures J Nerv Ment Dis 184:468–474

Kuzniecky, R., et al. (1993) Magnetic resonance imaging in childhood intractable partial epilepsies: pathological correlations. Neurology 43:681–687

Kuzniecky, R., et al. (1997) Intrinsic epileptogenesis of hypothalamic hamartomas in gelastic epilepsy Ann Neurol 42:60–67

Kwan, P., and Brodie, M. J. (2000) Epilepsy after the first drug fails: substitution or add-on? Seizure. 9:468–468

Kwan, P., et al. (2009) Association of subjective anxiety, depression, and sleep disturbance with quality of life ratings in adults with epilepsy Epilepsia 50:1059–1066

Kyllerman, M., et al. (1991) Clinical and neurophysiological development of Unverricht-Lundborg disease in four Swedish siblings. Epilepsia 32:900–909

Labar, D. (2004) Vagus nerve stimulation for 1 year in 269 patients on unchanged antiepileptic drugs. Seizure 13:392–398

Labar, D., et al. (2008) Stimulation parameters after vagus nerve stimulator replacement. Neuromod. Tech. Neuro Inter 11:155–158

Labate, A., et al. (2005) Typical childhood absence seizures are associated with thalamic activation. Epileptic Disord. 7:373–377

Labate, A., et al. (2007) Usefulness of a morning routine EEG recording in patients with juvenile myoclonic epilepsy. Epil. Res. 77:17–21

Labovitz, D, Hauser, W., and Sacco, R. (2001) Prevalence and predictors of early seizure and status epilepticus after first stroke Neurology 57:200–206

Lagae, L. (2009) Rational treatment options with AEDs and ketogenic diet in Landau-Kleffner syndrome: still waiting after all these years. Epilepsia 50:59–62

Lagae, L., et al. (1998) Successful use of immunoglobulins in Landau-Kleffner sydrome Pediatr Neurol 18:165–168
Lahl, R., and Dorer, B. (1972) Subacute sclerosing inclusion body encephalitis with myoclonic astatic petit mal. Schw. Arch Neurol Psychiatr 111:1–13
Laing, J., et al. (2006) Intranasal administration of the growth compromised HSV2 vector DeltaRR prevents kainite induced seizures and neuronal loss in rats and mice. Mol. Therapy 13:870–881
Lammers, M., et al. (1995) Monotherapy or polytherapy for epilepsy revisited: a quantitative assessment Epilepsia 36:440–446
Landau, W., and Kleffner, F. (1957) Syndrome of acquired aphasia with convulsive disorder in children. Neurology 7:523–530
Lawler, J., et al. (2002) Direct antioxidant properties of creatine. Biochem Biophys Res Comm 290:47–52
Lawn, N., and Wijdicks, E. (2002) Status epilepticus: a critical review of management options. Can J Neurol Sci 29:206–215
Lee, C., et al. (2008) A new animal model of infantile spasms with unprovoked persistent seizures. Epilepsia 49:298–307
Lee, J., et al. (2001) Patterns of cerebral glucose metabolism in early and late stages of Rasmussen's syndrome J Child Neurol 16:798–805
Lee, K., and McCormick, D. (1997) Modulation of spindle oscillations by acetylcholine, cholecystokinin and 1S, 3R-ACPD in the ferret lateral geniculate nand perigeniculate nuclei in vitro. Neurosci. 77:335–350
Lee, K., et al. (1997) A genetic animal model of human neocortical heterotopia associated with seizures. J. Neurosci. 17:6236–6242
Lee, K., Haight, P., and Olejniczak, P. (2000) Epilepsia partialis continua in Creutzfeldt-Jakob disease Acta Neurol Scand 102:398–402
Lee, K., Jang, K., and Shon, Y. (2006) Chronic deep brain stimulation of subthalamic and anterior thalamic nuclei for controlling refractory partial epilepsy Acta Neurochir Suppl. 99:87–91
Lee, S., et al. (2001) Generalized tonic seizures associated with ganglioma: treatment with surgical resection Eur Neurol 46:225–226
Lega, B., et al. (2009) Cortical resection tailored to awake intraoperative ictal recordings and motor mapping in the treatment of intractable epilepsia partialis continua. Neorosurg. 64:onsE195-onsE196
Lehtinen, M., et al. (2009) Cystatin B deficiency sensitizes neurons to oxidative stress in progressive myoclonus epilepsy, EPM1 J. Neurosci 29:5910–5915
Lemmens, E., et al. (2008) Cytognesis in the dentate gyrus after neonatal hyperthermia induced seizures: what becomes of surviving cells. Epilepsia 49:853–860
Lennox, W. (1951) The heredity of epilepsy as told by relatives and twins Jam Med Assoc 146:529–536
Lennox, W. G., and Cobb, S. (1928) Epilepsy: from the standpoint of physiology and treatment. Medicine 7:105–290
Leppik, I., et al. (1993) Efficacy and safety of zonisamide: a new antiepileptic drug Epil Res 14:165–173
Lerman, P. (1986) Seizures induced or aggravated by anticonvulsants. Epilepsia 27:706–710
Lerner,J., et al. (2009) Assessment and surgical outcomes for mild type 1 and severe type 2 cortical dysplasia. Epilepsia 50:435–446
Lesser, R. (1999) Unexpected places: how did vagus nerve stimulation become a treatment for epilepsy?Neurology 52:1117–1118
Letts, V. (2005) Stargazer-a mouse to seize. Epil. Curr. 5: 161–165
Letts, V., et al. (1998) The mouse stargazer gene encodes a neuronal Ca2+ channel gamma subunit. Nature genetics 19:340–347
Leutmezer, F., Lurger, S., and Baumgartner, C. (2002) Focal features in patients with idiopathic generalized epilepsy. Epil. Res. 50:293–300
Leuzzi, V. (2002) Inborn errors of creatine metabolism and epilepsy: clinical features, diagnosis and treatment. Child Neurol. 17:3S89–3S97

Leuzzi, V., et al. (2000) Brain creatine depletion: Guanidinoacetate methyltransferase deficiancy (improving with creatine supplementation). Neurology. 55(9):1407–1409
LeVan, Q., et al. (2001) Anticipation of epileptic seizures from standard EEG recordings Lancet 357:183–188
Levy, H. L., and Ghavami, M., (1996) Maternal phenylketonuria: a metabolic teratogen. Teratology. 53(3):176–184
Levy, R., Bourgeous, B., and Hachard, H., (2007) Drug drug interactions. In Engel, J., and Pedley, T. eds., Epilepsy: a comprehensive textbook Phila. Lippincott.
Lewis, A., and Parsons, S. (2008) Understanding of epilepsy by Children and young people with epilepsy Euro J Special Needs Educ 23:321–335
Lewis, D., Bello, J., and Chan, S. (2005) FABSTAT study team. Hippocampal abnormalities subsequent to febrile status epilepticus. Findings on early postictal imaging. Epilepsia 46:52–53
Lewis, M., et al. (1991) Impact of the Children's Epilepsy Program. Epilepsia 32:365–374
Lhatoo, S., and Sander, J. (2001) The epidemiology of epilepsy and learning disorder Epilepsia 42:6–9
Li, J., et al. (2010) Aberrant glutamate receptor 5 expression in temporal lobe epilepsy lesions. Brain Res. 1311:166–174
Li, S., Wu, J., and Salzberg, M (2005) The western pacific region and the global campaign against epilepsy Epilepsia 46:64–65
Liberson, W. T., (1989) Reciprocal relationship between the major EEG rhythms and latencies of evoked potential intermediate components. A tentative explanation. Electromyogr Clin Neurophysiol. 29:425–432
Lieuallen, K., et al. (2001) Cystatin B deficient mice have increased expression of apoptosis and glial activation genes. Hum Mol Genet 10:1867–1871
Liew, C., et al. (2009) 18F-FCWAY and 18F-FDG-PET in MRI negative temporal lobe epilepsy Epilepsia 50:234–239
Lin, C. M., and Thajeb, P., (2007) Valproic acid aggravates epilepsy due to MELAS in a patient with an A3243G mutation of mitochondrial DNA. Metab Brain Dis. 22(1):105–9. Erratum in: Metab Brain Dis. 2007 22(3–4):425
Lin, E., et al. (2003) Recombinant AAV mediated expression of galanin in rat hippocampus suppresses seizure developmant. Eur J Neurosci 18:2087–2092
Lin, K., et al. (2009) Magnetic resonance spectroscopy reveals an epileptic network in juvenile myoclonic epilepsy. Epilepsia 50:1191–1200
Litt, B., and Baltuch, G. (2001) Brain stimulation for epilepsy Epil and behav. 2:s61–s67
Liukkonen, E., et al. (2010) Long term outcome of 32 children with encephalopathy with status epilepticus during sleep or ESESS syndrome. Epilepsia DOI:10.1111/j.1528-1167.2010.02578.x
Livingston, S. (1972) Dietary management of epilepsy. In: Livingston, S. Comprehensive management of epilepsy in infancy, childhood, and adolescense. Springfield, Il. Charles Thomas pp 378–405
Lockard, J., Congdon, W., and DuCharme, L. (1990) Feasibility and safety of vagal stimulation in monkey model. Epilepsia 31:s20–s26
Loddenkemper, T., et al. (2001) Deep brain stimulation in epilepsy J. Clin. Neurophysiol 18:514–532
Logothetis. N., et al. (2001) Neurophysiological investigation of the basis of the fMRI signal. Nature 412:150–157
Lombroso, C. (1983) Prospective study of infantile spasms: clinical and therapeutic correlations Epilepsia 24:135–158
Lombroso, C. (2000) Pavor nocturnus of proven epileptic origin. Epilepsia 41:1221–1226
Loscher, W. (2002) Basic pharmacology of valproate: a review after 35 years of clinical use for the treatment of epilepsy. CNS Drugs 16:669–694
Loscher, W., and Honack, D. (1993) Profile of ucb L059, a novel anticonvulsant drug, in models of partial and generalized epilepsy in mice and rats. Eur. J. Pharmacol. 232:147–158

Loscher, W., and Schmidt, D (1988) Which animal models should be used in the search for new antiepileptic drugs? A proposal based on experimental and clinical considerations. Epilepsy Res. 2:145–181

Lothman, E., and Collins, R.C. (1981) Kainic acid induced limbic seizures: metabolic, behavioral, electroencephalographic and neuropathological correlates. Brain Res. 218: 299–318

Lothman, E., et al. (1990) Recurrent spontaneous hippocampal seizures in the rat as a chronic sequela to limbic status epilepticus. Epil. Res. 6:110–118

Lou Smith, M., Elliot, I., and Lach, L. (2006) Memory outcome after pediatric epilepsy surgery. Child. Neuropsychol. 12:151–164

Lowenstein, D. (2008) Seizures and epilepsy. In Fauci, A., et al. eds Harrison's Principles of Internal Medicine. N.Y. McGraw Hill Medical 17th ED., 2490t, 2498–2510

Lowry, O., and Passonneau, J. (1972) A Flexible system of enzymatic analysis New York, Academic Press pp 151–156, 174–177

Lucarini, N., et al. (2007) Genetic polymorphisms and idiopathetic generalied epilepsies. Pediatr. Neurol. 37:157–164

Luders, H., et al. (1998) Semiological seizure classification. Epilepsia 39:1006–1013

Luhdorf, K., Hensen, L., and Plesner, A. (1986) Epilepsy in the elderly: incidence, social function and disability. Epilepsia 27:135–141

Lundgren, J., et al. (1998) Vagus nerve stimulation in 16 children with refractory epilepsy. Epilepsia 39:809–817

Lusardi, T. (2009) Adenosine Neuromodulation and traumatic brain injury Curr Neuropharmacol 7:228–237

Lust, W. D., et al. (1978) Changes in brain metabolites induced by convulsants or electroshock: effects of anticonvulsant agents. Molec. Pharmacol. 14:347–356

Luszczki, J. (2009) Interactions between pregabalin and phenobarbital in the mouse maximal electroshock induced seizure model. J. Pre-clin. Clin Res. 3:103–109

MacDonald, B., et al. (2000) The incidence and lifetime prevalence of neurological disorders in a prospective community based study in the UK Brain 123:665–676

Mackay, M., et al. (2004) Practice parameter: medical treatment of infantile spasms. Neurololy 62:1668–1681

Maeda, Y., et al. (2003) Rasmussen's syndrome: Multifocal spread of inflammation suggested from MRI and PET findings Epilepsia 44:1118–1121

Magistretti, P., and Pellerin, L. (1996) Cellular mechanisms of brain energy metabolism. Ann NY Acad Sci 777:380–387

Magnes, J., Moruzzi, G., and Pompeiano, O (1961) Synchronization of the EEG produced by low frequency electral stimulation of the region of the solitary tract. Arch. Ital. Biol. 99:33–67

Magnusson, A., Nakken, K., and Brubakk, E. (1996) Autosomal dominant frontal epilepsy. Lancet 347:1191–1192

Maher, J., and McLachlan, R. (1995) Febrile convulsions. Is seizure duration the most important predictor of temporal lobe epilepsy? Brain 118:1521–1528

Makulkin, R., Novytskyi, S., and Korniienko, T. (1992) Role of globus pallidus in mechanisms of antiepileptic caudate-cortical effects. Fiziol Zh 38:3–9

Malla, B., et al. (1999) Acute postoperative seizures following anterior temporal lobectomy for intractable partial epilepsy. Neurosurg. Focus 2:article 2

Malmgren, K., et al. (2008) Seizure outcome after resective epilepsy surgery in patients with low IQ. Brain 131:535–542

Mancardi, M., et al. (2007) Severe epilepsy in X-linked creatine transporter defect (CRTR-D) Epilepsia 48:1211–1213

Manchanda R, Freeland A, Schaefer B, McLachlam RS, Blume WT. 2000. Auras, seizure focus, and psychiatric disorders. Neuropsychiatry Neuropsychol Behav Neurol. 13(1):13–9.

Mancini, G., et al. (2005) Two novel mutations in SLC6A8 cause creatine transporter defect and distinctive X-linked mental retardation in two unrelated Dutch families. Amer J Med Gen 132:288–295

Mandel, H., Tirosh, E., and Berant, N. (1989) Sandifer syndrome reconsidered Acta Pediatr Scand 78:797–781

Mandelbaum, D., and Burack, G. (1997) The effect of seizure type and medication on cognitive and behavioral functioning in children with idiopathetic epilepsy Dev Med Child Neurol 39:731–735

Manford, M., et al. (1992) National general practice study of epilepsy (NGPSE) : partial seizure patterns in a general population. Neurology 42:1911–1917

Manford, M., Fish, D., and Shorvon, S. (1996) Startle provoked epileptic seizures: features in 19 patients. J. Neurol. Neurosurg. Psychiatr. 61:151–156

Manganotti, P., et al. (2011) Highly focal BOLD activation on functional MRI in a patient with progressive myoclonic epilepsy and diffuse giant somatosensory evoked potentials Epil. Behav. PMID 21339087 Feb 19

Manni, R., Terzaghi, M., and Repetto, A. (2008) The FLEP scale in diagnosing nocturnal frontal lobe epilepsy, NREM and REM parasomnias: data from a tertiary sleep and epilepsy unit. Epilepsia 49:1581–1585

Maquet, P., Smith, C., and Stickgold, R., eds. (2003) Sleep and brain plasticity. Oxford, Oxford Univ Press

Marciani MG, Santone G, Sancesario G, Massa R, Stanzione P, Bernardi G. 1991. Epileptic activity following cerebral ischemia in Mongolian gerbils is depressed by CPP, a competitive antagonist of the N-methly-D-aspartate receptor. Neurosci Lett 129:306–310.

Maremmani, C., et al. (1991) Descriptive epidemiological study of epilepsy syndromes in a district of northwest Tuscany, Italy Epilepsia 32:294–298

Marescaux, C., and Vergnes, M. (1995) Genetic absence epilepsy in rats from Strasbourg (GAERS) Ital. J. Neurol Sci. 16:113–118

Marescaux, C., et al. (1982) Stuporous episodes during treatment with sodium valproate: report of 7 cases. Epilepsia 23:297–305

Marini C, King MA, Archer JS, Newton MR, Berkovic SF. 2003. Idiopathic generalized epilepsy of adult onset: clnical syndromes and genetics. J Neurol Neurosurg Psychiatry 74:192–196.

Marini, C., et al. (2003) Childhood absence epilepsy and febrile seizures: a family with a GABA a receptor mutation. Brain 126:230–240

Marini, C., et al. (2003) Idiopathetic generalized epilepsy of adult onset: clinical syndromes and genetics. J. Neurol. Neurosurg. Psychiat. 74:192–196

Marini, C., et al. (2007) Idiopathic epilepsies with seizures precipitated by fever and SCN1A abnormalities. Epilepsia 48:1678–1685

Markand, O. (1977) Slow spike wave activity in EEG and associated clinical features. Neurology 27:746–757

Markland, O. (2003) Lennox-Gastaut syndrome (childhood epileptic encephalopathy). J Clin Neurophysiol. 20:426–441

Marrosu, F., et al. ((2003) Correlation between GABA A receptor density and vagus nerve stimulation in individuals with drug resistant partial epilepsy. Epilepsy Res. 55:59–66

Marson, A., et al. (2007) The SANAD study of effectiveness of carbamazepine, gabapentin, lamotrigine, oxcarbazepine, or topiramate for treatment of partial epilepsy: an unblinded randomized controlled trial. Lancet 369:1000–1015

Martin, P., Maestu, F., and Sola, R. (2002) Effects of surgical treatment on intellectual performance and memory in a Spanish sample of drug resistant partial onset temporal lobe epilepsy patients. Seizure 11:151–156

Martinez-Juarez, I., et al. (2006) Juvenile myoclonic epilepsy subsyndromes. Brain 129:1269–1280

Martinovic, Z., and Jovic, N (1997) Seizure recurrence after a first generalized tonic-clonic seizure in children, adolescents, and young adults Seizure 6:461–465

Martynyuk, A., et al. (2007) Epilepsy in phenylketonuria: A complex dependence on serum phenylalanine levels Epilepsia 48:1143–1150

Marzec, M., et al. (2003) Effects of vagus nerve stimulation on sleep-related breathing in epilepsy patients. Epilepsia. 44(7):930–935

Mascetti, L., et al. (2009) Some facts about sleep relevant for Landau-Kleffner syndrome Epilepsy 50:43–46

Masino, and Geiger, J. (2008) Are purines mediators of the anticonvulsant/neuroprotective effects of ketogenic diets? Trends Neurosci 31:273–278

Masino, S., et al. (2009) Adenosine, ketogenic diet and epilepsy: The emerging therapeutic relationship between metabolism and brain activity Curr Neuropharm 7:257–268

Mathern, G., et al. (1996) The pathogenic and progressive features of chronic human hippocampal epilepsy. Epil Res 26:151–161

Mathiak, K., et al. (2010) Quality of life in childhood epilepsy with lateralized epileptogenic foci BMC Neurol 10:69–75

Matsumoto M, Inoue T, Nagata E, Tanaka K, Takano H, Minowa O, Kuno J, Sakakibara S, Yamada M, Yoneshima H, Miyawaki A, Fukuuchi Y, Furuichi T, Okano H, Mikoshiba K, Noda T. 1996. Ataxia and epileptic seizures in mice lacking type 1 inositol 1,4,5-trisphosphate receptor. Nature 379:168–171.

Matsumoto, A., et al. (1981) Infantile spasms: etiological factors, clinical aspects, and long term prognosis in 200 cases Eur J Pediatr 135:239–244

Matsumoto, M., et al. (1995) Ataxia and epileptic seizures in mice lacking type 1 inositol 1, 4, 5-triphosphate receptor. Nature 379:168–171

Mattson, R., and Cramer, Joyce (1988) Crossover from polytherapy to monotherapy in primary generalized epilepsy Amer. J. Med. 84:23–28

Mattson, R., et al. (1985) Comparison of carbamazepine, phenobarbital, phenytoin, and primidone in partial and secondary generalized tonic clonic seizures. N. Eng. J. Med. 313:145–151

Max, J., et al. (2010) Pediatric stroke: plasticity, vulnerability, and age of lesion onset J Neuropsychiatr Ciln Neurosci 22:30–39

May, R., and Sunder, T. (1993) Hematologic manifestations of long term valproate therapy. Epilepsia 34:1098–1101

May, T., and Pfafflin, M. (2005) Psychoeducational Programs for patients with epilepsy Dis Manag Health Outcomes 13:185–198

Maydell, B., et al. (2002) Efficacy of the ketogenic diet in focal versus generalized seizures. Pediatr. Neurol. 25:208–212

Mayer, S., et al. (2002) Refractory status epilepticus Arch Neurology 59:205–210

Maytal, J., et al. (1989) Low morbidity and mortality of status epilepticus in children. Pediatrics 83:323–331

Maytal., and Shinnar, S. (1990) Febrile status epilepticus. Pediatr. 86:611–616

McCandless DW, DeFrance JF, Dworsky S, Presley-Zimmer E. (1986) Status epilepticus-induced changes in primate cortical energy metabolism. Am J Physio 251:C774–779.

McCandless, D. (2010) Thiamine deficiency and associated clinical disorders. Springer, N. Y.

McCandless, D. (2011) Jaundice in malaria. In Kernicterus. D. McCandless, ed., Springer, New York, NY. pp 121–132

McCandless, D. W., et al. (1979) Metabolite levels in brain following experimental seizures: effects of isoniazid and sodium valproate in cerebellar and cerebral cortical layers. J. Neurochem. 32: 755–760

McCandless, D. W., et al. (1979) Metabolite levels in brain following experimental seizures: the effects of maximal electroshock and phenytoin in cerebellar layers. J. Neurochem:32:743–753

McCandless, D. W., et al. (1979) Sparing of metabolic stress in Purkinje cells after maximal electroshock. PNAS 76:1482–1484

McCandless, D. W., et al. (1979) Sparing of metabolic stress in Purkinje cells after maximal electroshock. Proc. Natl. Acad. Science 76:1482–1484

McCandless, D. W., et al. (1986) Status epilepticus induced changes in primate cortical energy metabolism. Amer. J. Physiol. 251:C774-C779

McCandless, D. W., et al. (1986) Status epilepticus-induced changes in primate cortical energy metabolism. Am J Physiol 251 (5 Pt 1) C774–779

McCandless, D. W., Mintun, M., and Collins, R. C. (1986) Comparison of 6–14C-glucose and 14C DG quantitative autoradiography. Trans. Amer. Soc. Neurochem. 17:173

McCandless, D.W. and FineSmith, R. (1992) Animal models of experimental seizures. In: Animal models of neurological disease II. Bolton, A., Baker, G., and Butterworth, R., eds. Humana Press, Baltimore, pp 133–150

McCandless, D.W., et al. (1987) Pentylenetetrazole induced changes in cerebral energy metabolism in Tupaia Glis. Epilepsia 28:184–189

McCandless, J., and McCandless, D. (2011) Gene therapy for hyperbilirubinemia, In McCandless, D, (Editor) Kernicterus, Springer, pp 235–244

McCormick, D, and Contreras, D. (2001) On the cellular and network bases of epileptic seizures. Ann Rev. Physiol. 63:815–846

McCormick, D. (2002) Cortical and subcortical generators of normal and abnormal rhythmicity Int. Rev. Neurobiol. 22:2323–2334

McDade, G., and Brown, S. (1992) Nonepileptic seizures : management and predictive factors of outcome Seizure 1:7–10

McDonald, S. (2000) Exploring the cognitive basis of right hemisphere pragmatic language disorders Brain and Language 75:82–107

McFarland, R., Taylor, R., and Turnbull, D. (2002) The neurology of mitochondrial DNA disease. Lancet Neurol. 1:343–351

McGregor, A., and Wheless, J. (2006) Pediatric experience with sudden unexplained death in epilepsy at a tertiary epilepsy center. J. Child. Neurol. 21:782–787

McKay, K., and Halperin, J. (2001) ADHD, aggression, and antisocial behavior across the lifespan. Interactions with neurochemical and cognitive function Ann NY Acad Sci 931:84–96

McLachlan, R. (1993) Suppression of interictal spikes and seizures by stimulation of the vagus nerve. Epilepsia 34:918–923

McLellan, A., et al. (2005) Psychopathology in children with epilepsy before and after temporal lobe resection. Dev. Med. Child. Neurol. 47:666–672

McNally, K., and Blumenfeld, H. (2004) Focal network involvement in generalized seizures: new insights from electroconvulsive therapy. Epil. And Behav. 5:3–12

Meador, K., et al. (2009) Cognitive function at 3 years of age after fetal exposure to antiepileptic drugs New Eng J of Med 360:1597–1605

Meencke, H., and Veith, H. (1992) Migration disturbances in epilepsy. Epil. Rs. 9:31–40

Meeren, H., et al. (2002) Cortical focus drives widespread corticothalamic networks during spontaneous absence seizures in rats. J. Neurosci 22:1480–1495

Meeren, H., et al. (2005) Evolving concepts on the pathophysiology of absence seizures. Arch Neurol 63:371–376

Melander, T., Hokfelt, T., and Rokaeus, A. (1986) Distribution of galanin like immunoreactivity in the rat central nervous system. J Comp Neurol 248:475–517

Melchior, J. C., (1977) Infantile spasms and early immunization against whooping cough. Danish survey from 1970 to 1975. Arch Dis Child. 52(2):134–137

Meldrome, B. (1993) Excitotoxicity and selective neuronal loss in epilepsy Brain Path. 3:405–412

Meldrum, B, Vigoroux, R., and Brierly and Brierley, J. (1973) Systemic factors and epileptic brain damage: Prolonged seizures in paralyzed artificially ventilated baboons Arch Neurol 28:82–87

Meldrum, B. in Status epilepticus-mechanisms of brain damage and treatment. Ed Delgado, A., et al. (1983) New York Raven

Meldrum, B. S., (1993) Excitotoxicity and selective neuronal loss in epilepsy. Brain Pathol. 3:405–412

Meldrum, B., and Brierly, J. (1978) Prolonged epileptic seizures in primates: ischemic cell change and its relation to ictal physiological events. Arch Neurol. 28:10–17

Meldrum, B., and Rogawski, M. (2007) Molecular targets for antiepileptic drug development. Neurother. 4:18–61

Meldrum, B., Nanji, N., and Cornell, R. (1989) Lack of effect of aspartame or of phenylalanine on photically induced myoclonus in the baboon, Papio papio Epil Res 4:1–7

Mellanby, J., et al. (1977) Epileptiform syndrome in rats produced by injecting tetanus toxin into the hippocampus. J. Neurol. , Neurosurg., Psychiatr. 40:404–414

Menuz K, Nicoll RA. 2008. Loss of inhibitory neuron AMPA receptors contributes to ataxia and epilepsy in stargazer mice. J Neurosci 42:10599–10603.

Merkenschlager, A., et al. (2009) Development of hippocampal sclerosis after a complex febrile seizure. Eur. J. Pediatr. 168:487–490

Merren, H., et al. (2002) Cortical focus drives widespread corticothalamic networks during spontaneous absence seizures in rats. J. Neurosci 22:1480–1495

Merrick, J., Aspler, S., and Schwartz, G. (2003) Phenylalanine restricted diet should be life long: A case report on long term follow up of an adolescent with untreated phenylketonuria Int J Adolesc Med Health 15:165–168

Metrakos, K., and Metrakos, J. (1961) Genetics of convulsive disorders. II. Genetics and electroencephalographic studies in centrencephalic epilepsy. Neurology 11:474–483

Metz-Lutz, M-N., and Filippini, M. (2006) Neuropsychological findings in Rolandic epilepsy and Landau-Kleffner syndrome Epilepsia 47:71–75

Mikami, M., et al. (1999) Localization of a gene for benign adult familial myoclonic epilepsy to chromosome 8q23.3-q24.1. Am J Hum Gen 65:745–751

Mikati, M., and Shamseddine, A. (2005) Management of Landau-Kleffner syndrome Pediatr. Drugs 7:377–389

Mikati, M., et al. (2004) Effects of creatine and cyclocreatine supplimentation on kainate induced injury in pre pubescent rats. Brain Inj. 18:1229–1241

Millati, N., et al. (2003) The clinical spectrum of epilepsy in patients with hypothalamic hamartoma has also been explored. Epilepsia 44:1310–1319

Miller, S., et al. (2002) Seizure associated brain injury in term newborns with perinatal asphyxia. Neurology 58:542–548

Milner, T., et al. (1997) Selective changes in hippocampal Neuropeptide Y neurons following removal of the cholinergic septal inputs. J Comp Neurol 386:46–59

Minton, M., et al. (1986) Increases in intracranial pressure from succinylcholine: prevention by prior non depolarizing blockade. Anesth. 65:165–169

Mintun, M., et al. (2001) Blood flow and oxygen delivery to human brain during functional activity: theoretical modeling and experimental data. Proc. Natl. Acad. Sci. 98:6859–6864

Mirza, W., Credeur, L., and Penry, J. (1993) Results of antiepileptic drug reduction in patients with multiple handicaps and epilepsy Drug Invest 5:320–326

Mitchll, W., et al. (2000) Adherence to treatment in children with epilepsy: who follows doctors orders? Epilepsia 41:1616–1625

Mitsufuji N, Ikuta H, Yoshioka H, Sawada T. 1996. Lennox-Gastaut syndrome associated with leukoencephalopathy. Pediatr Neurol 15:63–65.

Miyauchi, T., et al. (1988) Positron emmision tomography in 3 cases of Lennox-Gastaut syndrome. Jpn J Psychiatr Neurol. 42:795–804

Mizrahi, E. and Kellaway, P. (1987) Characterization and classification of neonatal seizures. Neurology 37:1837–1844

Mizuno, Y., et al. (1985) Chronic localized encephalitis with focal cerebral seizures revisited J. Neuropath Exp Neurol 44:351–356

Modica, P., Tempelhoff, R., and White, P. (1990) Pro- and anticonvulsant effects of anesthetics. Anesth. And Analgesia 4:433–444

Mohit, A., Martin, J., and Miller, C. (1995) p493F12 kinase: a novel MAP kinase expressed in a subset of neurons in the human nervous system. Neuron 14:67–78

Molinari, F., et al. (2005) Impaired mitochondrial glutamate transport in recessive neonatal myoclonic epilepsy. Am J Hum Genet 76:334–339

Molyneux, M. (2000) Impact of malaria on the brain and its prevention. Lancet 355:671–672

Montagna, P. (1992) Nocturnal paroxysmal dystonia and nocturnal wandering. Neurology 42:61–67

Montouris, G., et al. (2000) Nonfocal generalized tonic-clonic seizures: response during long term topiramate treatment. Epilepsia 41:s77-s81

Moreau, Th., et al. (1998) Epilepsy in patients with multiple sclerosis: radiological clinical correlations. Epilepsia 39:893–896

Morell, et al. (2008) Long term safety and efficacy of the RNSTM system in adults with medically intractible partial onset seizures Epilepsia 49:480

Moretti, P., et al. (2008) Brief report: autistic symptoms, developmental regression, mental retardation, epilepsy, and dyskinesias in CNS folate deficiency 38:1170–1177

Morgan, G., et al. (2006) Anesthesia for neurosurgery. Clin. Anesthesia (pp 631–646). Boston: McGraw-Hill

Morita, T., et al. (2002) Cliniconeuropathologic findings of familial frontal lobe epilepsy in Shetland sheepdogs. Can. J. Vet Res. 66:35–41

Morrell, F., et al. (1995) Landau-Kleffner syndrome: treatment with subpial intracortical transection Brain 118:1529–1546

Morrell, F., Whisler, W., and Bleck, T. (1989) Multiple subpial transection: a new approach to the surgical treatment of focal epilepsy. J. Neurosurg. 70:231–239

Morrell, F., Whistler, W., and Smith, M. (1991) Multiple subpial transection in Rasmussen's encephalitis. In Andermann, F., ed. Chronic encephalitis and epilepsy. Ramussen's syndrome Boston Butterworth-Heinemann pp 219–233

Morrell, M., et al. (2008) Long term safety and efficacy of the RNS system in adults with medically intractable partial onset seizures Epilepsia 49:480–481

Morris, G., and Mueller, W. (1999) The vagus nerve stimulation EO1-EO5: long term treatment with vagus nerve stimulation in patients with refractory epilepsy. Neurology 53:1731–1735

Moruzzi, G., and Magoun, H. (1949) Brain stem reticular formation and activation of the EEG. Electroenceph. Clin Neurophys. 1:455–473

Muayqil. T., Rowe, B., and Ahmed, S. (2007) Treatment adherence and outcomes in the management of convulsive status epilepticus in the emergency room. Epil. Discord. 9:43–50

Mula, M., et al. (2003) Psychiatric adverse events during levetiracetam therapy. Neurology 61:704–706

Mulley, J., et al. (2005) SCN1A mutations and epilepsy. Hum Mut 25:535–542

Mulley, J., et al. (2006) A new molecular mechanism for severe myoclonic epilepsy in infancy: exonic deletions in SCN1A. Neurology 67:1094–1095

Munari, C., et al. (1995) Role of the hypothalamic hamartoma in the genesis of gelastic fits Electro Clin Neurophys 95:154–160

Munuz, K., and Nicoll, R. (2008) Loss of inhibitory neuron AMPA receptors contributes to ataxia and epilepsy in stargazer mice. J. Neurosci. 28:10599–10603

Murakami, N., et al. (1996) Mesial temporal lobe epilepsy in childhood. Epilepsia 3:525–526

Murphy, J., et al. (2003) Vagal nerve stimulation in refractory epilepsy. Arch Pediatr adolesc med 157:560–564

Muzykewicz, D., et al. (2009) Infantile spasms in tuberous sclerosis complex: prognostic utility of EEG. Epilepsy 50:290–296

Nabbout, R., et al. (2003) Spectrum of SCN1A mutations in severe myoclonic epilepsy of infancy. Neurology 60:1961–1967

Nadkarni, S., Lajoie, J., and Devinsky, O. (2005) Current treatments of epilepsy. Neurol. 64:s2–s11

Nair, R., and Thomas, S. (2004) Genetic liability to epilepsy in Kerala State, India. Epil. Res. 62:163–170

Nair, S., et al. (2008) Absence seizures as resetting mechanisms of brain dynamics. Cyber Sys Anal 44:664–672

Nakasu, S., et al. (1997) Serial magnetic resonance imaging findings in Rasmussen's encephalitis Neurol Med Chir 37:924–928

Nakatomi, H., et al. (2002) Regeneration of hippocampal pyramidal neurons after ischemic brain injury by recruitment of endogenous neural progenators. Cell 110:429–441

Nakken, K., Magnusson, A., and Steinlein, O. (1999) Autosomal dominant nocturnal frontal lobe epilepsy: an electroclinical study of a Norwegian family with ten affected members. Epilepsia 40:88–92

Naritoku, D., Terry, W., and Helfert, R. (1995) Regional induction of fos immunoreactivity in the brain by anticonvulsant stimulation of the vagus nerve. Epil Res 22:53–58
Naylor, D. (2010) Glutamate and GABA in the balance: Convergent pathways sustain seizures during status epilepticus. Epilepsia 51:106–109
Naylor, D., Liu, H., and Wsterlain, C. (2005) Trafficking of GABAa receptors, loss of inhibition, and a mechanism for pharmacoresistance in status epilepticus. J. Neurosci. 25:7724–7733
Neal, E., et al. (2009) A randomized trial of classical and medium chain triglyceride ketogenic diets in the treatment of childhood epilepsy. Epilepsia 50:1109–1117
Nei, M., et al. (1999) The EEG and prognosis in status epilepticus. Epilepsia 40:157–163
Nei, M., et al. (2006) Refractory generalized seizures: response to corpus callosotomy and vagal nerve stimulation. Epilepsia 47:115–122
Nelson, K., and Ellenberg, J. (1976) Predictors of epilepsy in children who have experienced febrile seizures. N. Eng. J. Med. 295:1029–1033
Nersesyan, H., et al. (2004) Dynamic fMRI and EEG recordings during spike wave seizures and generalized tonic-clonic seizures in WAG/Rij rats. J. Cereb Blood flow Met. 24:589–599
Nettekoven, S., et al. (2008) Effects of antiepileptic drug therapy on vitamin D status and biochemical markers of bone turnover in children with epilepsy Euro J Pediatr 167: 1369–1377
Neville, B., Chin, R., and Scott, R. (2007) Childhood convulsive status epilepticus: epidemiology., management and outcome. Acta Neurol Scand 115:21–24
Newmark, M. and Dubinsky, S. (2004) Zonisamide monotherapy in a multi group clinic Seizure 13:223–225
Ng, M., and Devinsky, O. (2004) Vagus nerve stimulation for refractory idiopathic generalized epilepsy. Seizure 13:176–178
Ng, Y., and Rekate, H. (2005) Emergency transcallosal resection of hypothalamic hamartoma for "status gelasticus" Epilepsia 46:592–594
Ng, Y., et al. (2006) Transcallosal resection of hypothalamic hamartoma for intractable epilepsy Epilepsia 47:1192–1202
Ngoungou, E., and Preux, P. (2008) Cerebral malaria and epilepsy Epilepsia 49:19–24
Ngoungou, E., et al. (2006) Epilepsy as a consequence of cerebral malaria in an area in which malaria is endemic in Mali, West Africa Epilepsia 47:873–879
Nickels, K., and Wirrell, E. (2008) Electrical status epilepticus in sleep Semin Pediatr Neurol 15:50–60
Nicoletti, A., et al. (2005) Epilepsy and neurocysticercosis in rural Bolivia: a population based survey Epilepsia 46:1127–1132
Nicoll, R., Tomita, S., and Bredt, D. (2006) Auxiliary subunits assist AMPA type glutamate receptors. Science 311:1253–1256
NIH Consensus Development Conference, (1990). Consensus statement: surgery for epilepsy. JAMA 264:729–733
Noachtar, S., et al. (2008) Levetiracetam for the treatment of idiopathic generalized epilepsy with myoclonic seizures. Neurology 70:607–616
Nobili, L., et al. (2004) Sleep related hyperkinetic seizures of temporal lobe origin. Neurology 62:482–485
Nobili, L., et al. (2007) Surgical treatment of drug resistant nocturnal frontal lobe epilepsy. Brain 130:561–573
Noe, F., et al. (2008) Neuropeptide Y gene therapy decreases chronic spontaneous seizures in a rat model of temporal lobe epilepsy. Brain doi:10.1093/brain/awn079
Noebels, J., et al. (1990) Stargazer: a new neurological mutant on chromosome 15 in the mouse with prolonged cortical seizures. Epilepsy Res. 7:129–135
Noh, H., et al. (2004) A cdna microarray analysis of gene expression profiles in rat hippocampus following a ketogenic diet. Brain Res. 129:80–87
Norden, A., and Blumenfeld, H. (2002) The role of subcortical structures in human epilepsy. Epil. Behav. 3:219–231

Nordi, D., and Pedley, T. (2008) The Use of Electroencephalography in the Diagnosis of Epilepsy in Childhood. In Pellock, J, Bourgeois, B., and Dodson, W. eds Pediatric Epilepsy, 3rd ed New York, Demos pp 213–226

Nordli, D., Kuroda, M., and Carroll, J. (2001) Experience with the ketogenic diet in infants. Pediatr. 108:129–133

Noron, Ana, et al. (2007) Prevalence and pattern of epilepsy treatment in different socioeconomic classes in Brazil Epilepsia 48:880–885

Noronha, A. L., et al. (2007) Prevalence and pattern of epilepsy treatment in different socioeconomic classes in Brazil. Epilepsia 48:880–885

Novak, G., et al. (2007) Carisbamate (RWJ333369). Neurother. 4:106–109

Nylen, K., et al. (2005) A comparison of the ability of a 4:1 ketogenic diet and a 6.3:1 ketogenic diet to elevate seizure thresholds in young and adult rats. Epilepsia 46:1198–1204

Nylen, K., et al. (2009) The effects of a ketogenic diet on ATP concentrations and the number of hippocampal mitochondria in Aldh5a1 mice. Biochem. Biophys Acta 123:208–212

Oakley, J., et al. (2009) Temperature and age dependent seizures in a mouse model of severe myoclonic epilepsy in infancy. Proc Nat Acad Sci 106:3994–3999

Oakley, J., et al. (2009) Temperature and age dependent seizures in a mouse model of severe myoclonic epilepsy in infancy. Proc. Natl. Acad. Sci. 106:3994–3999

Obeid, T. (1994) Clinical and genetic aspects of juvenile absence epilepsy. J. Neurol. 241:487–491

Ochs, R., et al. (1987) Effect of generalized spike wave discharge on glucose metabolism measured by positron mission tomography. Ann. Neurol. 21:458–464

Odajima, H., et al. (2003) Occurrence of convulsions after administration of theophylline in a large Japanese pediatric population with asthma. Ped Asth Aller Immunol 16:163–173

Oguni H, Uehara T, Imai K, Osawa M. 1997. Atonic epileptic drop attacks associated with generalized spike-and-slow wave complexes: video-polygraphic study in two patients. Epilepsia 38:813–818.

Oguni, H., et al. (1991) Anterior callosotomy in the treatment of medically intractable epilepsies. Ann. Neurol. 30:357–364

Oguni, H., et al. (1992) A video EEG analysis of drop seizures in myoclonic astatic epilepsy of early childhood (Doose syndrome) Epilepsia 33:805–813

Oguni, H., et al. (1992) The syndrome of chronic encephalitis and epilepsy: a study based on the MNI series of 48 cases. Adv Neurol 57:419–433

Oguni, H., et al. (1994) Effect of anterior callosotomy on bilaterally synchronous spike and wave and other EEG discharges. Epilepsia 35:505–513

Oguni, H., et al. (1997) Atonic epileptic drop attacks associated with generalized spike-and-slow wave complexes: video polygraphic study in two cases. Epilepsia 38:813–813

Oguni, H., et al. (1997) Video-EEG analysis of drop seizures in myoclonic astatic epilepsy of early childhood. Epilepsis 33:805–813

Ohr, T., et al. (2007) Lacosamide, a novel anti-convulsant drug shows efficacy with a wide safety margin. Epil. Res. pil. Res. 74:147–154

Oldani, A., et al. (1998) Autosomal dominant nocturnal frontal lobe epilepsy,. A video polysomnographic and genetic appraisal of 40 patients and deliniation of the epileptic syndrome. Brain 121:205–233

Oldani, Z., et al. (1998) Autosomal dominant nocturnal frontal lobe epilepsy: a video polysomnographic and genetic appraisal of 40 patients and deliniation of the epileptic syndrome. Brain 21:205–223

Olsson, I., and Campenhausen, G. (1993) Social adjustment in young adults with absence epilepsies. Epilepsia 34:846–851

Olsson, I., Steffenburg, S., and Gillberg, C. (1988) Epilepsy in autism and autistic like conditions Arch Neurol 45:666–668

Oommen, J., Morrell, M., and Fisher, R. (2005) Experimental electrical stimulation therapy for epilepsy Curr Treat Neurology 7:261–271

Oostrom, K., et al. (2005) Three to four years after diagnosis: Cognition and behavior in children with epilepsy only Brain 128:1546–1555
Opeskin, K., et al. (2000) Idiopathic generalized epilepsy: lack of significant microdysgenesis. Neurology 55:1101–1106
Osorio, I., and Reed, R. (1989) Treatment of generalized tonic clonic status epilepticus with pentobarbital anesthesia after high dose phenytoin. Epilepsia 30:464–471
Ott, D., et al. (2001) Measures of psychopathology in children eith complex partial seizures and primary generalized epilepsy with absence J Amer Acad Child Adol Psychiatr 40:907–914
Ottman, R., et al. (1989) Seizure risk in offspring of parents with generalized vs. partial epilepsy. Epilepsia 30:157–161
Ounsted, C., Lindsay, J., and Norman, R. (1966) Biological factors in temporal lobe epilepsy. Clin Dev Med 22:31–33
Owen, O., et al. (1967) Brain metabolism during fasting. J. Clin. Invest. 46:1589–1595
Paetau, R. (2009) Magnetoencephalography in Landau-Kleffner syndrome Epilepsia 50:51–54
Pal, D. (2003) Epilepsy control in the twenty first century: hidden impact on children and families Child Care, Health and Development 29:233–236
Pal, D., et al. (2002) Help seeking patterns for epilepsy in rural India: implications for service delivery Epilepsia 43:904–911
Pal, D., et al. (2003) BRD2 is a probable major susceptibility gene for common juvenile myoclonic epilepsy. J. Hum. Genet. 73:261–270
Palmini, A., and Gloor, P. (1992) The localizing value of auras in partial seizure: a prospective and retrospective study. Neurology 52:801–808
Palmini, A., et al. (2005) Status gelasticus after temporal lobectomy: ictal FDG-PET findings and the question of dual pathology involving hypothalamic hamartomas Epilepsia 46:1313–1316
Pan, J., et al. (1999) Ketosis and epilepsy: 31P spectroscopic imaging at 4.1T Epilepsia 40:703–707
Pan, J., et al. (2002) C2-beta-hydroxybutyrate metabolism in human brain. J. Cereb. Blood Flow Metab. 22:890–898
Panayiotopoulos, C. (1999) Typical absence seizures and their treatment. Arch. Dis. Child. 81:351–355
Panayiotopoulos, C. (2001) Treatment of typical absence seizures and related epileptic syndromes. Pediatr Drugs 3:379–403
Panayiotopoulos, C.(1997) Absence epilepsies. In: Engel, JJ, and Pedley, T. eds. Epilepsy: a comprehensive textbook. Phil., Lippincott pp 2327–2346
Panayiotopoulos, C., et al. (1997) Idiopathic generalized epilepsy in adults manifested by phantom absences, generalized tonic-clonic seizures, and frequent absence status. J. Neurol. Neurosurg, Psychiatr. 63:622–667
Panayiotopoulos, C., Obeid, T., and Tahan, A. (1994) Juvenile myoclonic epilepsy: a 5 year prospective study. Epilepsia 35:258–296
Paolicchi, J. (2002) The spectrum of nonepileptic events in children Epilepsia 43:60–64
Papayannis, C., et al. (2008) Clinical spectrum and difficulties in management of hypothalamic hamartoma in a developing country Acta Neurol Scand 118:313–319
Pardo, C., et al. (2004) The pathology of Rasmussen's syndrome: stages of cortical involvement and neuropathological studies in 45 hemispherectomies Epilepsia 45:516–526
Parekh, M., et al. (2010) Early MR diffusion and relaxation changes in the parahippocampal gyrus precede the onset of spontaneous seizures in an animal model of chronic limbic epilepsy. Exper. Neurol. Doi:10.1016/j
Park, S-P., et al. (2007) Long term efficacy and safety of zonisamide monotherapy in epilepsiy patients. J. Clin Neurology 3:175–180
Parvizi, J., and Damasio, A. (2003) Neuroanatomical correlates of brainstem coma. 126:1524–1536
Pascual, O., et al. (2005) Astrocytic purinergic signaling coordinates synaptic networks. Science 310:113–116
Passaro, E., and Jobe, K. (2010) Identification of Potential Epilepsy Surgery Candidates. eMedicne 1–19 (http://emedicine.medscape.com?article/251449 1185203

Passonneau, J., and Lowry, O. (1993) Enzymatic Analysis: a practical guide. Humana Press, Totowa N.J.

Passonneau, J., Lust, W.D., and McCandless, D.W. (1980) Fixation and extraction of viable tissues. In: Techniques in Metabolic Research. C. Pogson, ed. Elsevier Pub. Co. pp 1–27

Patel, H., et al. (2006) Focal cerebral pathology presenting as tonic seizures. J. Child. Neurol. 21:813–816

Patrick, J., et al. (2007) Proposal of an algorithm for diagnosis and treatment of neonatal seizures in developing countries Epilepsia 48:1158–1164

Patwardhan, R., et al. (2000) Efficacy of vagal nerve stimulation in children with medically refractory epilepsy Neurosurg. 47:1353–1358

Pearce, J. (2004) A note on gelastic epilepsy Eur Neurol 52:172–174

Pearl, P., et al. (2009) Succinic semialdehyde dehydrogenase deficiency: lessons from mice and men. J. Inhert. Metab. Dis. 32:343–352

Pedersen, H., and Birket-Smith, E (1974) Neurological abnormalities in phenylketonuria Acta Neurol Scand 50:589–594

Pedley, T., and Guilleminault, C. (1977) Episodic nocturnal wanderings responsive to anticonvulsant drug therapy. Ann. Neurol. 2:30–35

Pedro, V., and Leisman, G. (2005) Hemispheric integrative therapy in Landau-Kleffner syndrome: Applications for rehabilitation sciences Int. J. Neurosci. 115:1227–1238

Peeters, B., et al. (1988) The WAG/Rij model for absence epilepsy: anticonvulsant drug evaluation. Neurosci. Res. Comm. 2:93–97

Pellock, J. (2004) Understanding co-marbidities affecting children with epilepsy. Neurology 62:s17-s26

Pellock, J., and Brodie, M. (1997) Felbamate: 1997 update. Epilepsia 38:1261–1264

Pellock, J., et al. (2010) Infantile spasms: A U.S. consensus report. Epilepsia 51:2175–2189

Pendl, G., et al. (1999) Corpus callosotomy with radiosurgery. Neurosurg. 45:303–307

Penfield, W. (1958) Functional localization in temporal lobe and deep Sylvian areas Res Pub Assoc Res Nerv Ment Dis 36:210–226

Penfield, W., and Jasper, H (1954) in: Epilepsy and the functional anatomy of the human brain. Boston, Little Brown & Co.

Penfield, W., and Jasper, H. (1954) Epilepsy and the Functional Anatomy of the Human Brain. Little Brown, Boston

Penfield, W., and Paine, K. (1955) Results of surgical therapy for focal epileptic seizures. Can Med Assoc J. 73(7):515–531

Penfield, W., and Perot, P. (1963) The brain's record of auditory and visual experience: a final summary and discussion. Brain 86:595–696

Pennacchio, L., et al. (1996) Mutations in the gene encoding cystatin B in progressive myoclonus epilepsy (EPM1) Science 271:1731–1734

Pennacchio, L., et al. (1998) Progressive ataxia, myoclonic epilepsy, and cerebellar apoptosis in cystatin B deficient mice. Nature 20:251–258

Penry, J. and Dean, J. (1990) Prevention of intractable partial seizures by intermittant vagal stimulation in humand: preliminary results. Epilepsia 31:s40–s43

Perlman, J., and Volpe, J. (1983) Seizures in the preterm infant: effects on cerebral blood flow velocity, intracranial pressure, and arterial blood pressure. J. Pediatr.102:288–293

Perucca, E., et al. (1998) Antiepileptic drugs as a cause of worsening seizures. Epilepsia 39:5–17

Petroff, O., et al. (2002) Glutamate-glutamine cycling in the epileptic human hippocampus Neurology 60:1646–1651

Phillips, H., et al. (1995) Localization of a gene for autosomal dominant nocturnal frontal lobe epilepsy to chromosome 20q13.2 Nat Genet 10:117–118

Phillips, H., et al. (1998) Autosomal dominant nocturnal frontal lobe epilepsy: genetic heterogeneity and evidence for a second locus at 15q24. Am. J. Hum Genet 63:1108–1116

Phillips, H., et al. (2000) A de novo mutation in sporadic nocturnal frontal lobe epilepsy. Ann Neurol. 48:264–267

Piazzini, A., et al. (2008) Frontal cognitive dysfunction in juvenile myoclonic epilepsy. Epilepsy 49:657–662
Picard, F., et al. (2006) Intracerebral recordings of nocturnal hyperkinetic seizures: demonstration of a longer duration of the preseizure sleep spindle. Clin. Physiol. 118:928–939
Picard, F., et al. (2007) Intracerebral recordings of nocturnal hyperkinetic seizures: demonstration of a longer duration of the pre-seizure sleep spindle. *Clin Neurophysiol.* 118(4):928–39. Epub 2007 Feb 20
Pitt, D., and Danks, D (1991) The natural history of untreated phenylketonuria over 20 years J Pediatr Child Health 27:189–190
Placencia, M., et al. (1994) The characteristics of epilepsy in a largely untreated population in rural Ecuador J. Neurol, Neurosurg., and Psychiatr. 57:320–325
Polak, P., et al. (2009) Inactivation of the somatosensory cortex prevents paroxysmal oscillations in cortical and related thalamic neurons in a genetic model of absence epilepsy. Cerebral Cortex 19:2078–2091
Ponds, R., and Hendriks, M. (2006) Cognitive rehabilitation of memory problems in patients with epilepsy. Seizure 15: 267–273
Power, C., et al. (1990) Cytomegalovirus and Rasmussen's encephalitis Lancet 336:1282–1284
Praline, J., et al. (2003) Outcome at adulthood of the continuous spikewaves during slow sleep and Landau-Kleffner Syndromes Epilepsia 44:1434–1440
Pranzatelli, M (1989) In vivo and in vitro effects of adrenocortico trophic hormone on serotonin receptors in neonatal brain. Dev Pharmacol Ther 12:49–56
Prayson, R., and Frater, J. (2002) Rasmussen encephalitis Am J Clin Path 117:776–782
Prevey, M., et al. (1996) Effect of valproate on cognitive functioning. Comparison with carbamazepine Arch Neurol 53:1008–1016
Prevost, J., et al. (2006) Nonlesional frontal lobe epilepsy of childhood: clinical presentation, response to treatment and comorbidity. Epilepsia 47:2198–2201
Prince, D. (1968) The depolarization shift in epileptic neurons. Exp. Neurol. 21:467–485
Provini, F., et al. (1999) Nocturnal frontal lobe epilepsy. A clinical and polygraphic overview of 100 consecutive cases. Brain 122:1017–1031
Prueter, C., et al. (2002) Dissociative and associated psychopathological symptoms in patients with epilepsy, pseudoseizures, and both seizure forms Epilepsia 43:188–192
Pulsifer, M., et al. (2004) The cognitive outcome of hemispherectomy in 71 children Epilepsia 45:243–254
Pung, T., Schmitz, B. (2006) Circadian rhythm and personality profile in juvenile myoclonic epilepsy. Epilepsia 47:111–114
Putnam, T., and Merritt, H. (1937) Experimental determination of the anticonvulsant properties of some phenyl derivatives. Science 85:525–526
Qiao, X., and Noebels, J. (1993) Developmental Analysis of hippocampal mossy fiber outgrowth in a mutant mouse with inherited spike wave discharges. J. Neurosci. 13:4622–4635
Qiao, X., et al. (1998) Cerebellar brain derived neurotrphic factor TrkB defect associated with impairment of eyeblink conditioning in stargazer mutant mice. J. Neurosci. 18:319–324
Qu, L., and Leung, L. (2008) Mechanisms of hyperthermia induced depression of GABAergic synaptic transmission in the immature rat hippocampus. J. Neurochem. 106:2158–2169
Quiske, A., et al. (2006) Cognitive functions in juvenile and adult patients with gelastic epilepsy due to hypothalamic hamartoma Epilepsia 47:153–158
Radvanyi-Bouvet,M., et al. (1985) Seizures and electrical discharges in premature infants. Neuroped. 16:143–148
Raedt, R., and Boon, P. (2005) Cell therapy for neurological disorders: a comprehensive review. Acta Neurol. Belg. 105:158–170
Raffo, E., et al. (2008) Calorie-restricted ketogenic diet increases thresholds to all patterns of pentylenetetrazole induced seizures: critical importance of electroclinical assessment Epilepsia 49:320–328

Rahman, M., et al. (2010) Disrupting abnormal electrical activity with deep brain stimulation: is epilepsy the next frontier? Neurosurg Focus 29:E7
Rahul, M., et al. (2000) Auras, seizure focus, and psychiatric disorders Neuropsych., Neoropsyhology, and Behav. Neurol.13:13–19
Raisinghani, M., and Faingold, C. (2005) Pontine reticular formations are implicated in the neuronal network for generalized clonic seizures which is itensified by audiogenic kindling. Brain Res. 1064:90–97
Ramaekers, V., et al. (2002) Psychomotor retardation, spastic paraplegia, cerebellar ataxia and dyskinesia associated with low 5-methyltetrahydrofolate in cerebrospinal fluid: a novel neurometabolic condition responding to folinic acid substitution. Neuropediatrics 33(6):301–308
Ramaratnam, S., and Narasimha, M. (2005) Prevalence and patterns of epilepsy in India Epilepsia 46:89–90
Rambo, M., et al. (2009) Additive anticonvulsant effects of creatine supplimentation and physical exercise against pentylenetetrazole induced seizures. Neurochem Intern 55:333–340
Ramsay, R., et al. (1992) Efficacy and safety of valproic acid versus phenytoin as sole therapy for newly diagnosed tonic-clonic seizures. J. Epil. 5:55–60
Randich, A., and Gebhart, G. (1992) Vagal afferent modulation of nociception. Brain Res 17:77–99
Rane, C., et al. (2002) A pharmacoeconomic analysis of the impact of therapeutic drug monitoring in adult patients with generalized tonic-clonic epilepsy. J. Clin. Pharmacol. 52:193–195
Rantala, H., and Putkonen, T. (1999) Occurrence, outcome, and prognostic factors of infantile spasms and Lennox-Gastaut syndrome. Epilepsia 40:286–289
Raol, Y., Budreck, E., and Brooks-Kayal, A. (2003) CaM kinase 2 is required for seizure evoked reduction of hyperpolarization activated cyclic nucleotide gated channel expression. Soc. Neurosci. Abst. 34:377
Raol, Y., et al. (2006) Enhancing GABA A receptor alpha 1 subunit levels in hippocampal dentate gyrus inhibits epilepsy development in an animal model of temporal lobe epilepsy. J. Neurosci. 26:11342–11346
Rasmussen, T. (1978) Further observations on the syndrome of chronic encephalitis and epilepsy. Appl. Neurophys. 41:1–12
Rasmussen, T. (1991) Tailoring of cortical excisions for frontal lobe epilepsy. Can. J. Neurol. Sci. 18:606–610
Rasmussen, T., and Andermann, F. (1989) Update on the syndrome of "chronicencephalitis" and eplepsy Clev Clin J Med 56:181–184
Raspall-Chaure, M. et al. (2007) The epidemiology of convulsive status epilepticus in children: a critical review. Epilepsia 48:1652–1663
Reasoner, D., et al. (1994) The incidence of pneumocephalus after supratentorial craniotomy-observations on the disappearance of intracranial air. Anesthesiology 80:1008–1012
Redrobe, J., et al. (1999) Multiple receptors for Neuropeptide Y in the hippocampus: putative roles in seizures and cognition. Brain Res. 848:153–166
Rees, M., et al. (1994) Linkage analysis of idiopathic generalized epilepsy in families of probands with juvenile myoclonic epilepsy and marker loci in the region of EPM 1 on chromosome 21 q. Neuroped 25:20–25
Reid, S. (1990) Surgical technique for implantation of the neurocybernetic prothesis. Epilepsia 31:s38–s39
Reimers, T., et al. (2003) Cognitive deficits in long term survivors of childhood brain tumors. Med Pd Med Ped oncol 40:26–34
Rener-Primec, Z., Stare, J., and Neubauer, D. (2006) The risk of lower mental outcome in infantile spasms increases after three weeks of hypsarrhythmia duration. Epilepsia 47:2202–2205
Reppert, S., and Eeaver, D. (2002) Coordination of circadian timing in mammals. Nature 418:935–941
Restak, R. (1995) Complex partial seizures present diagnostic challenge. Psychiat Tim. 12:34–40
Reuber, M., and Elger, C. (2003) Psychogenic nonepileptic seizures: review and update Epil. Behav. 4:205–216

Reynolds, E., Chadwick, D., and Shorvon, S. (1976) One drug (phenytoin) for treatment of epilepsy. Lancet 1:923–926
Rho, J., and Rogawski, M. (2007) The ketogenic diet: stroking the powerhouse of the cell. Epil. Curr. 7:58–60
Rho, J., et al. (2002) Acetoacetate, acetone, and dibenzylamine exhibit direct anticonvulsant actions in vivo. Epilepsia 43:358–361
Richardson, K., et al. (2003) In vivo modulation of hippocampal epileptiform activity with radial electric fields Epilepsia 44:768–777
Richichi, C., et al. (2004) Anticonvulsant and antiepileptic effects mediated by adenoassociated virus vector Neuropeptide Y expression in the rat hippocampus. J Neurosci 24: 3051–3059
Riggio, S., and Harner, R. (1995) Repetitive motor activity in frontal lobe epilepsy. Adv Neurol 66:153–164
Rigoulot, M., Boehrer, A., and Nehlig, A. (2003) Effects of topiramate in two models of genetically determined generalized epilepsy, the GAERS and the audiogenic Wistar AS Epilepsia 44:14–19
Ring, H., et al. (2000) A SPECT study of the effect of vagal nerve stimulation on thalamic activity in patients with epilepsy. Seizure 9:380–384
Riter, F., et al. (1993) Efficacy of felbamate in childhood epileptic encephalopathy. N Eng J Med 328:29–33
Rizzi, M., et al. (2003) Glia activation and cytokine increase in rat hippocampus by kainic acid induced status epilepticus during postnatal development. Neurobiol. Dis. 14: 494–503
Robitaille, Y. (1991) Neuropathological aspects of chronic encephalitis. In Andermann, F. ed Chronic encephalitis and Epilepsy, Rasmussen's syndrome Stoneham, Butterworth-Heinemann pp 79–110
Rodenburg, R., et al. (2006) Family predictors of psychopathology in children with epilepsy Epilepsia 47:601–614
Rodin, E. (1968) Prognosis for seizure control. In Rodin, E. , ed, The prognosis of patients with epilepsy. Springfield Illinois, Char. Thomas pp 179–262
Rodin, E., and Ancheta, O. (1987) Cerebral electrical fields during petit mal absences. Elect Clin Neurophys 66:457–466
Rogers, S., et al. (1994) Autoantibodies to glutamate receptor GluR3 in Rasmussen's encephalitis Science 265:648–651
Roks, G., et al. (1999) Effects of polytherapy compared with monotherapy in antiepileptic drugs: an animal study J. Pharm. Exp. Thera. 288:472–477
Rosenow, F., and Luders, H. (2000) Startle induced seizures. In Luders, H, Naochtar, S., eds. Epileptic seizures,. Pathophysiology and clinical semiology. N.Y. Churchill livingston pp 585–592
Rosenow, F., Hamer, H., and Knake, S. (2007) The epidemiology of convulsive and nonconvulsive status epilepticus. Epilepsia 48:82–84
Rossetti, A., and Bromfield, E. (2005) Levetiracetam in the treatment of status epilepticus in adults: a study of 13 episodes. European nerrology 54:34–38
Rossetti, A., et al. (2005) Simple partial seizures with hemisensory phenomena and dysgeusia: an insular pattern. Epilepsia 46:590–591
Rossi, P., et al. (1999) Landau-Kleffner syndrome: long term follow-up and links with electrical status epilepticus during sleep (ESES) Brain Dv 21:90–98
Rowan, A., et al. (1983) Valproate-ethosuximide combination therapy for refractory absence seizures. Arch. Neurol. 40:797–802
Royes, L., et al. (2003) Creatine protects against the convulsive behavior and lactate production elicited by the intrastriatal injection of methylmalonate. Neurosci. 118:1079–1090
Royes, L., et al. (2006) Effectiveness of creatine monohydrate on seizures and oxidative damage induced by methylmalonate. Pharm. Biochem. Behav. 83:136–144
Rubenstein, J. (2008) Use of the ketogenic diet in neonates and infants. Epilepsia 49:30–32
Rubenstein, J., et al. (2005) Experience in the use of the ketogenic diet as early therapy. J. Child. Neurol. 20:31–34

Rulolf, G., et al. (2009) From Rolandic epilepsy to continuous spike-and-waves during sleep and Landau-Kleffner syndromes: insights into possible genetic factors Epilepsia 50:25–28
Rutter, M., Grahm, P., and Yule, W. (1970) A neuropsychiatric study in childhood Phila Lippincott
Rutter, N., and Metcalfe, D. (1978) Febrile convulsions-what do parents do? Br. Med. J. 2:1345–1346
Sabatino, M., et al. (1986) Relations between basal ganglia and hippocampus: Action of substantia nigra and pallidum. Rev Electroenceph. Neurophys Clin 16:179–190
Sabaz, M., et al. (2001) The health related quality of life of children with refractory epilepsy: a comparison of those with and without intellectual disability Epilepsia 42:621–628
Sachdeo, R., et al. (1999) A double blind randomized trial of topiramate in Lennox-Gastaut syndrome. Neurology 52:1882–1887
Sackellares, D., and Sackellares, J. (2001) Impaired motor function in patients with psychogenic pseudoseizures Epilepsia 42:1600–1606
Sackellares, I., et al. (2000) Epilepsy-When chaos fails. In: Lehnertz, K., et al. eds. Chaos in the brain. Singapore, World Sci. pp 112–133
Saddichha, S., and Akhtar, S. (2010) Quality of life and marital adjustment in epilepsy and comparisons with psychiatric illnesses Int J Psychosoc Rehab 14:105–112
Saemundsen, E., et al. (2007) Autism spectrum disorders In children with seizures in the first year of life-a population study Epilepsia 48:1724–1730
Saemundsen, E., Ludvigsson, P., and Rafnsson, V. (2008) Risk of autism spectrum disorders after infantile spasms: a population based study nested in a cohort with seizures in the first year of life. Epilepsia 49:1865–1870
Sagduyu, A., Tarlaci, S., and Sirin, H. (1998) Generalized tonic clonic status epilepticus: causes, treatment, complications and predictors of case fatality. J. Neurol. 245:640–646
Said, R., Andrade, A., and Arnold, S. (2006) Pseudoseizures in twelve children under the age of 12 years old Epilepsia 47:S4
Saillet, S., et al. (2009) Manipulating the epileptic brain using stimulation: a review of experimental and clinical studies Epil. Disorders 11:100–112
Salanova, V., Markland, O., and Worth, R. (2005) Temporal lobe epilepsy: analysis of failures and the role of reoperation. Acta Neurol. Scand. 111:126–133
Salek-Haddadi, A., et al. (2002) Imaging absence seizures using fMRI. Epilepsia (abstracts) 43: (suppl) 123
Salek-Haddadi, A., et al. (2003) Functional magnetic resonance imaging of human absence seizures. Ann Neurolm 53:663–667
Saliba, R., Annegers, J., and Mizrahi, E. 1996) Incidence of clinical neonatal seizures. Epilepsia 37:13–14
Salmenpera, R., et al. (2001) Hippocampal and amygdala damage in partial epilepsy-a cross sectional MRI study of 241 patients. Epil Res 46:69–82
Salovey, P., and Mayer, J. (1990) Emotional intellegence Imag Cogn Pers 9:185–211
Sampath, H., and Ntambi, J. (2005) Polyunsaturated fatty acid regulation of genes of lipid metabolism. Ann. Rev. Nutr. 25:317–340
Sanchez, R., et al. (2001) Decreased glutamate receptor 2 expression and enhanced epileptogenesis in immature rat hippocampus after perinatal hypoxia-induced seizures. J. Neurosci. 21: 8154–8163
Sanchez-Carpintero, R., and Neville, B. (2003) Attentional ability in children with epilepsy Epilepsia 44:1340–1349
Sander, T., et al. (1998) The gene encoding the alpha 1a voltage dependent calcium channel (CACN1A4) is not a candidate for causing common subtypes of idiopathetic generalized epilepsy. Epil. Res. 29:115–122
Sansig, G., et al. (2001) Increased seizure susceptibility in mice lacking metabotropic glutamate receptor 7. J. Neuroscience 21:8734–8745
Sanz, E. (2003) Concordance and childrens use of medicines Brit Med J 327:858–860
Sardo, P., and Ferraro, G. (2007) Modulatory effects of nitric oxide active drugs on the anticonvulsant activity of lamotrigine in an experimental model of partial complex epilepsy in the rat. BMC Neurosci. 8:47

Sardo, P., et al. (2009) In the rat maximal dentate activation model of partial complex epilepsy, the anticonvulsant activity of levetiracetam is modulated by nitric oxide active drugs. J. Neural. Trans. 116:831–839

Sasa, M., et al. (1988) Effects of antiepileptic drugs on absence like and tonic seizures in the spontaneously epileptic rat, a double mutant rat. Epilepsia 29:505–508

Sato, S. (1983b) Generalized seizures: absence. In Dreyifuss, F., ed Pediatric Epileptology. Little Brown, MA pp 65–91

Sato, S., et al. (1983) Long term follow up of absence seizures. Neurology 33:1590–1595

Satow, T., et al. (2002) Partial epilepsy manifesting atonic seizure: Report of two cases. Epilepsia 43:1425–1431

Savic, I., et al. (2000) MR spectroscopy shows reduced frontal lobe concentrations of N-acetyl asparate in patients with juvenile myoclonic epilepsy. Epilepsia 41:290–296

Sawrie, S., et al. (2001) Verbral retention lateralizes patients with unilateral temporal lobe epilepsy and bilateral hippocampal atrophy. Epilepsia 42:651–659

Scala, E., et al. (2005) CDKL5/STK9 is mutated in Rett syndrome variant with infantile spasms. J. Med. Genet. 42:103–107

Scantlebury, M., Velisek, L., and Moshe, S. (2007) Catastrophic epilepsies of infancy: from bedside to the bench and back. Neurol. Asia. 12:7–9

Schachter, S., and Saper, C. (1998) Vagus nerve stimulation. Epilepsia 39:677–686

Schallier, A., et al. (2009) vCLUT2 heterozygous mice show more susceptibility to clonic seizures induced by pentyltetrazolol. Neurochem. Int. 55:41–44

Scharfman, H., Goodman, J., and Sollas, A. (2000) Granule like neurons at the hilar/CA3 border after status epilepticus and their synchrony with area CA3 bpyramidal cells: functional implications of seizure induced neurogenesis. J. Neurosci. 20:6144–6158

Scheffer, I., and Berkovic, S. (1997) Generalized epilepsy with febrile seizures plus. Brain 120:479–490

Scheffer, I., et al. (1994) Autosomal dominant frontal epilepsy misdiagnosed as sleep disorder. Lancet 343:•515–517

Scheffer, I., et al. (1995) Autosomal dominant frontal epilepsy: a distinctive clinical disorder. Brain 118:61–63

Scheffer, I., et al. (2007) Temporal lobe epilepsy and GEFS+ phenotypes associated with SCN1B mutations. Brain 130:100–109

Scheffer, I., et al. (2008) Epilepsy and mental retardation limited to females: an under recognized disorder. Brain 131:918–927

Scher, M., et al. (1993) Electrographic seizures in preterm and full term neonates Pediatrics 91:128–134

Schlesinger, K., Schreiber, R., and Griek, B. (1969) Effects of experimentally induced phenylketonuria on seizure susceptibility in mice J Comp Physiol Psych 67:149–155

Schlienger, R., and Shear, N. (1998) Antiepileptic drug hypersensitivity syndrome Epilepsia 39: s3–s7

Schmidt, D., Baumgartner, C., and Loscher, W. (2004) Seizure recurrence after planned discontinuation of anti epilepsy drugs in seizure free patients after epilepsy surgery. Epilepsia 45:179–186

Schmitz, B., and Wolf, P. (1991) Psychosis in elderly. In Devinsky, O., and Theodore, W., eds. Liss-Wiley, Boston pp. 97–128

Schmutzhard, E., and Gerstenbrand, F. (1984) Cerebral malaria in Tanzania: its epidemiology, clinical symptoms and neurological long term sequelae in the light of 66 cases. Trans R Soc Trop Med Hyg 78:351–353

Schoenfeld, J., et al. (1999) Neuropsychological and behavioral status of children with complex partial seizures Dev Med Child Neurol 41:724–731

Scholtes. F., Hendriks, M., and Renier, W (2005) Cognitive deterioration and electrical status epilepticus during slow sleep. Brain Dev 6:167–173

Schreiner, A., et al. (1995) Epileptic seizures in subcortical vascular encephalopathy J Neurol Sci 130:171–177

Schridde, U., and van Luijtelaar, G. (2004) Corticosterone increases spike wave discharges in a dose and time dependent manner in WAG/Rij rats. Pharmacol. Biochem. Behav. 78:369–375

Schrijvers, C., Coebergh, J., and Mackenbach, J. (1997) Socioeconomic status and comorbidity among newly diagnosed cancer patients Cancer 80:1482–1488

Schuchat, A., et al. (1997) Bacterial meningitis in the United States in 1995. N. Eng. Med. 337:970–976

Schuchmann, S., et al. (2008) Pronounced increase in breathing rate in the "hair Dryer model" of experimental febrile seizures. Epilepsia 49:926–928

Schuchmann, S., Vanhatalo, S., and Kaila, K. (2009) Neurobiological and physiological mechanisms of fever related epileptiform syndromes. Brain Dev. 31:378–382

Schuh, L., and Drury, I. (1997) Intraoperative electrocorticography and direct cortical electrical stimulation Seminars in Anesthesia, Perioperative Medicine and Pain Volume 16, Issue 1, 16:46–55

Schuler, V., et al. (2001) Epilepsy, hyperalgesia, impaired memory and loss of pre and postsynaptic GABA B responses in mice lacking GABA B Neuron 31:47–58

Schulz, R., et al. (1995) Amnesia of the epileptic aura. Neurology 45:231–235

Schulze, A. (2003) Creatine deficiency syndromes Mol Cell Biochem 244:143–150

Schulze-Bonhage, A., et al. (2009) Seizure control resulting from intrahippocampal depth electrode insertion J Neurol Neurosurg Psychiatr 81: 352–353

Schwartzkroin, P. (1999) Mechanisms underlying the anti-epileptic efficacy of the ketogenic diet. Epil. Res. 37:171–180

Scott, R. (2010) Adverse outcomes following convulsive status epilepticus in children: relationship with hippocampal injury. Epilepsia 51:178–181

Scott, R., et al. (2001) Quantitative magnetic resonance characterization of mesial temporal sclerosis in childhood. Neurol. 56:1659–1665

Scott, R., et al. (2002) Magnetic resonance imaging findings within 5 days of status epilepticus in childhood. Brain 125:1951–1959

Scott, R., et al. (2006) Prolonged febrile seizures are associated with hippocampal vasogenic edema and developmental changes. Epilepsia 47: 1493–1498

Scott, R., Lhatoo, S., and Sander, W. (2001) The treatment of epilepsy in developing countries: where do we go from here? Bull. Of World Health Org. 79:344–351

Scoville, W., and Milner, B. (1957) Loss of recent memory after bilateral hippocampal lesions. J. Neural. Neurosurg. Psychiatr. 20: 11–21

Seidenberg, M., et al. (1987) Neuropsychological correlates of academic achievement in children with epilepsy J. Epil 1:23–30

Seino, S., and Miki, T. (2003) Physiological and pathological roles of atp sensitive k+channels. Prog. In Biophys. And Mol. Biol. 81:133–176

Selkirk, M., et al. (2008) Clinical differences between patients with nonepileptic seizures who report antecedent sexual abuse and those who do not Epilepsia 49:1446–1450

Selwa, L., et al. (1994) Serial Cognitive Testing in Temporal Lobe Epilepsy. Epilepsia 35:743–749

Seo, J., et al. (2007) Efficacy and tolerability of the ketogenic diet according to lipid: nonlipid ratios-comparison of 3:1 with 4:1 diet. Epilepsia 48:801–805

Serikawa, T., et al. (1991) Ontogeny of absence like and tonic seizures in the spontaneously epileptic rat. Lab Animals 25:216–221

Shah, A., et al. (2007) Seizures lead to elevation of intracranial pressure in children undergoing invasive EEG monitoring. Epilepsia 48:1097–1103

Shamansky, S., and Glaser, G. (1979) Socioeconomic characteristics of childhood seizure disorders in the New Haven area: an epidemiological study Epilepsia 19:457–474

Shank, R., et al. (1991) Investigation of the mechanism of topiramate's anticonvulsant activity. Epilepsia 32:7–8

Shannon, P., et al. (2002) Neuropathological changes in a mouse model of progressive myoclonus epilepsy: cystatin B deficiency and Unverricht-Lundborg disease. J. Neuropath. Exp. Neurol. 61:1085–1091

Sharma, A., et al. (1984) Mesial temporal lobe epilepsy: Pathogenesis, induced rodent models and lesions. Toxicol. Path. 22:932–943
Sharma, K. (2005) Genetic epidemiology of epilepsy: a twin study Neurol India 53:93–98
Sharov, V., et al. (1987) Protection of ischemic myocardium by exogenous phosphocreatine. J. Thor Cardiov. Surg. 94:749–761
Shaw, P., et al. (2007) A prospective study of the effects of anterior temporal lobectomy on emotion recognition and theory of mind. Neuropsych 45:2783–2790
Shehab, S., et al. (2006) Experimental manipulations of the subthalamic nucleus fail to suppress tonic seizures in the electroshock model of epilepsy. Exp Brain Res 173:274–281
Sherman, E., Slick, D., and Eyrl, K. (2006) Executive dysfunction is a significant predictor of poor quality of life in children with epilepsy Epilepsia 47:1936–1942
Shetty, A., and Turner, D. (2000) Fetal hippocampal glutamate decarboxylase positive interneuron numbers in a rat model of temporal lobe epilepsy. J. Neurosci. 20:8788–8801
Shi, L., et al. (2006) Deep brain stimulation of the substantia nigra pars reticularis exerts long lasting suppression of amygdala kindled seizures Br Res 1090:202–207
Shih, T., and Hirsch, L. (2003) Tonic-absence seizures: an underrecognized seizure type Epilepsia 44:461–465
Shinnar, S., and Pellock, J. (2002) Update on the epidemiology and prognosis of pediatric epilepsy J Child Neurol 17:s4–s17
Shinnar, S., et al. (1996) The risk of seizure recurrence after a first unprovoked afebrile seizure in childhood: an extended follow-up Pediatr 98:216–225
Shinnar, S., et al. (2001) How long do new onset seizures in children last? ANN Neurology 49:659–664
Shinnar, S., et al. (2001) Language regression in childhood Pediatric Neurology 24:183–189
Shinnar, S., et al. (2001) Neuroimaging abnormalities in children with an apparent first unprovoked seizure Epil Res 43:261–269
Shorvon, S., and Walker, M. (2005) Status epilepticus in idiopathic generalized epilepsy. Epilepsia 46:73–79
Shorvon, S., Reynolds, E. (1979) Reduction in polytherapy for epilepsy Brit Med J 2:474–476
Shuer, L., et al. (2009) Epilepsy surgery. In R. Jaffe (Ed). Anesthesiologists Manual of Surgical Procedures (pp 81–85) Phila: Lippincott, Williams & Wilkins
Siegel, A., et al. (2004) Resective reoperation for failed epilepsy surgery: seizure outcome in 64 patients. Neurol. 63:2298–2303
Sijben, A., et al. (2009) Does a SCN1A gene mutation confer earlier age of onset of febrile seizures in GEFS+? Epilepsia 50:953–956
Sillanpaa, M., et al. (1998) Long term prognosis of seizures with onset in childhood N Eng J Med 338:1715–1722
Silva-Barrat, C., et al. (1986) Multiunitary activity analysis of cortical and subcortical structures in paroxysmal discharges and grand mal seizures in photosensitive baboons. Electroenceph. Clin. Neurophys. 64:455–468
Singh, G., and Prabhakar, S. (2008) The association between CNS infections and epilepsy: Epidemiological approaches and microbiological and epileptological perspectives. Epilepsia 49:2–7
Singh, N., et al. (2008) Mouse models of human KCNQ2 and KCNQ3 mutations for benign familial neonatal convulsions show seizures and neuronal plasticity without synaptic reorganization. J. Physiol. 586:3405–3423
Singh, N., et al. (2009) A role of SCN9A in human epilepsies as a cause of febrile seizures and as a potential modifer od Dravet syndrome. Plos Genet. 5: doi:10.1371 1000649
Sirvanci, S., et al. (2003) Immunocytochemical analysis of glutamate and GABA in hippocampus of genetic absence epilepsy rats (GAERS). Brain Res 988:180–188
Sirven, , J., et al. (2007) Second generation antiepileptic drugs' impact on balanes: a meta analysis Mayo Clin Proc 82:40–47
Sloviter, R. (1999) Status epilepticus induced neuronal injury and network reorganization. Epilepsia 40:S34-S39

Smith, D., et al. (1987) Results of a nationwide Veterans Administration Cooperative Study comparing the efficacy and toxicity of carbamazepine, phenobarbitol, phenytoin. and primidone Epilepsia 28:s50–s58
Smith, D., et al. (2006) Recovery of function after vagus nerve stimulation initiated 24 hours after percussion brai injury. J. Neurotrauma 23:1549–1553
Smith, M., Elliott, I., and Lach, L. (2004) Cognitive, psychosocial, and family function one year after pediatric epilepsy surgery Epilepsia 45:650–660
Smith, S., et al. (2008) A novel transcutaneous vagus nerve stimulation leads to brainstem and cerebral activations measured by functional MRI. Biomedical Engineering 53 June
Smith, W. (1945) The functional significance of the rostal cingulate cortex as revealed by its response to electrical excitation. J. Neurophysiol. 8:241–255
Smith,M., Elliot, I., and Lach, L. (2002) Cognitive skills in children with intractable epilepsy: Comparison of surgical and nonsurgical candidates Epilepsia 43:631–637
Snead, O. (2001) Surgical treatment of medically refractory epilepsy in childhood Brain Dev 23:199–207
Snead, O. (2004) Basic mechanisms of generalized absence seizures. Ann. Neurol. 37:146–157
Sokic, D., et al. (2001) Seizures in multiple sclerosis. Epilepsia 42:72–79
Sokoloff, L., et al. (1977) The 14C deoxyglucose method for the measurement of local cerebral utilation: theory, procedure, and normal values in the conscious and anesthetized albino rat. J. Neurochem. 28:897–916
Sorensen, A., et al. (2008) NPY gene transfer in the hippocampus attenuates synaptic plasticity and learning. Hippocampus 18:564–574
Sorensen, A., et al. (2009) NPY gene transfer attenuates seizures without affecting epilepsy induced impairment of LTP. Exper. Neurol. 215:328–333
Sorensen, J., et al. (1997) Corpus callosotomy for medically intractable seizures. Pediatr Neurosurg 27:260–267
Soria, C., et al. (2007) Quality of life in children with epilepsy and cognitive impairment: a review and a pilot study Develop Neurorehab 10:213–221
Specchio, N., et al. (2008) Effects of levetiracetam on EEG abnormalities in juvenile myoclonic epilepsy. Epilepsia 49:663–669
Spence, T., Biederman, J., and Mick, E. (2007) Attention deficit hyperactivity disorder. diagnosis, lifespan, comorbidities, and neurobiology Ambul Pediatr 7:73–81
Spencer, D. (1991) Anteromedial temporal lobectomy: directing the surgical approach to the pathologic substrate. In Surgery For Epilepsy. Spencer, S., and Spencer, S., Eds Blackwell Sci. Pub. Boston PP 129–137
Spencer, S., et al. (2005) Predicting long term seizure outcome after resective epilepsy surgery. Neurol. 65:912–918
Spencer, S., et al. (2007) Health related quality of life over time since resective epilepsy epilepsy surgery Ann Neurol 62:327–334
Spuck, S., et al. (2008) Right sided vagus nerve stimulations in humans: an effective therapy? Epil. Res. 82:232–234
Sridharan, R., and Murthy, B. (1999) Prevalence and pattern of epilepsy in India Epilepsia 40:631–636
St Louis, E., Rosenfeld, W., and Bramley, T. (2009) Antiepileptic drug monotherapy: the initial approach in epilepsy management. Current Neuropharm. 7:1875–1890
Stafstrom, C. (2004) Information processing during spike wave seizures. Epil. Curr. 4:15–16
Stavem, K., et al. (2000) The health related quality of life of patients with epilepsy compared with angina pectoris, rheumatoid arthritis, asthma, and chronic obstructive pulmonary disease. Quol. Life Res. 9:865–871
Steering Committee on Quality Improvement and Management, Subcommittee on Febrile Seizures. (2008) Febrile Seizures. Pediatr. 121:1281–1286
Stefan, H., et al. (2002) Autonomic Auras: Left hemispheric predominance of epileptic generators of cold shivers and goose bumps. Epilepsia 43:41–45
Stefan, H., et al. (2006) Onset of action of levetiracetam: an RCT trial using therapeutic intensive seizure analysis Epilepsia 47:3

Stefanovic, B., Warnking, J., and Pike, G. (2004) Hemodynamic and metabolic responses to neuronal inhibition. Neuroimage 22:771–778

Steffens, M., et al. (2005) Unchanged glutamine synthetase activity and increased NMDA receptor density in epileptic human neocortex: implications for the pathophysiology of epilepsy Neurochem int 47:379–384

Steinmeyer, K., et al. (1991) Inactivation of muscle chloride channel by transposition insertion in myotonic mice. Nature 354:304–308

Steriade, M. (2003) Neuronal substrate of sleep and epilepsy. Cambridge Univ. Press, Ny pp 215–227

Steriade, M. (2005) Sleep, epilepsy and thalamic reticular inhibitory neurons. Trends in Neurosci. 28: 317–324

Steriade, M. (2006) Grouping of brain rhythms in corticothalamic systems Neurosci. 137:1087–1106

Steriade, M., and Amzica, F. (2003) Sleep oscillations developing into seizures in corticothalamic systems. Epilepsia 44:9–20

Stern, R., et al. (1996) Influence of cognitive reserve on neuropsychological functioning in asymptomatic human immunodeficiency virus-1 infection. Arch. Neurol. 53: 148–153

Stillwagen, D., et al. (2005) Differential regulation of AMPA receptor and GABA receptor trafficking by tumor necrosis factor-alpha J Neurosci 25:3219–3228

Stobbe, G., et al. (2006) Use of levetiracetam in children with autism spectrum disorders and EEG epileptiform abnormalities Epilepsia 47:S4-S5

Stockler, S., et al. (1994) Creatine deficiency in the brain: a new treatable inborn error of metabolism. Pediatr Res. 36:409–413

Stoelting, R., et al. (2006) Antiepileptic drugs. Pharmacol. Phys. In Anesth. Practice (pp 570) Phila: Lippincott, Wiliams & Wilkins

Stohr, T., et al. (2007) Lacosamide, a novel anti-convulsant drug, shows efficacy with a wide safety margin in rodent models for epilepsy. Epil. Res. 74:147–154

Stores, G. (1999) Practitioner review: recognition of pseudoseizures in children and adolescents J Child Psychol Psychiatr 40:851–857

Stores, G., et al. (1992) Psychological effects of sodium valproate and carbamazepine in epilepsy Arch Dis Childhood 67:1330–1337

Stores, G., Zaiwalla, Z., and Bergel, N. (1991) Frontal lobe complex partial seizures in children: a form of epilepsy at particular risk of misdiagnosis Dev Med Child Neurol.

Striano, P., et al. (2003) Epilepsia partialis continua as a first symptom of multiple sclerosis: electrophysiological study of one case. Mult. Scl. 9:199–203

Striano, P., et al. (2007) An open label trial of levetiracetam in severe myoclonic epilepsy of infancy. Neurology 69:250–254

Striano, S., et al. (2005) The clinical spectrum and natural history of gelastic epilepsy-hypothalamic hamartoma syndrome Seizure 14:232–239

Striano, S., et al. (2009) The syndrome gelastic seizures-hypothalamic hamartoma: severe potentially reversible encephalopathy Epilepsia 50:62–65

Strine, T., et al. (2005) Psychological distress, comorbidities, and health behaviors among US adults with seizures: results from the 2002 National Health interview Survey. Epilepsia 46: 1133–1139

Strohl, K., et al. (2007) Sleep related epilepsy in the A/J mouse. Sleep 30: 169–176

Stroink, H., et al. (1998) The first unprovoked untreated seizure in childhood: a hospital based study of the accuracy of the diagnosis, rate of recurrence, and long term outcome. Dutch study of epilepsy in childhood J Neurol Neurosurg Psychiatr 64:595–600

Stypulkowski, P., Giftakis, J., and Billstrom, T. (2011) Development of a large animal model for investigation of deep brain stimulation for epilepsy Stereotax Funct Neurosurg 89:111–122

Suchomelova, L., et al. (2006) Treatment of experimental status epilepticus in immature rats: dissociation between anticonvulsant and antiepileptic effects. Pediatr Res 59:237–243

Suhy, L., et al. (2002) H-1 MRSIn predicts surgical outcome in MRI negative temporal lobe epilepsy. Neurol. 58:821–823

Sullivan, P., et al. (2004) The ketogenic diet increases mitochondrial uncoupling protein levels and activity. Ann. Neurol. 55:576–580

Sum Ping, S., Mefta, M., and Symreng, T (1991) Reliability of capnography in identifying esophageal intubation with carbonated beverage or antiacid in the stomach. Anesthesia And Analgesia 73:333–337

Sun, H., et al. (2008) SCN1A, SCN1B, and GABRG2 gene mutation analysis in Chinese families with generalized epilepsy with febrile seizures plus. J. Hum. Gen. 53:769–774

Sun, M., et al. (2009) Comparison of add on valproate and primidione in carbamazepine unresponsive patients with partial epilepsy. Seizure 18:90–93

Sutula, T., et al. (1981) Intensive monitoring in refractory epilepsy Neurology 31:243–247

Sutula, T., et al. (1989) Mossy fiber synaptic reorganization in the epileptic human temporal lobe. Ann. Neurol. 26:321–330

Suzuki, J. (1976) Paroxysmal discharges in the electroencephalogram of the EL mouse. Experiment. 15: 336–338

Suzuki, J., Nakamoto, Y., and Shinkawa, Y. (1983) Local cerebral glucose utilization in epileptic seizures of the mutant E1 mouse Brain Res 266:359–363

Swartz, B., et al. (1996) Visual working memory in primary generalized epilepsy: an 18FDG-PET study. Neurology 47:1203–1212

Syed, T., et al. (2008) Do observer and self reports of ictal eye closure predict psychogenic nonepileptic seizures? Epilepsia 49:898–904

Szabo, C., et al. (1999) Epilepsy surgery in children with pervasive developmental disorder. Pediatr. Neurol. 20:349–353

Szaflarski, M., et al. (2006) Quality of life in medication resistant epilepsy: the effects of patients age, age at seizure onset, and disease duration. Epil. And Behav. 8:547–551

Szucs, A., et al. (2008) The risk of paradoxical levetiracetam effect is increased in mentally retarded patients. Epilepsia. 49(7):1174–1179

Tae, W., et al. (2008) Cortical thickness abnormality in juvenile myoclonic epilepsy. J. Neurol. 255:561–566

Takahashi, Y., et al. (2009) A substantial number of Rasmussen's syndrome patients have increased IgG, CD4+T cells, TNFalpha, and Granzyme B in CSF Epilepsia 50:1419–1431

Tamburrini, G., et al. (2004) Prolonged intracranial pressure (ICP) monitoring in non traumatic pediatric neurosurgical diseases. Med. Sci. Monit. 10:mt53-mt63

Tanaka, E., (1999) Clinically significant pharmacokinetic drug interactions between antiepileptic drugs J Clin Pharm Therapy 24:87–92

Tanriverdi, T., et al. (2009) Long term seizure outcome after corpus callosotomy: a retrospective analysis of 95 patients. J. Neurosurg. 110:332–342

Tassinari, C., et al. (1992) Epilepsy with continuous spikes and waves during slow sleep-otherwise described as ESES. In Roger, J., Bureau, M., Dravet, C., eds. Epileptic Syndromes in Infancy, Childhood, and Adolescence. 2nd ed London Libby

Tassinari, C., et al. (1997) Gelastic seizures. In Tuxhorn,I., Hothausen, H., and Boenigk, H. eds. Pediatric Epilepsy Syndromes and their Surgical Treatments. London, Libby, pp 429–446

Tassinari, C., et al. (2003) Transcranial magnetic stimulation and epilepsy Clin Neurophysiol 114:777–798

Tassinari, C., et al. (2005) Epilepsy with myoclonic absences. In: Roger, J.et al. Eds. Epileptic Syndromes in infancy, childhood, and adolescence, 2nd ed. London: john Libbey publ., pp 313–328

Tatemoto, K. (1982) Neuropeptide Y: complete amino acid sequence of the brain peptide. Proc NatlAcad Sci 79:5485–5489

Tatemoto, K., et al. (1983) Galanin, a novel biologically active peptide from porcine intestine. FABS Lett 164:124–128

Tatiana, F., et al. (2006) Cognitive dysfunctions in juvenile myoclonic epilepsy. Epilepsia 47:102

Tatum, W., et al. (2000) New antiepileptic drugs Arch Family Med 9:1135–1141

Taylor, D. (1972) Mental state and temporal epilepsy: a correlative account of 100 patients treated surgically. Epilepsia 13:2345–2351
Taylor, D., and Diner, B. (2007) Evidence based medicine: do anti epileptic drugs following acute brain injury prevent seizures and/or affect morbidity and mortality? Isr J Emer Med 7:36–39
Taylor, D., and Lochery, M. (1987) Temporal lobe epilepsy: origin and significance of simple and complex auras. J. Neurol. Neurosurg. Psychiatr. 50:673–681
Taylor, L. (1991) Neuropsychologic assessment of patients with chronic encephalitis. In Andermann, F. ed. Chronic Encephalitis and Epilepsy, Rasmussen's Syndrome. Stoneham Butterworth-Heinemann pp 111–121
Tedeschi, D., and Swinyard, E. (1958) Effects of anticonvulsant drugs on recovery time from tonic clonic seizures in mice and rats. J. Pharmacol. Exp. Ther. 124:161–164
Tellez-Zenteno, J., Dhar, R., and Wiebe, S. (2005) Long term seizure outcomes following epilepsy surgery: a systematic review and meta analysis. Brain 128:1188–1198
Tellez-Zenteno, J., et al. (2006) Hippocampal electrical stimulation in mesial temporal lobe epilepsy Neurology 66:1490–1494
Tellez-Zenteno, J., Soofi, A., and Wiebe, S. (2007) Age related somatic comorbidity in people with epilepsy in the general population Neurology 68: A359
Temkin, N (2002) Prophylactic Anticonvulsants after neurosurgery. Epil. Currents 2:105–107
Tenney, J., et al. (2003) Corticothalamic modulation during absence seizures in rats: A functional MRI assessment. Epilepsia 44:1133–1140
Tenney, J., et al. (2004) fMRI of brain activation in a genetic rat model of absence seizures. Epilepsia 45:576–582
Tenney, J., et al. (2004b) FMRI of generalized absence status epilepticus in conscious marmoset monkeys reveals corticothalamic activation. Epilepsia 45:1240–1247
Teper, Y., et al. (2007) Nicotine induced dystonic arousal complex in a mouse line harboring a human autosomal dominant nocturnal frontal lobe epilepsy mutation. J. Neurosci. 27: 10128–10142
Tergau, F., et al. (1999) Low frequency repetitive transcranial magnetic stimulation improves intractable epilepsy Lancet 353:2209–2212
Terry, R., Tarver, W., and Zabars, J. (1990) An implantable neurocybernetic prosthesis system. Epilepsia 31:s33–s37
Testa, S., et al. (2007) Mood, Personality, and health related quality of life in epileptic and psychogenic seizure disorders Epilepsia 48:973–982
Than, K., et al. (2005) Can you predict an immediate complete, and sustained response to the ketogenic diet? Epilepsia 46:580–592
Theodore, W., et al. (1987) Cerebral glucose metabolism in the Lennox-Gastaut syndrome. Ann Neurol 21:14–21
Theodore, W., et al. (1999) Hippocampal atrophy, epilepsy duration, and febrile seizures in patients with partial seizures. Neurology %2: 132–136
Theodore, W., et al. (2006) Epilepsy in North America: A report prepared under the auspices of the global campaign against epilepsy, the International Bureau for Epilepsy, the International League Against Epilepsy, and the World Health Organization. Epilepsia 47:1700–1722
Thio, L., et al. (2000) Ketone bodies do not directly alter excitatory or inhibitory hippocampal synaptic transmission. Neurology 54:325–331
Thom, S., et al. (2002) Cytoarchitectural abnormalities in hippocampal sclerosis. J. Neuropath Exp Neurol 61:510–519
Thomas, J., Reagan, T., and Klass, D. (1977) Epilepsia partialis continua: A review of 32 cases. Arch. Neurol. 34:266–275
Thomas, P., et al. (1999) Nonconvulsive status epilepticus of frontal origin. Neurology 52:1174–1183
Thomas, P., Valton, L., and Genton, P., (2006) Absence and myoclonic status epilepticus precipitated by antiepileptic drugs in idiopathic generalized epilepsy. Brain 129:1281–1292
Tian, G., et al. (2005) An astrocytic basis of epilepsy Nat Med 11:973–981

Tibussek, D., et al. (2006) Proven startle provoked epileptic seizures in childhood semiologic and electrophysiologic variability. Epilepsia 47:1050–1058

Tien, R., et al. (1992) Rasmussen's encephalitis A. J. R. 158:1329–1332

Tigaran, S., et al. (2003) Acute postoperative seizures after frontal lobe cortical resection for intractable epilepsy. Epilepsia 44:831–835

Tinsa, F., et al. (2010) Role of lumbar puncture for febrile seizure among infants under one year old. Tunis Med. 88(3):178–183

Tinuper, P., et al. (1996) Partial epilepsy of long duration: changing semiology with age Epilepsia 37:162–164

Tinuper, P., et al. (2006) Diagnosis of nocturnal frontal lobe seizures: the diagnostic accuracy of clinical history. Epilepsia 47:S4 266

Tober, C., et al. (1996) D-23129: a potent anticonvulsant in the amygdala kindling model of complex partial seizures. Euro. J. Pharm. 303:163–169

Toda, H., et al. (2008) The regulation of adult rodent hippocampal neurogenesis by deep brain stimulation J. Neurosurg 108:132–138

Tolmacheva, E., and van Luijtelaar, G. (2007) Absence seizures are reduced by the enhancement of GABA-ergic inhibition in the hippocampus in WAG/Rij rats. Neurosci. Lett. 38:17–21

Tolmacheva, E., et al. (2004) Cortical and limbic excitability in rats with absence epilepsy. Epilepsy Res. 62:189–198

Tortella, F., et al. (1986) Neuroprotection (focal ischemia) and neurotoxicity studies in rats with AHN649- a 3 aino analog of dextro methorwhan and low affinity N methyl daspartate antagonist J Pharm exp Ther 291:399–408

Toth, Z., et al. (1998) Seizure induced neuronal injury: vulnerability to febrile seizures in an immature rat model. J. Neurosci. 18:4285–4294

Towne, A., et al. (1994) Determinants of mortality in status epilepticus. Epilepsia 35:27–34

Tran, D., et al. (2006) Prevalence of epilepsy in a rural district in Lao PDR Neuroepid. 26:109–116

Tran, D., et al. (2008) The challenge of epilepsy control in deprived settings: Low compliance and high fatality rates during a community based phenobarbitol program in rural Laos Epilepsia 49:539–540

Tran, T., et al. (2000) Rasmussen's Syndrome CNS Drugs 14:343–354

Trevethan,E. (2004) Seizures and epilepsy among children with language regression and autistic spectrum disorders. J. Child. Neurol. 19: S49–S57

Tsai, J., et al. (2006) Efficacy and safety of levetiracetam in Taiwanese patients with refractory partial seizures Epilepsia 47:1

Tsubota, Y., et al. (2003) The Wakayama epileptic rat (WER), a new mutant exhibiting tonic-clonic seizures and absence like seizures. Exp. Anim. 52:53–62

Tuchman, R. (1994) Epilepsy, language, and behavior: clinical models in childhood J. Child Neurology 9:95–102

Tuchman, R. (2000) Treatment of seizure disorders and EEG abnormalities in children with autistic spectrum disorders. J Autism Dev Disord 20:485–489

Tuchman, R. (2004) AEDs and psychotropic drugs in children with autism and epilepsy. Ment. Retard. Dev. Disabil. Res. Rev. 10:135–138

Tuchman, R., and Rapin, I. (2002) Epilepsy in autism. Lancet Neurol. 6:352–358

Tucker, D., et al. (2007) Discharges in ventromedial frontal cortex during absence spells. Epilepsy behav. 11:546–557

Tudur Smith, C., Marson, A., and Williamson, P. (2009) Phenytoin versus valproate monotherapy for partial onset seizures and generalized onset tonic-clonis seizures. Cochrane Database of Systematic Reviews. DOI 10.1002/14651858.CDC001769

Turanli, G., et al. (2006) Outcome and long term follow up after corpus callosotomy in childhood onset intractable epilepsy. Childs. Nerve Sys. 22:1322–1327

Turk, J., et al. (2009) Autism disorder in children with and without epilepsy Acta Paediatr. 98:675–681

Turk, V., and Bode, W. (1991) The cystatins: protein inhibitors of cysteine proteinases. FASEB Lett. 285:213–219

Turnbull, D., et al. (1985) Which drug for the adult epileptic patient: phenytoin or valproate? Brit. Med. J. 290:815–819

Turnbull, D., et al. (1985) Which drug for the adult epileptic patient: phenytoin or valproate? Brit. Med. J. 290:8150–8159

Turski, W., et al. (1983) Cholinomimetics produce seizures and brain damage in rats. Experentia 39:1408–1411

Tutuncuoglu, S., Serdaroglu, G., and Kadioglu, B. (2002) Landau-Kleffner syndrome beginning with stuttering: a case report J Child Neurol 17:785–788

Uhlemann, E., and Neims, A. (1972) Anticonvulsant properties of the ketogenic diet in mice. J. Pharm. Exp. Ther. 180:231–238

Upton,A., et al. (1985) Suppression of seizures and psychosis of limbic system origin by chronic stimulation of anterior nucleus of the thalamus. Int. J. Neurol. 19:223–230

Urbanska, E., et al. (1998) Mitochondrial toxin 3-nitropropionic acid evokes seizures in mice. Eur. J. Pharmacol. 369:55–58

Uthman, B., et al. (2004) Effectiveness of vagus nerve stimulation in epilepsy patients. Neurology 63:1124–1126

Vajda, F., and Eadie, M. (2005) Maternal valproate dosage and fetal malformation Acta Neurol Scand 112:137–143

Van Cott, A., et al. (2010) Suicide related behaviors in older patients with new anti-epileptic drug use: data from the VA hospital system. BMC Med 8:4–12

van Gassen, K., et al. (2008) Characterization of febrile seizures and febrile seizure susceptibility in mouse inbred strains. Gener, Brain, and Behav. 7:578–586

Van Laere, K., et al. (2000) Vagus nerve stimulation in refractory epilepsy: SPECT activation study. J. Nucl. Med. 41:1145–1154

van Luijtelaar, E. and Coenen, A. (1986) Two types of electrocortical paroxysms in an inbred strain of rats. Neurosci. Lett. 70:393–397

Van Luijtelaar, E., and Coenen, A. (1988) Circadian rhythmicity in absence epilepsy in rats. Epil. Res. 2:331–336

van Rooij, L., et al. (2007) Neurodevelopment outcome in term infants with status epilepticus detected with amplitude integrated electroencephalography. Pediatr 120:e354-e363

Van Roost, D., Boon, P., Vonck, K., and Caemaert, J (2007) Neurosurgical aspects of temporal deep brain stimulation for epilepsy Acta Neurol Supp 97: 333–336

Van Stuijvenberg, M., et al. (1998) Randomized controlled trail of ibuprofen syrup administered during febrile illness to prevent febrile seizure recurrences. Pediatr. 102 E51

Vander, T., Medvedovsky, M., and Herishanu, Y. (2003) Epilepsia partialis continua possibly caused by cerebellar lesion. The Cerebellum 2:126–128

Vandesteene, A., et al. (1988) Effect of propofol on cerebral blood flow and metabolism in man. Anesth. 43:42–43

Vanhatalo, S., et al. (2002) Visual field constriction in 91 Finnish children treated with vigabatrin. Epilepsia 43:748–756

Varghese, G. (2009) Clinical use of ictal SPECT in secondarily generalized tonic-clonic seizures. Brain: J. of Neurol. 132:2102–2113

Vasconcellos, E., et al. (2001) Mental retardation in pediatric candidates for epilepsy surgery: the role of early seizure onset. Epilepsia 42:268–274

Vazzani, A., and Baram, T. (2007) New roles for interleukin-1 beta in the mechanisms of epilepsy. Epil. Curr. 7:45–50

Veech, R., et al. (2001) Ketone bodies, potential therapeutic uses. IUBMB Life 51:241–247

Velasco, A., et al. (2000) Subacute and chronic electrical stimulation of the hippocampus on intractible temporal lobe seizures: preliminary report. Arch Med Res. 31:316–328

Velasco, A., et al. (2006) Neuromodulation of the centromedian thalamic nuclei in the treatment of generalized seizures and the improvement of the quality of life in patients with Lennox-Gastaut syndrome. Epilepsia 47:1203–1212

Velasco, A., et al. (2007) Electrical stimulation of the hippocampal epileptic foci for seizure control: a double blind long term follow up study. Epilepsia 48:1895–1903

Velasco, F., et al. (1987) Electrical stimulation of the centromedian thalamic nucleus in the treatment of convulsive seizures Epilepsia 28:421–439

Velasco, F., et al. (2005) Double blind randomized controlled pilot study of bilateral cerebellar stimulation for treatment of intractable motor seizures. Epilepsia 46:1–11

Velasco, F., et al. (2010) Deep brain stimulation for epilepsy in Epilepsy, Rho, J., Sankar, R., and Stafstrom, C. eds., Epilepsy. CRC Press, Boca Raton

Velisek, L., Vondrickova, R., and Mares, P. (1993) Models of simple partial and absence seizures in freely moving rats Pharm. Biochem. Behav. 45:889–896

Vercueil, L., et al. (1998) High frequency stimulation of the subthalamic nucleus suppresses absence seizures in the rat. Epil. Res. 31:39–46

Verhey, L. (2009) Quality of life in childhood epilepsy: What is the level of agreement between youth and their parents? Epil Behav 14:407–410

Verity, C., Ross, E., and Golding, J. (1993) Outcome of childhood status epilepticus and lengthy ferbrile convulsions: findings of national cohort study. Br Med J 307:225–228

Vespa, P., and Nuwer, M. (2000) Post traumatic seizures CNS drugs 13:129–138

Vespa, P., et al. (1997) Incidence of nonconvulsive and convulsive seizures in the ICU following traumatic brain injury Crit Care Med 1:A120

Vezzani, A., and Sperk, G. (2004) Overexpression of NPY and Y2 receptors in epileptic brain tissue: an endogeneous neuroprotective mechanism in temporal lobe epilepsy? Neuropep 38:245–252

Vezzani, A., Sperk, G., and Colmers, W. (1999) Neuropeptide Y: emerging evidence for a functional role in seizure modulation. Trends Neurosci. 22:25–30

Vibha, P., et al. (2010) Quality of life (Qol) and marital adjustment in epilepsy and comparisons with psychiatric illnesses. Int. J. Psychosocial Rehabilitation 14:105–112

Viehaber, S., et al. (2003) Correlation of hippocampal glucose oxidation capacity and interictal FDG-PET in temporal lobe epilepsy. Epilepsia 44:193–199

Vielhaber, S., et al. (2003) Hippocampal N-acetyl aspartate levels do not mirror neuronal cell densities in creatine supplimented epileptic rats. Euro J Neurosci. 18:2292–2300

Vielhaber, S., et al. (2008) Subfield specific loss of hippocampal N-acetyl aspartate in temporal lobe epilepsy. Epilepsia 49:40–50

Vignatelli, L., et al. (2007) Interobserver reliability of video recording in the diagnosis of nocturnal frontal lobe seizures. Epilepsia 48:1506–1511

Villasana, D., et al. (1989) Neurological deterioration in adult phenylketonuria. J Inherit Metab Dis. 1989;12(4):451–457

Villeneuve, N., et al. (2009) The ketogenic diet improves recently worsened focal epilepsy. Dev. Med. Child Neurol. 51:276–281

Vining, E. (2006) Struggling with Rasmussen's syndrome Epil Curr 6:20–21

Vining, E. (2008) Long term health consequences of epilepsy diet treatments. Epilepsia 49: 27–29

Vining, E. (2010) Ethosuximide in childhood absence epilepsy-older and better. N Eng J Med 362:843–845

Vining, E., et al. (2002) Growth of children on the ketogenic diet. Dev. Med. Child. Neurol. 44:796–802

Vinters, H., Wang, B., and Wiley, C. (1993) Herpes viruses in chronic encephalitis associated with intractable childhood epilepsy Hum Path 24:871–879

Visudtibhan, A., et al. (2004) Treatment of infantilen spasms with vigabatrin: an 8 year experience in Thai children in a referral hospital. Neurol Asia 9:117–118

Vlaini J, Perici D. 2009. Effects of acute and repeated zolpidem treatment on pentylenetetrazole –induced seizure threshold and on locomotor activity: comparsion with diazepam. Neuorpharmacology June 56(8):1124–1130. Epub. 2009 Apr.

Vlainic, J., and Pericic, D. (2010) Zolpidem is a potent anticonvulsant in adult and aged mice. Brain Res. 1310:181–188

Volterra, A., and Meldolesi, J. (2005) Astrocytes, from brain glue to communication elements : the revolution continues Nat Rev Neurosci 6:626–640
Walczak, T., et al. (2001) Incidence and risk factors in sudden unexplained death in epilepsy. Neurology 56:519–525
Waldau, B., et al. (2009) Few isolated neurons in hypothalamic hamartomas may cause gelastic seizures Ped Neurosurg 45:225–229
Walker, A. (1940) The hypothalamus and pilomotor regulation. Assoc Res Ment Dis Proc. 20:400–415
Walker, A., Johnson, H., and Kollros, J. (1945) Penicillin convulsions: the convulsant effects of penicillin applied to the cerebral cortex of monkey and man. Surg. Gynecol. Obstet. 81:692–703
Wallace, R., et al. (2003) Sodium channel alpha 1 subunit mutation in severe myoclonic epilepsy of infancy and infantile spasms. Neurology 61:765–769
Walpole, P., Isaac, C., and Reynders, H. (2008) A comparison of emotional and cognitive intelligence in people with and without temporal lobe epilepsy Epilepsia 49:1470–1474
Walsh, L., et al. (1999) Acute pentylenetetrazole injection reduces rat GABA A receptor mRNA levels. J pharm Exp Ther 289:1626–1633
Walter, G., and Renella, R. (1989) Epstein-Barr virus in brain and Rasmussen's encephalitis Lancet 1:279–280
Warrell, D., Molyneaux, M., and Beasles, P. (1990) Severe and complicated malaria Trans R Soc Trop Med Hyg 84:1–65
Waterhouse, E., et al. (1998) Synergistic effect of status epilepticus and ischemic brain injury on mortality Epil Res 29:175–183
Watts, A. (1989) A model for managing epilepsy in a rural community Brit Med J 298:805–807
Wauquier, A., and Zhou, S (1996) Topiramate: a potent anticonvulsant in the amygdala kindled rat. Epilepsia 27:73–77
Weiss, S., et al. (1995) Quenching: inhibition of development and expression of amygdala kindled seizures with low frequency stimulation Neuroreports 6:2171–2176
Welty, T. (2006) Juvenile myoclonic epilepsy Pediatr Drugs 303:303–310
Wen, X., et al. (2010) Is antiepileptic drug use related to depression? Epilepsy and Behavio 19:454–500
West, W. (1841) On a peculiar form of infantile convulsions. Lancet 1:724–725
Westerveld, M., et al. (2000) Temporal lobectomy in children: cognitive outcome. J Neurosurg. 92:24–30
Wheless, J. (2006) Intractable epilepsy: a survey of patients and caregivers Epil Behavior 8:756–764
White, A., et al. (2010) EEG spike activity precedes epilepsy after kainate induced status epilepticus. Epilepsia 51:371–383
White, H., et al. (1998) The NIH anticonvulsant drug development program. In French J., Leppik, I., and Dichter, , M. eds Antiepileptic Drug Development Adv in Neurology Phila., Lippincott-Raven Publ pp 29–39
White, H., et al. (2006) The novel investigational neuromodulator plays a broad spectrum anticonvulsant profile in rodent seizure and epilepsy models. Epilepsia 47:200–201
WHO (1978) Application of advances in neurosciences for control of neurological disorders. Tech Report Series 629:9–20
WHO (2004) Malaria incidence estimates at country level for the year 2004-proposed estimates and draft report Geneva, Roll Back Malaria. WHO
Widdess-Walsh, P., et al. (2007) Multiple auras: clinical significance and pathophysiology. Neurology 69:755–761
Wiebe, S., and Hesdorffer, D. (2007) Epilepsy: being ill in more ways than one Epil Currents 7:145–148
Wiebe, S., et al. (1999) Burden of epilepsy: the Ontario health survey Canad J Neurol Sci 26: 263–270
Wiebe, S., et al. (2001) A randomized controlled trial of surgery for temporal lobe epilepsy. N. Engl. J. Med. 345:311318

Wieser, H. (1997) Simple partial status epilepticus. In Engel, J., and Pedley, T. eds. Epilepsy: a comprehensive textbook. Phila. Lippincott-Raven pp 709–723

Wieser, H., et al. (2003) Long term seizure outcomes following amygdalohippocampectomy. J. Neurosurg. 98:751–763

Wiest, M., and Nicolelis, M. (2003) Behavioral detection of tactile stimuli during 7–12 Hz cortical oscillations an awake rats. Nat Neurosci 6:913–914

Wild, J., et al. (1999) Characteristics of a unique visual field defect attributed to vigabatrin. Epilepsia 40:1784–1794

Wilder, R. (1921) The effects of ketonemia on the course of epilepsy. Proc. Mayo Clin. 2:307–308

Wilder-Smith, E., et al. (2005) The NORSE (new onset refractory status epilepticus) syndrome: defining a disease entity. Ann Acad Med Singapore 34:417–420

Wilfong, A., and Schultz, R. (2006) Vagus nerve stimulation for treatment of epilepsy in Rett syndrome. Dev. Med. Child Neurology 48:683–686

Williams, A. J., et al. (1999) Antiepileptic drug treatment of nonconvulsive seizures induced by experimental focal brain ischemia. J Pharmacol Exp Therapeutics. 111:220–227

Williams, A., et al. (2004) Antiepileptic drug treatment of nonconvulsive seizures induced by experimental focal brain ischemia J. P. E. T. 311:220–227

Williams, D. (1953) A study of thalamic and cortical rhythms in petit mal . Brain:76:50–69

Willmore, J., et al. (2009) Vigabatrin: 2008 update. Epilepsia 50:163–173

Wilmot-Lee, B. (2008) Improving medicine taking in epilepsy Pediatr Nur. 20:37–43

Winawer, M., and Shinnar, S. (2005) Genetic epidemiology of epilepsy or what do we tell the families? Epilepsia 46:24–30

Winawer, M., et al. (2003) Evidence for distinct genetic influences on generalized and localization related epilepsy Epilepsia 44:1176–1182

Wirrell, E. (2003) Natural history of absence epilepsy in children. Canad J neurological Sci 30:184–188

Wiznitzer, M. (2004) Autism and tuberous sclerosis J. Child Neurol 19:738–748

Woermann, F., et al. (1998) Quantitative MRI in patients with idiopathic generalized epilepsy. Evidence of widespread cerebral structural changes. Brain 121:1661–1667

Woermann, F., et al. (1999) Abnormal cerebral structure in juvenile myoclonic epilepsy demonstrated with voxel based analysis of MRI. Brain 122:2101–2108

Wohlrab, G., et al. (2009) Vigabatrin therapy in infantile spasms: solving one problem and inducing another. Epilepsia 50:2003–2009

Woldbye, D., et al. (1997) Powerful inhibition of kainic acid seizures by Neuropeptide Y via Y5 like receptors. Natl. Med. 3:761–764

Wolfson, B., and Freed, B. (1980) Influence of alcohol on anesthetic requirements and acute toxicity. Anesthesia & Analgesia 59:826–833

Wolfson, L., Sakurada, O., and Sokoloff, L. (1977) Effects of gamma butyrolactone on local cerebral glucose utilization in the rat. J. Neurochem. 29:777–783

Wong, et al. (2005) Complications of intracranial electrodes monitoring in refractory partial epilepsy. Epilepsia 46:259

Wong, R., and Prince, D. (1979) Dendritic mechanisms underlying penicillin induced epileptiform activity. Science 204:1228–1231

Wong, V., et al. (2006) Tuberous sclerosis complex: correlation of MRI findings with comorbidities J Child Neurol. 21:99–105

Woodbury, J., and Woodbury, D. (1991) Vagal stimulation reduces the severity of maximal electroshock seizures in intact rats: use of a cuff electrode for stimulating and recording. Pacing Clin Electrophysiol 14:94–107

Woolf, L., et al. (1958) Dietary treatment of phenylketonuria Acta Dis Child 33:167–173

Wu, J. (2007) Electrophysiological properties and subunit composition of GABA A receptors in patients with gelastic seizures and hypothalamic hamartoma J. Neurophys 98:5–15

Wurtman, R. (1985) Aspartame: possible effect on seizure susceptibility Lancet II:1060–1064

Wyllie, E., et al. (1985) Postictal serum creatine kinase in the diagnosis of seizure disorders Arch Neurology 42:123–126

Wyllie, E., et al. (1996) Epilepsy surgery in infants. Epilepsia 37:625–637

Wyllie, E., et al. (1999) Psychiatric features of children and adolescents with pseudoseizures Arch Pediatr Adolesc Med 153:244–248

Yamamoto, H., and Tang, H. (1996) Preventive effect of melatonin against cyanide induced seizures and lipid peroxidation in mice. Neurosci. Letters 207:89–92

Yang, D., et al. (1997) Absence of excitotoxicity induced apoptosis in the hippocampus of mice lacking the Jnk3 gene. Nature 389:865–870

Yang, Y., et al. (2007) Complex seizures disorder caused by Brunol4 deficiency in mice PLoS Genet. 3:175–1284

Yoshikawa, H. (2007) First line therapy for theophylline associated seizures. Acta Neurol Scand 115:57–61

Yoshinaga, H., et al. (2001) Utility of the scalp recorded ictal EEG in childhood epilepsy Epilepsia 35:471–476

You, S., Ahn, H., and Ko, T. (2006) Vigabatrin and visual field defects I pediatric epilepsia patients J. Korean Med Sci 21:728–732

Young, G., Jordan, K., and Doig, G. (1996) An assessment of nonconvulsive seizures in the intensive care unit using continuous EEG monitoring: an investigation of variables associated with mortality Neurology 47:83–89

Younkin, D., et al. (1986) Cerebral metabolic effects of neonatal seizures measured with in vivo 31P NMR spectroscopy. Ann Neurol. 20:513–519

Yu, F., et al. (2006) Reduced sodium current in GABAergic interneurons in a mouse model of severe myoclonic epilepsy in infancy. Nat. Neurosci. 9:1142–1149

Yudkoff, M., et al. (2005) Response of brain amino acid metabolism to ketosis. Neurochem. Inter. 47:119–128

Yule, W. (1980) Educational achievement. In : Kulig, B., Meinardi, H. and Stores, C., eds. Epilepsy and behavior. Lisse, The Netherlands, Swets and Zeitlinger

Zaman, V., and Shetty, A. (2001) Fetal hippocampal CA3 cell grafts transplanted to lesioned CA3 region of the adult hippocampus exhibit long term survival in a rat model of temporal lobe epilepsy. Neurobiol. Dis. 8:943–952

Zarrelli, M., et al. (1999) The incidence of epileptic syndromes in Rochester Minnesota Epilepsia 40:1708–1714

Zhao, J., Afra, P., and Adamolekun, B. (2010) Partial epilepsy presenting as focal atonic seizure: a case report. Seizure: the J of the Brit Epil. Assoc. 19:326–329

Zhu, G., et al. (2008) Rats harboring S284L CHRNA4 mutation show attenuation of synaptic and extrasynaptic GABAergic transmission and exhibit the nocturnal frontal lobe epilepsy phenotype. J. Neurosci. 28:12465–12476

Zifkin, B., Andermann, E., and Andermann, F. (2005) Mechanisms genetics, and pathogenesis of juvenile myoclonic epilepsy. Curr Opin Neurology 18:147–153

Zini, S., et al. (1993) Galanin reduces release of endogenous excitatory amino acids in the rat hippocampus. Eur. J Pharmacol 245:1–7

Zucconi, M., and Ferini-Strambi, L. (2000) NREM parasomnias: arousal disorders and differentiation from nocturnal frontal lobe epilepsy. Clin Neurophysiol. 111:S129-S135

Zuddas, A., et al. (2002) Neuroepileptic determinants of autism spectrum disorders in tuberous spectrum disorders in tuberous sclerosis complex Brain 125:1247–1255

Zupanc, M. (2003) Infantile spasms Exp Opin Pharmacother 4:2039–2048

Index

D